Geophysical Monograph Series

175 **A Continental Plate Boundary: Tectonics at South Island, New Zealand** *David Okaya, Tim Stem, and Fred Davey (Eds.)*
176 **Exploring Venus as a Terrestrial Planet** *Larry W. Esposito, Ellen R. Stofan, and Thomas E. Cravens (Eds.)*
177 **Ocean Modeling in an Eddying Regime** *Matthew Hecht and Hiroyasu Hasumi (Eds.)*
178 **Magma to Microbe: Modeling Hydrothermal Processes at Oceanic Spreading Centers** *Robert P. Lowell, Jeffrey S. Seewald, Anna Metaxas, and Michael R. Perfit (Eds.)*
179 **Active Tectonics and Seismic Potential of Alaska** *Jeffrey T. Freymueller, Peter J. Haeussler, Robert L. Wesson, and Göran Ekström (Eds.)*
180 **Arctic Sea Ice Decline: Observations, Projections, Mechanisms, and Implications** *Eric T. DeWeaver, Cecilia M. Bitz, and L.-Bruno Tremblay (Eds.)*
181 **Midlatitude Ionospheric Dynamics and Disturbances** *Paul M. Kintner, Jr., Anthea J. Coster, Tim Fuller-Rowell, Anthony J. Mannucci, Michael Mendillo, and Roderick Heelis (Eds.)*
182 **The Stromboli Volcano: An Integrated Study of the 2002–2003 Eruption** *Sonia Calvari, Salvatore Inguaggiato, Giuseppe Puglisi, Maurizio Ripepe, and Mauro Rosi (Eds.)*
183 **Carbon Sequestration and Its Role in the Global Carbon Cycle** *Brian J. McPherson and Eric T. Sundquist (Eds.)*
184 **Carbon Cycling in Northern Peatlands** *Andrew J. Baird, Lisa R. Belyea, Xavier Comas, A. S. Reeve, and Lee D. Slater (Eds.)*
185 **Indian Ocean Biogeochemical Processes and Ecological Variability** *Jerry D. Wiggert, Raleigh R. Hood, S. Wajih A. Naqvi, Kenneth H. Brink, and Sharon L. Smith (Eds.)*
186 **Amazonia and Global Change** *Michael Keller, Mercedes Bustamante, John Gash, and Pedro Silva Dias (Eds.)*
187 **Surface Ocean–Lower Atmosphere Processes** *Corinne Le Quèrè and Eric S. Saltzman (Eds.)*
188 **Diversity of Hydrothermal Systems on Slow Spreading Ocean Ridges** *Peter A. Rona, Colin W. Devey, Jérôme Dyment, and Bramley J. Murton (Eds.)*
189 **Climate Dynamics: Why Does Climate Vary?** *De-Zheng Sun and Frank Bryan (Eds.)*
190 **The Stratosphere: Dynamics, Transport, and Chemistry** *L. M. Polvani, A. H. Sobel, and D. W. Waugh (Eds.)*
191 **Rainfall: State of the Science** *Firat Y. Testik and Mekonnen Gebremichael (Eds.)*
192 **Antarctic Subglacial Aquatic Environments** *Martin J. Siegert, Mahlon C. Kennicut II, and Robert A. Bindschadler*
193 **Abrupt Climate Change: Mechanisms, Patterns, and Impacts** *Harunur Rashid, Leonid Polyak, and Ellen Mosley-Thompson (Eds.)*
194 **Stream Restoration in Dynamic Fluvial Systems: Scientific Approaches, Analyses, and Tools** *Andrew Simon, Sean J. Bennett, and Janine M. Castro (Eds.)*
195 **Monitoring and Modeling the *Deepwater Horizon* Oil Spill: A Record-Breaking Enterprise** *Yonggang Liu, Amy MacFadyen, Zhen-Gang Ji, and Robert H. Weisberg (Eds.)*
196 **Extreme Events and Natural Hazards: The Complexity Perspective** *A. Surjalal Sharma, Armin Bunde, Vijay P. Dimri, and Daniel N. Baker (Eds.)*
197 **Auroral Phenomenology and Magnetospheric Processes: Earth and Other Planets** *Andreas Keiling, Eric Donovan, Fran Bagenal, and Tomas Karlsson (Eds.)*
198 **Climates, Landscapes, and Civilizations** *Liviu Giosan, Dorian Q. Fuller, Kathleen Nicoll, Rowan K. Flad, and Peter D. Clift (Eds.)*
199 **Dynamics of the Earth's Radiation Belts and Inner Magnetosphere** *Danny Summers, Ian R. Mann, Daniel N. Baker, and Michael Schulz (Eds.)*
200 **Lagrangian Modeling of the Atmosphere** *John Lin (Ed.)*
201 **Modeling the Ionosphere-Thermosphere** *Jospeh D. Huba, Robert W. Schunk, and George V Khazanov (Eds.)*
202 **The Mediterranean Sea: Temporal Variability and Spatial Patterns** *Gian Luca Eusebi Borzelli, Miroslav Gacic, Piero Lionello, and Paola Malanotte-Rizzoli (Eds.)*
203 **Future Earth - Advancing Civic Understanding of the Anthropocene** *Diana Dalbotten, Gillian Roehrig, and Patrick Hamilton (Eds.)*
204 **The Galápagos: A Natural Laboratory for the Earth Sciences** *Karen S. Harpp, Eric Mittelstaedt, Noémi d'Ozouville, and David W. Graham (Eds.)*
205 **Modeling Atmospheric and Oceanic Flows: Insightsfrom Laboratory Experiments and Numerical Simulations** *Thomas von Larcher and Paul D. Williams (Eds.)*
206 **Remote Sensing of the Terrestrial Water Cycle** *Venkat Lakshmi (Eds.)*
207 **Magnetotails in the Solar System** *Andreas Keiling, Caitríona Jackman, and Peter Delamere (Eds.)*
208 **Hawaiian Volcanoes: From Source to Surface** *Rebecca Carey, Valerie Cayol, Michael Poland, and Dominique Weis (Eds.)*
209 **Sea Ice: Physics, Mechanics, and Remote Sensing** *Mohammed Shokr and Nirmal Sinha (Eds.)*
210 **Fluid Dynamics in Complex Fractured-Porous Systems** *Boris Faybishenko, Sally M. Benson, and John E. Gale (Eds.)*
211 **Subduction Dynamics: From Mantle Flow to Mega Disasters** *Gabriele Morra, David A. Yuen, Scott King, Sang Mook Lee, and Seth Stein (Eds.)*
212 **The Early Earth: Accretion and Differentiation** *James Badro and Michael Walter (Eds.)*
213 **Global Vegetation Dynamics: Concepts and Applications in the MC1 Model** *Dominique Bachelet and David Turner (Eds.)*
214 **Extreme Events: Observations, Modeling and Economics** *Mario Chavez, Michael Ghil, and Jaime Urrutia-Fucugauchi (Eds.)*
215 **Auroral Dynamics and Space Weather** *Yongliang Zhang and Larry Paxton (Eds.)*
216 **Low-Frequency Waves in Space Plasmas** *Andreas Keiling, Dong-Hun Lee, and Valery Nakariakov (Eds.)*
217 **Deep Earth: Physics and Chemistry of the Lower Mantle and Core** *Hidenori Terasaki and Rebecca A. Fischer (Eds.)*

Geophysical Monograph 218

Integrated Imaging of the Earth

Theory and Applications

Max Moorkamp
Peter G. Lelièvre
Niklas Linde
Amir Khan
Editors

This Work is a co-publication between the American Geophysical Union and John Wiley and Sons, Inc.

AGU
American Geophysical Union

WILEY

This Work is a co-publication between the American Geophysical Union and John Wiley & Sons, Inc.

Published by John Wiley & Sons, Inc., Hoboken, New Jersey
Published simultaneously in Canada

Library of Congress Cataloging-in-Publication Data

ISBN: 978-1-118-92905-6

Cover image: Gravity anomalies (left), magnetic anomalies (middle) and snapshot of global seismic waves travelling through the Earth (right). The wavefield calculations were performed using AxiSEM; the source is a strike-slip earthquake located in the Apennines, Italy; image courtesy of Martin van Driel. The magnetic and gravity anomalies were plotted using the data and scripts described in Bezděk, A., and J. Sebera (2013), Matlab script for 3D visualizing geodata on a rotating globe, *Computers & Geosciences, 56*, 127–130, doi:10.1016/j.cageo.2013.03.007.

Printed in the United States of America

10 9 8 7 6 5 4 3 2 1

CONTENTS

CONTRIBUTORS

Juan Carlos Afonso
CCFS—Department of Earth and Planetary Sciences
Macquarie University
Sydney, New South Wales, Australia

Miguel Bosch
Applied Physics Department, Engineering Faculty
Universidad Central de Venezuela
Caracas, Venezuela

Knud Skou Cordua
Solid Earth Physics
Niels Bohr Institute
University of Copenhagen
Copenhagen, Denmark

Paul Cupillard
Laboratoire GeoRessources
Université de Lorraine, CNRS
Vandoeuvre-lès-Nancy, France

Joseph Doetsch
Swiss Competence Center for Energy Research
Supply of Electricity (SCCER-SoE), ETH Zurich
Zurich, Switzerland

Colin G. Farquharson
Department of Earth Sciences
Memorial University of Newfoundland
St. John's, Newfoundland and Labrador
Canada

Javier Fullea
Institute of Geosciences (CSIC, UCM)
Madrid, Spain and
Dublin Institute for Advanced Studies
Dublin, Ireland

Luis A. Gallardo
Earth Science Division
CICESE
Ensenada, Mexico

Björn Heincke
Geomar, Helmholtz Centre for Ocean Sciences
Kiel, Germany
Present address: Geological Survey of
Denmark and Greenland
Copenhagen, Denmark

Richard W. Hobbs
Department of Earth Sciences
Durham University, Science Labs
Durham, United Kingdom

Marion Jegen
Geomar, Helmholtz Centre for Ocean Sciences
Kiel, Germany

Amir Khan
Institute of Geophysics
Swiss Federal Institute of Technology
Zürich, Switzerland

Peter G. Lelièvre
Department of Earth Sciences
Memorial University of Newfoundland
St. John's, Newfoundland and Labrador, Canada

Niklas Linde
Applied and Environmental Geophysics Group
Institute of Earth Sciences
University of Lausanne
Lausanne, Switzerland

Thomas Mejer Hansen
Solid Earth Physics
Niels Bohr Institute
University of Copenhagen
Copenhagen, Denmark

Max A. Meju
Exploration Technical Services Division
Petronas Upstream
Kuala Lumpur, Malaysia

Max Moorkamp
Department of Geology
University of Leicester
Leicester, United Kingdom

Klaus Mosegaard
Solid Earth Physics
Niels Bohr Institute
University of Copenhagen
Copenhagen, Denmark

Hendrik Paasche
UFZ-Helmholtz Centre for Environmental Research
Department Monitoring and Exploration Technologies
Leipzig, Germany

Alan W. Roberts
Department of Earth Sciences
Durham University, Science Labs
Durham, United Kingdom
Present address: Geospatial Research Limited
Durham, United Kingdom

Andrea Zunino
Solid Earth Physics
Niels Bohr Institute
University of Copenhagen
Copenhagen, Denmark

FOREWORD

Geophysics is all about being able to see the unseen and to make the subsurface as transparent as the atmosphere. With this goal in mind, integrated imaging of the Earth and the monitoring of its processes occupy a central position in the field of geophysics. Geophysical inversion has rapidly evolved in the last three decades, and especially in the last few years, to include more and more information in the inverse problem. Although classic Tikhonov regularization is still used a lot, more physically based regularization operators are currently employed in deterministic inverse problems. The use of stochastic methods is another avenue to merging various types of information while respecting the physics of the processes and their uncertainties. It is important to remember that the subsurface is not a random structure. It obeys rules governed by sedimentation processes and erosion and is shaped by tectonic forces and mineralogical transformations. Incorporating these "geological" and "geochemical" data in the geophysical inverse problem is currently a hot topic. The joint inversion of geophysical data, especially with different sensitivities to subsurface properties, is also a new frontier. This can be done using petrophysics-based approaches, structural approaches such as the cross-gradient technique, and geostatistical methods. Finally, time-lapse imaging (e.g., dynamic tomography), which can be used to monitor subsurface processes, is an area of fertile research where a variety of approaches are currently being developed. Among them is the fully coupled inversion approach where the geophysical observables are directly tied to the processes we try to image. This brings me to the role of petrophysics and mineralogy in geophysical imaging. Historically, petrophysics has been loosely used to interpret geophysical models in terms of parameters of interest. Nowadays, petrophysics is more and more used upfront, integrated directly in the inversion process. This can be done, for instance, through petrophysical clustering (in a deterministic way or by including some randomness in the petrophysical relationships for each facies of the subsurface) or by filling the gap between the variables describing a given process and the geophysical observables. Finally, some novelties include the way we parameterize the inverse problem. Geophysical images exhibit usually some sparsity, which one can take advantage of. This is often crucial when using stochastic methods because of their high computational cost.

Geophysical inversion has been treated in a number of books, so we may wonder why another book on the topic is needed. The originality of "Integrated Imaging of the Earth" is that it covers, perhaps for the first time, the topics of joint and cooperative inversion of geophysical data. In a fast growing field, this book summarizes in a very didactic way the current state-of-knowledge in combining different types of geophysical data and information to better image the subsurface. In this quest for information, to paraphrase Tarantola and Valette, the readers will find an extensive discussion of the joint inversion problem together with innovative solution strategies. In addition to developing the background theory, as expected for such a book, the authors have paid special attention to providing many detailed examples covering a broad spectrum of applications, from the shallow subsurface to the Earth's mantle. At each scale and for each application, different amounts of information may be integrated in the inverse problem, and this book illustrates exciting strategies to do so. It will propel the readers to what geophysics should be in the twenty-first century.

André Revil
Directeur de Recherche CNRS, Le Bourget-du-Lac

PREFACE

The idea for this volume emerged from a successful session on integrated imaging approaches at the AGU Fall meeting in 2012 that was organized by three contributors of this book. This session showcased a variety of different methods to unify geophysical data and demonstrated how such approaches can be applied successfully in a variety of settings. Thus when one of us was subsequently contacted by Wiley with the proposal to work on an edited volume on this topic, it was easy to see the potential content for such a book. Furthermore, we were not aware of any similar efforts that compile the current state-of-the-art in integrated Earth imaging in a concise and comprehensive manner.

Now, three years later, we are happy to see the shape that this project has taken. The chapters cover all important aspects of integrated imaging from basic theory to current areas of application. This broad view on the topic should provide newcomers with a good introduction and provide experienced practitioners with new ideas on how to advance the field. We hope that this volume becomes a valuable resource to you, as it has become to us.

Max Moorkamp
Peter G. Lelièvre
Niklas Linde
Amir Khan
Leicester, St. John's, Lausanne, and Zürich

ACKNOWLEDGMENTS

The editors would like to thank the contributing authors for their efforts in writing the individual chapters. Without their expertise and knowledge, this volume would not be what it is now. The reviewers of the various chapters helped to ensure that each contribution adheres to the highest scientific standards.

1
Introduction

Max Moorkamp,[1] Peter G. Lelièvre,[2] Niklas Linde,[3] and Amir Khan[4]

Reliable and detailed information about the Earth's subsurface is of crucial importance throughout the geosciences. Improved descriptions of the Earth's internal structure and composition are fundamental to better understand and predict physical processes within the Earth. Earth models are generally more reliable and practical if they unify multiple sources of information. This unification of geophysical, petrophysical, geological, and geochemical aspects ranging from theory, field measurements, and laboratory experiments forms a truly multidisciplinary challenge. In this volume, we consider primarily the combination of complementary, yet possibly disparate, types of geophysical data in presence of geological and petrophysical constraints. The literature on this subject encompasses a wide variety of analysis methods from joint inversion, cooperative inversion, and statistical post-inversion analysis methods, which come with different assumptions, advantages, and challenges. We use the term *integrated imaging of the Earth* to not only designate this broad range of different approaches, but also as a possible name for this emerging branch of solid-Earth geophysics that has recently gained considerable attention within the geosciences.

[1] *Department of Geology, University of Leicester, Leicester, United Kingdom*
[2] *Department of Earth Sciences, Memorial University of Newfoundland, St. John's, Newfoundland and Labrador, Canada*
[3] *Applied and Environmental Geophysics Group, Institute of Earth Sciences, University of Lausanne, Lausanne, Switzerland*
[4] *Institute of Geophysics, Swiss Federal Institute of Technology, Zürich, Switzerland*

This book reviews and synthesizes a variety of approaches, successes, and challenges related to integrated imaging of the Earth. The aim is to promote further understanding of the science involved, provide a coherent framework for practitioners and students, and outline promising avenues for future research. The book covers the fundamental theory and a broad range of applications at spatial scales that range from meters to hundreds of kilometers. In the remainder of this chapter we discuss some of the issues common to all integrated approaches and provide definitions of key terms. We then give a short overview of the content of the different chapters in this book and conclude with a brief look to the future. Given the already extensive literature on integrated approaches, we will not provide references for all aspects discussed in this introductory chapter. Instead we refer to individual chapters in this book which provide extensive references to the current literature.

1.1. SOME DEFINITIONS

A key question in integrated Earth imaging is how to best combine various geophysical methods and data to produce robust and consistent Earth models. Presently, there exists no consistent and widely adopted terminology to describe and classify different methodologies designed for this task. Here we provide a set of definitions that we hope will enable a more consistent usage in the literature. We have chosen these definitions to best fit with the usages preferred by the contributing authors. In some cases, these choices represent a trade-off between historical and current usage trends.

Integrated Imaging of the Earth: Theory and Applications, Geophysical Monograph 218, First Edition.
Edited by Max Moorkamp, Peter G. Lelièvre, Niklas Linde, and Amir Khan.

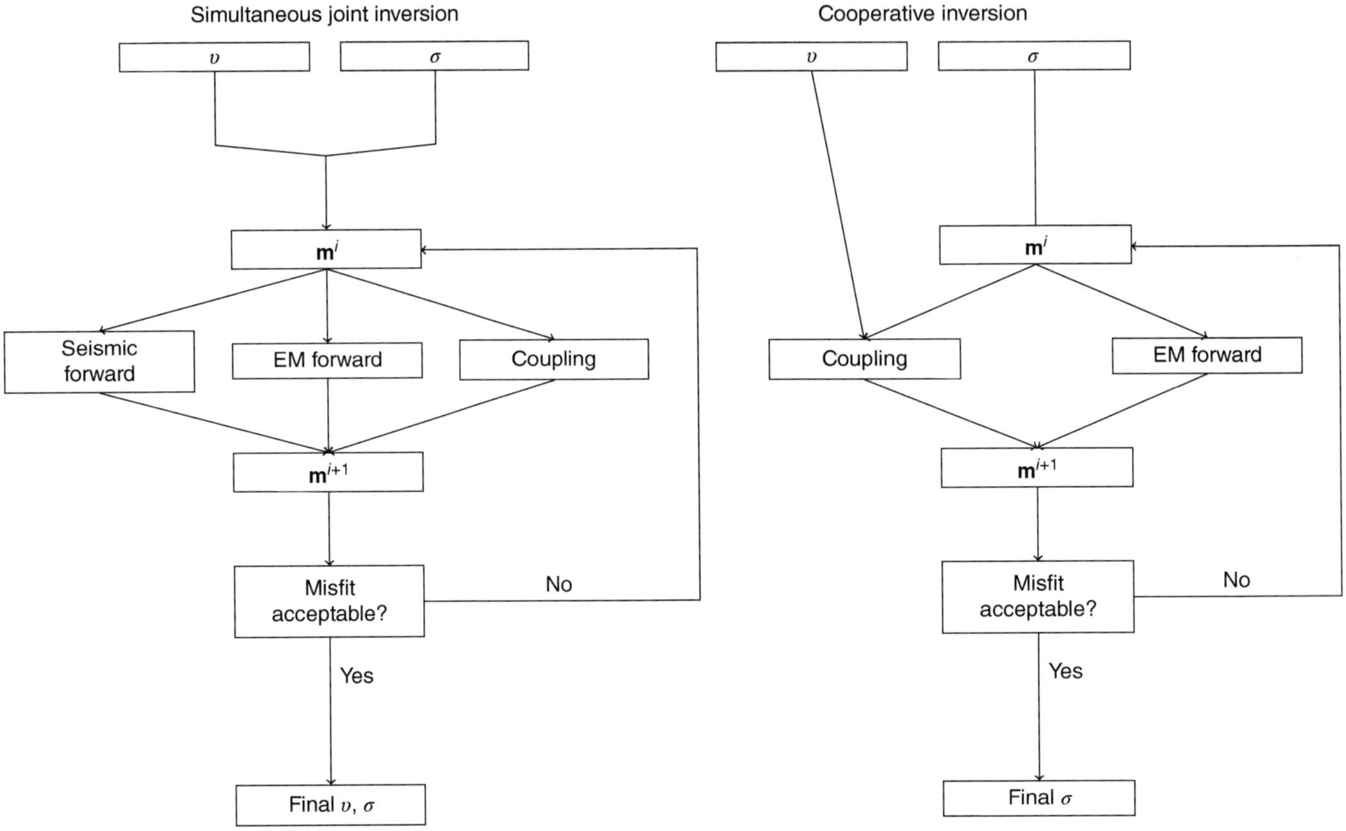

Figure 1.1 Generalized and simplified flowcharts for typical joint inversion algorithms **(left)** and cooperative algorithms **(right)** highlighting the main differences between the two approaches. We choose inversion for seismic velocity ν and conductivity σ as an example. For simplicity we do not show regularization and how exactly the seismic velocities and conductivities are related to the model vector **m** in the inversion. This will depend on the type of coupling chosen, and some examples can be found in the application chapters. The flowchart for the cooperative inversion is typically part of a larger algorithm that exchanges the roles of ν and σ after each step or several iterations.

Joint inversion, sometimes also termed *simultaneous inversion* (although we prefer the former), refers to approaches where different data types are inverted within a single algorithm, with a single objective function, and where all model parameters describing the property fields are adjusted concurrently throughout the inversion process. This stands in contrast to *cooperative inversion* approaches where single dataset inversions are performed, sequentially or in parallel, and information is shared between the different inversions. Note that this definition of cooperative inversion differs from the one offered by *Lines et al.* [1986], who used the term cooperative inversion to encompass both joint inversion and what we define as cooperative inversion. However, we argue that our definition is most consistent with contemporary usage. We suggest the term *coupled inversion* to encompass both joint and cooperative inversion approaches.

Both joint and cooperative inversion share the same goal; that is, that the combination of multiple datasets will lead to improved resolution and more consistent inference of Earth properties. Currently there is no clear evidence to suggest that one of the two approaches is universally superior to the other, and the approach needs to be chosen based on the context of the study at hand. Figure 1.1 shows generalized flow charts for joint and cooperative inversion algorithms that highlight the most important differences between the two approaches.

Cooperative inversion has the advantage that there is no need to explicitly define the relative weights of different data types, and a cooperative approach may potentially have better convergence properties than joint inversions. Cooperative inversions may also be more practical when dealing with legacy data, or when it would be difficult to construct a feedback loop for one dataset—for example, when incorporating seismic reflection or ground-penetrating radar in the inversion. In this sense, many cooperative inversion strategies can be considered special cases of incorporating complex geological priors into the inversion (see also Chapter 6).

The main advantage of joint inversion is that the information provided by all data is considered simultaneously. This can potentially help to avoid inversion artifacts because all methods contribute to the model evolution. Therefore, spurious features are unlikely to appear in regions that several datasets are sensitive to. Also, approaches that parameterize the inversion in terms of petrophysical properties (for example, mineralogy or porosity), that simultaneously influence several geophysical parameters become joint inversion problems as changing one model parameter will influence the misfit for all datasets.

Within joint inversion approaches, we can identify different subcategories. *Single-property joint inversions* combine different types of geophysical data that are sensitive to the same physical parameters. For example, receiver functions and surface waves are both sensitive to seismic velocity. This type of joint inversion is simplified because the different methods are naturally coupled through their common sensitivity to the same single physical property and no additional mathematical coupling measures are required. However, several complications can arise in practice. For example, controlled source electromagnetics (CSEM) and magnetotellurics (MT) are both sensitive to electrical conductivity within the Earth. In the MT case, the plane wave nature of the source leads to current flow that is largely horizontal in the absence of major lateral changes in conductivity. In contrast, CSEM dipole sources produce significant vertical components in the electric field and anisotropy must be considered. Similar arguments also apply to electrical resistivity tomography. Thus, a naive approach that neglects this complication and assumes an isotropic conductivity distribution may produce questionable results.

For *multi-property joint inversions* that combine, for example, electrical resistivity and seismic velocity, the manner in which these very different properties are coupled will have a large impact on the joint inversion and its ability to resolve different structures within the Earth. For this reason, a multitude of different coupling methods have been developed and we further divide multi-property joint inversion approaches into *structurally coupled* and *property coupled* approaches. Structural and property-based coupling approaches can also be applied in cooperative inversion strategies.

Property coupled joint inversion approaches directly link the different physical parameters that are inverted for. The coupling relationship can be obtained from local site-specific data (e.g., collected from boreholes; see Chapters 8 and 9) or by common petrophysical parameters such as porosity (see Chapter 9) or rock mineralogy (see Chapters 10 and 11 for examples). The theoretical foundations of these approaches are discussed in Chapter 3. Property coupled approaches must be tailored to the specifics of the dataset and region under investigation. As such, they require a detailed analysis of the data before the inversion can be performed and careful assessment of the inversion results. The reward for this is potentially improved resolution and direct information on quantities of geological interest.

Structurally coupled approaches focus on producing coincident boundaries or gradients within the different physical property models. The assumption of coincident physical property changes is often appropriate but is not generally valid (see Chapter 8 for an example). Chapter 4 describes different approaches for structural coupling in detail, and many of the application chapters contain examples that use structural coupling. One of the main reasons for their popularity is their versatility: They need little or no adjustment for different applications, while still providing sufficient coupling between the geophysical methods. However, certain studies have shown that structural coupling may be insufficient for specific scenarios (see Chapters 8 and 9) and stronger property coupled approaches may be required.

Post-inversion analysis methods are completely decoupled from the inversion process. They can be applied to the different models recovered by joint or cooperative inversion, or to co-located models obtained independently from single dataset inversions. They can even be applied to models generated by different groups of researchers. For independently generated models, it is necessary to ensure that they have comparable parameterizations and resolution such that the features in each model can be related to one another. Chapter 5 gives an overview of different approaches for post inversion analysis of multiple models.

All the integrated imaging methods mentioned above move away from qualitative, subjective comparisons of different geophysical models. Instead they seek to utilize the different geophysical methods in a quantifiable and reproducible manner by defining mathematical relationships between different geophysical quantities. The difficulty in defining relationships between disparate geophysical parameters, such as seismic velocity and electrical conductivity, is reflected in the variety of different approaches that have been proposed to couple the different methods. Defining meaningful relationships is arguably the most important scientific challenge in integrated imaging of the Earth. As the examples presented in the different chapters of this book demonstrate, these relationships largely determine to what extent a given approach is applicable to a certain region and how much improvement we can expect compared to individual analyses.

1.2. OVERVIEW OF CHAPTER CONTENTS

This book contains contributions from leading researchers and covers the main aspects of integrated Earth imaging. The book is divided into two parts: the first deals with theory and the second deals with applications. The theory chapters contain the underlying mathematical and conceptual foundations. In Chapter 2, Mosegaard and Hansen consider the foundations of inversion and optimization methods that form the basis of most integrated imaging approaches with particular emphasis on statistical aspects of inverse problems. For deterministic inversion approaches, *Nocedal and Wright* [1999] and *Menke* [2012] provide excellent introductions. It is shown how inverse problems can be formulated within a probabilistic framework using probability distributions that describe the various states of information or current knowledge about the system being studied (e.g., information on model parameters, data, physical relationships between model and data). Of particular importance when solving inverse problem is the choice of parameterization, and it is discussed how solutions can be obtained that are independent of the choice of parameterization.

The discussion of inverse theory leads directly to the next necessary ingredient for joint and cooperative inversion approaches: the definition of a suitable coupling between the different methods. The mathematical background of different classes of coupling methods are covered in Chapters 3 and 4, along with ways to incorporate them into integrated imaging approaches.

In Chapter 3, Bosch introduces the general framework of lithological tomography to integrate multiple data sources and prior knowledge through statistical relationships. Formulating the posterior probability density function can be highly challenging when dealing with hierarchical models and multiple datasets. Bosch shows how direct acyclic graphs can greatly simplify this task. He then introduces the reader to typical lithological tomography problems in exploration and global-scale studies together with solution strategies based on either sampling (Markov chain Monte Carlo) or optimization (gradient) methods.

In Chapter 4, Meju and Gallardo motivate and trace the history of structurally based joint inversion. After a short discussion about data fusion and a review of alternative measures of common model structure, they focus on algorithms that use the cross-gradients function to enforce structural similarity between disparate physical properties. Field examples are provided for both near-surface and deeper exploration targets.

In Chapter 5, Paasche reviews different approaches for post-inversion integration of geophysical models obtained from either independent or coupled inversion. The various approaches can help identify structural features and classify relationships between physical properties. The post-inversion integration approaches of Chapter 5 ultimately ease the lithological interpretation of multiple geophysical models and substantially improve the information extraction from those models. Paasche provides the mathematical fundamentals for the approaches and demonstrates their application through illustrative examples.

Chapter 6 by Hansen et al. rounds off the theory section, where the problem of inferring information about the Earth is described as a probabilistic data integration problem. Probabilistic data integration requires that information is quantifiable in the form of a probability distribution. This can be achieved either (a) directly through the specification of an analytical description of a probability distribution or (b) indirectly through an algorithm that samples a typically unknown probability distribution. In this contribution, methods are described for characterizing different kinds of information pertinent to Earth science and for making inferences from probability distributions that combine all available information. Methods are discussed that are capable of dealing with complex data integration problems.

While the theoretical foundation for joint inversion is common across applications, each specific application area faces its own set of challenges. The application chapters are roughly sorted by depth of investigation. We start with near-surface and hydrogeophysical applications (Chapter 7) that typically investigate the upper tens to hundreds of meters and have a long history of multimethod experiments. Linde and Doetsch provide a review of the coupled inverse modeling approaches that have played important roles in pushing hydrogeophysical applications towards increasingly challenging targets. Joint inversion with cross-gradient structural coupling is often favored and has been applied successfully to a wide range of near-surface applications and geophysical data types. They also present significant work where the coupling involves hydrological flow and transport simulators combined with petrophysical relationships that link hydrological state variables and geophysical properties. They conclude with a discussion of important challenges and suggest future research avenues. While the focus of Chapter 7 is on hydrogeophysical applications, Linde and Doetsch speak briefly on the potential of joint geophysical inversion to aid in archaeological investigations, civil engineering applications, and unexploded ordinance detection and discrimination.

Mineral exploration (Chapter 8) and hydrocarbon exploration (Chapter 9) focus on the upper hundreds of meters to few kilometers of the subsurface. Both areas of exploration have started to embrace integrated imaging approaches because of the potentially high economic benefits associated with new and improved techniques. The integrated imaging methods employed can be quite different in these two applications because of the different geological environments in which mineral deposits and hydrocarbon reservoirs are typically found. Imaging the sedimentary basins where hydrocarbons are found is traditionally performed with seismic methods, and thus virtually all joint inversion approaches in this area include seismics. In contrast, seismic methods are costly and have proven more problematic for the hard-rock geology typically encountered in mineral exploration scenarios; there the focus lies mainly on electromagnetic and potential field methods.

In Chapter 8, Lelièvre and Farquharson outline the aspects of mineral exploration that lead to difficulties with geophysical inversion and the need for integrated imaging approaches. They provide recommendations to practitioners for overcoming those challenges and they focus their review on joint and cooperative inversion approaches used in the field. They point out a great variety of possible geological scenarios, geophysical and geological data combinations available, and exploration questions posed; these provide ample interesting exploration problems where integrated imaging methods can be applied to potentially improve the success of mineral exploration projects. They emphasize that a solid understanding of physical property information is critical for directing joint and cooperative inversion strategies. They conclude by encouraging academic, government, and industry cooperation towards further research and development of coupled inversion methods and software.

In Chapter 9, Moorkamp et al. describe the current state of the art in joint inversion for hydrocarbon exploration. An extensive amount of work has focused on maximizing the level of detail gained by the inversion by combining full-waveform and controlled source electromagnetic inversion approaches. Also, when investigating oil and gas reservoirs, the porosity and permeability of the source rock are quantities of prime interest. Thus, petrophysical approaches that formulate the inversion in terms of these quantities have been explored in some detail. The authors also present two detailed case studies associated with current areas of exploration interest. These illustrate practical issues associated with joint inversion and provide some recipes that can potentially be used in other application areas as well.

In Chapters 10 (lithosphere) and 11 (upper mantle and transition-zone), Afonso et al. and Zunino et al. jointly consider seismic, potential field, and electromagnetic methods with the goal of making inferences about the deep Earth. As a means of obtaining information about the fundamental parameters of interest to basic Earth science, the studies invert directly for thermochemical and petrological quantities. The link between geophysical observables and thermochemical state is provided by thermodynamic concepts. The studies presented here rely on a self-consistent thermodynamic formalism for computing mantle mineral phase equilibria and physical properties from which quantitative inferences about the underlying processes that produce the observed variations in physical properties (e.g., seismic wave speeds, density, electrical conductivity) can be drawn. This ensures that temperature, composition, physical properties, and discontinuities associated with mineral phase transformations are anchored by laboratory-measured data of mantle minerals while permitting the use of inverse methods to sample a range of profiles of physical properties matching geophysical data. The advantage of this approach is that the inverse problem is formulated in terms of quantities of direct interest for geological interpretation and understanding Earth's geodynamic evolution. The main issues these approaches are facing are the high computational costs, the accuracy of laboratory measurements at high temperature and pressure, and sparser sampling of the deep Earth by surface measurements. Nevertheless, results are encouraging and allow for direct prediction of additional quantities of interest such as surface topography and heat flow.

One important area of application for integrated imaging approaches is absent from this compilation: Exploration for geothermal resources is a field of strong research activity and high significance for securing energy supplies for the future. At present, integrated imaging of any form—be it joint inversion, cooperative inversion, or integrated analysis—is still quite sparsely used in this area. The limited work performed in this direction in the context of geothermal applications is highly innovative [e.g., *Bauer et al.*, 2012; *Revil et al.*, 2015] and highlights the advantages of integrated approaches. However, it seems that the methods have not yet been widely adopted by the community. We therefore consider imaging geothermal resources as a promising field for future developments.

1.3. FUTURE DEVELOPMENTS

Both academic institutions and commercial providers are actively developing methodology and software for integrated Earth imaging. This dual interest is fundamental to ensure the development and implementation of new ground-breaking approaches. The interaction of scientific curiosity and commercial interests ensures significant funding and a balance between theoretically interesting approaches and direct applicability to concrete applications of Earth imaging. We argue that the collaboration between academia and industry is a prominent factor for past and future success of integrated imaging. However, the strong commercial interest in these methods also has its downsides. There is no doubt that important work remains unpublished for reasons of confidentiality and competitive advantage. Furthermore, there exists presently a multitude of patents for joint inversion of virtually any possible combination of geophysical datasets. We know from experience that the threat from these patents is not purely hypothetical: Patents have presented, and continue to present, barriers to the publication of open source joint inversion software. We consider this most unfortunate as the exchange of computer codes helps to advance the field more rapidly. In the long term, such an exchange would also benefit the commercial providers, since any advances that evolve through this type of exchange could also be integrated into their products.

Overall, we see a bright future for integrated imaging approaches. As illustrated by the wide range of applications displayed in this book, coupled inversion and joint interpretation methods have established themselves in virtually all areas of solid-Earth geophysics. Various companies offer integrated interpretation and joint inversion as commercial services. This indicates that industry has recognized the added value that these approaches offer and is willing to collect a variety of data and bear the additional cost over traditional subsurface imaging methods. In academia, several large recent initiatives, such as US-Array and Sinoprobe, are collecting several types of geophysical data that enable integrated analyses. Furthermore, in recent years we have seen a shift in the types of publications on integrated approaches: Whereas earlier studies focused on the theoretical properties of the algorithms and used real data mostly as illustrative examples, various recent studies focus on the interpretation of the results and consider integration as a useful tool to achieve better results. This is a clear sign that integrated imaging has reached a certain level of maturity.

These developments do not imply that there is little to do for the future. Even though some coupling approaches, such as cross-gradient coupling, are widely used and significant practical experience has been gained from these, there are still only few systematic studies that rigorously compare the theoretical properties of different methods. Unfortunately, formal resolution analysis for complex models is still beyond reach due to the computational complexity of the problem. However, with the ever-increasing speed of computer hardware, it is only a question of time before this can be performed for complex models and multiple datasets. For petrophysical coupling approaches, only a very limited number of formulations has been explored. In exploration of sedimentary environments, for example, Archie's law is usually employed to link electrical conductivity to porosity and permeability even though it is known to be restricted to clean sandstones. Alternative and more complete formulations that consider other factors such as clay content should be explored. Similar arguments hold for other application areas where experience has so far been limited to few geological settings.

Based on the growing interest in the field and a larger number of researchers conducting integrated studies, we are optimistic that these issues and many others will be investigated. Being a truly multidisciplinary field, these additional researchers ideally include not only Earth imaging specialists, but also geochemists, petrologists, and geodynamicists. If we can combine all our knowledge of Earth in a consistent and systematic way, we have a unique opportunity to unravel some of the mysteries of our home planet.

REFERENCES

Bauer, K., G. Muñoz, and I. Moeck (2012), Pattern recognition and lithological interpretation of collocated seismic and magnetotelluric models using self-organizing maps, *Geophys. J. Int.*, *189* (2), 984–998.

Lines, L. R., A. K. Schultz, and S. Treitel (1986), Cooperative inversion of geophysical data, *Geophysics*, *53* (1), 8–20.

Menke, W. (2012), *Geophysical Data Analysis: Discrete Inverse Theory Matlab Edition*, Academic Press, Waltham, MA.

Nocedal, J., and S. J. Wright (1999), *Numerical Optimization*, Springer-Verlag, New York.

Revil, A., S. Cuttler, M. Karaoulis, J. Zhou, B. Raynolds, and M. Batzle (2015), The plumbing system of the Pagosa thermal Springs, Colorado: Application of geologically constrained geophysical inversion and data fusion, *J. Volcanol. Geotherm. Res.*, *299*, 1–18, doi:http://dx.doi.org/ 10.1016/j.jvolgeores.2015.04.005.

Part I
Theory

2

Inverse Methods: Problem Formulation and Probabilistic Solutions

Klaus Mosegaard and Thomas Mejer Hansen

ABSTRACT

Inverse problems are problems where physical data from indirect measurements are used to infer information about unknown parameters of physical systems. Noise-contaminated data and prior information on model parameters are the basic elements of any inverse problem. Using probability theory, we seek a consistent formulation of inverse problems; and from our fully probabilistic results we can, in principle, answer any question pertaining our state of information about the system when all information has been integrated. A practical way of implementing probabilistic inversion methods for nonlinear problems with complex prior information is to use Monte Carlo methods. However, under certain circumstances, as when the problem is linear or mildly nonlinear and uncertainties are Gaussian, deterministic or even analytical solutions are available.

2.1. INTRODUCTION

Inverse problems arise when information about structures and processes of a physical system is inferred from empirical data. The history of inverse problems has been intimately connected to the development of mathematical geophysics, where the Earth's interior is studied from surface measurements or space observations. The advent of modern digital computers in the middle of the twentieth century enabled numerical solutions to partial differential equations and large systems of equations, and this development encouraged mathematicians to analyze the sensitivity of solutions to inaccuracies in data. The concept of ill-posedness, originally introduced by *Hadamard* [1923], formalized the study of sensitivity, and after investigations by *Tikhonov* [1943, 1963] it led to a series of methods that, up to this day, dominates linear inversion practice. The important work of *Backus and Gilbert* [1967, 1968, 1970] proposed a linear inverse theory, focusing on problems where an unknown function is derived from discrete data.

Keilis-Borok and Yanovskaya [1967] introduced Monte Carlo inversion methods into geophysics. Their work offered a way of dealing with the nonuniqueness of solutions, but it also paved the way for later development of computational methods for probabilistic inversion [*Mosegaard and Tarantola*, 1995; *Sambridge*, 1998; *Tarantola*, 2005]. Following the rapid growth of digital computing, probabilistic methods are playing and increasing role. In combination with deterministic methods, Monte Carlo methods are, in the coming years, likely to enable probabilistic analysis of even large-scale inverse problems. For this reason we focus this review entirely on probabilistic inversion.

2.2. INVERSE THEORY

Computational reconstruction or forecasting of Earth systems from data must be based on a mathematical framework and some principles for expressing Earth

Solid Earth Physics, Niels Bohr Institute, University of Copenhagen, Copenhagen, Denmark

Integrated Imaging of the Earth: Theory and Applications, Geophysical Monograph 218, First Edition.
Edited by Max Moorkamp, Peter G. Lelièvre, Niklas Linde, and Amir Khan.

structures or processes quantitatively. It is important that every step in the selected computational procedure can, in principle, be documented and agreed upon between analysts, thereby securing a high degree of objectivity. Unfortunately, quantification alone is not sufficient to guarantee that analysts will agree on the solution to the same inverse problem. Internal consistency of our methods will therefore be a principal theme in this exposition.

The mathematical framework of probabilistic data analysis is not a complete procedure for computing probabilities of, for example, Earth structure. It is a method for mapping input probabilities describing information given in advance (probability densities describing noise on the data, and prior probability densities for Earth properties) into output (posterior) probability densities for Earth parameters. In the following we will briefly describe the mathematical framework of probabilistic data analysis, but a main focus will be on the most critical issue, namely the basic formulation of the problem, including parameterization of the Earth, assignment of prior probability densities over data and Earth parameters, and calculation of posterior probabilities. It is important to emphasize that input probabilities must be meaningful and used in a consistent way to ensure correct and unbiased conclusions.

2.2.1. Parameterization of the Complex Earth

A complete, overall description of the inversion process may look like this:

1. Parameterize the Earth structure m:

$$\boldsymbol{m} = \boldsymbol{f}(m) \tag{2.1}$$

to obtain a finite set of *model parameters* $\boldsymbol{m}$.

2. Solve an *inverse problem*

$$\boldsymbol{d} = \boldsymbol{g}(\boldsymbol{m}) \tag{2.2}$$

to infer information about $\boldsymbol{m}$ from data $\boldsymbol{d}$.

3. Go backwards from the parameters $\boldsymbol{m}$ to arrive at statements about the Earth structure m.

$$\boldsymbol{m} \rightarrow m. \tag{2.3}$$

(see Figure 2.1).

Most discussions about modeling/inverse problems is about calculation of Earth parameters $\boldsymbol{m}$ from data. However, it must be remembered that the choice of subsurface parameterization may have a large impact on the final result. A straightforward parameterization where the Earth is represented by a fine, regular grid will in general work well; but if the number of parameters must be reduced (e.g., for computational reasons), care must be taken to avoid spurious structure in the computed model.

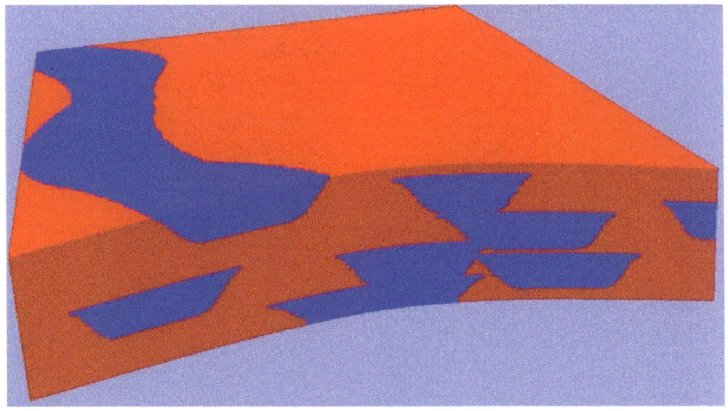

Figure 2.1 Earth structure constructed from a parameterized model.

2.2.2. Model Parameters and Observable Parameters

Any physical system—the Earth as a whole, a hydrological system of rivers, lakes, and aquifers, or a small rock crystal under a microscope—may be described by a set of parameters, of which some, the *data*,

$$\boldsymbol{d} = (d_1, d_2, \ldots, d_N) \tag{2.4}$$

are directly observable, and others, *the model parameters*,

$$\boldsymbol{m} = (m_1, m_2, \ldots, m_M) \tag{2.5}$$

are not directly observable. To compute $\boldsymbol{m}$, we take advantage of results from mathematical physics which allow us to associate to any model $\boldsymbol{m}$ the theoretical data $\boldsymbol{d}$,

$$d_n = f_n(m_1, m_2, \ldots, m_M) \quad \textbf{for} \quad n = 1, \ldots, N \tag{2.6}$$

or, for short,

$$\boldsymbol{d} = \boldsymbol{g}(\boldsymbol{m}). \tag{2.7}$$

It is in fact this expression that separates the total set of parameters $\boldsymbol{x}$ into the subsets $\boldsymbol{d}$ and $\boldsymbol{m}$, although sometimes there is no difference in physical nature between the parameters in $\boldsymbol{d}$ and the parameters in $\boldsymbol{m}$.

In the general case where data cannot be computed uniquely from model parameters through the function $\boldsymbol{d} = \boldsymbol{g}(\boldsymbol{m})$, we describe the relation between $\boldsymbol{d}$ and $\boldsymbol{m}$ through an implicite equation $\boldsymbol{h}(\boldsymbol{d}, \boldsymbol{m}) = 0$, where $\boldsymbol{h}$ is a function defined over the combined data–model space.

2.2.3. What Is Data?

In natural sciences, virtually all knowledge about phenomena in Nature is based on *interpretation* of *data*. In everyday language the word *data* can be loosely defined as "facts and statistics collected together for reference or

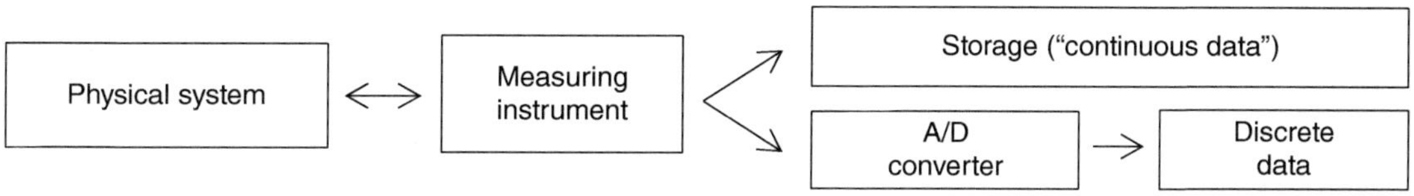

Figure 2.2 The data generation process.

analysis" (*New Oxford American Dictionary*), but specialized definitions are used in, for example, computer science and philosophy.

Our definition of data will be rooted in physics. In general we have the following scenario (see Figure 2.2): A *primary physical system* under study (Earth, a geological layer, an atom, a star, a star cluster, etc.) interacts (through force fields or particle exchange) with another physical system, the *measuring instrument* (a telescope with a CCD, a seismograph, the human eye, a voltmeter, etc.). The physical state (position, illumination, etc.) of one or several parts of the measuring instrument (a pointer/needle, capacitors in a CCD, etc.) is sensitive to the state of the primary system, and it may be recorded (saved) to a *storage* (seismogram paper, photographic paper, etc.) whose physical state is *stable over a long period of time*. Furthermore, the state of the measuring instrument may interact with a third physical system (an *analog-to-digital converter*) that will assign digital numbers to physical states of the measuring instrument.

The word *data* stands for either (1) the stable state of the storage system (for instance, the curves drawn on a seismograph paper, or text in a book) or (2) the digital numbers produced by the analog-to-digital conversion. In the following, the word *data* will denote the finite set of digital numbers produced by an A/D conversion. This means that our data is a finite (ordered) set $\boldsymbol{d}$ of numbers, each represented by a finite number of digits.

2.2.4. Parameterization

Any computation aimed at reconstruction or forecasting of Earth systems requires that the Earth be represented as numbers. This process of mathematization, often referred to as *parameterization*, proceeds formally by establishing a mapping $\boldsymbol{m} = f(m)$ from the Earth structure m to a finite set of *model parameters* $\boldsymbol{m}$ which are believed to represent the Earth "with the required accuracy." Reconstruction or forecasting is performed by algorithms whose operations are based on mathematical and physical laws, and the result of the computation is statements about the model parameters $\boldsymbol{m}$. But how can we go backward from the parameters $\boldsymbol{m}$ to statements about the model Earth m?

As an example, let us consider a model of some domain V in the Earth describing the spatial concentration $m(x, y, z)$ of a certain constituent, say quartz. We use Cartesian coordinates (x, y, z) to describe positions in space. $m(x, y, z)$ is a (positive) real function and is considered to be a complete model of the quartz concentration. In order to handle it in a digital computation, we need to represent $m(x, y, z)$ by a finite set of numbers. This is technically done by considering $m(x, y, z)$ to be an element in a *Hilbert Space* (see, e.g., *Conway* [1990]), in which we define an infinite set of orthonormal basis functions $\varphi_1(x, y, z)$, $\varphi_2(x, y, z)$, $\varphi_3(x, y, z)$, ... and parameters m_1, m_2, ... such that $m(x, y, z)$ can be expressed as

$$m(x,y,z) = \sum_{n=1}^{\infty} m_n \varphi_n(x,y,z). \tag{2.8}$$

The coefficients m_n in this expansion are the model parameters. They are infinite in number, and together they fully represent $m(x, y, z)$. However, for practical reasons, we chose to work with a finite subset of parameters $m_1, m_2, \ldots, m_M$; and if the set of basis functions $\varphi_n(x,y,z)$ is *complete*, then, if M is sufficiently large, the truncated expansion

$$m(x,y,z) \approx \sum_{n=1}^{M} m_n \varphi_n(x,y,z) \tag{2.9}$$

approximates $m(x, y, z)$ as well as we wish. No matter which complete set of base functions we choose, we are able to approximate (reconstruct) $m(x, y, z)$ arbitrarily well, given that M is large enough.

The latter property is important. It shows that two different analysts, using different choices of base functions, will arrive at consistent results, because they the can both represent the Earth exactly (in practice, the approximation error can be made as small as desired; see Figures 2.3 and 2.4).

2.2.5. Parameters as Random Variables

Classic mathematical physics makes unique predictions of data from a physical system observed by given instruments. However, experience shows that, with very few exceptions, if we repeat an experiment with the same instrumental setup, data will usually not repeat itself. There will be smaller or larger deviations between different data realizations, and the deviations will be unpredictable. Observational data is *not* a unique set of numbers.

Nonprobabilistic data analysis suffers from the problem that when noisy data are not fully reproducible, data inversions are not reproducible either. In fact, inversion of two instances of data from the same physical system under the same conditions may differ dramatically.

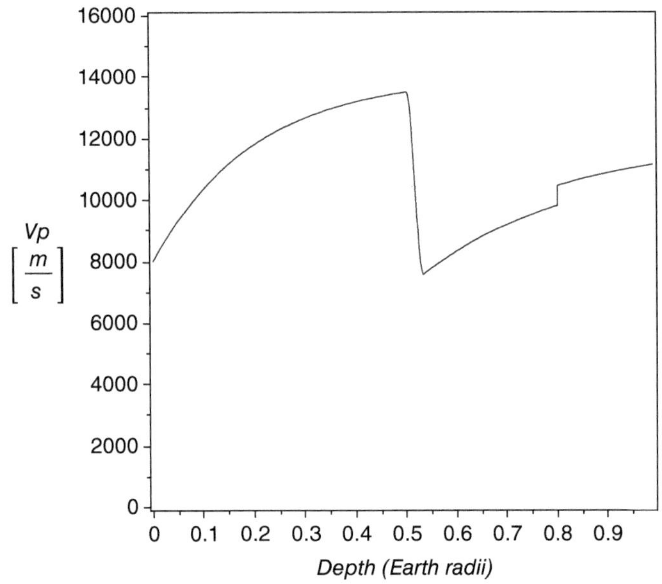

Figure 2.3 A model not unlike the P-wave velocity in the Earth's interior.

This problem, called instability, is of course well known in nonprobabilistic data analysis, and a number of *ad hoc* methods have been invented to defeat it. The most common of these methods is *regularization*, which produces smoothed solutions that are less sensitive to noise. Unfortunately, the choice of smoothing in regularization methods is arbitrary and based on a subjective choice, without reference to quantitative, empirical observations.

If the considered physical system is characterized by parameters $\boldsymbol{m} = (m_1, m_2, \ldots, m_M)$ and mathematical physics predicts data through the function $\boldsymbol{d}_{\mathbf{pred}} = \boldsymbol{g}(\boldsymbol{m})$, the observed data $\boldsymbol{d}$ deviates from $\boldsymbol{d}_{\mathbf{pred}}$ by an unpredictable amount, usually termed "noise." The noise $\boldsymbol{n} = \boldsymbol{d} - \boldsymbol{d}_{\mathbf{pred}}$ can originate from a number of factors:

• The measuring instrument is sensitive to disturbances ("additive noise")which are independent of the considered physical system. An example is thermal noise in an electronic measurement instrument.

• The measuring instrument is sensitive to structure or processes in the considered physical system ("modelization noise"), not accounted for by the model parameters $\boldsymbol{m} = (m_1, m_2, \ldots, m_M)$.

• The method used to calculate $\boldsymbol{d}_{\mathbf{pred}}$ is inaccurate.

• The A/D-conversion introduces round-off errors when representing the original signal by a finite number of digits.

The presence of noise means that $\boldsymbol{d}$ cannot be derived from $\boldsymbol{d}_{pred}$ with certainty. In probabilistic inverse theory we

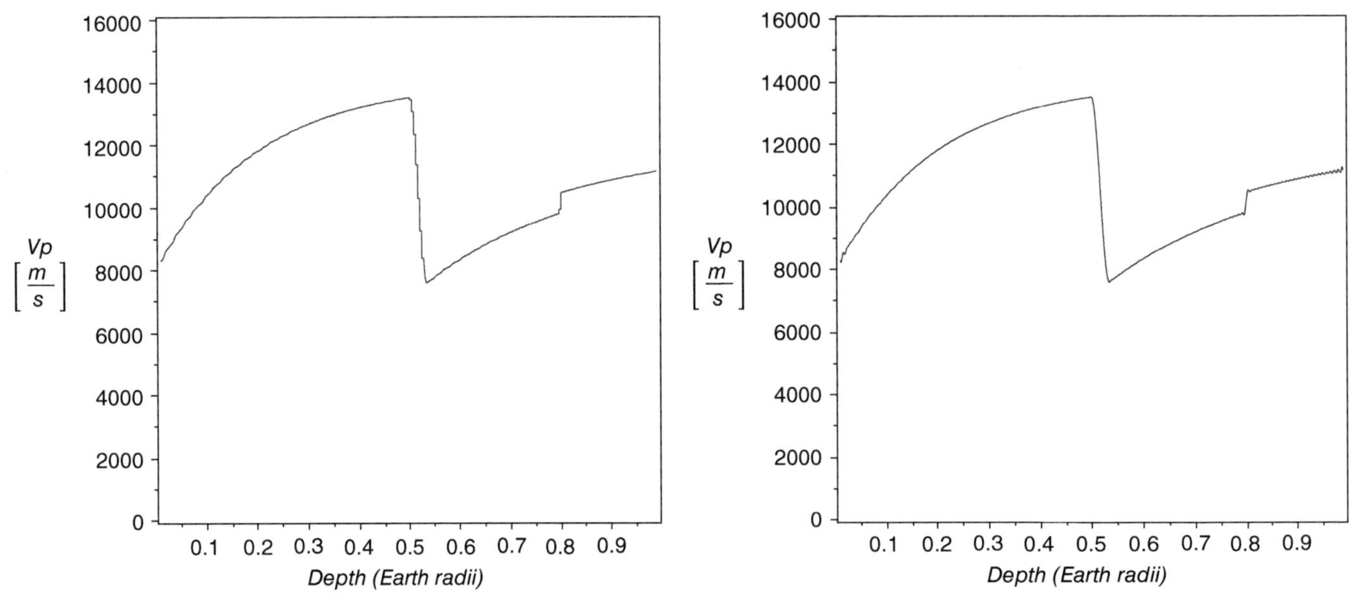

Figure 2.4 Two representations of the P-wave velocity function in Figure 2.1. **Left:** Expansion through 256 Haar-basis functions (piecewise constant). **Right:** Expansion through 128 Fourier (sin/cos)-basis functions. The figure illustrates how expansions through two different, complete sets of basis functions will, in the limit of many basis functions, approximate the function very accurately. For this reason, if two analysts use different complete sets of basis functions, they will still obtain consistent results.

therefore describe $\boldsymbol{d}$ by a joint probability density $f_D(\boldsymbol{d})$ with mean (expectation) $\boldsymbol{d}_{pred}$ and a width (dispersion, standard deviation) quantifying the uncertainty of $\boldsymbol{d}$. The density $f_D(\boldsymbol{d})$ describes not only uncertainties on individual data values, but also possible dependencies (correlations) between data uncertainties.

Probabilistic inversion is an attempt to remove unsupported choices in data analysis. The starting point of the fully probabilistic approach (see, e.g., *Tarantola* [2005]) is to acknowledge that information about parameters of a physical system, both model parameters and data, is almost always incomplete and will remain incomplete after data analysis. For this reason, the parameters are not represented by real numbers (or integers), but by *random numbers* and their corresponding probability distributions. This, however, raises a new problem which should not be taken lightly, namely the problem of finding the input probability distributions of data and model parameters (the latter often termed the *prior probability distributions*). In the following section we shall discuss this problem.

2.3. THE ORIGIN OF PROBABILITY

Probability can be seen from two viewpoints: a purely mathematical perspective and a heuristic perspective. We shall start with Kolmogorov's mathematical definition of probability, although it appeared last on the scene.

2.3.1. Kolmogorov's Axioms of Probability

Consider a set of "possible outcomes" D. We define the *probability* that an outcome x belongs to a subdomain (event) A of D as the value of a real function P(A) with the three properties [*Kolmogorov*, 1933]:

1. For any event A of D,

$$0 \le P(A) \le 1. \tag{2.10}$$

2.

$$P(D) = 1. \tag{2.11}$$

3. Let A_1, A_2, ... be a countable (possibly countably infinite) sequence of pairwise disjoint events of D. Then,

$$P\left(\bigcup_{i=1}^{\infty} A_i\right) = \sum_{i=1}^{\infty} P(A_i). \tag{2.12}$$

The function $P(A)$ is often termed a *probability function* or a *probability distribution*. Kolmogorov's axioms apply to probabilities over discrete or continuous spaces. Here, we shall consider probabilities over spaces of physical parameters, and they are usually continuous spaces (manifolds). In this case, a probability distribution is represented by an integrable *probability density function* $f(\boldsymbol{x})$ over the parameter space, such that

$$P(A) = \int_A f(x)dx \quad \text{for} \quad A \subseteq X \tag{2.13}$$

and

$$\int_X f(x)dx = 1. \tag{2.14}$$

Thus, a probability density $f(x)$ is a compact way of representing a probability function $P(A)$ in a given coordinate system. The value of P on a subset A is simply given as the integral of f over A.

It is important to realize that Kolmogorov's axioms are just requirements that any probability distribution must satisfy. Hence the question remains: Given a physical parameter, how do we find its probability distribution? Below, we shall briefly discuss the basic methods that are used for this purpose.

2.3.2. Probability Densities from Frequencies

One way of assigning probabilities is to use empirical frequencies obtained from experiments. This method is the core of statistical sciences, to which we shall refer for more details. Here we will restrict ourselves to some comments on the underlying assumptions and problems in this approach.

An example of the method is illustrated in Figure 2.5. It shows a histogram of rock densities obtained from samples collected worldwide. Such histograms may serve as an "approximation" to an (unknown) probability density of observing a given mass density of a randomly sampled rock (shown as a black, solid curve).

Traditional statistics takes the following viewpoint: Considering a given type of distribution (in Figure 2.5 the proposed probability distribution is a log-normal distribution), to what degree is the histogram and the distribution consistent? We shall refer to the vast statistics literature for an answer to this type of question, where a particular kind of probability density is assumed from the outset.

2.3.3. Probability Densities as Predictions

A probability density can be seen as the limit of a histogram for an infinitely large number of sample items. In this way, probabilities are *predicted fractions* of an infinitely large sample. For instance, in the example shown in Figure 2.5, we predict that if we collect a very large number of rock samples, the fraction of samples having a density between 5.0 g/cm^3 and 5.1 g/cm^3 will be close to the area under the solid curve between these two

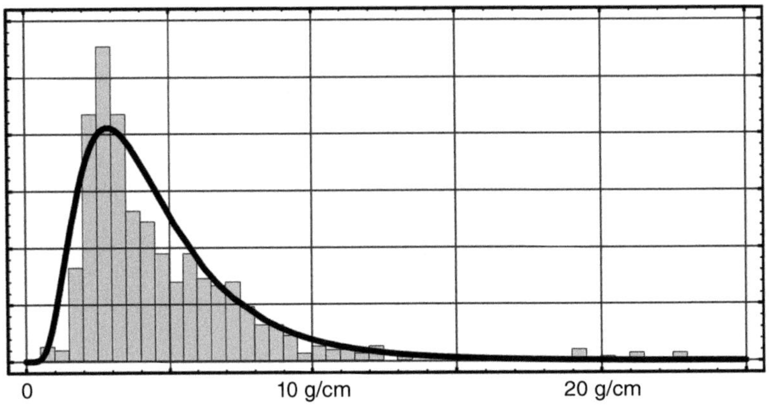

Figure 2.5 Probability density as the limit of a histogram of mass densities. The histogram is built from 571 different known minerals in the Earth's crust [*Johnson and Olhoeft*, 1984].

densities. This definition is not even using probability, or the notion of independent samples. On the other hand, the probability density defined in this way can be used to compute probabilities through expression (2.13).

As any other prediction, this interpretation of probability density is based on some critical assumptions:

• Two different histograms from the same random process must, in the limit of infinitely many sample items, approach the same limit. In order to obtain large sample sizes, it is in practice necessary to use sample items from a wide range of points in space and time. It is therefore required, but often difficult, to make sure that these sample items are satisfying this assumption.

• In practice, sample sizes are often limited. We must have a way of measuring how close a given histogram is to its limit probability density.

2.3.4. Getting More Samples from Stationarity Assumptions

One of the most important methods for finding a probability density is to make histograms. But this requires a fair number of samples to be useful, and in Earth science it is usually difficult to obtain many repetitions of the same process. One useful method is to use *stationarity* where we assume that signals or structure occurring at different times or at different locations in space follow the same probability distribution. One example from geoscience is the assumption of temporal and/or spatial stationarity of noise in seismic reflection data. Figure 2.6 shows a case where noise statistics of near-vertical-incidence seismic reflection data was found by assuming (1) stationarity in time for each seismic trace and (2) horizontal layering resulting in identical data from trace to trace, except for the noise. The noise covariance could then be found by analyzing differences between data in different traces.

2.3.5. Testing the Probability Distribution: Is the Histogram a Likely Outcome of a Sampling Experiment?

For a probability distribution to be an acceptable prediction, our histogram must be a likely result of the sampling process. The probability of getting the histogram $\pi_1, \ldots, \pi_K$ with N counts from a discrete (or discretized) probability distribution $p_1, \ldots, p_K$ is given by the *Multinomial Distribution*:

$$P(\pi_1,\ldots,\pi_K) = \frac{N!}{\pi_1!,\ldots,\pi_K!} p_1^{\pi_1} \ldots p_K^{\pi_K}. \tag{2.15}$$

In practice, this distribution is inconvenient to work with. For this reason, simplified methods have been developed, including the well-known χ^2-test. Note that if a histogram has a high probability under the assumed probability density, it does not guarantee that the probability distribution is the correct limit for the histogram.

2.3.6. Probability Densities from Degrees of Belief

In some cases, probabilities are assigned from judgment, based on (nonquantitative) experience. An example of this non-empirical method is when we guess the magnitude of a physical quantity for which we have no data, say the viscosity of the Earth's inner core, and provide a number, expressing the uncertainty we feel should be attached to our guess. Probabilities based on degrees of belief are often called *subjective* probabilities. If we need consistent, objective inversion results, such probabilities should be avoided. If, on the other hand, we emphasize that the result of our analysis (only) reflects the state of knowledge of the analyst, the use of such probabilities are acceptable seen from a Bayesian viewpoint.

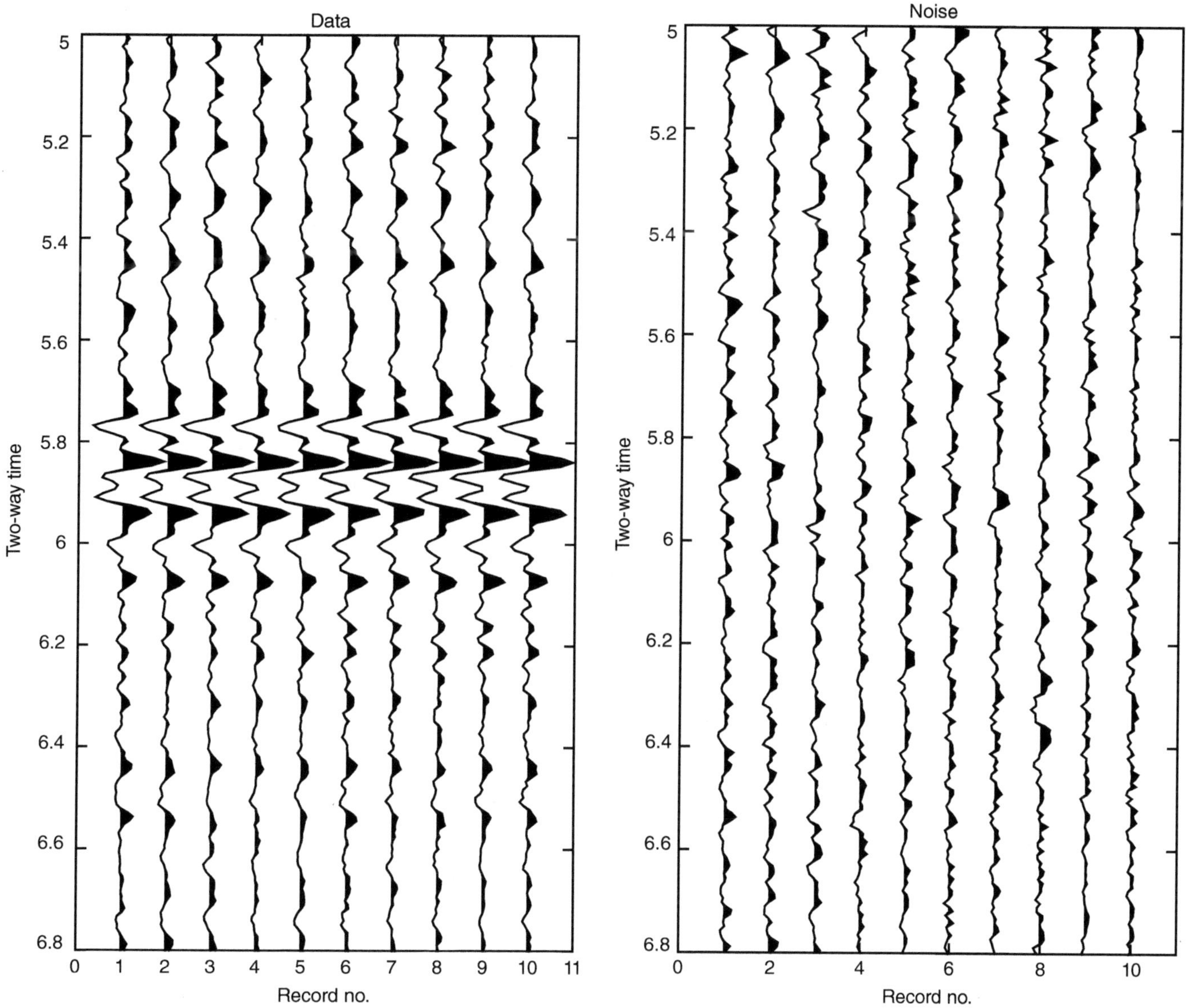

Figure 2.6 Left: A selection of deep seismic reflections from the Earth's lithosphere (DRUM profile [*BIRPS*, 1984]). **Right:** The noise found by assuming horizontal stratification and temporal and spatial stationarity of noise in the data.

2.3.7. Probabilities Based on Symmetry Arguments

The strongest non-empirical arguments for assigning probabilities are based on symmetries. The most well known example of this method is the dice which has a sixfold symmetry and therefore has a probability of 1/6 to land on each side. A fair coin (a two-sided dice!) has two-fold symmetry and therefore has a probability of 1/2 to land on each side. The argument is simple: "There is no reason" for one side to have higher probability than others, and the probabilities must add up to 1.

A common example from inverse problem practice is when we wish to express that all values of the model parameters $\mathbf{m} = (m_1, m_2, \ldots, m_M)$ are a priori equally likely (that we have *no reason to prefer* some values over others). The intuitive way to express this is through the probability density $f_M(\mathbf{m}) = constant$. This answer seems reasonable, but should be used with great care. If a probability density has a certain shape in one parameterization, it may not have the same shape when using another parameterization in the same space. We shall discuss this problem in more detail below.

2.3.8. Inconsistent Probabilities from Non-empirical Sources

The use of belief-based and symmetry-based probabilities carries a considerable risk of producing results that depend on the personal choices of the analyst. The risk is present both when probability models are chosen for the noise on the data and when prior probabilities for model parameters are chosen. Sometimes conflicts between different choices of different analysts are obvious, but in other cases they may be well hidden. This happens, for instance, when two analysts choose different parameterizations to describe exactly the same physical entities.

One example is a conductivity. This parameter, like most other inherently positive parameters, has a commonly used reciprocal parameter, in this case the resistivity:

$$\text{Resistivity}\,\rho=\frac{1}{\sigma}\leftrightarrow\text{Conductivity}\,\sigma=\frac{1}{\rho}. \quad (2.16)$$

Another example of reciprocal positive parameter pairs with equal applicability is the period T, and its corresponding frequency $\nu=1/T$. A third, more complicated example is seen in Hooke's law, relating stress σ_{ij} to strain ε_{ij}, which can be expressed as

$$\sigma_{ij}=\sum_{kl}c_{ijkl}\varepsilon_{kl} \quad (2.17)$$

introducing the stiffness tensor c_{ijkl}, or as

$$\varepsilon_{ij}=\sum_{kl}d_{ijkl}\sigma_{kl} \quad (2.18)$$

introducing the compliance tensor d_{ijkl}, the reciprocal of the stiffness tensor.

Let us take a look at the conductivity–resistivity pair. Consider the following thought experiment:

One analyst, Alice, has measured a resistivity ρ_{obs} with uncertainty s_ρ. She expresses this through the probability distribution (Figure 2.7)

$$f_A(\rho)=\frac{1}{s_\rho\sqrt{2\pi}}\exp\left(-\frac{(\rho-\rho_{obs})^2}{2s_\rho^{\,2}}\right). \quad (2.19)$$

Bob makes the same measurement, but prefers to work with conductivity: $\sigma_{obs}=1/\rho_{obs}$. He expresses this through the probability distribution (Figure 2.8)

$$g_B(\sigma)=\frac{1}{s_\sigma\sqrt{2\pi}}\exp\left(-\frac{(\sigma-\sigma_{obs})^2}{2s_\sigma^{\,2}}\right). \quad (2.20)$$

Alice and Bob decide to compare their distributions. Bob re-computes his distribution to resistivity (Figure 2.9)

$$f_B(\rho)=\left|\frac{d\sigma}{d\rho}\right|g_B(\sigma)=\frac{1}{\rho^2}\frac{1}{s_\sigma\sqrt{2\pi}}\exp\left(-\frac{\left(\frac{1}{\rho}-\sigma_{obs}\right)^2}{2s_\sigma^{\,2}}\right). \quad (2.21)$$

and they compare. His result is clearly different from Alice's distribution, having a long tail to the positive side.

Because Alice and Bob had chosen different parameters to describe the same physical entity, they did not realize that they assigned two different probabilities to this entity. Incidentally, it turns out that if they instead

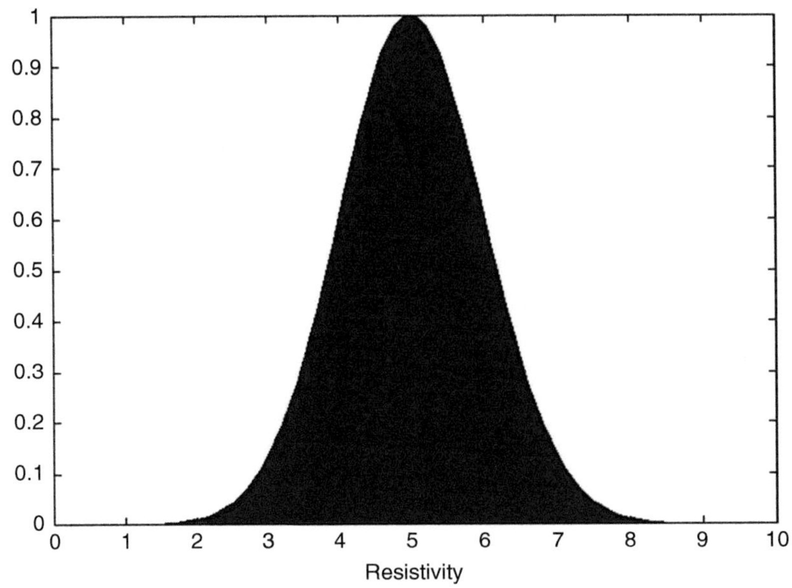

Figure 2.7 A Gaussian distribution of resistivity.

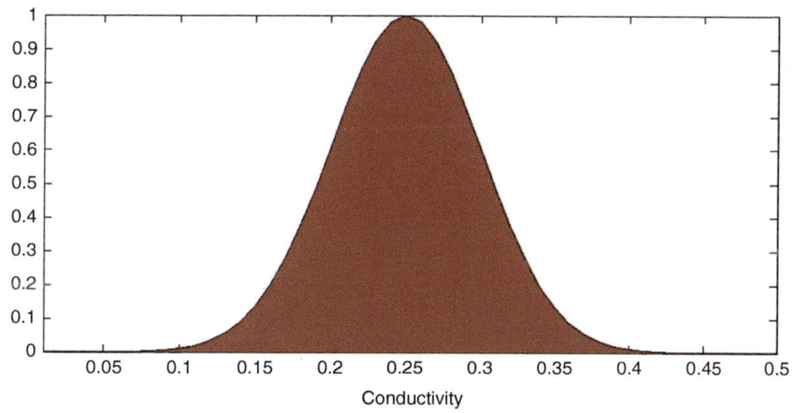

Figure 2.8 A Gaussian distribution of conductivity with mean and standard deviation roughly compatible with the resistivity distribution in Figure 2.7.

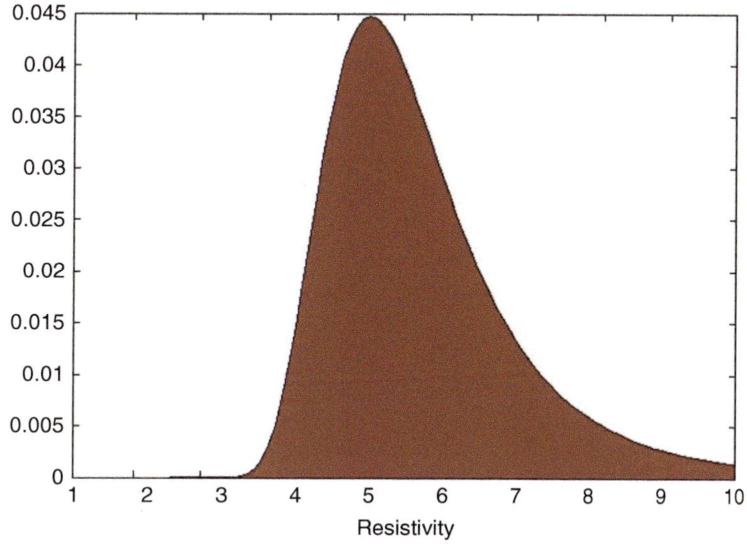

Figure 2.9 Resistivity distribution derived from the conductivity distribution in Figure 2.8.

had chosen the logarithmic parameters $\log(\rho/\rho_0)$ and $\log(\sigma/\sigma_0)$, where ρ_0 and σ_0 are reference values, they had not run into contradictions.

The above example shows the inherent risks associated with the use of non-empirical probabilities. The inconsistencies come from the fact that such probabilities are not objective, in the sense that they are not derived from external information that the analysts can agree upon. In the following presentation of the rules of probabilistic inversion, we shall therefore advocate the use of empirically based distributions whenever possible.

2.4. PROBABILISTIC INVERSION

A characteristic property of most inverse problems is that their solutions are nonunique. Usually there are infinitely many values of the model parameters that satisfy equation (2.2) in the sense that the magnitude of $\boldsymbol{d} - \boldsymbol{g}(\boldsymbol{m})$ does not exceed the magnitude of the noise. This means that we must deal with a vast space of solutions, from which we must choose how to extract the desired information. This choice is inherently subjective and can be seen as an act of *interpretation*. This is a dominating method in nonprobabilistic inversion where the most well known method for choosing a preferred solution is *regularization* where additional constraints are applied to the model parameters (e.g., smoothness constraints) to single out one particular solution.

In a probabilistic inversion, however, the approach is fundamentally different. The ambition here is to reduce the role of the subjective interpretation element as much as possible. This is done by propagating the data uncertainties (the data probability density) back through the equation $\boldsymbol{d} = \boldsymbol{g}(\boldsymbol{m})$ and thereby obtain a likelihood for

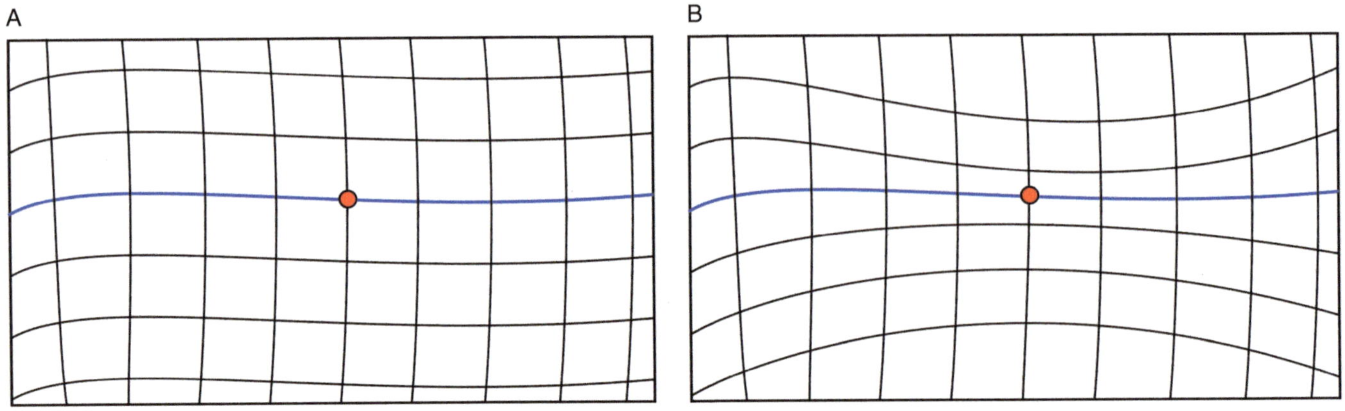

Figure 2.10 Two different coordinate grids in the same space. A probability distribution assigning equal probabilities to equal volumes is defined over the space. In coordinate system A the gridline spacing, and hence the probability density, is almost constant. In coordinate system B the spacing betweenhorizontal gridlines is smaller around the red mark, and hence the probability density has a minimum in this area.

each $\boldsymbol{m}$. In addition to this, it is recognized that some data-independent information about $\boldsymbol{m}$ may be available—the so-called a priori information on $\boldsymbol{m}$—and this information must be integrated with the likelihood to give a combined probability density known as the *posterior probability density* over $\boldsymbol{m}$. The advantage of this approach is that, given the data distribution and prior information on $\boldsymbol{m}$, all possible solutions $\boldsymbol{m}$ are weighted in an objective way. We can now pick the $\boldsymbol{m}$ with the highest value of the posterior probability density (the so-called maximum a posteriori (MAP) model) and compute the uncertainty of $\boldsymbol{m}$, or we can extract other probabilistic information from this distribution.

In the following, we shall discuss the methods of probabilistic inversion in more detail. We have made extensive use of the word *probability density*; but to simplify the exposition, we shall, in the following, assume that they are unnormalized. Thus, strictly speaking, they are just *densities* (or *measures*). Once calculated, they can be normalized when needed.

However, as pointed out by *Tarantola* [2005], probabilistic rules formulated in terms of probability densities are, in general, not invariant (do not preserve the same form) under coordinate transformations. A serious consequence of this is the so-called *Borel Paradox* where conditional probability densities calculated on the same subspace, but in different coordinate systems, may be different.

2.4.1. The Borel Paradox

Consider Figure 2.10 where two different coordinate systems A and B are defined over the same two-dimensional space. The horizontal and vertical coordinate gridlines of system A and the vertical gridlines of system B are approximately equidistant, but the horizontal gridlines of system B are not. A one-dimensional subspace (the horizontal blue line) located in the middle of the figure coincides with a horizontal gridline. Because of the fact that the vertical gridlines are the same in system A and B, the local coordinate grid in the one-dimensional subspace is identical in the two systems.

Let us now assume that we wish to describe a probability distribution over the entire space assigning equal probabilities to equal volumes (areas). Since the gridline spacing in coordinate system A is almost constant, the corresponding probability density in A will be almost constant. In coordinate system B, the spacing between horizontal gridlines is smaller in the middle (around the red mark), and hence the probability density in B has a minimum in this area. This is due to the fact that, if the coordinates are called x and y, the physical volume/area between $x+dx$ and $y+dy$ is smallest in B; and considering that $f(x,y)\,dx\,dy$ is the probability between $x+dx$ and $y+dy$, this means that $f(x,y)$ must be smaller in B to ensure that "equal volumes have equal probabilities" in the physical space.

Figure 2.11 shows the conditional probability densities on the one-dimensional subspace computed in systems A and B. Although the basic two-dimensional probability distribution expressed in the two systems is the same, and despite the fact that the local coordinate grid in the one-dimensional subspace remains the same in A and B, the two conditional distributions are different. What we see here is the so-called *Borel Paradox* where the probability density, restricted to a subspace, is influenced by a deformation of gridlines, not in the subspace but in the surrounding higher-dimensional space in which it is embedded!

However, as pointed out by *Mosegaard and Tarantola* [1995] and *Tarantola* [2005], there is a way of avoiding the

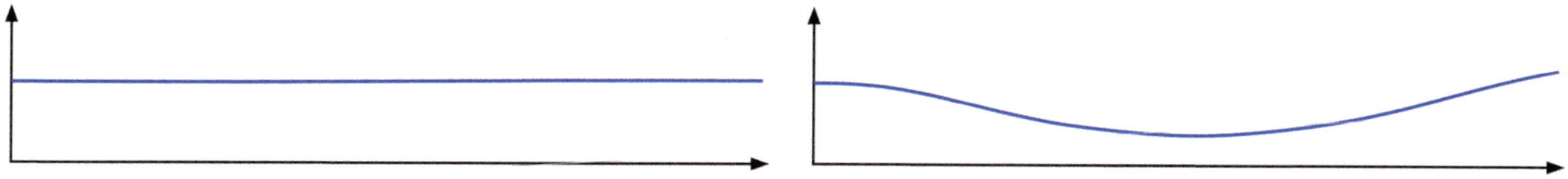

Figure 2.11 Conditional probability densities defined on the blue 1D line in Figures 2.10A and 2.10B. **Left:** The conditional density calculated in coordinate system A. **Right:** The conditional density calculated in coordinate system B. Note how volume variations between gridlines in the two-dimensional space influence the calculated conditionals on the blue line, despite the fact that (1) the basic two-dimensional distributions in both systems express a constant volume distribution of probability and (2) the grid system on the 1D line is identical in the two systems. This is an example of the so-called Borel Paradox.

contradictions. Replacing probability densities $\hat{f}(\boldsymbol{x})$, $\boldsymbol{x} \in \mathcal{X}$, with $f(\boldsymbol{x}) = \hat{f}(\boldsymbol{x}) / \mu(\boldsymbol{x})$ where $\mu(\boldsymbol{x})$ is a nonzero *volume density*, renders $f(\boldsymbol{x})$ invariant and removes the paradox. $\hat{f}(\boldsymbol{x})d\boldsymbol{x}$ is, per definition of $\hat{f}$, the probability that $\boldsymbol{x}$ is in the infinitesimal "cube" $d\boldsymbol{x}$; and $\mu(\boldsymbol{x})d\boldsymbol{x}$ is, per definition of μ, the physical volume in $d\boldsymbol{x}$. Hence, $\hat{f}(\boldsymbol{x}) / \mu(\boldsymbol{x})$ is the (coordinate independent) probability per physical unit volume near $\boldsymbol{x}$.

Dividing with $\mu(\boldsymbol{x})$ corresponds to the well-known multiplication with an appropriate Jacobian when changing coordinates, and for Cartesian coordinates, $\mu(\boldsymbol{x})$ is constant. In the following exposition we will work with volume-compensated densities $f(\boldsymbol{x}) = \hat{f}(\boldsymbol{x}) / \mu(\boldsymbol{x})$.

2.4.2. Constructing a Probabilistic Solution

Given a choice of model parameters $m_1 \dots m_M$ and data parameters $d_1 \dots d_N$ for a physical system, an exhaustive probabilistic characterization would require specification of a joint probability density $f(m_1,\dots,m_M,d_1,\dots,d_N) = f(\boldsymbol{m},\boldsymbol{d})$. This distribution, however, is usually not directly available in practice, so below we will explain how it may be constructed by integrating different sources of information.

A very important piece of information comes from the laws of physics allowing us to calculate predicted (theoretical, noise free) data $\boldsymbol{d}_{pred}$ from the model parameters through a function $\boldsymbol{g}$:

$$\boldsymbol{d}_{\text{pred}} = \boldsymbol{g}(\boldsymbol{m}). \tag{2.22}$$

An example of this is when predicted seismic data $\boldsymbol{d}_{\text{pred}}$ are uniquely calculated from elastic parameters $\boldsymbol{m}$ of the earth through partial differential equations, appropriate initial and boundary conditions, and source parameters.

Another significant piece of information is the probabilistic description of noise $\boldsymbol{n}$ that contaminates the observed data $\boldsymbol{d}$:

$$\boldsymbol{d} = \boldsymbol{d}_{\text{pred}} + \boldsymbol{n}. \tag{2.23}$$

Assuming that $f_{\text{n}}(\boldsymbol{n})$ is the density of the noise, we obtain the following conditional density of $\boldsymbol{d}$ for given $\boldsymbol{m}$:

$$f(\boldsymbol{d} \mid \boldsymbol{m}) = f_n(\boldsymbol{d} - \boldsymbol{g}(\boldsymbol{m})). \tag{2.24}$$

A commonly occurring noise density is the normal (Gaussian) distribution with zero mean and variance σ^2:

$$f_n(\boldsymbol{n}) = \exp\left(-\frac{\boldsymbol{n}^2}{2\sigma^2}\right) \tag{2.25}$$

giving

$$f(\boldsymbol{d} \mid \boldsymbol{m}) = \exp\left(-\frac{(\boldsymbol{d} - \boldsymbol{g}(\boldsymbol{m}))^2}{2\sigma^2}\right) \tag{2.26}$$

(remember that we ignore normalizations). If we can somehow (see below) obtain the marginal probability density $f(\boldsymbol{m})$, often termed the *prior density*, we will be able to reconstruct the complete joint distribution

$$f(\boldsymbol{m},\boldsymbol{d}) = f(\boldsymbol{d} \mid \boldsymbol{m}) f(\boldsymbol{m}). \tag{2.27}$$

Once we have performed a measurement, we have a concrete realization $\boldsymbol{d}_{\text{obs}}$ of $\boldsymbol{d}$ and we can compute

$$p(\boldsymbol{m}) = f(\boldsymbol{m},\boldsymbol{d}_{\text{obs}}) \tag{2.28}$$

known as the *posterior distribution* of $\boldsymbol{m}$. This density is considered to be the complete, probabilistic solution to the inverse problem.

2.4.3. Empirically Based Prior Information

Some priors $f(\boldsymbol{m})$ are data-independent and are derived from symmetry arguments or subjective beliefs. We will denote this category of priors *native priors*. Such priors are often subjective and depend on the analyst's preferences. The literature contains numerous examples of native priors formulated as Gaussian distributions with means and standard deviations chosen without reference to objective information. If consistent, objective results are needed, such priors should be used with great care, or preferably avoided.

Another category of priors is based on empirical background knowledge, which may or may not depend on the measurements $\boldsymbol{d}$. If the empirical prior information $f(\boldsymbol{m})$ is based on background information $\boldsymbol{b}$ that is independent of $\boldsymbol{d}$, we can use the formalism from the previous section. If, however, the background information $\boldsymbol{b}$ depends on $\boldsymbol{d}$, we can incorporate it into our formalism as "data" by considering the joint density

$$f(\boldsymbol{m},\boldsymbol{d},\boldsymbol{b}), \tag{2.29}$$

allowing a complete probabilistic description of our system. Assuming that $f(\boldsymbol{m})$ describes a native prior (e.g., from symmetry considerations), we can write

$$f(\boldsymbol{m},\boldsymbol{d},\boldsymbol{b}) = f(\boldsymbol{d},\boldsymbol{b} \mid \boldsymbol{m}) f(\boldsymbol{m}). \tag{2.30}$$

This allows us to compute the *posterior* density for $\boldsymbol{m}$,

$$f(\boldsymbol{m},\boldsymbol{d}_{\mathrm{obs}},\boldsymbol{b}_{\mathrm{bg}}), \tag{2.31}$$

where $\boldsymbol{b}_{\mathrm{bg}}$ is the concrete background information on which our prior information is based.

2.4.4. Linear Gaussian Inverse Problems

The simplest inverse problems are the linear problems with additive and Gaussian noise, and with Gaussian prior probability density for the model parameters. These problems can be treated analytically and may even serve as (local) approximations to nonlinear inverse problems if the nonlinearity is mild.

For a linear Gaussian problem, we have in general

$$f(\boldsymbol{d} \mid \boldsymbol{m}) = \exp\left(-\frac{1}{2}(\boldsymbol{d}-\boldsymbol{Gm})^T \boldsymbol{C}_n^{-1}(\boldsymbol{d}-\boldsymbol{Gm})\right), \tag{2.32}$$

where $\boldsymbol{C}_n$ is the noise covariance matrix, and

$$f(\boldsymbol{m}) = \exp\left(-\frac{1}{2}(\boldsymbol{m}-\boldsymbol{m}_0)^T \boldsymbol{C}_m^{-1}(\boldsymbol{m}-\boldsymbol{m}_0)\right), \tag{2.33}$$

where $\boldsymbol{m}_0$ is the center (mean) of the prior density in the model space, and $\boldsymbol{C}_m$ is the prior model covariance matrix. From the above expression we get

$$p(\boldsymbol{m}) = \exp(-\boldsymbol{S}(\boldsymbol{m})), \tag{2.34}$$

where

$$\boldsymbol{S}(\boldsymbol{m}) = \frac{1}{2}\Big[(\boldsymbol{d}-\boldsymbol{Gm})^T \boldsymbol{C}_n^{-1}(\boldsymbol{d}-\boldsymbol{Gm}) + (\boldsymbol{m}-\boldsymbol{m}_0)^T \boldsymbol{C}_m^{-1}(\boldsymbol{m}-\boldsymbol{m}_0)\Big]. \tag{2.35}$$

It can be demonstrated (see, e.g., *Tarantola* [2005]) that $p(\boldsymbol{m})$ is a Gaussian density with mean

$$\boldsymbol{m}_{post} = \boldsymbol{m}_0 + \left(\boldsymbol{G}^T\boldsymbol{C}_n^{-1}\boldsymbol{G}+\boldsymbol{C}_m^{-1}\right)^{-1}\boldsymbol{G}^T\boldsymbol{C}_n^{-1}(\boldsymbol{d}-\boldsymbol{Gm}_0) \tag{2.36}$$

and covariance

$$\boldsymbol{C}_{post} = \left(\boldsymbol{G}^T\boldsymbol{C}_n^{-1}\boldsymbol{G}+\boldsymbol{C}_m^{-1}\right)^{-1}. \tag{2.37}$$

This is the analytical and fully probabilistic solution for the linear, Gaussian inverse problem.

2.4.5. Example: Inversion of Seismic Reflection Data

Consider again Figure 2.6 (left) where seismic reflection data are measured at the surface, with the aim of reconstructing subsurface structure in a vertical plane below the recording profile. The subsurface is parameterized by the logarithm of the acoustic impedance m_{ij} on a regular two-dimensional grid (x_i, z_j) in the plane, and the data are measured at a one-dimensional grid x_i at the surface. At the ith surface point we measure the seismogram

$$\boldsymbol{d}_i = \begin{pmatrix} d_{i1} \\ \vdots \\ d_{iN} \end{pmatrix} = \boldsymbol{Gm}_i, \tag{2.38}$$

where $\boldsymbol{m}_i$ is the ith column in the matrix $\boldsymbol{m} = \{m_{ij}\}$. $\boldsymbol{G}$ is given by $\boldsymbol{G} = \boldsymbol{WD}$, where $\boldsymbol{D}$ performs a differentiation of $\boldsymbol{m}_i$ to obtain an approximate reflectivity (valid when reflection coefficients are small), and $\boldsymbol{W}$ is a matrix that convolves the reflectivity with the source signal (see Figure 2.12, top left). This approximate expression for the seismogram is the well-known convolution model.

Inspection of the seismic data records gives us an impression of the subsurface structure. In this particular example with data taken from the DRUM dataset [*BIRPS*, 1984; *Mosegaard et al.*, 1997], the reflections from the Moho (around 6 s two-way time) are approximately the same for all horizontal positions, and this leads us to assume that the Earth is near-horizontally stratified at this location. Based on this assumption, we obtain an approximation to the noise by computing the average of all 10 data records and subtracting it from each record. Figure 2.6 (right) shows the noise $\boldsymbol{n}$ derived in this way. A histogram of the noise values

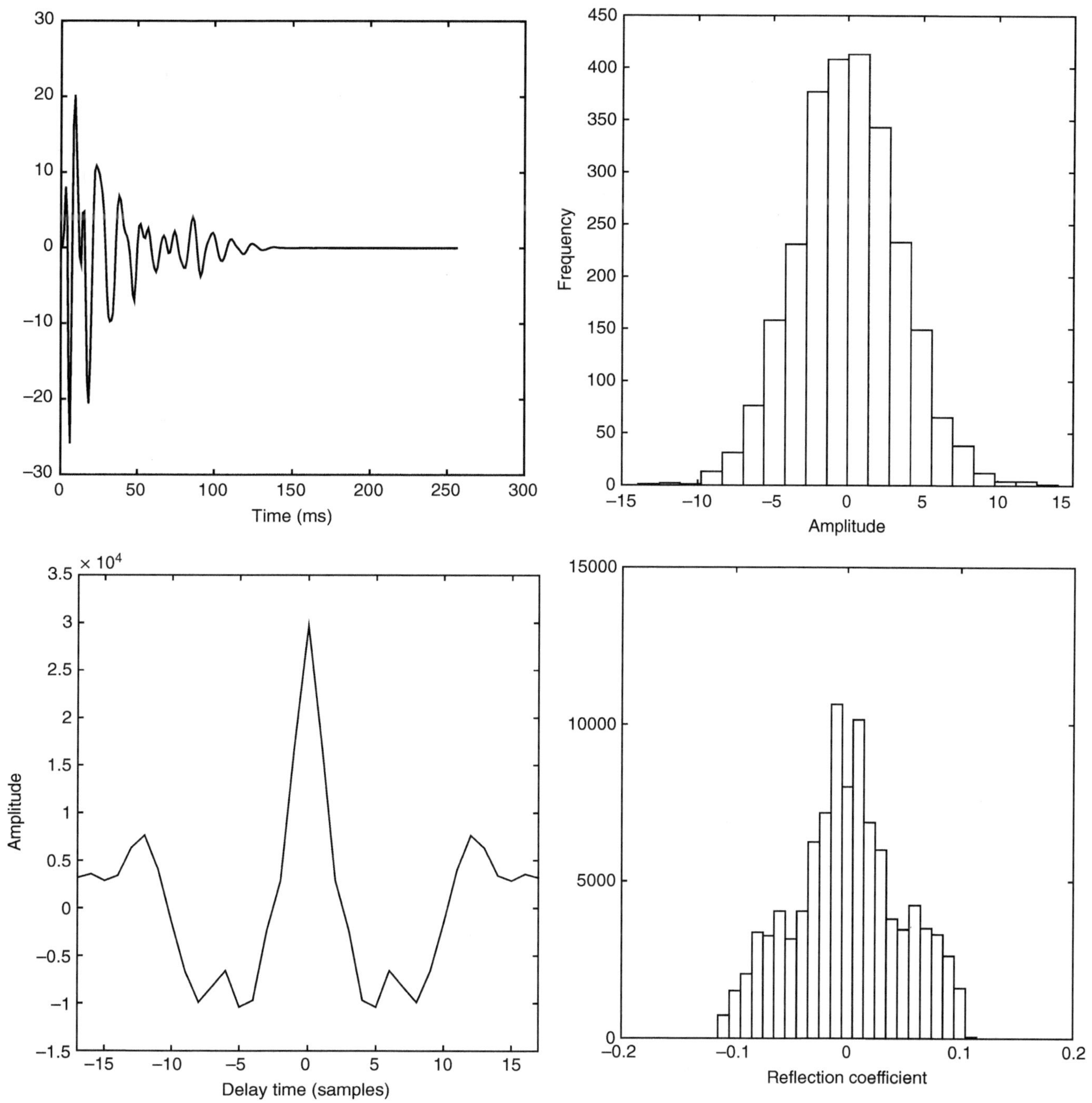

Figure 2.12 Top left: Recorded wavelet. **Top right:** Histogram of noise values estimated from data. **Bottom left:** Estimate of the (temporal) covariance function of the noise. **Bottom right:** Histogram of reflection coefficients derived from field measurements of rock properties.

(Figure 2.12, top right) suggests that we should use a Gaussian distribution with zero mean to describe $\boldsymbol{n}$. An analysis of the 10 records of $\boldsymbol{n}$ results in an estimate of the temporal, normalized covariance function shown in Figure 2.12 (bottom left). From this information and an estimated S/N ratio of 2.0, we can construct the noise covariance matrix $\boldsymbol{C}_n$.

Our prior information on acoustic impedances was derived from the data-independent empirical information shown in Figure 2.12 (bottom right), which is a histogram of reflection coefficients obtained from field studies of igneous intrusions of Rum, Scotland, and Great Dyke, Zimbabwe [*Singh and McKenzie*, 1993]. Assuming that a Gaussian distribution with zero mean and standard

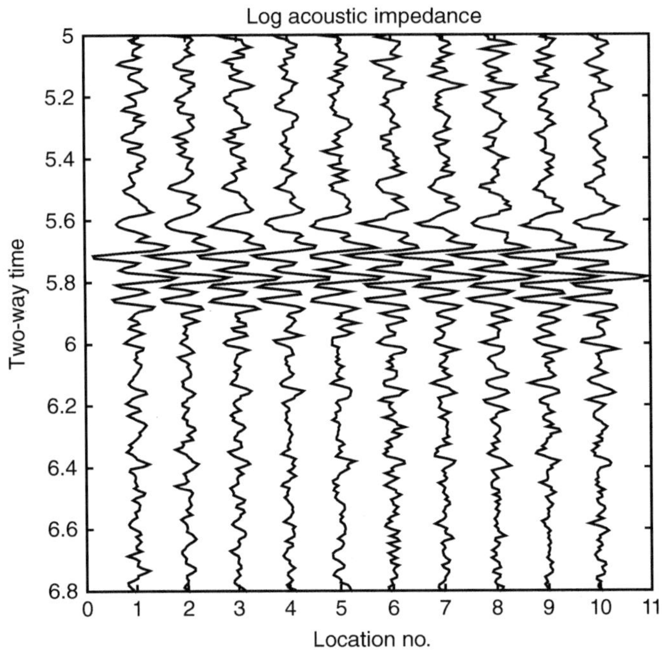

Figure 2.13 A posteriori mean model obtained from a linear, Gaussian inversion of the data shown in Figure 2.6 (left). The figure shows a plot of $\log(\boldsymbol{I}/\boldsymbol{I}_0)$ where $\boldsymbol{I}$ is the acoustic impedance and $\boldsymbol{I}_0$ is a (here arbitrary) reference value of $\boldsymbol{I}$.

deviation 0.047 can be used to represent this information, and assuming uncorrelated impedances, we can construct the a priori covariance matrix $\boldsymbol{C}_m$. We can now use expression (2.36) to compute the mean posterior model, shown in Figure 2.13. The posterior uncertainties are represented by the posterior covariance matrix (not shown), obtained from expression (2.37).

2.4.6. Gaussian, Mildly Nonlinear Problems Solved Through Linearization

A nonlinear inverse problem

$$\boldsymbol{d} = \boldsymbol{g}(\boldsymbol{m}) \tag{2.39}$$

may in some cases, when the nonlinearity of $\boldsymbol{g}$ is not too severe, be solved iteratively by a local approximation to a linear problem:

$$\boldsymbol{m}_{k+1} = \boldsymbol{m}_k + \epsilon_k \left(\boldsymbol{G}_k^T \boldsymbol{C}_n^{-1} \boldsymbol{G}_k + \boldsymbol{C}_m^{-1}\right)^{-1} \left(\boldsymbol{G}_k^T \boldsymbol{C}_n^{-1} \boldsymbol{d} - \boldsymbol{g}(\boldsymbol{m}_k)\right) \\ -\boldsymbol{C}_m^{-1}\left(\boldsymbol{m}_k - \boldsymbol{m}_0\right), \tag{2.40}$$

where

$$\boldsymbol{G}_k = \left(\frac{\partial g_i}{\partial m_j}\right)_{m=m_k}. \tag{2.41}$$

In the limit $k \to \infty$ we obtain the local posterior covariance (of the tangent Gaussian centered at $\boldsymbol{m}_\infty$):

$$\boldsymbol{C}_{post} \approx \left(\boldsymbol{G}_\infty^T \boldsymbol{C}_n^{-1} \boldsymbol{G}_\infty + \boldsymbol{C}_m^{-1}\right)^{-1}. \tag{2.42}$$

2.5. SAMPLING A PROBABILITY DENSITY

In probabilistic, nonlinear inversion with complex priors, Monte Carlo methods are often the only method that is able to characterize distributions in high-dimensional model spaces [*Mosegaard*, 2006]. The idea is to attempt to produce *independent* realizations from the posterior, but the problem of generating such realizations may be difficult to solve. In some cases it may even be practically unsolvable.

The difficulties mainly arise from the fact that, in many applications, the probability density f to be sampled is not fully known. Often, an explicit closed-form expression for the probability density is not available, and the only way to acquire information about f is to pick one point $\boldsymbol{x} \in X$ at a time and to evaluate $f(\boldsymbol{x})$ at the selected point using some numerical algorithm. We shall in the following use the terminology that "$f(\boldsymbol{x})$ can only be evaluated pointwise" for this important case.

2.5.1. Rejection Sampling

It is quite easy to design a so-called "perfect" sampler—that is, a method that draws perfectly independent sample points according to a probability density which can only be evaluated pointwise. A simple method, called the *rejection sampler*, is available if a number $M \geq max_x f(x)$ is given: Each iteration consist of two steps, where in the first step a point $\boldsymbol{x}$ is drawn uniformly at random from X and in the second step an acceptance probability of $f(\boldsymbol{x})/M$ is used to decide if the point is accepted. For problems with many parameters, however, his method is extremely inefficient, even when $M = max_x f(x)$. The reason is that in many practical applications, where the space X is high-dimensional, the probability density $f(\boldsymbol{x})$ is near-zero almost everywhere in X. Hence, the average waiting time between finding points with a high value of $f(\boldsymbol{x})$ is extremely large, and the algorithm is very slow. This method is only of practical interest in low-dimensional spaces.

A more efficient version of the rejection sampler is obtained if we already have a "candidate density" $h(\boldsymbol{x})$, which is an approximation to $f(\boldsymbol{x})$. In this case we can choose a number M such that

$$M \geq \max_x \frac{f(x)}{h(x)}.$$

Then each iteration will consist of two steps where a point $\boldsymbol{x}$ is first drawn randomly from X with probability density $h(\boldsymbol{x})$, after which an acceptance probability of $f(\boldsymbol{x})/(Mh(\boldsymbol{x}))$ is used to decide if the point is accepted. If $h(\boldsymbol{x})$ is a good approximation to $f(\boldsymbol{x})$, the number M can be chosen close to 1, leading to generally high acceptance probabilities.

2.5.2. The Metropolis Algorithm

An efficient and widely used sampling method is the *Metropolis Algorithm* [*Metropolis et al.*, 1953]. To sample a probability density f, this algorithm picks, in each iteration, the next realization from pair of points $\boldsymbol{x}_1$ and $\boldsymbol{x}_2$ (the current point and a 'candidate' point, respectively) with probabilities proportional to $f(\boldsymbol{x}_1)$ and $f(\boldsymbol{x}_2)$, respectively. This ensures, in the long run, a correct balance between sample densities at all points visited by the algorithm and therefore, asymptotically, produces a sample of $f(\boldsymbol{x})$, even if $f(\boldsymbol{x})$ is unnormalized and only can be evaluated pointwise. More precisely, the algorithm is, for our purposes, defined in the following way:

Algorithm *(Metropolis)*. *Given a (possibly unnormalized) probability density $f(\boldsymbol{x})$ over the manifold X, a random function $V(\boldsymbol{x})$ which samples a constant probability density if applied iteratively:*

$$\boldsymbol{x}^{(n+1)} = V\left(\boldsymbol{x}^{(n)}\right) \tag{2.43}$$

and a random function $U(0, 1)$ generating a uniformly distributed random number from the interval [0, 1]. *The random function W, which iteratively operates on the* current *parameter vector $\boldsymbol{x}^{(n)}$ and produces the next parameter vector $\boldsymbol{x}^{(n+1)}$:*

$$\boldsymbol{x}^{(n+1)} = W\left(\boldsymbol{x}^{(n)}\right) = \begin{cases} V\left(\boldsymbol{x}^{(n)}\right) & \text{if } U(0,1) \leq \min\left[1, \dfrac{f\left(V\left(\boldsymbol{x}^{(n)}\right)\right)}{f\left(\boldsymbol{x}^{(n)}\right)}\right] \\ \boldsymbol{x}^{(n)} & \text{otherwise} \end{cases} \tag{2.44}$$

asymptotically samples the probability density $Cf(\boldsymbol{x})$, where C is a normalization constant.

The word "asymptotically" means in this case that the set of points $\boldsymbol{x}^{(1)}, \ldots, \boldsymbol{x}^{(n)}$ visited in n successive steps by the algorithm converges towards a sample of f as n goes to infinity.

A possible extension of the Metropolis Algorithm would be to pick the next realization, not from a pair of points $\boldsymbol{x}_1$ and $\boldsymbol{x}_2$ (where $\boldsymbol{x}_1$ is the current point) but from a large collection of points $\boldsymbol{x}_1, \boldsymbol{x}_2, \boldsymbol{x}_3, \ldots$ (including the current point) with probabilities proportional to $f(\boldsymbol{x}_1)$, $f(\boldsymbol{x}_2)$, $f(\boldsymbol{x}_3)$, ..., respectively. This would again ensure, in the long run, a correct balance between sample densities in all regions of X visited by the algorithm and therefore asymptotically produce a sample of $f(\boldsymbol{x})$. An example of this strategy is the sampling algorithm introduced by *Geman and Geman* [1984], the so-called *Gibbs Sampler*, where the next realization is picked from a large collection of equidistant points $\boldsymbol{x}_1, \boldsymbol{x}_2, \boldsymbol{x}_3, \ldots$ located along an axis-parallel line (through the current point) in the parameter space. In a typical implementation of this method, all axis directions for $\boldsymbol{x}_1, \boldsymbol{x}_2, \boldsymbol{x}_3, \ldots$ are visited cyclically.

The Metropolis algorithm and its relatives are so-called *Markov chains*. For a Markov chain, the probability of visiting a point in X in a given iteration is only conditionally dependent on the point visited in the previous iteration (and not on any of the earlier iterations). In this sense a Markov chain has the shortest possible "memory," and this is an advantage from the point of view of simplicity and required memory for the algorithm. However, the fact that a Markov chain algorithm discards information on the probability density to be sampled will obviously limit its efficiency.

2.5.3. Genetic Algorithms

Another notable category of Monte Carlo techniques are the genetic algorithms. They were originally proposed by *Fogel et al.* [1966], who established the concept of evolutionary computation, but it was *Holland* [1975] who developed them to powerful search techniques for global optimization. Genetic algorithms are not samplers in the same sense as the rejection sampler and the Metropolis algorithm, because they do not generate a set of points in sampling space whose sampling density function is proportional to f. They do, however, generate points from areas of high probability and can provide initial input points to a resampling algorithm (e.g., a Metropolis algorithm) whose outputs are approximate samples of f. The efficiency of there sampling will depend critically on how close the point density of its input is to f.

A characteristic of genetic algorithms is that they work with a population of several models simultaneously. The population is initially generated randomly, but in each iteration it is altered by three different actions named *mutation*, *recombination,* and *selection*. Each action mimics, and is named after, a genetic process that is known in Nature: Mutation randomly changes a parameter in one or several models, recombination randomly swaps parameter values between individual models, and selection allows models with a poor datafit to be replaced by copies of models with a better datafit. In each iteration, the misfit for each model is evaluated by solving the forward problem for each of the models, and the purpose of the genetic algorithm is then to seek out models in parameter space with an acceptable fit. For more information about genetic algorithms, see, for example, *Davis* [1987], *Goldberg* [1989], *Rawlins* [1991], *Whitley* [1994], *Gallagher and Sambridge* [1994], and *Winter et al.* [1995].

2.5.4. The Neighborhood Algorithm

The Metropolis Algorithm is extensively used for sampling in analysis of inverse problems. However, it is well known amongst practitioners of this approach that, when the number of sample points is too limited, there is a danger that certain parts of the parameter space remain unexplored, and important solutions are overlooked.

The *neighborhood algorithm* [*Sambridge*, 1998, 1999a, 1999b] seeks to avoid this problem. An extensive exploration of the parameter space with probability density f is performed by generating a set of points, in successive generations, whose point density is an approximation to f for high values of f and is derived from all previous models. The point density is approximately constant on nearest-neighbor (Voronoi) cells about each of the previous models. The approximation is regularly updated, and the generated points are then concentrated in several regions of high probability. The number of vertices of Voronoi cells grows exponentially with the number of model parameters, but the neighborhood algorithm does not require their explicit calculation and hence remains practical in spaces with many dimensions.

Like a genetic algorithm, the neighborhood algorithm is not a sampler, but needs to be supplemented with a resampling procedure to produce samples with the given target density f.

2.5.5. Recent Developments

Methods for Probabilistic/Bayesian sampling and optimization is a vast and fast developing field in computational statistics. A recent example of a method that has been introduced in geoscience is the Parallel Tempering algorithm (see *Sambridge* [2014] for a review and some examples). In this method, originally proposed by *Swendsen and Wang* [1986], N randomly initialized copies of the sampling are run in parallel at different temperatures (noise levels). Based on the Metropolis acceptance rule, configurations (sets of model parameters) at different temperatures can be exchanged, and in this way configurations at high temperatures are made available to simulations at low temperatures. The result is an algorithm that has performed very well in a number of applications.

A recent and completely different algorithm for probabilistic inference, currently receiving considerable attention, is the method of Optimal Maps [*El Moselhy and Marzouk*, 2012]. The method is not even based on Monte Carlo calculations. Instead, it is using an optimization method to construct a mapping from the prior distribution to the posterior distribution. The algorithm avoids many of the computational difficulties known from Markov chain Monte Carlo. Among the benefits are analytical expressions for moments of the posterior distribution, and an ability to generate an arbitrary number of independent samples of the posterior distribution without additional likelihood evaluations. The future will show what this type of algorithm has to offer to geoscience applications.

2.6. INVERSE THEORY IN PRACTICE: LIMITATIONS AND INTERPRETATIONS

2.6.1. The Dilemma of Sparse Parameterizations

Representing the Earth with a large number of orthogonal basis functions and corresponding model parameters not only allows us to make close approximations to the real Earth, it also ensures that different analysts, using different basis functions, will work with the same Earth model, except for a small error.

In some cases, however, the computational burden of data inversion is so large that it is tempting, or even necessary, to use what is known as *sparse methods*. These methods attempt at representing the Earth with a limited number of parameters, thereby paving the way for inversion of large-scale datasets. However, when using sparse parameterizations (sparse-spike inversion, compressed sensing, transdimensional methods, wavelet transform methods, etc.) it is important to be aware of the inherent risk of this approach: Different analysts may arrive at different representations of the Earth, depending on their initial choice of basis functions.

Nonprobabilistic sparse methods are commonly based on the following idea: Imagine an inverse problem with many unknown parameters—so many that the problem is very underdetermined by the given data. Assume that the presence of noise leads us to conclude that the resolving power of the data is insufficient to allow meaningful computation of all the parameters. Now suppose that we remove one parameter at a time and that we, in each step, solve the inverse problem. In the beginning we can fit the data perfectly, due to the many parameters, but when the number of parameters are fewer than the data, the problem may become overdetermined, and the misfit becomes positive. It grows for each parameter we remove, and at some point the misfit is so large that we fit the data "barely within the error bars." Now we are satisfied, because we have avoided overfitting. We stop the process, and what we have left is a *sparse solution* given as M model parameters and the value of these parameters.

However, this approach introduces a dilemma: Every time we select a set of discrete parameters m_1, m_2, ... for an inverse problem, we are implicitly choosing a set of basis functions. In this process we have a choice, because there are many complete sets of basis functions available: Some are smooth like sines and cosines, others are rough (discontinuous and/or have discontinuous derivatives) like "sawtooth" functions or indicator functions (constant in a range, 0 outside). They can all approximate equally well, given that the number of basis functions is sufficiently large (Figures 2.3 and 2.4). However, if you use a sparse inversion technique, you will strive towards using a small number of parameters, and hence a small number of basis functions, to build your solution. Consequently, the "optimal" M depends (strongly) on the chosen set of basis functions. A model that is "simple" (in the sense that it is described by few parameters) when using one set of basis functions is "complex" (requires many parameters) when using a different set of basis functions. This means that different analysts who choose different basis functions will, in general, disagree about the simplicity of an Earth model.

As seen in Figure 2.14, sparse methods will give your solution a strong imprint of the basis functions: If you

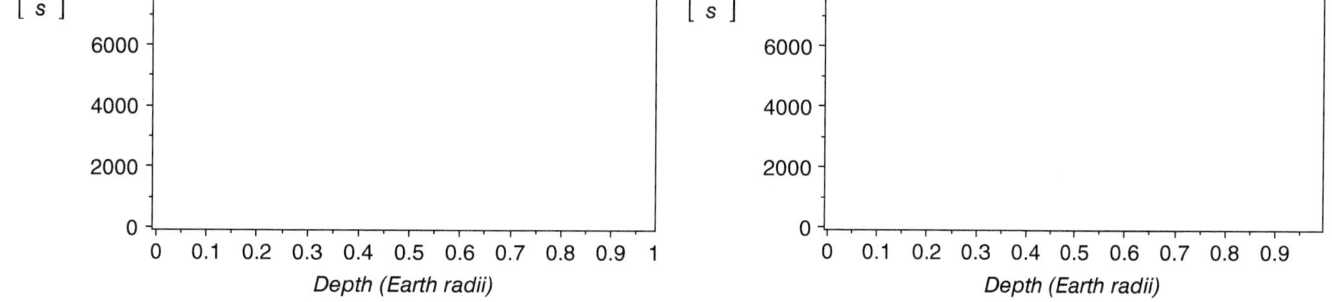

Figure 2.14 Two sparse representations of the true P-wave velocity function in Figure 2.1, having approximately the same misfit (in this case: L_2-distance to the true function). **Left:** Expansion through 16 Haar-basis functions. **Right:** Expansion through four Fourier (sin/cos)-basis functions. It is clear that the different character of the two sets of basis functions makes a strong imprint on the shape of their expansions: for example, one analyst finds marked discontinuities, but the other analyst finds no discontinuities. For this reason, if two analysts use different sets of basis functions, their results will be inconsistent (contradictory).

select smooth functions, it will be smooth; if you choose "saw tooth" functions, it will be serrated; if you choose indicator functions, it will be piecewise constant; and so on. Your choice of basis functions (which is dictating your choice of parameters) determines the solution's appearance.

Another important lesson can be learned from the example in Figure 2.14: The sparsest model with a given accuracy in the Haar-basis needs 16 model parameters, whereas the sparsest model in the Fourier-basis (with approximately the same accuracy) needs only four parameters. In other words, "simplicity" (the number of parameters N) is not an objective property of a model, and certainly not a physical property.

In recent years, a probabilistic formulation of sparse inversion, named *transdimensional probabilistic sampling*, has been introduced into geosciences (see, e.g., *Malinverno* [2002], *Sambridge et al.* [2006], or *Bodin et al.* [2012]). This method treats not only parameter values, but also the number of parameters as random variables, and allows sampling in this extended space, either for a single class of basis functions or for several classes of basis functions. In contrast to nonprobabilistic, sparse techniques, this method does not strive to minimize the number of parameters, but rather samples a wide set of model structures, allowing a more general exploration of models than is possible with deterministic methods. In this way the dependence on particular sets of basis functions is reduced.

It should be noted that, in a probabilistic formulation of a problem with sparse parameterization, care should be taken when defining prior information about the number of parameters N. When N is related to (or can be derived from) physical information—for instance, when N is the number of layers in an Earth model—it may be possible to define prior probabilities. However, if N is a pure mathematical entity, determined through an arbitrary choice of the analyst, this is hardly meaningful.

In cases where the computational workload is manageable, a safe and consistent choice is to use as many parameters as practically possible and to ensure sensible models through the definition of hard or probabilistic prior constraints on these parameters.

2.6.2. Inverse Theory and Statistics

Is inverse theory physics, mathematics or statistics? This question is usually left unanswered, resulting in a series of misunderstandings not without consequences for practical applications. We shall try to give a partial answer to this question. First, it is important to understand the differences between statistics and inverse theory. Statistics is the science of constructing probability distributions from empirical data, and to this aim, parameterized probability distributions are normally assumed. Statistical theory is then used to estimate these parameters. Two schools of thought dominate in statistics: The first, so-called "frequentist" thinking, sees parameters of probability distributions (mean values, standard deviations, etc.) as plain numbers to be determined from frequencies of outcomes from the distributions. The other school of thought, the Bayesian school (see, e.g., *Bernardo and Smith* [2000]), sees the parameters as random variables with their own probability distributions. These so-called prior distributions must be provided by the analyst, and to frequentists the "subjectivity" of the prior is seen as a weakness of the Bayesian approach and has been heavily criticized over the years.

Despite this criticism, Bayesian statistics has been the backbone of what we now call probabilistic inversion (often named "Bayesian inversion"). This has been a very fertile development, but it also has its problems. A major problem is the statistical concept of a "parameter." In inverse theory, a parameter is describing a physical entity, and importantly, this means that prior information about such a parameter can be obtained from physical observations or physical arguments (e.g., symmetry considerations). This is in sharp contrast to traditional statistical parameters like covariances or medians, or the number of parameters N used to parameterize a distribution. Often, such parameters are unphysical (cannot be related to empirical data) and therefore have no physically meaningful prior distribution.

The existence of these two distinct classes of parameters sometimes creates a problem when statistical thinking is used uncritically in inverse theory. It arises because probability distributions used in probabilistic inversion are both functions of the traditional statistical parameters and functions of physical parameters describing the Earth. If no distinction is made between the two parameter categories, it is tempting to estimate not only the physical Earth parameters, but also unphysical, statistical model parameters, such as a priori standard deviations of Earth parameters or the number of parameters used to characterize Earth structure. The problem with this is twofold: Firstly, no meaningful prior can usually be assigned to unphysical parameters, and secondly, computing such parameters from the data violates the information flow that characterizes inverse problems. In this flow, the noise distribution and the prior information is input information.

2.6.3. Inverse Theory and Mathematics

Inverse problem theory belongs to mathematical physics. Superficially, it looks like a purely mathematical topic: Parameters are identified, probability densities are defined, and equations relating the individual parameters are used to solve the problem. But it is important to remember that inverse theory—like other branches of mathematical physics—must obey significant external constraints: It must be physically meaningful. Not all equations in inverse theory that have a mathematical solution have a physically meaningful solution.

Physical meaningfulness is closely related to the single most important principle of mathematical physics, *the principle of invariance (covariance)*:

• Consequences of a physical theory must be independent of the chosen parameterization (reference frame).

This principle asserts that two different, independent analysts are free to choose their own parameterization of a problem, but the physical laws must be formulated such that, when they have finished their analysis, and one of the analysts transforms his/her result to the parameterization of the other analyst, there should be no disagreement between their results. Otherwise, objectivity is lost.

As we have seen, special difficulties occur if we switch between parameter systems with incompatible distance/volume measures. A transformation from resistivity to conductivity gave conflicting results when trying to define Gaussian probability densities for their uncertainty.

Failure to apply the fundamental principle of invariance when analyzing inverse problems may give mathematically acceptable results, but physically inconsistent models.

ACKNOWLEDGMENTS

The authors wish to thank Andrea Zunino, Knud Cordua, associate editor Amir Khan, and reviewers Malcolm Sambridge and Miguel Bosch for constructive comments and criticism about the manuscript and about inverse problems and the use of probabilistic methods in general.

REFERENCES

Backus, G., and F. Gilbert (1967), Numerical applications of a formalism for geophysical inverse problems, *Geophys. J. R. Astron. Soc.*, *13*, 247–276.

Backus, G., and F. Gilbert (1968), The resolving power of gross Earth data, *Geophys. J. R. Astron. Soc.*, *16*, 169–205.

Backus, G., and F. Gilbert (1970), Uniqueness in the inversion of inaccurate gross Earth data, *Philos. Trans. R. Soc. London*, *266*, 123–192.

Bernardo, J. M., and Smith, A. F. M. (2000), *Bayesian Theory*, 1st ed., John Wiley & Sons, Hoboken, NJ.

Conway, J. B. (1990), *A Course in Functional Analysis*, 2nd ed., XVI, Springer Graduate Texts in Mathematics, Vol. *96*. Springer, New York, 400 pages.

Davis, L. (1987), *Genetic Algorithms and Simulated Annealing*, Research Notes in Artificial Intelligence, Pitman, London.

El Moselhy, T. A., and Marzouk, Y. M. (2012), Bayesian inference with optimal maps, *J. Comput. Phys.*, *231*, 7815–7850.

Fogel, L.J., A. J. Owens, and M. J. Walsh (1966), Artificial Intelligence through Simulated Evolution, John Wiley, New York.

Gallagher, K., and M. Sambridge (1994), Genetic algorithms: A powerful tool for large-scale non-linear optimization problems, *Comput. Geosci.*, *20*(7/8), 1229–1236.

Geman, S., and D. Geman (1984), Stochastic relaxation, Gibbs distributions and the Bayesian restoration of images, *IEEE Trans. Pattern Anal. Mach. Intell.*, *6*, 721–741.

Goldberg, D. E. (1989), *Genetic Algorithms in Search, Optimization, and Machine Learning*, Addison-Wesley, Reading, MA.

Hadamard, J. (1923), *Lectures on Cauchy's Problem in Linear Partial Differential Equations*, Yale University Press, New Haven.

Keilis-Borok, V. J., and T. B. Yanovskaya (1967), Inverse problems in seismology (structural review), *Geophys. J. R. astr. Soc.*, *13*, 223–234.

Kolmogorov, A. (1933), Grundbegriffe der Wahrscheinlichkeitsrechnung (in German), Berlin: Julius Springer.

Metropolis, N., M. N. Rosenbluth, A. W. Rosenbluth, A. H. Teller, and E. Teller (1953), Equation of state calculations by fast computing machines, *J. Chem. Phys. 21*, 1087–1092.

Mosegaard, K., and A. Tarantola (1995), Monte Carlo sampling of solutions to inverse problems, *J. Geophysi. Res.*, *100*, B7, 12431–12447.

Mosegaard, K., S. C. Singh, D. Snyder, and H. Wagner (1997), Monte Carlo analysis of seismic reflections from Moho and the W-reflector, *J. Geophys. Res. B*, *102*, 2969–2981.

Mosegaard, K. (2006), Monte Carlo Analysis of Inverse Problems, Doctoral thesis, University of Copenhagen, ISBN 87-991228-0-4.

Rawlins, G. J. E. (ed.), (1991), *Foundations of Genetics Algorithms*, Morgan Kaufmann, Burlington, MA.

Sambridge, M. (1998), Exploring multi-dimensional landscapes without a map, *Inverse Probl.*, *14*(3), 427–440.

Sambridge, M. (1999a), Geophysical inversion with a neighborhood algorithm, I, Searching a parameter space, *Geophys. J. Int.*, *138*, 479–494.

Sambridge, M. (1999b), Geophysical inversion with a neighborhood algorithm, II, Appraising the ensemble, *Geophys. J. Int.*, *138*, 727–746.

Sambridge, M. (2014), A Parallel tempering algorithm for probabilistic sampling and multimodal optimization, *Geophys. J. Int.*, *196*, 357–374.

Singh, S. C., and D. P. McKenzie (1993), Layering in the lower crust, *Geophys. J. Int.*, *113*, 622–628.

Swendsen, R. H., and Wang, J. S. (1986), Replica Monte Carlo simulation of spin glasses, *Phys. Rev. Lett.*, *57*, 2607–2609.

Tarantola, A. (2005), *Inverse Problem Theory and Methods for Model Parameter Estimation*, Society of Industrial and Applied Mathematics, Philadelphia.

Tikhonov, A. N. (1943), On the stability of inverse problems, *C. R. (Doklady) Acad. Sci. URSS (N.S.)*, *39*, 176–179.

Tikhonov, A. N. (1963), Solution of incorrectly formulated problems and the regularization method, *Doklady Akad. Nauk USSR 151*, 501–504.

Whitley, D. L. (1994), A genetic algorithm tutorial, *Stat. Comput.*, *4*, 65–85.

Winter, G., J. Periaux, M. Galan, and P. Cuesta, eds., *Genetic Algorithms in Engineering and Computer Science*, John Wiley & Sons, New York.

3

Inference Networks in Earth Models with Multiple Components and Data

Miguel Bosch

ABSTRACT

The integration of information for the inference of earth structure and properties can be treated in a probabilistic framework by considering a posterior probability density function (PDF) that combines the information from a new set of observations and a prior PDF. To formulate the posterior PDF in the context of multiple datasets, the data likelihood functions are factorized assuming independence of uncertainties for data originating across different surveys. A realistic description of the earth medium requires the modelization of several properties and other structural parameters, which relate to each other according to dependency and independency notions. Thus, conditional probabilities across model components also factorize. The relationships across model components can be described via a direct acyclic graph. The basic rules for factorization of the posterior PDF are easily obtained from the graph organization. Once the posterior probability has been formulated, realizations can be obtained following a sampling approach or searching for a maximum posterior probability earth medium configuration. In the first case, sampling algorithms will adapt to the factorized structure of the posterior PDF. In the second case, iterative second- or first-order approximations of the objective function conduce to the solution of a system of equations for the model update.

3.1. INTRODUCTION

The appraisal of solid Earth structure and properties requires modeling the medium's heterogeneous composition and lithotypes, the morphology of geological bodies, pore fluids, fractures, knowledge, and information that are linked at various scales, and the medium's physical response to field measurements. This implies the description of various types of (1) medium properties and structural parameters, (2) observations provided by surveys, well-logs, and rock sample studies, and (3) knowledge to establish relationships across the model properties and observations. The full multiplicity of components and intervening relationships could be considered too large to be workable. Nevertheless, given the goal of a specific study and the available information, a relevant set of such components and relationships can be retained for modeling, while the rest is neglected. The appropriate selection conforms to the pertinence of the model component to the phenomenon observed and the goal of the study.

Integration of multiple data, information, and knowledge has been considered a key issue in natural sciences, and particularly in Earth sciences. The relevant information is heterogeneous in its nature (properties, objects, phenomena, scales) and available at diverse treatment levels: The data are the raw support of the information (processed and interpreted data), which is understood in terms of knowledge (a successful theory). With the advent of larger computational possibilities in the past decades, the quantitative treatment of multicomponent

Applied Physics Department, Engineering Faculty, Universidad Central de Venezuela, Caracas, Venezuela

Integrated Imaging of the Earth: Theory and Applications, Geophysical Monograph 218, First Edition.
Edited by Max Moorkamp, Peter G. Lelièvre, Niklas Linde, and Amir Khan.

complex models and observations is in progress to provide (a) more realism to the description of the Earth media and (b) more accuracy and precision to the estimates [*Linde et al.*, 2006a; *Bosch et al.*, 2010; *Torres-Verdin et al.*, 2012]. Examples of joint inversion of multiple geophysical datasets, in various Earth sciences inference contexts, are described in the works by *Lines et al.* [1988], *Haber and Oldenburg* [1997], Bosch [1999], *Bosch et al.* [2001, 2004], *Tiberi et al.* [2003], *Gallardo et al.* [2003], *Gallardo and Meju* [2004], *Linde et al.* [2006b], *Guillen et al.* [2007], *Khan et al.* [2007], *Alpak et al.* [2008], *Doetsch et al.* [2010], *Buland and Kolbjomsen* [2012], and *Chen and Hoversten* [2012].

The present work focuses on the formulation of inverse problems in Earth sciences, for the case of models configured with multiple spatially distributed properties, subjected to different types of physical observations and/or embodying multiple relationships across properties and observations. The nature of the formulation and approaches to solve inverse problems is not different for this scenario. The issues to solve consists in (1) how to compose relationships and information across the model components and (2) how to draw from the composed information realizations of the joint model, or maximum posterior probability model configurations, according to the solution approach followed. To explain these issues, I will not follow a rigorous deductive path. I will mention the basic principles and illustrate their formulation for various common examples of inferential interest in earth sciences.

The first section of this chapter describes the formulation of posterior probability densities for complex models, for the case where their components are structured in hierarchical layers, and for multiple sets of data. The second section unfolds the structure of the posterior probability density for less structured models, via the support of direct acyclic graphs [*Pearl*, 1986; *Thulasiraman and Swamy*, 1992] that describe the relations across the model components and datasets. The third section describes the stochastic approach to the solution of the inverse problem by sampling the posterior density with Markov Chains, following its factor structure. The fourth section describes the optimization approach to the solution of the inverse problem, which consists of searching for the model configuration that maximizes the posterior density. Finally, the discussion and conclusion sections close the chapter.

3.2. MULTIPLE PHYSICAL OBSERVATIONS AND MODEL COMPONENTS

The general formulation of inverse problems is outlined here in a probabilistic framework, within the scope of Bayesian inference. The term Bayesian refers to the interpretation of probabilities as a description of the information, knowledge, or uncertainty on parameter spaces that support a model of a natural object or phenomenon. A state of information about the modeled object is described by the *probability density function* (PDF) defined over the model parameter space or, equivalently, by the *cumulative distribution function* (CDF).The variables in the parameter space are considered *random* as they take different values each time they are evaluated and represent the corresponding state of information. Each time the model parameters are evaluated, their outcome is drawn in proportion to the corresponding PDF realizing a different configuration of the modeled earth medium. The discussion section expands the analysis of the relationship between the model parameter space and the modeled object space.

The common formulation of inverse problems considers a *prior* state of information, a set of observations related with the modeled object, and the improved *posterior* state of information resulting from the combination of the prior and the new information provided by the data interpretation. The *posterior* probability density is given by

$$\sigma(\mathbf{m}) = c\rho(\mathbf{m}) L_{\text{data}}(\mathbf{m}), \tag{3.1}$$

where **m** is a multivariate random variable in the model parameter space, c is a normalization constant, ρ (**m**) is the prior PDF, and $L_{\text{data}}(\mathbf{m})$ is the data likelihood function, which embodies the new information provided by the observations. Normalization constants will be included in the following equations of the posterior PDF but not further identified.

As we know, observations are not commonly made directly on the model parameters, but in terms of additional related parameters that we refer to as *data*. The knowledge of the relationship between the data and the model parameters, including the associated uncertainties, provides the means to transform the former into information of the latter—that is, interpret the data in terms of the modeled object information. The formal derivation of the data likelihood function in (3.1) depends on the formulation of the relationship between the model and data spaces and the associated uncertainties. In the general case [*Tarantola*, 2005], the information provided by the theory and the observations are modeled independently and combined in a joint model-data space. The data likelihood function is calculated as a marginal non-normalized probability in the model parameter space,

$$L_{\text{data}}(\mathbf{m}) = \int \theta_{\text{theory}}(\mathbf{d} \mid \mathbf{m}) \rho_{\text{obs}}(\mathbf{d}) d\mathbf{d}. \tag{3.2}$$

Above, **d** are the true data parameters here considered with uniform homogeneous probability density. It is the data that would have been observed in the absence of observational and data processing errors. The probability density $\rho_{\text{obs}}(\mathbf{d})$ describes the information on the true data provided by the observation experience, $\mathbf{d} = \mathbf{d}_{\text{obs}} + \Delta\mathbf{d}_{\text{obs}}$, via the corresponding observed data $\mathbf{d}_{\text{obs}}$ and the associated uncertainties $\Delta\mathbf{d}_{\text{obs}}$. The conditional probability density $\theta_{\text{theory}}(\mathbf{d} \mid \mathbf{m})$ describes the data–model relationship, based on theoretical or empirical knowledge, including the data modeling uncertainties. The homogeneous probability density [*Mosegaard*, 2011; *Mosegaard and Tarantola*, 2002; *Tarantola*, 2005] describes the state of null information about the parameter. The homogeneous PDF of the data should be included in the denominator of the integrand in (3.2), in the case where it is not modeled as uniform (constant).

It is also common to model the observational uncertainties within the data–model conditional, instead of providing the independent observational PDF, $\rho_{\text{obs}}(\mathbf{d})$, present in the general expression (3.2). In this case the data likelihood function can be formulated straight forwardly as

$$L_{\text{data}}(\mathbf{m}) = \Theta_{\text{theory}}(\mathbf{d}_{\text{obs}} \mid \mathbf{m}), \tag{3.3}$$

where $\mathbf{d}_{\text{obs}}$ are the observed data. Both formulations are equivalent if uncertainties are appropriately modeled. The reader is referred to the work by *Tarantola* [2005] for a general derivation of the above expressions from the theory of combination of information states, generalized for parameters with nonuniform homogeneous probability densities. Equations (3.1)–(3.3) are the common basis for statistical inference, although derived from different theoretical approaches.

3.2.1. Likelihood Function Factorization

Let us now consider a model with various inner components of model parameters, $\mathbf{m} = \{\mathbf{m}_1, \mathbf{m}_2, \ldots, \mathbf{m}_K\}$, influenced by various types of data observations, $\mathbf{d} = \{\mathbf{d}_1, \mathbf{d}_2, \ldots, \mathbf{d}_N\}$, as required in integrated data modeling. The inner components of the model parameters, $\mathbf{m}_k$, could correspond to different property fields, the same property fields at different scales, geometric boundaries of geological objects, and other subsets of the model parameters specific case. Each of these multiple components is commonly of high dimensionality, such as a property field distributed on a spatial grid. The data components are partitioned according to different observational phenomenon (gravity, seismic, electric), derived data (seismic travel times, amplitudes, frequencies), or field surveys. We can decompose the observational PDF in the joint data space by the product of marginals,

$$\begin{aligned}\rho_{\text{obs}}(\mathbf{d}) &= \rho_{\text{obs}N}(\mathbf{d}_N)\rho_{\text{obs}N-1}(\mathbf{d}_{N-1})\ldots\rho_{\text{obs}1}(\mathbf{d}_1)\\ &= \prod \rho_{\text{obs}n}(\mathbf{d}_n),\end{aligned} \tag{3.4}$$

under the assumption of independent observational data uncertainties across the surveys. Recall that the observational PDF, $\rho_{\text{obs}}(\mathbf{d})$, only embodies information about the measurement process and does not anticipate the posterior information in data space—that is, as present in the posterior marginal of (3.1). The assumption is well justified for different surveys, which commonly use different instrumentation and field teams, and for different observational phenomenon (rock samples, well-logs, seismic experiment, gravity measurement). Nevertheless, in a strict sense, Eq. (3.4) is an approximation, as measurements across different surveys may be affected by commonly used information (a common digital elevation model, seasonal terrain conditions, or other factors). We assume herein that possible correlated factors are minor compared with the total observational uncertainty. The above decomposition of the observational PDF can in some cases be applied to data components derived from the same survey. As an example, various types of partial data can be obtained from seismic surveys, such as phase travel times, reflection amplitudes, and frequency content, which are commonly interpreted independently assuming unrelated uncertainties.

Similarly, the theoretical conditional probability is composed by the product of conditional marginals,

$$\begin{aligned}\Theta_{\text{theory}}(\mathbf{d} \mid \mathbf{m}) &= \Theta_{\text{theory}N}(\mathbf{d}_N \mid \mathbf{m})\Theta_{\text{theory}N-1}(\mathbf{d}_{N-1} \mid \mathbf{m})\ldots\\ &\Theta_{\text{theory}1}(\mathbf{d}_1 \mid \mathbf{m}) = \prod \Theta_{\text{theory}n}(\mathbf{d}_n \mid \mathbf{m}),\end{aligned} \tag{3.5}$$

when assuming independence of the modeling uncertainties. Modeling of different phenomenon (seismic, electric, gravity) and different components of the same phenomenon (seismic travel times, amplitude reflections, frequency decay) are based on different types of theoretical knowledge and are likely to have independent modeling uncertainties, thereby supporting the stated assumption. Again, the expression is an approximation that neglects possible related factors emerging from common modeling choices (e.g., common spatial property discretization for instance).

By substitution of (3.4) and (3.5) in (3.2) followed by integration, we obtain the joint likelihood function as the product of the data likelihood functions for each of the data components,

$$L_{\text{data}}(\mathbf{m}) = L_{\text{data}1}(\mathbf{m})L_{\text{data}\,2}(\mathbf{m})\ldots L_{\text{data}N}(\mathbf{m}) = \prod L_{\text{data}n}(\mathbf{m}), \tag{3.6}$$

with

$$L_{\text{data}n}(\mathbf{m}) = \int \Theta_{\text{theory}n}(\mathbf{d}_n \mid \mathbf{m})\rho_{\text{obs}n}(\mathbf{d}_n)d\mathbf{d}_n. \tag{3.7}$$

A similar result is obtained by substitution of (3.5) in (3.3). The factorization of the joint data likelihood according to the data subsets is equivalent to the addition of data objective function terms in the framework of deterministic solutions to the inverse problem, as will be explained in the corresponding section of this chapter.

3.2.2. Prior PDF Factorization

At the multiple components model, the prior PDF on the model parameters can be decomposed by following the rule of conditional probabilities:

$$\rho(\mathbf{m}) = \rho(\mathbf{m}_K \mid \mathbf{m}_{K-1},\ldots,\mathbf{m}_1) \times \rho(\mathbf{m}_{K-1} \mid \mathbf{m}_{K-2},\ldots,\mathbf{m}_1)\ldots\rho(\mathbf{m}_1). \tag{3.8}$$

Given the knowledge on the relationships across the model components, some of these conditionals could be simplified. Causal or empirical statistical relationships impose the relevant dependencies, and independencies, enforcing a hierarchy to the model components. In figure 3.1, we present a common scheme for a multicomponent model with multiple data observations, structured in two layers of properties: *primary*, $\mathbf{m}_{\text{pri}} = \{\mathbf{m}_1, \mathbf{m}_{2,} \ldots, \mathbf{m}_M\}$, and *secondary*, $\mathbf{m}_{\text{sec}} = \{\mathbf{m}_{M+1}, \mathbf{m}_{M+2,} \ldots \mathbf{m}_K\}$, and one observed data layer. In this setting, the secondary properties are dependent on the primary properties, while the observed data are dependent on the secondary properties. The data are not directly (explicitly) dependent on the primary properties. With the imposed hierarchy and model layers, the prior information can be satisfactorily decomposed by

$$\rho(\mathbf{m}) = \rho(\mathbf{m}_{\text{sec}} \mid \mathbf{m}_{\text{pri}})\rho(\mathbf{m}_{\text{pri}}) = \rho(\mathbf{m}_{M+1},\ldots,\mathbf{m}_K \mid \mathbf{m}_M,\ldots,\mathbf{m}_1)\rho(\mathbf{m}_M,\ldots,\mathbf{m}_1), \tag{3.9}$$

and the posterior PDF for the situation in Figure 3.1 takes the form

$$\sigma(\mathbf{m}) = c\,\rho(\mathbf{m}_{M+1},\mathbf{m}_K \mid \mathbf{m}_M,\ldots,\mathbf{m}_1)\rho(\mathbf{m}_M,\ldots,\mathbf{m}_1) \times L_{data1}(\mathbf{m}_{M+1})L_{data2}(\mathbf{m}_{M+2})L_{data3}(\mathbf{m}_{M+2},\mathbf{m}_K), \tag{3.10}$$

with the factorization of the likelihood function as previously explained and including the explicit dependency of each data component on the corresponding secondary model component as indicated in Figure 3.1. It is common to define an objective function proportional to the logarithm of the posterior PDF, as will be described in the optimization approach section of this chapter. The factorization of the posterior PDF, shown in expression (3.10), is in this setting equivalent to the addition of terms in the objective function, each one corresponding to a particular factor of the posterior PDF.

In Figure 3.1, I have drawn separate data component boxes to indicate independence across the data

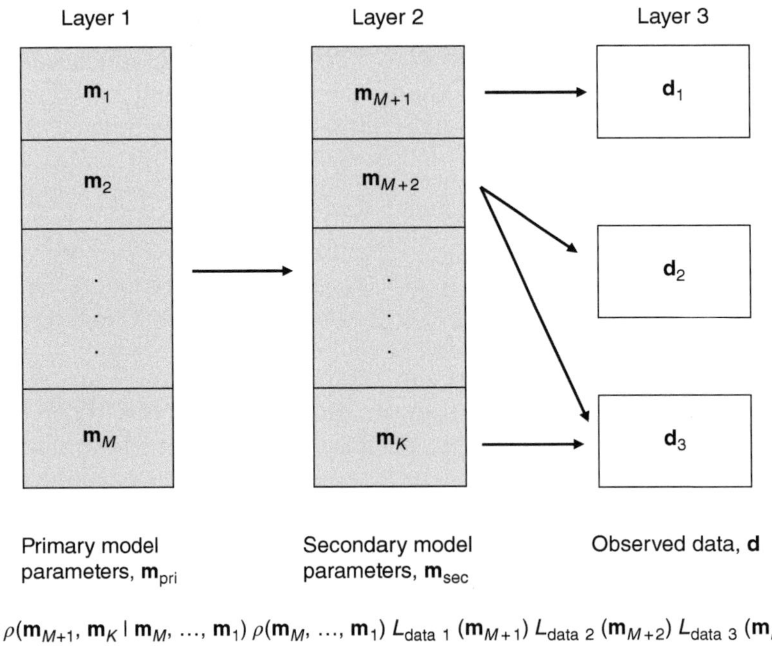

Figure 3.1 Random variables organized in hierarchical layers describing model parameters and data in an inference problem. Bold arrows indicate dependencies across random variables and its modeling sense. Variables in common blocks are modeled jointly. Gray boxes indicate model parameters describing Earth medium properties and structure, while white boxes indicate parameters describing experimental observations and measurements. The composition of the posterior PDF is shown at the bottom as a product of data likelihood functions, priors, and conditional PDFs.

components with respect to the observational and modeling data uncertainties. The juxtaposition of boxes shown for model components at the primary and secondary model layers indicate retention of dependencies across these components in the prior, $\rho(\mathbf{m}_M, \ldots, \mathbf{m}_1)$, and conditional, $\rho(\mathbf{m}_{M+1}, \mathbf{m}_K \mid \mathbf{m}_M, \ldots, \mathbf{m}_1)$; these properties should be modeled jointly considering their cross-relations, and spatial relations could be accounted for.

3.2.3. Examples in Common Inferential Settings

When an inferential problem is analyzed, the first step is to define the model components, their internal relations, and their data. This should be done by an expert, or a team of experts, in order to ensure the pertinence of the model and data to satisfy the goal of the inference. The network design introduced above involves retaining model components, their relevant relations, their sets of data, and their relations with the model components. Dependencies and independencies need to be defined.

In this section, I will illustrate the structure of common inferential problems in Earth sciences and the appropriate formulation of the posterior PDF. In Figures 3.2 and 3.3, I show layered multicomponent models that are useful in inverse problems at local, regional and planetary scales. In Figure 3.2 the setting for an integrated description of a siliciclastic sedimentary medium is depicted by four parameter layers: three model parameter layers and one data layer.

In sedimentary basins, the spatial statistical characteristics of the medium properties is at large scope heterogeneous, whereas within the same formation or units the statistics can be analyzed as spatially homogeneous. Thus, for appropriate statistical modeling, a primary space describing the formation delineation and their sequence is needed. This information can be parameterized by the formation category sequence (formation identification) and a geometrical framework delimiting the statistically homogeneous medium regions by the corresponding horizons. Prior information on these primary parameters, $\mathbf{m}_{\text{formation}}$, is usually obtained via interpreted seismic horizons, well-log data, and geological knowledge of the area.

Within formations, several types of lithology can be present (carbonates, igneous intrusions, siliciclastic

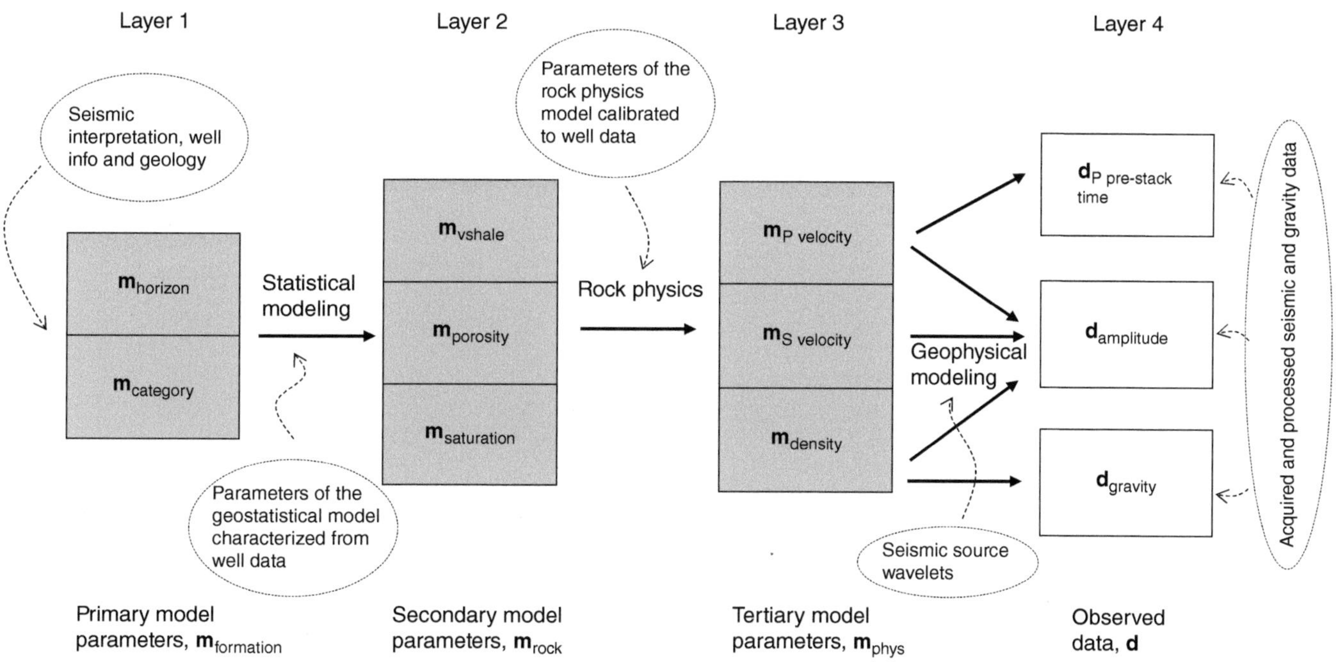

$$\sigma(\mathbf{m}) = \rho(\mathbf{m}_{\text{formation}})\, \rho(\mathbf{m}_{\text{vshale}}, \mathbf{m}_{\text{porosity}}, \mathbf{m}_{\text{saturation}} \mid \mathbf{m}_{\text{formation}})\, \rho(\mathbf{m}_{\text{P velocity}}, \mathbf{m}_{\text{S velocity}}, \mathbf{m}_{\text{density}} \mid \mathbf{m}_{\text{vshale}}, \mathbf{m}_{\text{porosity}}, \mathbf{m}_{\text{saturation}})$$
$$\times L_{\text{amplitude}}(\mathbf{m}_{\text{P velocity}}, \mathbf{m}_{\text{S velocity}}, \mathbf{m}_{\text{density}})\, L_{\text{P pre-stack time}}(\mathbf{m}_{\text{P velocity}})\, L_{\text{gravity}}(\mathbf{m}_{\text{density}})$$

Figure 3.2 Example of model parameter structure in Earth science inference settings: the case of siliciclastic sedimentary basin description based on seismic reflection amplitudes, seismic pre-stack P arrival times and gravity observations. Dependencies across random variables and their hierarchies are shown by bold arrows. Nonrandom parameters required for the conditionals, priors, and likelihoods are indicated in dashed ellipses. Bold boxes show the random parameters for description of the earth medium (gray) and observations (white). The corresponding structure of the posterior PDF is shown as the bottom.

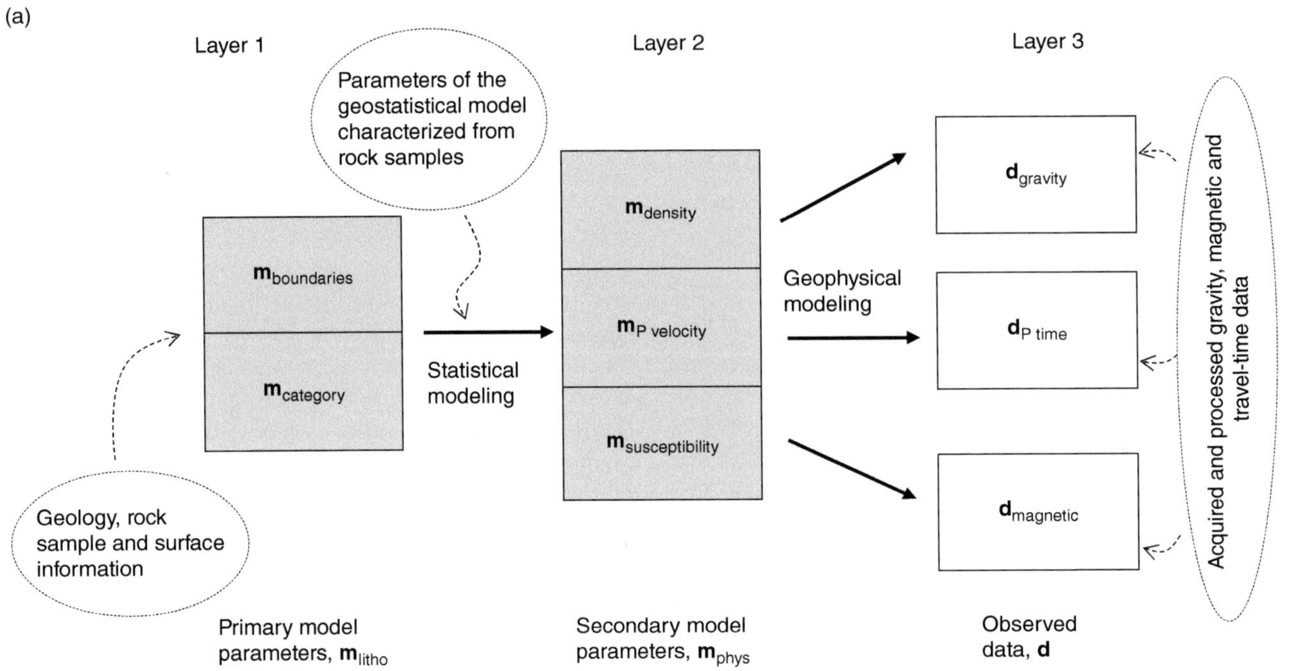

$$\sigma(\mathbf{m}) = \rho(\mathbf{m}_{\text{litho}})\, \rho(\mathbf{m}_{\text{density}}, \mathbf{m}_{\text{P velocity}}, \mathbf{m}_{\text{susceptibility}} \mid \mathbf{m}_{\text{litho}})\, L_{\text{gravity}}(\mathbf{m}_{\text{density}})\, L_{\text{P time}}(\mathbf{m}_{\text{P velocity}})\, L_{\text{magnetic}}(\mathbf{m}_{\text{susceptibility}})$$

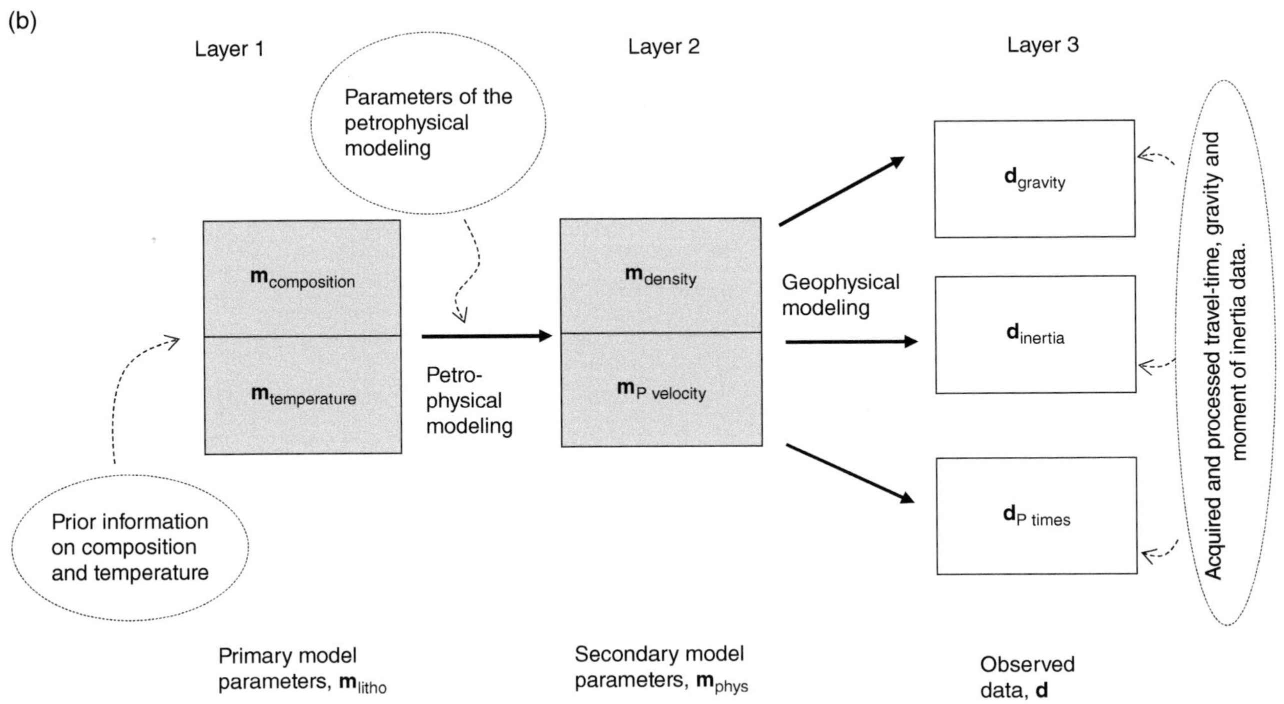

$$\sigma(\mathbf{m}) = \rho(\mathbf{m}_{\text{litho}})\, \rho(\mathbf{m}_{\text{density}}, \mathbf{m}_{\text{P velocity}} \mid \mathbf{m}_{\text{litho}})\, L_{\text{gravity}}(\mathbf{m}_{\text{density}})\, L_{\text{P time}}(\mathbf{m}_{\text{P velocity}})\, L_{\text{inertia}}(\mathbf{m}_{\text{density}})$$

Figure 3.3 Example of model parameter structure in earth science inference settings: **(a)** Lithotype geobody description at crustal regional scale based on gravity, magnetic, and seismic travel-time data. **(b)** Planet scale composition and temperature description constrained by seismic travel times, gravity data, and the inertia moment. Symbols are the same as in previous figures. The corresponding structure of the posterior PDF is shown at the bottom of each figure.

sedimentary rocks). We will consider here the case of siliciclastic sedimentary rocks, where lithology can be described by the shale volume fraction. Another secondary parameter is the total porosity that influences the elastic medium properties and density, and finally the pore fluid volume fraction (saturation) is important in systems with two or more fluids. Conditioned to the formation and well-log information, geostatistical parameters describing the rock matrix and fluid properties are commonly characterized to model this secondary layer of parameters, $\mathbf{m}_{\text{rock}} = \{\mathbf{m}_{\text{vshale}}, \mathbf{m}_{\text{porosity}}, \mathbf{m}_{\text{saturation}}\}$. In this case we refer to the use of well data to calibrate spatially homogeneous property statistics (means, covariances). An example with spatially localized well-log information will be considered in the next section of this chapter.

According to the rock matrix and fluid configurations, rock physical models are used to calculate properties that characterize the mechanical behavior of the medium, like compressional seismic velocity and shear seismic velocity as well as the mass density, $\mathbf{m}_{\text{phys}} = \{\mathbf{m}_{\text{P velocity}}, \mathbf{m}_{\text{S velocity}}, \mathbf{m}_{\text{density}}\}$. This set of parameters represents the third layer of model parameters.

Finally, Figure 3.2 includes in the fourth layer data from common geophysical surveys that provide information to interpret sedimentary basin stratification, depth, and structure. Interpretation of the seismic data can be in various ways, either in full wave form (full data) or by separating data components. It is common to interpret in a separate manner the pre-stack P-wave travel time for major well-identified reflectors and the reflection amplitudes after migration (spatial repositioning of the seismic data). We conform the data parameter layer in this example with these seismic partial data and the gravity data, $\mathbf{d} = \{\mathbf{d}_{\text{amplitude,}}\ \mathbf{d}_{\text{P pre-stack time,}}\ \mathbf{d}_{\text{gravity}}\}$.

Notice that seismic data subsets of different nature contribute at the right-hand side of the figure to the observed data and at the left-hand side to the prior information. The prior information is based on interpreted horizons in migrated and stacked data, whereas the data to be modeled at the right-hand side correspond to (1) seismic reflection amplitudes and (2) travel times of major events in pre-stack domain. There is no redundancy or cyclicity in the problem definition.

According to the layered model and the data relations in Figure 3.2, and applying the previous concepts for composition of the posterior density, we have

$$\begin{aligned}\sigma(\mathbf{m}) = {} & c\,\rho\left(\mathbf{m}_{\text{formation}}\right)\rho\left(\mathbf{m}_{\text{vshale}}, \mathbf{m}_{\text{porosity}}, \mathbf{m}_{\text{saturation}} \mid \mathbf{m}_{\text{formation}}\right) \\ & \times \rho\left(\mathbf{m}_{\text{P velocity}}, \mathbf{m}_{\text{S velocity}}, \mathbf{m}_{\text{density}} \mid \mathbf{m}_{\text{vshale}}, \mathbf{m}_{\text{porosity}}, \mathbf{m}_{\text{saturation}}\right) \\ & \times L_{\text{amplitude}}\left(\mathbf{m}_{\text{P velocity}}, \mathbf{m}_{\text{S velocity}}, \mathbf{m}_{\text{density}}\right) L_{\text{P pre-stack time}} \\ & \times \left(\mathbf{m}_{\text{P velocity}}\right) L_{\text{gravity}}\left(\mathbf{m}_{\text{density}}\right). \end{aligned} \tag{3.11}$$

where, $\rho(\mathbf{m}_{\text{formation}})$ is the prior PDF on the formation sequence and delimiting horizon boundaries, $\rho(\mathbf{m}_{\text{vshale}}, \mathbf{m}_{\text{porosity}}, \mathbf{m}_{\text{saturation}} \mid \mathbf{m}_{\text{formation}})$ is the PDF of the rock matrix and fluid parameters conditioned by the formation, and $\rho(\mathbf{m}_{\text{P velocity}}, \mathbf{m}_{\text{S velocity}}, \mathbf{m}_{\text{density}} \mid \mathbf{m}_{\text{vshale}}, \mathbf{m}_{\text{porosity}}, \mathbf{m}_{\text{saturation}})$ is the PDF of the physical rock properties conditioned by the rock matrix and fluid parameters. The likelihood functions are identified according to the set of observations and the related physical model argument. Figure 3.2 also shows some of the information needed for the definition of the priors, conditionals, and likelihoods, which is employed in the modeling as nonrandom parameters: the seismic source wavelets, the parameters of the rock physics model, and parameters of the conditional geostatistical models. These parameters will not vary in the inference and are previously estimated from the analysis of the data and additional information.

Similar components to the example shown in Figure 3.2 can be found in the papers by *Bosch* [2004], *Larsen et al.* [2006], *Bosch et al.* [2007], *Bosch et al.* [2009], *Grana and Della Rosa* [2010], and *Grana et al.* [2012] with details on how to model the specific prior, conditionals, and observational PDFs.

Considering now a larger scale, Figure 3.3a shows a setting for the inference of the geological structure at the crust, similar to the one employed by *Bosch* [1999], *Bosch et al.* [2001, 2004] and *Guillen et al.* [2007]. In this case, the primary frame is given by the description of the geometry of major lithotype geobody boundaries (e.g., gabbro, granite, sediments). The physical medium properties are modeled conditioned to the geobody lithotype by an empirical joint physical property density that is derived based on laboratory rock measurements for each lithotype. Finally, the observed data corresponds to common survey observations that provide information at large regional/crustal scales: gravity, magnetic, seismic refraction, and/or earthquake travel times. The resulting posterior density according to the model and data structure is given in the figure.

A similar parameter structure, shown in Figure 3.3b, was used at global satellite and planetary scale by *Khan et al.* [2006] to infer the thermal and compositional structure of the moon from available data on P-wave travel times, gravity observations, and inertia moment. In this case the primary parameters were the temperature and composition, and the secondary parameters were the seismic compressional velocity and mass density. A petrological model, based on computations of mineral phase proportions in the mantle, was used for the prediction of the mass density and seismic velocity conditioned to the mantle composition and temperature. The same approach was applied, with differences in the constraining geophysical data, to infer the composition and temperature of Mars [*Khan and Connolly*, 2008] and the Earth's mantle [*Khan et al.*, 2008].

3.2.4. The Role of Rock Physics and Dynamic Models as Coupling Information

In the examples presented above, an important role is given to relationships across model components, which are described by inner model conditional PDFs. Multiple properties defined at the same points are naturally linked. In solid Earth models, relationships are imposed by rock physics, but also geology, sedimentology, mineralogy, and chemistry can provide relational information depending on the setting.

Approaches to model the conditionals between the physical rock properties (e.g., elastic moduli, seismic velocities, mass density, viscosity, electrical resistivity) from basic rock frame constitution and fluids (e.g., matrix lithology, porosity, fluid types, and fractions) are multiple. I will refer below to empirical and rock physics model-based approaches.

An *empirical approach* to the formulation of the conditional of physical medium parameters to lithotype categories can be illustrated, for example, by the work of *Bosch* [1999] and *Bosch et al.* [2001]. The characterization of the mass density and magnetic susceptibility was based on laboratory rock sample measurement data for each one of the involved geobody lithotypes of the studied area. An empirical spatial (geostatistical) simulation model was elaborated for the conditional $\rho(\mathbf{m}_{\text{density}}, \mathbf{m}_{\text{susceptibility}} \mid \mathbf{m}_{\text{litho}})$ by using mixtures of multivariate Gaussian functions. The PDF for the mass density and magnetic susceptibility in a node depended on the lithotype of the geobody (categorical variable) and the mass density and magnetic susceptibility values at the other nodes within the geobody, according to the mentioned model. Because of the scarcity of data in some applications, the covariance (or equivalently semivariogram) ranges need to be assumed; in the work by *Bosch et al.* [2001] it was based on the geostatistical characterization of similar areas from field measurements [*Bourne*, 1993]. Additional examples of the application of an empirical approach to formulate probabilities of the physical rock properties, conditioned to lithology and reservoir properties, are described in the works by *Mukerji et al.* [2001], *Larsen et al.* [2006], and *Ulvmoen et al.* [2010].

Relationships between the elastic moduli, mass density, and other physical rock properties have been studied for various types of rocks and Earth media, within the domain of rock physics. Common models of rock physics for relating acoustic and elastic properties to rock matrix and fluid components are described in detail by *Mavko et al.* [2003] and *Hilterman* [2001]. For sedimentary rocks, and particularly for the most common siliciclastic sedimentary rocks, a large set of modeling tools are available. Predictive models for elastic and other physical properties for mantle material have also been studied. A second type of approach, less dependent on a specific set of data than the empirical approach, consists in using appropriate rock physics models for the prediction of the physical properties from rock matrix and fluid properties,

$$\mathbf{m}_{\text{phys}} = \mathbf{f}\left(\mathbf{m}_{\text{rock}}\right) + \Delta\mathbf{m}_{\text{phys}}. \tag{3.12}$$

where, $\mathbf{f}(\mathbf{m}_{\text{rock}})$ is a rock physics function that models the involved physical rock properties and $\Delta\mathbf{m}_{\text{phys}}$ are the deviations from the prediction. It is important to mention that no rock physics model has universal validity. Hence, all rock physics models should be evaluated and calibrated against actual property data from the application area. The statistics for the deviations $\Delta\mathbf{m}_{\text{phys}}$ can be characterized by comparing the rock physics model prediction and property measurements for the area (usually well-log or core data). The statistical model for the deviations,

$$\rho\left(\Delta\mathbf{m}_{\text{phys}}\right) = \rho\left(\mathbf{f}\left(\mathbf{m}_{\text{rock}}\right) - \mathbf{m}_{\text{phys}}\right) = \rho\left(\mathbf{m}_{\text{phys}} \mid \mathbf{m}_{\text{rock}}\right), \tag{3.13}$$

conduce straightforwardly to the conditional PDF for the physical model parameters. In addition to the rock physics model deviations, measurement uncertainties could also be accounted for in $\rho(\Delta\mathbf{m}_{\text{phys}})$ when relevant.

Examples of the *rock physics-based* approach outlined above for modeling the dependency of acoustic and elastic properties from rock matrix and fluid properties are described by *Mavko and Mukerji* [1998], *Bosch* [2004], *Bosch et al.* [2007], *Spikes et al.* [2007], *Bosch et al.* [2009], *Grana et al.* [2012], *Suman and Mukerji* [2013], and *Grana* [2014] for modeling the dependency of acoustic and elastic properties from rock matrix and fluid properties.

At the planetary scale, physical medium properties, such as seismic velocities and density, are a function of the mantle composition and temperature. The problem is nonlinear due to mineral phase changes. Nevertheless, it is fully tractable via petrophysical models, such as the one described by *Connolly* [2005] that is based on the minimization of the free energy associated with the mixture of mantle minerals. This approach was followed by *Khan et al.* [2006] and *Khan and Connolly* [2008] to model petrophysical conditionals for inferring the thermal and compositional configuration of the moon and Mars. Also, *Hacker et al.* [2003] elaborated a model for the prediction of the compressional velocities and mass density in the Earth mantle.

When modeling phenomena evolving in time, dynamic models are the natural link between the various time-lapse observations or velocity observations. In the case of the mantle description, flow equations can be useful to link temperature fields and mantle kinematics. In the case of time-lapse seismic in reservoirs under production, fluid flow modeling can be also used as inner link between the time-lapse configurations. Applications under various approaches are shown in the papers by *Huang* [2001], *Mezgahni et al.* [2004], and *Dadashpour et al.* [2009].

In addition to rock physics and dynamic models, structural features such as the location of geologic boundaries [*Lines et al.*, 1988; *Haber and Oldenburg*, 1997; *Bosch et al.*, 2001; *Gallardo et al.*, 2003; *Guillen et al.*, 2007] or the requirement of similar directions for property spatial gradients [*Gallardo and Meju*, 2004; *Doetsch et al.*, 2010] are also used for model conditioning across medium property fields.

3.3. GRAPHS AND POSTERIOR PROBABILITY DENSITIES

As shown in the previous section, the information about parameter components and dependencies are easily presented in graphical form, as in Figures 3.1–3.3, facilitating a straightforward definition of the posterior probability densities. So far we have presented examples with hierarchical model parameters layers and a final data layer: Priors are given for the first model layer, conditionals at the intermediate model layers and data likelihoods at the final layer. However, more heterogeneous networks of model and data subspaces can be analyzed for inference following the underlying principles.

A graphical structure known as *direct acyclic graph* (DAG) [*Thulasiraman and Swamy*, 1992] is useful to describe relationships across model parameters that are less structured than a hierarchical sequence of model layers. The DAG is defined by a set of nodes, which are here the model and data components, and a set of directed arrows that link the nodes, which will represent direct dependency relationships. In the DAG, it is required that no closed directed path exists in the graph. If a parameter subspace $\mathbf{m}_n$ points in the graph to a subspace $\mathbf{m}_k$, the latter is considered a *descendant* of the former, and the former an *ascendant* of the latter. Acyclicity warrants that no node can be its own descendant or ascendant; Figure 3.4 shows an example of a DAG relating model and data components. Notice that we have data that are dependent on different model component generations and direct influences (arrows) across model components separated by more than one generation; data nodes do not have descendants. A given DAG and the PDF defined over the joint parameter space defines what is called a *Bayesian network*, sometimes also called belief network or simply inference network [*Pearl*, 1986, 1994; *Ben-Gal*, 2007; *Griffiths et al.*, 2008].

The same principles applied in (3.6) and (3.9) produce the following rules for the factorization of the posterior PDF over the DAG:

1. Each model node with no ascendants introduces a prior PDF factor.
2. Each arrow across model nodes contributes with a conditional PDF factor.
3. Each data node contributes with a likelihood function factor.

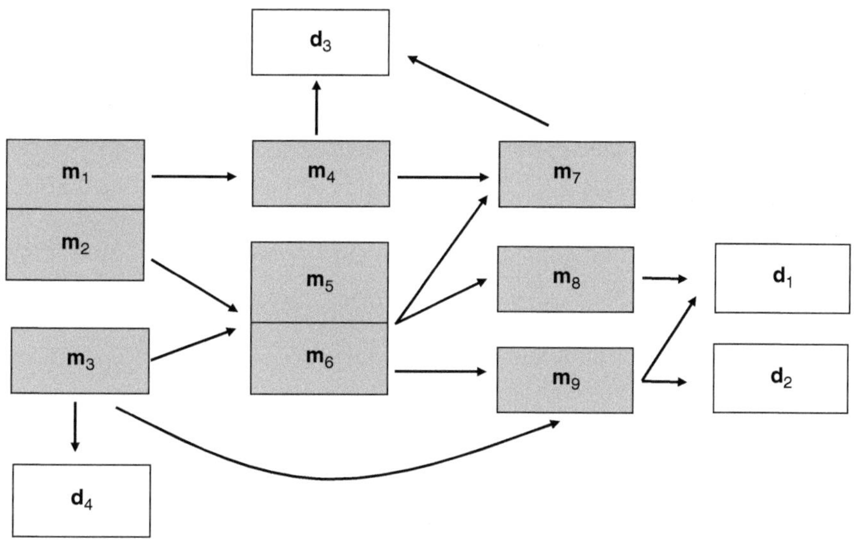

$$\sigma(\mathbf{m}) = \rho(\mathbf{m}_1, \mathbf{m}_2)\, \rho(\mathbf{m}_3)\, \rho(\mathbf{m}_5, \mathbf{m}_6|\mathbf{m}_2, \mathbf{m}_3)\, \rho(\mathbf{m}_4|\mathbf{m}_1)\, \rho(\mathbf{m}_7|\mathbf{m}_4, \mathbf{m}_5, \mathbf{m}_6)\, \rho(\mathbf{m}_8|\mathbf{m}_5, \mathbf{m}_6)$$
$$\times\, \rho(\mathbf{m}_9|\mathbf{m}_3, \mathbf{m}_6)\, L_{\text{data 1}}(\mathbf{m}_8, \mathbf{m}_9)\, L_{\text{data 2}}(\mathbf{m}_9)\, L_{\text{data 3}}(\mathbf{m}_4, \mathbf{m}_7)\, L_{\text{data 4}}(\mathbf{m}_3)$$

Figure 3.4 Example of an inference network defined over a direct acyclic graph (DAG). Dependencies across random variables are shown by bold arrows. The corresponding structure of the posterior PDF is shown at the bottom of the figure. Model components within common boxes are jointly modeled. Gray boxes indicate model parameters describing the earth medium properties and structure, while white boxes indicate model parameters describing experimental observations and measurements.

According to these rules, the posterior PDF for the DAG in Figure 3.4 is

$$\begin{aligned}\sigma(\mathbf{m}) = c\rho(\mathbf{m}_1,\mathbf{m}_2)\rho(\mathbf{m}_3)\rho(\mathbf{m}_5,\mathbf{m}_6 \mid \mathbf{m}_2,\mathbf{m}_3)\\ \times\rho(\mathbf{m}_4 \mid \mathbf{m}_1)\rho(\mathbf{m}_7 \mid \mathbf{m}_4,\mathbf{m}_5,\mathbf{m}_6)\rho(\mathbf{m}_8 \mid \mathbf{m}_5,\mathbf{m}_6)\\ \times\rho(\mathbf{m}_9 \mid \mathbf{m}_3,\mathbf{m}_6)L_{\text{data}1}(\mathbf{m}_8,\mathbf{m}_9)L_{\text{data}2}(\mathbf{m}_9)\\ \times L_{\text{data}3}(\mathbf{m}_4,\mathbf{m}_7)L_{\text{data}4}(\mathbf{m}_3).\end{aligned} \tag{3.14}$$

The sense of some of the DAG relations depends on the modeling decisions made. When the relationships are based on theoretical models, the nature of the theory commonly imposes the sense of the direct (easier) modeling, complying with a causality notion. Examples of Bayesian model networks used for oil reservoir description, with illustration of the specific graph structure defined, are described in the works by *Eidsvik et al.* [2004], *Bosch et al.* [2007], *Rimstad et al.* [2012] and *Chen and Hoversten* [2012]. Examples of similar networks applied to decision making can be found in the work by *Bhattacharjya and Mukerji* [2006] and *Martinelli et al.* [2013].

Figure 3.5 shows an application example of the inference with a DAG relational description across model and data components, for sedimentary strata description. In the setting of Figure 3.5, the formational random parameters of Figure 3.2 are simplified to be a known (known horizons and formations in reflection seismic time) part of the prior information. Additional data have been included consisting of well-log observations in given locations (well paths) for the porosity, shale fraction, water saturation, and elastic medium parameters (P-wave and S-wave velocities and mass density). The seismic source wavelet, considered a nonrandom parameter in Figure 3.3, has been randomized in order to adjust the seismic source

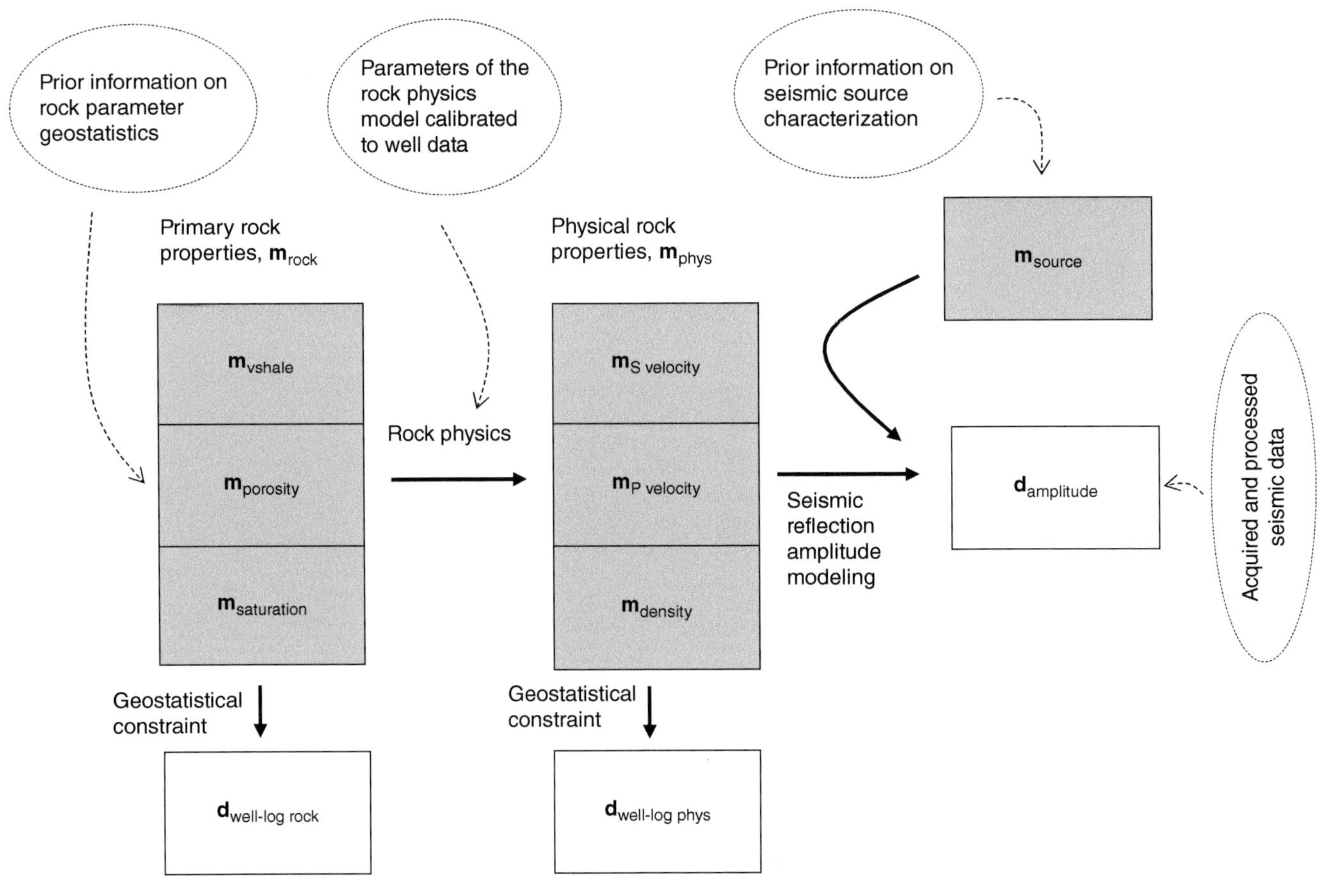

$$\sigma(\mathbf{m}) = \rho(\mathbf{m}_{\text{rock}})\,\rho(\mathbf{m}_{\text{source}})\,\rho(\mathbf{m}_{\text{P velocity}}, \mathbf{m}_{\text{S velocity}}, \mathbf{m}_{\text{density}} \mid \mathbf{m}_{\text{vshale}}, \mathbf{m}_{\text{porosity}}, \mathbf{m}_{\text{saturation}})$$
$$L_{\text{amplitude}}(\mathbf{m}_{\text{P velocity}}, \mathbf{m}_{\text{S velocity}}, \mathbf{m}_{\text{density}}, \mathbf{m}_{\text{source}})\,L_{\text{well-log rock}}(\mathbf{m}_{\text{rock}})\,L_{\text{well-log phys}}(\mathbf{m}_{\text{phys}})$$

Figure 3.5 Inference network for siliciclastic sedimentary basin description based on pre-stack seismic reflection data, including the estimation of the source wavelet and well-log data priors on the medium properties. Symbols are the same as in previous figures. The corresponding structure of the posterior PDF is shown at the bottom.

wavelet within the inference process. With the relationships shown in the figure, the posterior PDF is

$$\sigma(\mathbf{m}) = c\,\rho(\mathbf{m}_{rock})\rho(\mathbf{m}_{source})\rho(\mathbf{m}_{phys}\,|\,\mathbf{m}_{rock}) \times L_{amplitude}(\mathbf{m}_{phys}, \mathbf{m}_{source}) L_{well-log\,rock}(\mathbf{m}_{rock}) \times L_{well-log\,phys}(\mathbf{m}_{phys}), \quad (3.15)$$

where $\rho(\mathbf{m}_{source})$ is the prior information on the source wavelet, usually obtained by preliminary well to seismic tie, and $\rho(\mathbf{m}_{rock})$ is the prior information on the rock matrix and fluid parameters (total porosity, shale fraction, and water saturation). The seismic amplitude likelihood, $L_{amplitude}(\mathbf{m}_{phys}\ \mathbf{m}_{source})$, is now dependent on the source wavelet in addition to the elastic medium parameters, and there are likelihood functions, $L_{well\text{-}log\ rock}(\mathbf{m}_{rock})$ and $L_{well\text{-}log\ phys}(\mathbf{m}_{phys})$, corresponding to the well-log measurements of the rock and physical properties at specific well-path locations. More details about the source wavelet inference are given by *Bosch et al.* [2007], and for well-log data inclusion see *Bosch et al.* [2009].

Figure 3.6 shows another inference network defined over a DAG. I am showing in this figure a network proposal for the inference of mantle properties and dynamics, by coupling seismic tomography, gravity, and plate velocity observations. In this case, dynamic models for the mantle and plates are part of the inner coupling of the model components. Primary parameters are the mineralogical mantle composition, together with the temperature and pressure. A descendant set of parameters are the compressional seismic velocity, the shear velocity, the mass density, and the viscosity, which should be modeled by composing mineral fractions, temperature, and pressure dependencies. Another set of descendant parameters are the mantle dynamics given by stress and velocities, dependent on the mantle configuration and physical properties. Mantle velocity imposes anisotropy in the

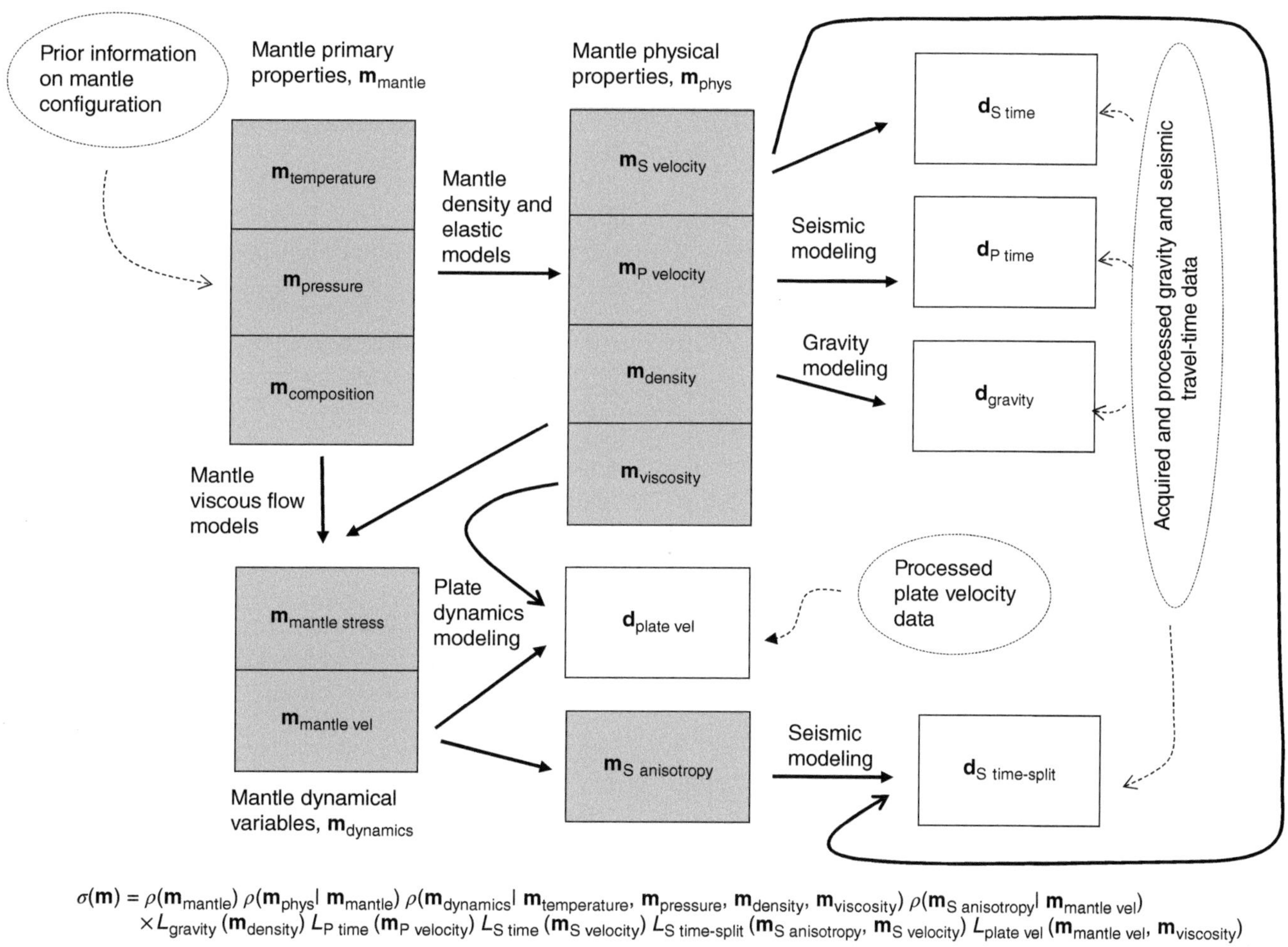

Figure 3.6 Example of a proposed inference network for mantle configuration coupling seismic tomography, mantle flow, plate dynamics, and gravity data constraints. The corresponding structure of the posterior PDF is shown at the bottom; compact notation $\mathbf{m}_{mantle}$, $\mathbf{m}_{phys}$, and $\mathbf{m}_{dynamics}$ have been used to abbreviate the expression. Same symbols as in previous figures.

propagation of seismic shear waves (birefringence phenomenon), which is measurable from seismological observations. Mantle velocities at the surface are also related to the plate kinematics, through dynamic plate models. The observed data includes travel-time seismological observations for P and S phases, for the travel-time split of the S phase due to anisotropy, plate measured velocities, and gravity data.

3.4 SAMPLING IN MODEL NETWORKS

Once the posterior PDF has been formulated in a model network, the solution of the inverse problem consists in drawing realizations from the posterior PDF, or alternatively solving for maximum posterior model configurations. We will discuss the first option in this section.

The structure of the model components, data components, and relationships is established via the factors of the combined PDF and, as explained before, satisfactorily depicted by the associated model network graph. The straightforward approach to sample from the posterior PDF is to construct a Markov Chain sampler following the sequence imposed by the graph, from the ascendants to the descendants. There are many techniques to sample from the priors, conditionals, and posterior PDFs [*Geyer*, 1992; *Smith and Roberts*, 1993; *Tierney*, 1994; *Gelfand and Smith*, 1990; *Liu*, 1998]. The procedure I recommend here is sufficiently general to successfully adapt to any model network, and it is efficient for sampling. The procedure can be separated in two major phases for (a) sampling the joint prior PDF and (b) sampling the joint posterior PDF.

3.4.1. Sampling the Joint Prior PDF

The prior PDF is an important factor of the posterior PDF and is equal to the posterior if the likelihood data factors are ignored (no observations or infinite data uncertainties). To sample a realization from the joint prior PDF:

1. Draw a realization from each one of the component priors included in the PDF.
2. According to the realization of the ascendant model parameters, draw realizations of the first generation of descendants according to the conditional PDFs.
3. Continue the procedure through all the *generations* until the last descendant has been realized. Recall that to generate a descendant realization, all its ascendants should be realized, as warranted by the acyclic graph configuration.

The appropriate technique to produce prior and conditional realizations varies according to the nature of each PDF. In the case of continuous variables, often the PDF can be formulated as a multivariate Gaussian function. In this case, standard sampling methods are well known, like the square root of the covariance matrix method or Gibbs sampling through Gaussian conditionals. Some parameters are not Gaussian distributed, but may be transformed to Gaussian after appropriate change of variables. For the case of categorical multivariate parameters or non-Gaussian continuous parameters, Gibbs sampling from the univariate conditionals is often convenient.

3.4.2. Sampling the Joint Posterior PDF

Likelihood function evaluations corresponding to geophysical observations are commonly costly (in terms of computation times), difficult to calculate (in terms of elaborated numerical nonlinear computations), and not represented by parameterized continuous PDF models. The Metropolis–Hastings sampler is an appropriate technique to account for this type of likelihood. In the inferential setting, this sampler uses as candidate outcome a realization of the prior PDF and proceeds by accepting or rejecting the realization by testing the likelihood function ratio between the candidate and the current realization in the chain. The recommended procedure is as follows:

1. Generate a candidate realization following the joint prior sampling chain rules described in the previous subsection.
2. Evaluate the joint data likelihood for this realization.
3. Calculate the joint data likelihood ratio between the candidate realization and the current realization.
4. Accept the candidate realization as the next step of the posterior chain with probability equal to the minimum between the likelihood ratio and one.
5. If the candidate is rejected, assign the current realization in the posterior chain.
6. Iterate the procedure from the first step.

The Metropolis sampler warrants the convergence of the chain to a sample of the posterior PDF in long enough runs. A description of the Metropolis sampler applied to posterior PDFs in geophysical inverse problems can be found in the work by *Mosegaard and Tarantola* [1995].

Likelihoods associated with property sampling in specific locations (e.g., well-logging or surface rock sampling) are less difficult to evaluate and can be in most cases related to geostatistical Gaussian spatial PDFs (Kriging and Gaussian simulation). These likelihoods may be either (a) included within the likelihood evaluated by the Metropolis sampler or (b) used as additional modeling constraint to the prior information chain as shown by *Bosch et al.* [2009]. For a review in statistical spatial models the reader is referred to the works by *Dubrule* [2003], *Chiles and Delfiner* [2009] and *Deutsch and Journel* [1992]. Multipoint statistics [*Caers et al.*, 2000; *Strebelle*, 2002] allows for prior PDF sampling with improved description of morphological features. Examples of their application to inversion of seismic data in complex models are described in the works by *González et al.* [2008] and *Grana et al.* [2012]. Object oriented modeling of fluvial sedimentary systems have been demonstrated by *Holden et al.* [1998] and *Deutsch and Wang* [1996].

3.4.3. Efficient Sampling Through Factorized Likelihoods

When several data likelihoods are present, the Metropolis sampler may be applied in cascade to each one of the likelihood factors; using partial posteriors as prior sampling PDFs for the consecutive data likelihood, as shown by *Bosch et al.* [2000] and *Tarantola* [2005]. The procedure is as follows:

1. Generate a candidate realization following the joint prior sampling chain procedure.
2. Evaluate one data likelihood ratio between the candidate and the current realization.
3. Retain the candidate realization for the next likelihood factor test with probability equal to the minimum between the likelihood ratio and one.
4. If the candidate is not retained, accept the current realization and go to 1.
5. If the candidate is retained, repeat from 2 following with the next likelihood factor. If retained after the last likelihood factor test, accept the model realization and go to 1.

Two criteria should be used when ordering the likelihood factors, leaving with preference at the beginning: (1) smoother likelihoods in terms of information (larger uncertainties) and (2) likelihoods with smaller computational cost. The former condition (smoothness) allows avoiding unwanted barrier problems (i.e., inability to mix the sampling across modes separated by very low probability zones) in the preliminary likelihood evaluations, which can potentially affect the efficiency of the method. In any case, the efficiency between various likelihood sequences should be tested for evaluation and should be compared with the option of single joint likelihood evaluation.

3.5. MAXIMUM POSTERIOR PROBABILITIES IN MODEL NETWORKS

Another alternative in realizing a solution to the inverse problem is searching for the configuration that maximizes the posterior probability density (MAP) and calculating the local posterior covariance matrix. The MAP search commonly converges to the nearest mode, although there are methods to search for the global MAP. In the case of multimodal PDFs a single MAP configuration is not a complete description of the problem solution; identification of major modes and the corresponding local MAP configuration may be an alternative. It is common in geophysical inference that the data likelihoods are complex multimodal functions. Nevertheless, if the prior information is monomodal and highly informative, and such that it circumscribes the posterior into the region of one of the modes of the data likelihood, the posterior is then close to monomodal and the MAP constitutes an acceptable description of the problem solution.

A classic method for searching a MAP configuration is the Gauss–Newton method [*Tarantola*, 2005], which requires the gradient and the approximate Hessian of the natural logarithm of the posterior PDF. By defining the *objective function*, $S(\mathbf{m}) = -\ln(\sigma(\mathbf{m}))$, we obtain

$$\sigma(\mathbf{m}) = \exp(-S(\mathbf{m})). \tag{3.16}$$

As the exponential is a positive monotonically increasing function, a MAP configuration corresponds to a minimal value of the objective function, and, neglecting third-order derivatives, the posterior local covariance matrix is the Hessian of the objective function evaluated at the MAP. The model parameters update, $\Delta\mathbf{m} = \mathbf{m}^{n+1} - \mathbf{m}^{n}$, for a step $n+1$ in the iterative search towards the $\mathbf{m}_{\mathrm{MAP}}$ satisfies

$$\mathbf{Hess}\left[S(\mathbf{m}^{n})\right]\Delta\mathbf{m} = -\mathbf{Grad}\left[S(\mathbf{m}^{n})\right], \tag{3.17}$$

where **Hess** symbolizes the Hessian operator and **Grad** the gradient operator. If multiplying by the prior model covariance matrix, $\mathbf{C}_{\mathrm{m}}^{\mathrm{prior}}$, the linear system matrix gets dimensionless and more stable for the numerical solution,

$$\underbrace{\mathbf{C}_{\mathrm{m}}^{\mathrm{prior}}\mathbf{Hess}\left[S(\mathbf{m}^{n})\right]}_{\text{Curvature}}\Delta\mathbf{m} = \underbrace{-\mathbf{C}_{\mathrm{m}}^{\mathrm{prior}}\mathbf{Grad}\left[S(\mathbf{m}^{n})\right]}_{\text{Steepest descent direction}}. \tag{3.18}$$

Notice that because the joint objective function is a logarithm of the posterior PDF, the factor structure in (3.10) transforms straightforwardly to the addition of objective function terms, each one accounting for the corresponding data likelihood, conditional probability, or prior probability.

The posterior model covariance is the inverse of the Hessian of the objective function, evaluated at the MAP configuration. It can be calculated by inverting the Hessian, but inverting the Curvature matrix (i.e., the product of the prior model covariance matrix and the Hessian of the objective function) and multiplying by the prior covariance is commonly a more stable procedure,

$$\mathbf{C}_{\mathrm{m}}^{\mathrm{posterior}} = \left(\mathbf{C}_{\mathrm{m}}^{\mathrm{prior}}\,\mathbf{Hess}\left[S(\mathbf{m}_{\mathrm{MAP}})\right]\right)^{-1}\mathbf{C}_{\mathrm{m}}^{\mathrm{prior}}. \tag{3.19}$$

To work out the linear equations to search the MAP configuration, the data likelihood, the conditional PDFs, and priors need to be explicitly formulated. I will use a simple setting of a two-layered model and one data layer, as shown in Figure 3.7, to illustrate the formulation of the linear system of equations (3.18). It corresponds to the case of inverting pre-stack seismic data for joint estimation

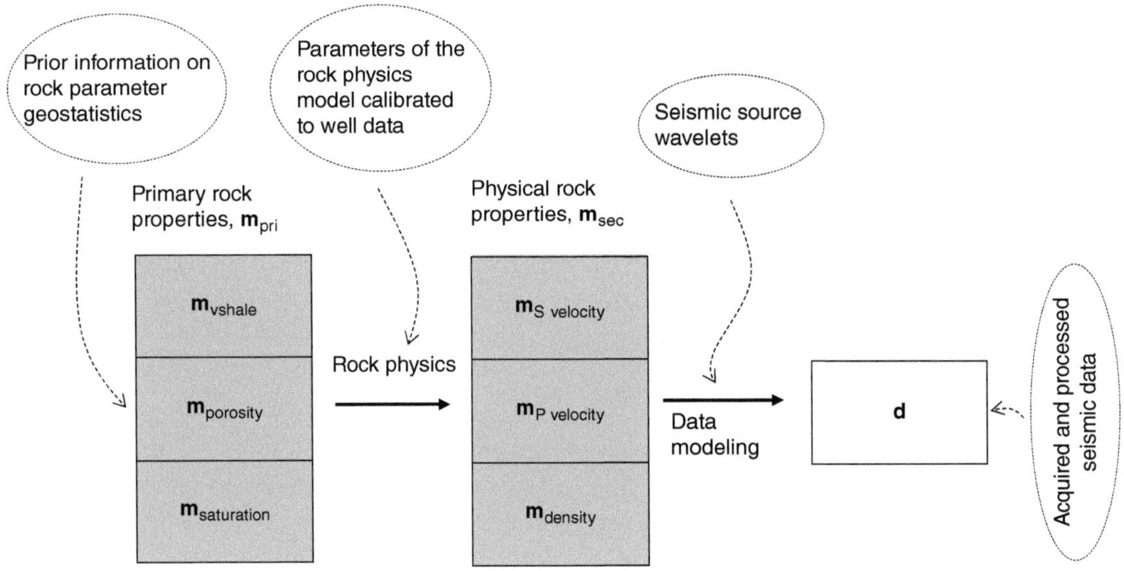

Figure 3.7 Inference network for a hierarchical two-layered model structure, illustrated with the case of a siliciclastic sedimentary basin description based on seismic data. Symbols are the same as in previous figures. The corresponding structure of the posterior PDF is shown at the bottom.

of the isotropic elastic medium parameters, and primary rock parameters in a siliciclastic sedimentary medium, describing the total porosity, the shale factor and pore fluid phase (water–hydrocarbon) fraction. The formulation is the same for any posterior PDF with a similar structure. For describing the isotropic medium, various combinations of three parameters (elastic moduli, impedances, mass density, seismic velocities) can be selected; a common choice is the mass density and the seismic P and S velocities. To be explicit, I will describe the medium by specifying primary, $\mathbf{m}_{prim} = \{\mathbf{m}_{vshale}, \mathbf{m}_{porosity}, \mathbf{m}_{saturation}\}$, and secondary, $\mathbf{m}_{sec} = \{\mathbf{m}_{P\,velocity}, \mathbf{m}_{S velocity}, \mathbf{m}_{density}\}$, model parameters, the joint model space being $\mathbf{m} = \{\mathbf{m}_{prim}, \mathbf{m}_{sec}\}$. Commonly, these parameters are specified at each point over a 3D grid.

The seismic data depends explicitly on the elastic medium configuration and parameters associated with the seismic survey experiment (source function and geometry). The seismic observations in this problem could be travel times (tomography problems), reflection amplitudes (reflectivity inversion), or the full wave field. In any of these cases, we can formulate a forward modeling of the data, $\mathbf{g}(\mathbf{m}_{sec})$, based on the seismic wave mechanical theory, such that

$$\mathbf{d}_{obs} = \mathbf{g}(\mathbf{m}_{sec}) + \Delta\mathbf{d}, \tag{3.20}$$

where $\mathbf{d}_{obs}$ is the observed data, and $\Delta\mathbf{d}$ are the deviations between the observed and the modeled data.

As already explained, rock physics is the natural link between the primary and secondary model parameters. After calibrating an appropriate rock physics model for the elastic parameters using local data if available, $\mathbf{f}(\mathbf{m}_{prim})$, we have

$$\mathbf{m}_{sec} = \mathbf{f}(\mathbf{m}_{prim}) + \Delta\mathbf{m}_{sec}, \tag{3.21}$$

where $\mathbf{m}_{sec}$ are the true medium elastic parameters and $\Delta\mathbf{m}_{sec}$ are the deviations between the true and modeled elastic parameters.

The posterior PDF in this problem according to what has been previously explained is

$$\sigma(\mathbf{m}) = \rho(\mathbf{m}_{prim})\,\rho(\mathbf{m}_{sec} \mid \mathbf{m}_{prim})\,L_{dat}(\mathbf{m}_{sec}). \tag{3.22}$$

I will model the three factors in the posterior formulation by multivariate Gaussian functions. The first PDF describes the prior information on the primary model parameters, which could be defined with

$$\rho(\mathbf{m}_{prim}) = c_1 \exp\left[-\tfrac{1}{2}\left(\mathbf{m}_{prim} - \mathbf{m}_{prim}^{prior}\right)^T \mathbf{C}_{prim}^{-1}\left(\mathbf{m}_{prim} - \mathbf{m}_{prim}^{prior}\right)\right], \tag{3.23}$$

with $\mathbf{m}_{prim}^{prior}$ being the expected prior primary parameters and $\mathbf{C}_{prim}$ the prior covariance matrix; commonly, $\mathbf{m}_{prim}^{prior}$ is spatially dependent and previously modeled according to the geological stratification and formation horizon information. The likelihood function and the conditional rock physics PDF are formulated by modeling the probability of the deviations $\Delta\mathbf{m}_{\text{sec}}$ and $\Delta\mathbf{d}$ in (3.20) and (3.21),

$$\rho\left(\mathbf{m}_{\text{sec}} \mid \mathbf{m}_{\text{prim}}\right) = c_2 \exp\left[-\tfrac{1}{2}\left(\mathbf{m}_{\text{sec}} - \mathbf{f}\left(\mathbf{m}_{\text{prim}}\right)\right)^{\text{T}} \mathbf{C}_{\text{sec|prim}}^{-1} \left(\mathbf{m}_{\text{sec}} - \mathbf{f}\left(\mathbf{m}_{\text{prim}}\right)\right)\right], \tag{3.24}$$

with $\mathbf{C}_{\text{sec|prim}}$ being the covariance matrix of the rock physics model deviations $\Delta\mathbf{m}_{\text{sec}}$. Similarly, the data likelihood function is

$$L_{\text{dat}}\left(\mathbf{m}_{\text{elas}}\right) = \exp\left[-\tfrac{1}{2}\left(\mathbf{d}_{\text{obs}} - \mathbf{g}\left(\mathbf{m}_{\text{sec}}\right)\right)^{\text{T}} \mathbf{C}_{\text{dat}}^{-1} \left(\mathbf{d}_{\text{obs}} - \mathbf{g}\left(\mathbf{m}_{\text{sec}}\right)\right)\right], \tag{3.25}$$

with $\mathbf{C}_{\text{dat}}$ being the data covariance, encompassing data observational and modeling uncertainties. By adding exponents of the three modeled factors of the posterior PDF, the full objective functions has each of the three information components,

$$\begin{aligned} S(\mathbf{m}) = & \underbrace{\tfrac{1}{2}\left[\left(\mathbf{m}_{\text{prim}} - \mathbf{m}_{\text{prim}}^{\text{prior}}\right)^{\text{T}} \mathbf{C}_{\text{prim}}^{-1} \left(\mathbf{m}_{\text{prim}} - \mathbf{m}_{\text{prim}}^{\text{prior}}\right)\right]}_{\text{Prior information term}} \\ & + \underbrace{\tfrac{1}{2}\left[\left(\mathbf{m}_{\text{sec}} - \mathbf{f}\left(\mathbf{m}_{\text{prim}}\right)\right)^{\text{T}} \mathbf{C}_{\text{sec|prim}}^{-1} \left(\mathbf{m}_{\text{sec}} - \mathbf{f}\left(\mathbf{m}_{\text{prim}}\right)\right)\right]}_{\text{Rock physics term}} \\ & + \underbrace{\tfrac{1}{2}\left[\left(\mathbf{d}_{\text{obs}} - \mathbf{g}\left(\mathbf{m}_{\text{sec}}\right)\right)^{\text{T}} \mathbf{C}_{\text{dat}}^{-1} \left(\mathbf{d}_{\text{obs}} - \mathbf{g}\left(\mathbf{m}_{\text{sec}}\right)\right)\right]}_{\text{Geophysical term}}. \end{aligned} \tag{3.26}$$

Expressions $\mathbf{g}(\mathbf{m}_{\text{sec}})$ and $\mathbf{f}(\mathbf{m}_{\text{prim}})$ are in general nonlinear, and hence the search of the model update needs successive iterations in the application of the Gauss–Newton's method described in expression (3.15). The first and second derivatives of the objective function, as well as algebraic simplifications, required to calculate the model update are detailed in the work by *Bosch* [2004]. The resulting model update for the secondary model parameters satisfies the linear system of equations,

$$\mathbf{A}\,\Delta\mathbf{m}_{\text{sec}} = \mathbf{b}, \tag{3.27}$$

with the left-hand side being

$$\mathbf{A} = \mathbf{I} + \left(\mathbf{C}_{\text{sec|prim}} + \mathbf{F}\mathbf{C}_{\text{prim}}\mathbf{F}^{\text{T}}\right)\mathbf{G}^{\text{T}}\,\mathbf{C}_{\text{dat}}^{-1}\mathbf{G}. \tag{3.28}$$

Above, matrices $\mathbf{G}$ and $\mathbf{F}$ are the Jacobian matrices of $\mathbf{g}(\mathbf{m}_{\text{sec}})$ and $\mathbf{f}(\mathbf{m}_{\text{prim}})$ correspondingly, and $\mathbf{I}$ is the identity matrix. The right-hand side of (27) is

$$\begin{aligned} \mathbf{b} = & \ \mathbf{f}\left(\mathbf{m}_{\text{prim}}\right) - \mathbf{m}_{\text{sec}} + \mathbf{F}\left(\mathbf{m}_{\text{prim}}^{\text{prior}} - \mathbf{m}_{\text{prim}}\right) \\ & + \left(\mathbf{C}_{\text{sec|prim}} + \mathbf{F}\mathbf{C}_{\text{prim}}\mathbf{F}^{\text{T}}\right)\mathbf{G}^{\text{T}}\,\mathbf{C}_{\text{dat}}^{-1}\left(\mathbf{d}_{\text{obs}} - \mathbf{g}\left(\mathbf{m}_{\text{sec}}\right)\right). \end{aligned} \tag{3.29}$$

After the model update for the secondary parameters is calculated, the model update for the primary parameters is obtained by [*Bosch,* 2004]

$$\begin{aligned} \Delta\mathbf{m}_{\text{prim}} = & \ \mathbf{m}_{\text{prim}}^{\text{prior}} - \mathbf{m}_{\text{prim}} \\ & + \mathbf{C}_{\text{prim}}\mathbf{F}^{\text{T}}\,\mathbf{G}^{\text{T}}\,\mathbf{C}_{\text{dat}}^{-1}\left(\mathbf{d}_{\text{obs}} - \mathbf{g}\left(\mathbf{m}_{\text{sec}}\right) - \mathbf{G}\,\Delta\mathbf{m}_{\text{sec}}\right). \end{aligned} \tag{3.30}$$

Once (3.27) and (3.30) are calculated, the model parameters are jointly updated and a new evaluation of (3.27)–(3.30) can be obtained by iterating towards proximity of the MAP configuration. Commonly, convergence monitoring the data residual evolution until substantial reduction and stabilization.

3.6. DISCUSSION

Once the corresponding posterior probability density is defined according to the models involved, the solution of the inverse problem relies in generating object outcomes that summarize the posterior information—that is, the combination of the various data likelihoods, conditionals, and prior densities. Two major approaches have been described here: (1) sampling object configurations in proportion to the posterior PDF and (2) searching maximum posterior probability object configurations. A comparison between the two options is a common subject of discussion.

A first point to mention is that the two approaches do not provide the same description of the posterior information. The sampling approach provides a complete description of the posterior PDF. Theoretically, with a large enough set of samples of the model parameters, all marginals, conditionals, and the joint posterior PDFs can be approximated by the sample statistics with arbitrarily small deviations. Also, expected values, standard deviations and frequency histograms of the modeled object parameters can be computed. Hence, the sampling approach produces a full solution of the inverse problem. The optimization approach, on the other hand, searches for the mode (the local MAP configurations) that is the closest to a starting search point. It is essentially local, as the model parameter space can be always divided in a series of local mode sectors. In the neighborhood of the mode the posterior covariance matrix can be calculated to describe the uncertainties and posterior correlations around the mode.

Another known limitation of the MAP configuration is that it is usually smooth in physical (3D plus time) space, because all the spatially distributed parameters align with their posterior expected values. Realizations produced from the sampling process better represent the true spatial variability of the parameters, and they are often more useful when a representation of the spatial heterogeneity is relevant. A typical case is for modeling the fluid flow in a permeable medium, where the driving flow locations correspond to the large permeability channels and not to the average permeability [*Dubrule*, 2003]. The MAP configuration commonly underestimates by large the fluid flow. This limitation can partly be overcome by producing posterior simulations centered at the MAP with superposed deviations generated according to the posterior covariance matrix.

The computational costs involved in the optimization and the sampling approaches are highly dependent on the specific case and objectives of the inference; a comparison requires a case-by-case analysis. In general, the number of iterations involved in the search of the MAP is much smaller than the number of model parameters (a few iterations), whereas sampling chains require a length (number of realizations) of several times the number of model parameters. However, the operations required in the computation of a single iteration for the optimization approach are much larger than the computations involved in generating one sampling step, and they increase faster with the number of model parameters. Direct matrix methods solving linear systems as expression (3.17) typically require operations in the order between the square and the cube of the number of model parameters. Nevertheless, methods that take advantage of the sparcity and/or structure of the system (3.17), or approximate iterative solvers such as the conjugate gradients method, decrease this dependency, commonly reaching performances with operations orders beneath the square of the number of model parameters. Such numerical methods are required for the solution of large to very large inverse problems with the optimization approach.

Another issue of discussion refers to the various spaces associated with the inference: model parameters, modeled object, modeled observations, and data parameters. Implicit applications link the model parameter space to the model object space, and they also link the modeled observation space to the data space. These applications are not commonly explicit in the general formulation of the inference problems, but need to be accounted to model the involved functions and PDFs.

The formulation presented herein relates data and model parameters with basis on a combination of multiple modeling processes, described by conditional probability densities and likelihood functions. The basic knowledge to establish these links across model components and data (physical laws, geostatistical relationships) is not given primarily on the model parameter space but on physical space (3D plus time plus modeled matter). For this reason, it is useful to comment on the difference between the space of model parameters, which here is given by random variables supporting a physical object model, and the space of physical modeled objects. A set of object modeling rules, sometimes identified as *parameterization* rules, are required to transform a model parameter outcome to a modeled object configuration in physical 3D space and in some cases time. These rules commonly involve the physical identification of the parameters and the construction of their outcome in space accordingly. Sometimes, they are straightforward, like in the case of assigning property values to a Euclidean three-dimensional grid, but in other cases can be more elaborated. Examples are parameters being coefficients of polynomials defining geological body surfaces, or when the model elements are defined over curved coordinate systems.

A realization of an object model configuration can be regarded as the process that combines (1) drawing an outcome of the random variables (in model parameter space) according to the correspondent PDF and (2) passing these parameters through the object modeling engine to end up with a configuration of the object in the modeled physical (3D and time) space. The modeled object configuration is the result of this realization process. Once the object modeling rules have been established and behave as a bijective function, each outcome of the model parameters is associated with a correspondent outcome of the modeled object configuration, and vice versa. An outcome of the model parameters implicitly indicates a realization of a modeled object configuration; hence they are sometimes treated as the same entity.

A similar distinction applies for the observation of experiments made on the studied object (geophysical surveys, well-log measurements) and the data, conceived as parameters that describe the observations. The data also requires a series of configuration rules to have physical meaning (in 3D space, time, and observation nature). Also, commonly the data used in the inverse problems involves processing from raw (lower level) field or instrumental data. The understanding of the differences between model parameters, modeled object, experimental observation, and data is useful for the formulation of the relations between data and model parameters and the complete description of the related uncertainties. In particular, uncertainties should involve the various modeling processes that are present.

The issue of modeling parameter probability densities is closely related with the object modeling transform, as different parameterizations produce different PDFs for the same state of information on the modeled object. In particular, homogeneous (i.e., null-information) parameter probability densities are straightforwardly related to the parameterization [see *Tarantola*, 2005]. The object and observational modeling transforms are implicit components in the corresponding conditionals and likelihoods, as well as in the geophysical (3.20) and petrophysical (3.21) functions used to model the objective function terms.

3.7. CONCLUSIONS

Inference problems in Earth sciences involve the integration of multiple types of knowledge, observations, and information, which ultimately is done by the expert and the scientific community at large by continuous processes of partial analysis and synthesis. To support these processes, methods for quantitative inference in complex models and multiple data are in progress. The inference formulation is done via the definition of probability densities over parameter spaces that model the object or phenomenon to be described. To model the posterior state of information, after a set of multiple observations have been included, the data likelihood functions can be factorized across surveys and observational methods, assuming independence of the observed data uncertainties.

To couple components of the object model that are responsible for diverse observations, the knowledge about inner relationships across the object model parameters need to be used as part of the prior information, entering as conditional probability densities between components of the object model. The identification of relevant dependencies and acceptable independencies across model components is an important issue for the pertinence of the model and the reliability of the inference. The presentation of these dependencies via direct acyclic graphs is useful, allowing a straightforward formulation of the posterior PDF.

Once the posterior PDF is modeled, the generation of the object posterior configurations can follow two lines: drawing multiple realizations from the posterior PDF (sampling approach) or searching maximum posterior PDF configurations and their local covariance, a procedure that commonly has only local validity depending on the modes of the PDF.

The theoretical capabilities of these methods are unlimited, in a mathematical sense, depending for their application on the computational capacities and the ability to reliably describe the object/phenomenon laws and inner relationships across their components.

ACKNOWLEDGMENTS

The author acknowledges the Universidad Central de Venezuela, the Institute de Physique du Globe de Paris, and the University of Cambridge, which housed different periods of the author's work in the subjects developed in the chapter. Special acknowledgment is due to Albert Tarantola, professor and friend, who promoted the interest in inverse problems and their probabilistic treatment in professional and academic Earth science community.

REFERENCES

Alpak, F., C. Torres-Vedin, and T. Habashy (2008), Estimation of in-situ petrophysical properties from wireline formation tester and induction logging measurements: A joint inversion approach, *J. Petroleum Sci. Eng.*, *63*, 1–17.

Bhattacharjya, D., and Mukerji, T. (2006), Using influence diagrams to analyze decisions in 4D seismic reservoir monitoring, *The Leading Edge*, *25*, 1236–1239.

Bosch, M. (1999), Lithologic tomography: From plural geophysical data to lithology estimation, *J. Geophys. Res.*, *104*, 749–766.

Bosch, M. (2004), The optimization approach to lithological tomography: Combining seismic data and petrophysics for porosity prediction, *Geophysics*, *69*, 1272–1282, doi: 10.1190/1.1801944.

Bosch, M., C. Barnes, and K. Mosegaard (2000), Multi-step samplers for improving efficiency in probabilisitic geophysical inference, in *Methods and Aplications of Inversion*, Springer, New York.

Bosch, M., L. Cara, J. Rodrigues, A. Navarro, and M. Díaz (2007), A Monte Carlo approach to the joint estimation of reservoir and elastic parameters from seismic amplitudes, *Geophysics*, *72*(6), O29–O39, doi: 10.1190/1.2783766.

Bosch, M., C. Carvajal, J. Rodrigues, A. Torres, M. Aldana, and J. Sierra (2009), Petrophysical seismic inversion conditioned to well-log data: Methods and application to a gas reservoir, *Geophysics, 74*(2), O1–O15, doi: 10.1190/1.3043796.

Bosch, M., Guillen, A., and Ledru, P. (2001), Lithologic tomography: An application to geophysical data from the Cadomian belt of northern Brittany, France, *Techtonophysics*, *331*, 197–227.

Bosch, M., R. Meza, R. Jimenez, and A. Honig (2004), Joint gravity and magnetic inversion in 3D using Monte Carlo methods, *Geophysics*, *71*, G153–G156, doi:10.1190/1.2209952.

Bosch, M., T. Mukerji, and E. Gonzalez (2010), Seismic inversion for reservoir properties combining statistical rock physics and geostatistics: A review, *Geophysics*, *75*, A165–A176.

Bourne, J. H. (1993), Use of magnetic susceptibility, density and modal mineral data as a guide to composition of granitic plutons, *Math. Geol.*, *25*, 357–375.

Buland, A., and O. Kolbjomsen (2012), Bayesian inversion of CSEM and magnetotelluric data: *Geophysics*, *77*, E33–E42.

Caers J., S. Srinivasan, and A. Journel (2000), Geostatistical quantification of geological information for a fluvial-type North Sea reservoir, *SPE Reservoir Eval. Eng.*, *3*, 457–467.

Chen, J., and G. M. Hoversten (2012), Joint inversion of marine seismic AVA and CSEM data using statistical rock-physics models and Markov random fields, *Geophysics*, *77*, R65–R80.

Chiles, J.-P., and P. Delfiner (2009), *Geostatistics: Modeling Spatial Uncertainty*, John Wiley & Sons, Hoboken, NJ.

Connolly, J. A. D. (2005), Computation of phase equilibria by linear programming tool for geodynamic modeling and an application to subduction zone decarbonation, *Earth Planet. Sci. Lett*, *236*, 524.

Dadashpour, M, D. Echeverria-Ciaurri, J. Kleppe, and M. Landro (2009), Porosity and permeability estimation by integration of production and time-lapse near and far offset seismic data: *J. Geophys. Eng.*, *6*, 325–344, doi: 10.1088/1742-2132/6/4/001.

Deutsch, C. R., and A. G. Journel (1992), *GSLIB, Geostatistical Software Library and User's Guide*, Oxford University Press, New York, 340 pages.

Deustch, C. V., and L. Wang (1996), Hierarchical object based stochastic modeling of fluvial reservoirs, *Math. Geol.*, *28*, 857–880.

Doetsch, J., N. Linde, I. Coscia, S. Greenhalgh, and A. Green (2010), Zonation for 3D aquifer characterization based on joint inversion of mutimethod crosshole geophysical data, *Geophysics*, G53–G64.

Dubrule, O. (2003), *Geostatistics for Seismic Data Integration in Earth Models*, SEG, New York.

Eidsvik J., P. Avseth, H. More, T. Mukerji, and G. Mavko (2004), Stochastic reservoir characterization using prestack seismic data, *Geophysics*, *69*, 978–993.

Gallardo, L., and M. Meju (2004), Joint two-dimensional DC resistivity and seismic travel-time inversion with cross-gradients constraints, *J. Geophys. Res.: Solid Earth*, *109*, B03311.

Gallardo, L, M. A. Perez, and E. Gomez (2003), A versatile algorithm for joint 3D inversión of gravity and magnetic data, *Geophysics*, *68*, 949–959.

Gelfand, A. E., and A. F. M. Smith (1990), Sampling-based approaches to calculating marginal densities, *J. Am. Stat. Assoc.*, *85*, 398–409.

Geyer, C. J. (1992), Practical Markov chain Monte Carlo, *Stat. Sci.*, *7*, 473–551.

González, E., T. Mukerji, and G. Mavko (2008), Seismic inversion combining rock physics and multiple-point geostatistics, *Geophysics, 73*, R11–R21.

Grana, D. (2014), Probabilistic approach to rock physics modeling, *Geophysics, 79*, D123–D143.

Grana, D., and E. Della Rossa (2010), Probabilistic petrophysical-properties estimation integrating statistical rock physics with seismic inversion, *Geophysics, 75*, O21–O37.

Grana, D., T. Mukerji, J. Dvorkin, and G. Mavko (2012), Stochastic inversion of facies from seismic data based on sequential simulations and probability perturbation method, *Geophysics, 77*, M53–M72.

Griffiths, T, C. Kemp, and J. Tenenbaum (2008), Bayesian models of cognition, in *The Cambridge Handbook of Computational Psychology*, R. Sun, ed., Cambridge University Press, New York.

Guillen, A., P. Calcagno, G. Courrioux, A. Joly, and P. Ledru (2007), Geological modelling from field data and geological knowledge, Part II—Modeling validation using gravity and magnetic data inversion, *Phys. Earth Planetary Interiors*, *171*, doi::10.1016/j.pepi.2008.06.014.

Haber, E., and D. Oldenburg (1997), Joint inversion: A structural approach, *Inverse Problems*, *13*, 63–77.

Hacker, B. R., G. A. Abers, and S. M. Peacock (2003), Subduction factory 1. Theoretical mineralogy, densities, seismic wave speeds and H2O contents, *J. Geophys. Res.*, *108*, doi:10.1029/2001JB001127.

Hilterman, F. J. (2001), *Seismic Amplitude Interpretation,* SEG, New York.

Holden, L., R. Hauge, O. Skare, A. Skorstad (1998), Modeling of fluvial reservoirs with object models, *Math. Geol.*, *30*, 473–496.

Huang, X. (2001), *Integrating Time-Lapse Seismic with Production Data: A Tool for Reservoir Engineering,* The Leading Edge, New York.

Khan, A., and J. A. D. Connolly (2008), Constraining the composition and thermal state of Mars from inversion of geophysical data, *J. Geophys. Res.*, *113*, E7003.

Khan, A., J. A. D. Connolly, J. Maclennan, and K. Mosegaard (2006), Joint inversion of seismic and gravity data for lunar composition and thermal state, *Geophys. J. Int.*, *168*, 243–258.

Khan, A., J. A. D. Connolly, and S. R. Taylor (2008), Inversion of seismic and geodetic data for the major element chemistry and temperature of the Earth's mantle, *J. Geophys. Res.*, *113*, B9308.

Larsen A., M. Ulvmoen, H. Omre, and A. Buland (2006), Bayesian lithology/fluid prediction and simulation on the basis of a Markov-chain prior model, *Geophysics, 71*, R69–R78.

Linde, N., J. Chen, M. Kowalsky, and S. Hubbard (2006a), Hydrogeophysical parameter estimation approaches for field scale characterization, In *Applied Hydrogeophysics*, H. Verrecken, A. Binley, G. Cassiani, A. Revil, and K. Titov, eds., Springer, New York.

Linde, N., A. Binley, A. Tryggvason, L. Pedersen, and A. Revil (2006b), Improved hydrogeophysical characterization using joint inversion of cross-hole electrical resistance and ground-penetrating radar traveltime data, *Water Resources Res.*, *42*, W12404.

Lines, L., A. Schultz, and S. Treitel (1988), Cooperative inversion of geophysical data, *Geophysics, 53*, 8–20.

Liu, S. J. (1998), Metropolized independent sampling with comparison to rejection sampling and importance sampling, *Stat. Comput.*, *6*, 1113–1119.

Martinelli, G., Eidsvik, J., Sinding-Larsen, R., Rekstad, S., and Mukerji, T. (2013), Building Bayesian networks from basin-modelling scenarios for improved geological decision making, *Petroleum Geosci.*, *19*(3), 289–304.

Mavko, G., and T. Mukerji (1998), A rock physics strategy for quantifying uncertainty in common hydrocarbon indicators, *Geophysics, 63*, 1997–2008.

Mavko, G., T. Mukerji, and J. Dvorkin (2003), *The Rock Physics Handbook,* Cambridge University Press, New York.

Mezghani, M., A. Fornel, V. Langlais, and N. Lucet (2004), *History Matching and Quantitative Use of 4D Seismic Data for an Improved Reservoir Characterization,* Society of Petroleum Engineers, Houston, TX, SPE-90420-MS, doi: 10.2118/90420.

Mosegaard, K. (2011), Quest for consistency, symmetry and simplicity—The legacy of Albert Tarantola, *Geophysics, 76*, W51–W61.

Mosegaard K., and A. Tarantola (1995), Monte Carlo sampling of solutions to inverse problems, *J. Geophys. Res.*, *100*, 12431–12447.

Mosegaard K., and A. Tarantola (2002), Probabilistic approach to inverse problems, in *International Handbook of Earthquake and Engineering Seismology*, W. Lee, H. Kanamori, P. Jennings, and C. Kisslinger, eds., Academic Press, New York.

Mukerji, T, A. Jorstad, P. Avseth, and J. R. Granli (2001), Mapping lithofacies and pore–fluid probabilities in a North Sea reservoir: Seismic inversions and statistical rock physics, *Geophysics, 66*, 988–1001.

Pearl, J. (1986), Fusion, propagation and structuring in belief networks, *Artifi. Intell.*, *29*, 241–288.

Pearl, J. (1994), Causal diagrams for empirical research, *Biometrika, 82*, 669–688.

Rimstad K., P. Avseth, and H. Omre (2012), Hierarchical Bayesian lithology/fluid prediction: A North Sea case study, *Geophysics, 77*, B69–B85.

Smith, A. F., and G. O. Roberts (1993), Bayesian computations via the Gibbs sampler and related Markov chain Monte Carlo methods, *J. Royal Stat. Soc.*, *55*, 3–23.

Spikes, K., T. Mukerji, and G. Mavko (2007), Probabilistic seismic inversion on rock-physics models, *Geophysics*, R87–R97.

Strebelle, S. (2002), Conditional simulation of complex geological structures using multiple-point statistics, *Math. Geol.*, *34*, 1–22.

Suman, A., and T. Mukerji (2013), Sensitivity study of rock-physics parameters for modeling time-lapse seismic response of Norne field, *Geophysics, 78*, D511–D523.

Tarantola, A. (2005), *Inverse Problem Theory and Methods for Model Parameter Estimation*, SIAM, Philadelphia.

Thulasiraman, K., and M. Swamy (1992), *Graphs: Theory and Algorithms*, John Wiley & Sons, New York.

Tiberi, C., M. Diament, J. Deverchere, C. Petit-Mariani, V. Mikhailov, S. Tikhostsky, and U. Achauer (2003), Deep structure of the Baikal rift zone by joint inversion of gravity and seismology, *J. Geophys. Res. Solid Earth*, *108*, 2133.

Tierney, L. (1994), Markov-chains for exploring posterior distributions, *Ann. Stat.*, *22*, 1702–1762.

Torres-Verdin, C., A. Revil, M. Oristaglio and T. Mukerji (2012), Multiphysics borehole geophysical measurements, formation evaluation, petrophysics and rock physics—Introduction, *Geophysics, 77*, WA1–WA2.

Ulvmoen M., H. Omre, and A. Buland (2010), Improved resolution in Bayesian lithology/fluid inversion from prestack seismic data and well observations, Part 2—Real case study, *Geophysics, 75*, B73–B82.

4

Structural Coupling Approaches in Integrated Geophysical Imaging

Max A. Meju[1] and Luis A. Gallardo[2]

ABSTRACT

Accurate reconstruction of the structure and physical property distribution within the Earth using data from individual geophysical methods is a difficult task due to geological complexity, inherent nonlinearity of most geophysical processes and modelling inaccuracies, and measurement uncertainties. It has emerged that structurally constrained inversion of multiple geophysical data sensing different physical properties of the Earth leads to improved imaging. This chapter presents an overview of structural coupling approaches used successfully in joint inversion of geophysical and other types of data.

4.1. INTRODUCTION

Multiple geophysical (or multiphysics) imaging is proving to be vital in a wide range of contemporary issues including landmine detection, the use of groundwater resources, natural hazard monitoring, methane storage in wetlands, carbon dioxide sequestration, and efficient extraction of fossil fuels (see, e.g., *Binley et al.* [2002], *Alpak et al.* [2008], *Doetsch et al.* [2010a], *Gallardo et al.* [2012], *Jardani et al.* [2013], and *Revil and Mahardika* [2013]). There is a need for combining information from data acquired using different geophysical methods. Combining observations of multiple physical phenomena on an object of investigation have potential for accurate predictions and hence risk reduction in decision making with data. The challenge in integrated imaging of the subsurface is how to combine correlated data from interrelated physical phenomena or disparate data from unrelated physical phenomena and taking into account the different support volumes of the data (Figure 4.1). In this chapter we describe the conceptual, physical, and computational frameworks for data fusion (or preconditioning) and structure-coupled inverse modeling of Earth observations.

[1]Exploration Technical Services Division, Petronas Upstream, Kuala Lumpur, Malaysia
[2]Earth Science Division, CICESE, Ensenada, Mexico

4.1.1. Combining Multiphysics Measurements on a Physical System: Data Fusion

Remote or invasive measurements of different properties of physical systems within the Earth are often made at different spatial and temporal scales (Figure 4.1). Such multidimensional and band-limited datasets are used to infer the structure and physical state of the system under investigation. In many cases, detecting and monitoring the presence or movement of fluids within such systems from multimodal signal sensing and imaging are a primary objective for several reasons, and they require accurate knowledge of the structure holding or confining the fluids. Accurate characterisation of material heterogeneity and its influence on fluid flow or storage in subterranean reservoirs is of crucial importance in the wide range of contemporary issues mentioned above (see, e.g., *Binley*

Integrated Imaging of the Earth: Theory and Applications, Geophysical Monograph 218, First Edition.
Edited by Max Moorkamp, Peter G. Lelièvre, Niklas Linde, and Amir Khan.

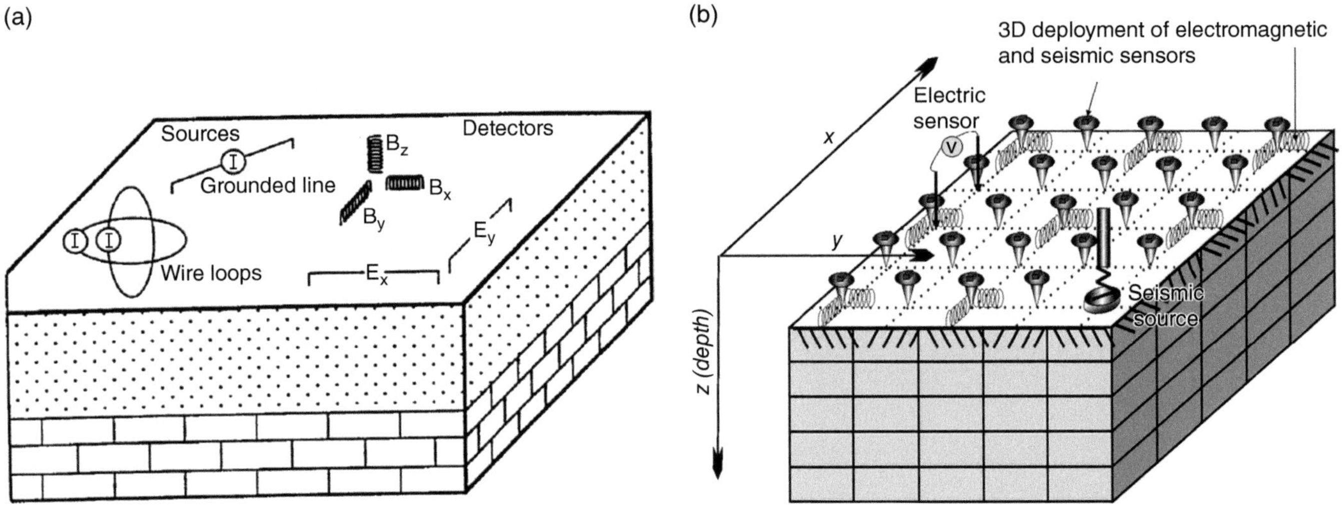

Figure 4.1 Illustration of the concept of multiphysics or multimodal remote sensing measurements. An array of sources and detectors are deployed within, above, or on the surface of a physical system such as the Earth. The detectors or sensors register data related to the object's response to physical excitations of the system by the different applied sources. **(a)** Interrelated physical phenomena, for example, are measured using various sources and detectors in electrical and electromagnetic methods. (After *Meju* [2002].) **(b)** Unrelated physical phenomena are, for example, measured by seismic and electromagnetic methods. (After *Gallardo and Meju* [2011].)

et al. [2002, 2004], *Jardani et al.* [2013], and *Revil and Mahardika* [2013]). The fluid and solid mineral components of subterranean materials have, for instance, electrical and mechanical signatures that can be sensed by geophysical methods and subsequently inverted to image the subsurface physical property distribution. However, the measured data are typically of limited bandwidth and corrupted by ambient noise and may thus be described as inaccurate, insufficient, and inconsistent [*Jackson*, 1972] for a unique characterization of the physical system. Natural physical systems are complex, and our physics-based models describing them are only as good as the limited information available about the natural world. So, how can we overcome these difficulties and generate more realistic images of a physical system? Multiphysics-based imaging has the potential to provide an answer to this question as is long known in geophysical inversion [*Vozoff and Jupp*, 1975; *Sasaki*, 1989], but model resolution can be significantly limited by (a) the lack of established procedures for characterization and propagation of uncertainties in the measured data and model predictions and (b) our often limited or incomplete knowledge of the physical system being imaged.

Until recently, the emphasis has been on combining and modeling measurements from physically related phenomena, for which there are established analytical relationships. An important consideration with such data is how to account for the different spatial scales (or footprints) of the numerous available methods of remote sensing, scanning, or depth sounding (Figure 4.1a). For example, to fuse the measured apparent resistivity data for coincidentally located frequency-domain magnetotelluric (MT) and in-line four-electrode direct current (DC) Schlumberger, Wenner, or dipole–dipole depth sounding arrays, one may use the approximate time–space scaling relationship (see, e.g., *Meju* [2005])

$$T = 2\pi\mu_0\sigma L^2, \tag{4.1}$$

where $\mu_0 = 4\pi \times 10^{-7}$ ($\Omega\cdot sm^{-1}$) is the magnetic permeability of free-space, σ is half-space electrical conductivity (siemens/m), T is the MT period in seconds, and L is the half-electrode array length (m) for DC resistivity sounding. Similarly, time-domain electromagnetic (TEM) and MT sounding data can be fused using the approximate relation $T = 4t$ (see, e.g., *Meju* [2005]), where t is the transient time in seconds. These empirical relations apply in any half-space medium and have been shown to also apply in general media (see, e.g., *Meju and Sakkas* [2007], *Bai and Meju* [2003], *Sakkas et al.* [2002], *Bai et al.* [2001], *Meju* [1998, 2002, 2005], and *Meju et al.* [1999, 2002, 2003]). Figure 4.2 shows the DC resistivity, TEM, and MT apparent resistivity sounding curves from the same location [*Meju et al.*, 1999]. It is obvious that the apparent resistivity sounding curves from the various methods are in excellent accord in the overlapping sections in this common spatiotemporal scale representation (see more examples in *Meju et al.* [1999] and *Meju* [2005]). The combined datasets are effectively broadband, image a more complete depth section than each separate method,

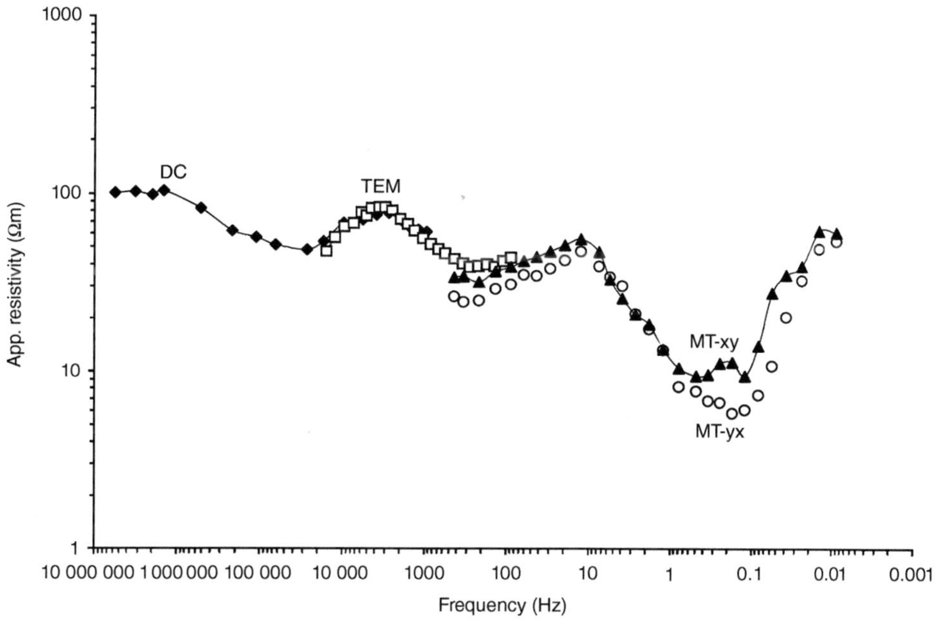

Figure 4.2 Example of relative space–time scaling of electrical and electromagnetic field data to yield broadband integrated Earth-resistivity soundings. Shown are Schlumberger DC resistivity, central-loop TEM, and MT apparent resistivity sounding curves from the same site in the Parnaiba Basin, Brazil. (After *Meju et al.* [1999].) For the DC case, the equivalent MT frequency is obtained via Eq. (4.1) while TEM transient times (t) are converted to frequencies (f) using the relation ($1/(4t) = f$) described in the text.

and the imaged resistivity-versus-depth variations are shown by *Meju et al.* [1999] to be in agreement with the known lithological variations in a borehole at the sounding location. This suggests the potential for improvement in subsurface imaging using more complete, consistent, and sufficient (i.e., broadband) geoelectromagnetic data.

Combining data from uncorrelated physical processes (Figure 4.1b) is more challenging and requires mathematical coupling in inverse modeling as explained later. A particularly challenging task is how to make two or more geophysical images of the subsurface structurally consistent if the geophysical methods themselves have different resolution characteristics or footprints (herein referred to as individual property fields). For example, seismic reflectivity and electrical resistivity may differ significantly in scale length. *Gallardo and Meju* [2003, 2004] introduced the cross-gradient criterion to link such disparate physical property fields when they spatially overlap. *Newman and Commer* [2010] and *Um et al.* [2014] propose an approach involving low-pass filtering (via Laplace–Fourier transformation) of the seismic traces [*Shin and Cha*, 2008, 2009; *Shin and Ha*, 2008] before they are jointly inverted with electromagnetic data. The proposed data transformation results in seismic data that have similar spatial resolution as electromagnetic data such that both datasets can be treated on equal terms during inversion, but this data homogenization approach will also lead to loss of high-frequency information (seismic reflectivity).

4.1.2. Structural Coupling of Multiphysics Inverse Problems: Model Fusion

Experimental data from coincidentally located multiple geophysical scanning of the Earth (Figure 4.1b) are commonly inverted separately since there is no established analytical relationship between the physical properties sensed by different depth-scanning methods. Separate data inversion may lead to inconsistent models for the same subsurface target under observation (see, e.g., *Gallardo et al.* [2005, 2012]). However, just as rock porosity and the presence and amount of fluids can influence the electrical and mechanical properties of porous media (see, e.g., *Pride* [1994], *del Rio and Whitaker* [2001], *Revil* [2013], and *Revil and Mahardika* [2013]), the structural attributes of the media can influence the interactions between these properties and any applied electrical, electromagnetic, and mechanical excitations. Thus in geophysical imaging, the rock structure associated with physical property variations (see, e.g., *Gallardo and Meju* [2003], *Haber and Oldenburg* [1997] or spatial homogeneity changes (see, e.g., *Bosch* [1999] and *Bosch et al.* [2002]) may remain the same to the multiphysics scanning methods (i.e., a common frame of reference) and serve for multiple property coupling and integration, which can help reduce the model uncertainty inherent in using the individual methods. This is the basis of structure-coupled joint inversion of data from non-physically related properties.

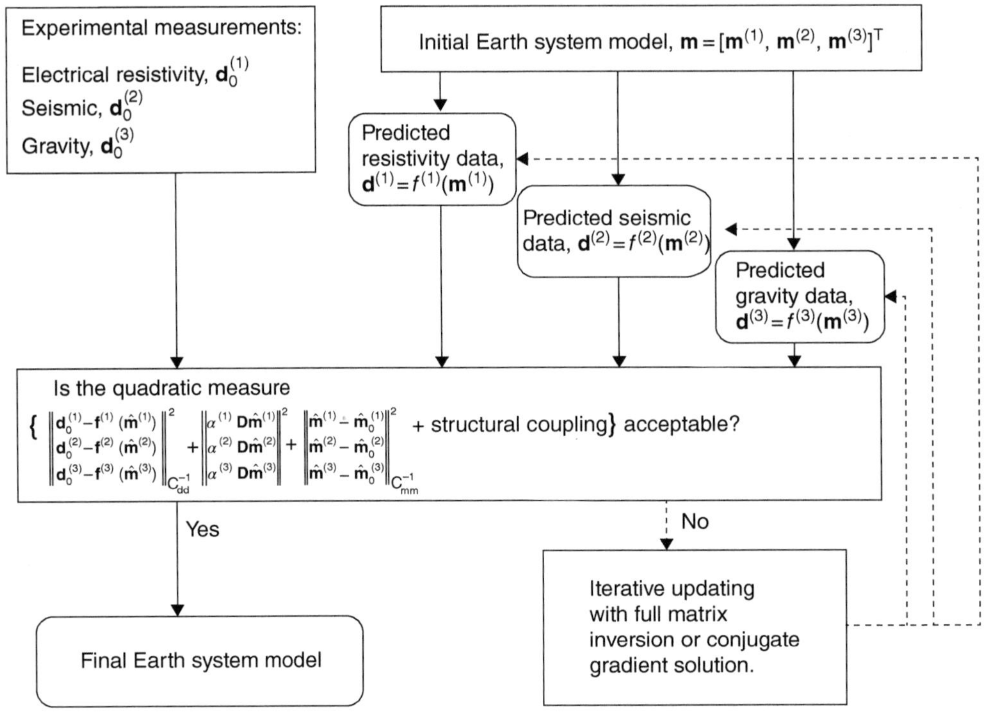

Figure 4.3 Flowchart of a basic structure-coupled joint geophysical imaging of the subsurface using multimodal measurements. (After *Gallardo and Meju* [2011].) Given multiphysics measurements over an Earth system, a quadratic measure is minimized under the imposed condition of structural similarity, enforced on the unknown models for the different physical property fields (electrical resistivity, seismic velocity, and density). The matrix **D** and the weighting factor α serve to regularize the problem while the available prior knowledge of the system parameters is contained in $\mathbf{m}_0$.

It is different from joint inversion of data from physics-coupled phenomena such as exploited in electrical and electromagnetic methods which have identical physical properties (see, e.g., *Sasaki* [1989] and *Jardani et al.* [2013]) or gravity and magnetic methods for which a transform relation exists (see, e.g., *Gallardo-Delgado et al.* [2003] and *Zhdanov et al.* [2012a, 2012b]. Structural coupling requires the imposition of mathematically driven matching conditions at structural boundaries in a numerically stable way (see *Gallardo and Meju* [2011], *Haber and Gazit* [2013], and references therein). An appropriate measure of data misfit is minimized subject to the conditions of structural similarity enforced on the sought physical models (Figure 4.3). Simultaneous multidimensional inverse modeling or "joint inversion" of multiphysics data with structural constraints has been shown to lead to models that are in better accord and closer to the true system property distribution [*Zhang and Morgan*, 1996; *Haber and Oldenburg*, 1997; *Berge et al.*, 2000; *Musil et al.*, 2003; *Gallardo* 2004; *Gallardo and Meju*, 2003; 2004; 2007; *Linde et al.*, 2006; 2008; *Bedrosian et al.*, 2007; *Gallardo*, 2007; *Hu et al.*, 2009; *Infante et al.*, 2010; *Moorkamp et al.*, 2011; *Zhu and Harris*, 2011; *Haber and Gazit*, 2013]. However, not all physical property distributions in the subsurface will be structurally coincident, and some flexibility in model reconstruction may be necessary in some geological environments. An obvious example is where there is strong geological heterogeneity together with variations in salinity or water content as partly discussed in *Linde et al.* [2006], and the limits of the structural approach thus need to be clarified. This is a key consideration for any realistic joint inversion scheme.

Following *Gallardo and Meju* [2011], we distinguish between structure-coupled joint inversion of phenomenologically different multiphysics data (see, e.g., *Gallardo and Meju* [2003]) and the alternative approach of traditional cooperative inversion [*Lines et al.*, 1988; *Oldenburg et al.*, 1997; *Paasche and Tronicke*, 2007] or structural cooperative inversion (see, e.g., *Lelievre* [2009] and *Jegen et al.* [2009]). In the iterative cooperative inversion of data from two geophysical methods, the respective datasets are inverted in separated steps, but the resulting model (i.e., structure and physical property distribution) of one iteration step is used as the starting model for the other step through an assumed relation between the two models (see, e.g., *Jegen et al.* [2009]). Some (compositional) correlation between the physical properties is assumed in the traditional cooperative inversion approach while it is assumed that the structural aspects of the physical property fields are spatially correlated in structural cooperative inversion (see, e.g., *Lelievre* [2009]). This approach may not lead to a unique solution but allows for the possibility that not all geophysical attributes are structurally coupled, is relatively easy to implement, and avoids an explicit weighting of the different datasets. It is possible to constrain the cooperative solution process to honor a priori structural information about the physical system, but the resulting knowledge of the system is often incomplete and may be inconsistent with the experimental data. There is thus a need for an approach to resolve or avoid such potential conflicts with measured data. It is emerging that appropriately coupled simultaneous inversion of multiphysics data with realistic *a priori* information (Figure 4.3) might provide a solution for such problems (see, e.g., *Gallardo and Meju* [2003, 2004, 2007, 2011], *Gallardo* [2007], *Cardiff and Kitanidis* [2009], *Lelievre and Oldenburg* [2009], *Gallardo et al.* [2012] and *Um et al.* [2014]). Note that the inversion of data from one geophysical method incorporating a priori information such as structural interpretation from existing higher-resolution surface geophysical or borehole data—akin to the "model fusion concept" of *Haber and Gazit* [2013] and the image-guided tomography of *Zhou et al.* [2014]—may not be strictly termed joint or multiphysics inversion. It is more appropriately termed integrated interpretation here. Note also that *Zhdanov et al.* [2012a, 2012b] use Gramian constraints to invert jointly gravity and magnetic data by enforcing (or enhancing *sensu stricto*) the correlations between the different model parameters or their attributes (e.g., spatial gradients). Importantly, the method assumes that a correlation between the different model parameters or their attributes exists, but the specific forms are unknown.

In the rest of this chapter, we will focus on the development of differential-geometry-based algorithms that permit the joint inversion of multiple geophysical and multiscale data for which the underlying physical properties may be uncorrelated. In these approaches, the normalized gradients of the intensity and directional property fields of the physical system provide a means to effectively couple unrelated as well as interrelated physical properties of a system. The gradient properties also provide a simple objective means of visualizing those components of the resulting images that are either resolved or unresolved by the combined multiphysics data (see, e.g., Figure 8 in *Gallardo and Meju* [2004]). They may thus serve as metrics for quickly appraising uncertainty in the reconstructed images. In Section 4.2 we seek to develop further understanding of methods for coupled simultaneous inversion of uncorrelated as well as interrelated multiplatform physical (e.g., electrical, electromagnetic, magnetic, gravity, sound wave, and fluid-flow) data using structural gradients-based approach for coupling fundamentally different physical fields. In Section 4.3 we discuss challenges and an instructive case study of integrated imaging. In Section 4.4 we suggest directions for future research in structure-coupled multiphysics imaging.

4.2. APPROACHES IN MULTIPHYSICS IMAGING BASED ON COMMON STRUCTURE

4.2.1. Gradient-Based Measures of Image Similarity

Morphological semblance between any two images can be defined in various ways [*Serra*, 1982]. Here, we use functions such as differences, sums, cross-products, and dot products of the image gradients. Such functions have been used to link different physical property fields in geophysical imaging (see, e.g., *Jilinski et al.* [2010]). Following *Gallardo and Meju* [2011], we group these functions into direction-dependent and direction-independent measures of semblance.

4.2.1.1. Direction-Dependent Measures

The similarities or differences in the gradients direction of any two images can be measured using a variety of multidimensional operators. A simple comparison of normalized directions is given by the normalized Gauss map function [*Droske and Rumpf*, 2003]

$$\hat{n} = \frac{\nabla m}{\|\nabla m\|}, \tag{4.2}$$

where ∇m is the gradient of an image. Function (4.2) describes the precise direction vector of the physical property fields; but for normalized gradients, singularities can occur when $\|\nabla m\| \to 0$ such as in homogenous media, local maxima-minima, and saddle points.

The simple difference between two normalized image gradients

$$\Delta\vec{\phi} = \hat{n}^{(1)} - \hat{n}^{(2)}, \tag{4.3}$$

provides a linear comparison of the structural resemblance of two images and has significant computational advantages. *Droske and Rumpf* [2003] show successful applications of this approach to medical image registration where careful attention was given to the singularity arising from (4.2) by a supervised selection of features with large property gradients that were used for the application of (4.3).

Another approach employs either angular or trigonometric functions as gradient measures. *Droske and Rumpf*

[2003] discuss product-related functions based on normalized Gauss maps, and they propose functions such as

$$\Delta\phi = \hat{n}^{(1)} \cdot \hat{n}^{(2)} \tag{4.4}$$

and

$$\Delta\vec{\phi} = \hat{n}^{(1)} - \hat{n}^{(2)}\left(\hat{n}^{(1)} \cdot \hat{n}^{(2)}\right). \tag{4.5}$$

The biggest drawback of these functions is the combination of the singularities associated with the normalized image gradients. Similarly, *Gallardo and Meju* [2004] propose the use of angles and vector products based on gradients, and they suggest functions such as

$$\Delta\phi = \cos^{-1}\left(\hat{n}^{(1)} \cdot \hat{n}^{(2)}\right) \tag{4.6}$$

and

$$\Delta\phi = \frac{1}{1 - \hat{n}^{(1)} \cdot \hat{n}^{(2)}}. \tag{4.7}$$

However, these not only face the problem of combined singularities of two Gauss map functions but also the additional nonlinearity of the trigonometric functions and their periodicity-related discontinuity.

Alternative un-normalized functions based on vector products have been proposed. Remarkably, the cross-gradient function [*Gallardo and Meju*, 2003] defined as

$$\Delta\vec{\phi} = \vec{\tau} = \nabla m^{(1)} \times \nabla m^{(2)} \tag{4.8}$$

has proven useful and stable, and its application has steadily grown in joint multiphysics inverse problems [*Gallardo and Meju* 2003, 2004, 2007, 2011; *Gallardo et al.* 2005, 2012; *Tryggvason and Linde* 2006; *Linde et al.*, 2006, 2008; *Gallardo*, 2007; *Hu et al.*, 2009; *Fregoso and Gallardo*, 2009; *Doetsch et al.*, 2010a, 2010b; *Infante et al.*, 2010; *Zhu and Harris* 2011; *Moorkamp et al.*, 2011; *Lochbühler et al.*, 2013; *Um et al.*, 2014]. The multiplicative character of this function does not demand additional normalizations/equalizations or angular transformations. It shows no discontinuities or singularities and can detect differences within large or small gradients. Figure 4.4 shows a three-dimensional map of this function for two 3D objects (ellipsoidal and spherical shells) with rotation geometry [*Fregoso and Gallardo*, 2009]. The direction of the cross-gradient vector is consistent with the circular geometry of the objects whereas the magnitude of the vector emphasizes the morphological differences due to conflicting gradients. Note that common structures are picked up when the magnitudes of the three components of the cross-gradient vector are zero. Also, the specific components of the cross-gradient vectors that are equal to zero show the structural similarity in a consistent geometrical direction. For instance, the radial geometry in Figure 4.4 reduces two components (radial and x-direction) of the cross-gradient vector to zero, whereas a two-dimensional geometry reduces the vector to just its transverse component (see Figure 1 in *Gallardo et al.* [2005]). In the general three-dimensional (3D) objects, if a coordinate basis can align parallel to the equivalent (equipotential) geometrical contours, it will reduce the whole cross-gradient vector to that of just one component, which is computationally advantageous. Unfortunately, the determination of this particular axis in models of multidimensional objects is a complex task. *Haber and Gazit* [2013], based on the vector identity

$$\left|\nabla m^{(1)} \times \nabla m^{(2)}\right|^2 = \left|\nabla m^{(1)}\right|^2 \left|\nabla m^{(2)}\right|^2 - \left|\nabla m^{(1)T} \nabla m^{(2)}\right|^2, \tag{4.9}$$

suggest the mathematical equivalence of the cross-product function (left-hand side) and the dot-product function (right-hand side terms) as measures of structural similarity. They also suggest that the dot-product function is

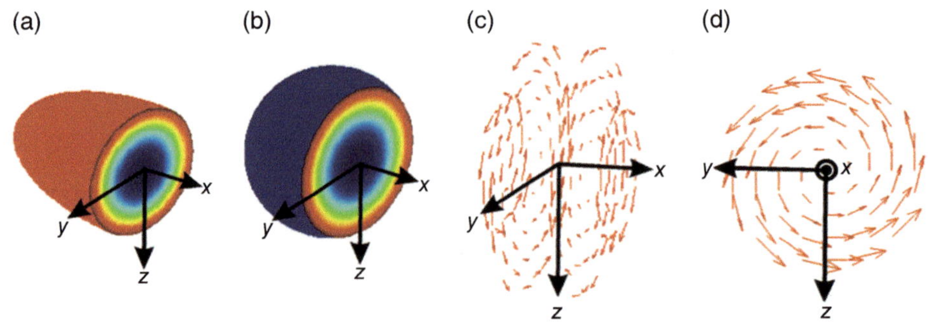

Figure 4.4 Comparison of ellipsoidal **(a)** and spherical **(b)** images to determine their 3D structural resemblance. **(c)** Perspective view of four selected vertical slices that show the cross-gradient vectors for these models. **(d)** Note the circular y–z behavior of the cross-gradients, and observe the partial 2D resemblance indicated by the null x-component of all cross-gradient vectors. (After *Fregoso and Gallardo* [2009].)

more accurate when discretizing using short differences since it only involves the products of normal derivatives.

4.2.1.2. Direction-Independent Measures

A different approach to measuring structural resemblance uses the magnitude of the gradient fields and ignores the directional characteristic of the property changes (see, e.g., *Cumani* [1991] and *Toivanen et al.* [2003]). When this strategy is applied to pattern recognition tasks such as edge detection, the gradients that are regarded as large are simply selected as the most appropriate to define the edges of the objects. This is measured by a threshold value which can be statistically regulated to guarantee that the amplitude of the change surpasses the noise level in the image. For geophysical imaging involving two different datasets, *Zhang and Morgan* [1996] and *Haber and Oldenburg* [1997] propose the use of model curvature as a measure of structure. *Zhang and Morgan* [1996] use a function of the form

$$\phi = \frac{\|\nabla^2 m\|}{\alpha^2} \tag{4.10}$$

while *Haber and Oldenburg* [1997] employ the function

$$\phi = \begin{cases} 0 & |\nabla^2 m| < \tau_1 \\ P_5\left(|\nabla^2 m|\right), & \tau_1 < |\nabla^2 m| < \tau_2 \\ 1 & \tau_2 < |\nabla^2 m| \end{cases} \tag{4.11}$$

as a structure operator. In Eqs. (4.10 and 4.11), P_5 is a one-dimensional polynomial of degree 5 that makes the structure operator twice Frechét differentiable, τ_1 and τ_2 define the interval within which this operator is twice differentiable, and α is an amplitude normalization factor. These functions principally detect the edges of objects in images, and the structural differences between the two images are given by

$$\Delta\phi = \phi^{(1)} - \phi^{(2)}. \tag{4.12}$$

A key feature of this approach is that it assumes that the coincident occurrence of sharp edges or prescribed smoothed edges is the main indicator of morphological resemblance and penalizes other variations such as ramp- or fuzzy-type boundaries. The presumed advantage is that $\Delta\phi$ is a linear operator and this facilitates its computation. However, it requires a model-dependent normalization (which actually suggests its nonlinear nature) and ignores direction-dependent morphological information that defines an object's shape.

It is noteworthy that *Günther and Rücker* [2006] introduced a direction-independent joint inversion scheme that adapts the regularization of one property model depending on the structure (gradient magnitude) of the other model. Recently, *Haber and Gazit* [2013] propose that any two physical models $\mathbf{m}^{(1)}$ and $\mathbf{m}^{(2)}$ can be coupled through the use of the square root function

$$\phi = \int_\Omega \sqrt{\left|\nabla m^{(1)}\right|^2 + \left|\nabla m^{(2)}\right|^2}\, dx \tag{4.13}$$

where $|\nabla m^{(1)}|$ and $|\nabla m^{(2)}|$ are the absolute values of their respective gradients. They termed this measure the joint total variation (JTV) of the two models, which has the useful property of being convex in both $\mathbf{m}^{(1)}$ and $\mathbf{m}^{(2)}$.

4.2.2. Joint Inversion Based on Gradient Direction of Resemblance

There are two approaches to structure-coupled joint inversion. One approach seeks exact structural resemblance between the sought images. In the other approach, structural resemblance is encouraged, rather than imposed, by minimizing a weighted norm of the structural-resemblance constraint. The advantages and drawbacks of each approach are discussed below in the context of cross-gradients inversion.

4.2.2.1. Inversion for Exact Structural Resemblance

Gallardo and Meju [2003] use the cross-gradient function (4.8) in multiphysics imaging and consider two geophysical images—electrical resistivity ($\mathbf{m}^{(1)}$) and seismic velocity ($\mathbf{m}^{(2)}$)—of the same object to be structurally similar when

$$\Delta\vec{\phi}_i = \vec{\tau}_i = \nabla m^{(1)}(x_i,y_i,z_i) \times \nabla m^{(2)}(x_i,y_i,z_i) = \vec{0}. \tag{4.14}$$

The inverse problem is posed as the search for structurally similar resistivity and velocity models, as gauged by Eq. (4.14), which satisfies the measured electrical resistivity and seismic travel-time data. The data are assumed to contain random noise whose distribution is given by an uncorrelated Gaussian distribution with zero mean and variance σ^2. They regularize the problem using smoothness assumptions, and they also require the images to honor any available a priori information ($\mathbf{m}_0$) about the sought objects. The errors in the a priori parameter estimates $\mathbf{m}_0$ are assumed to have a Gaussian distribution with covariance matrix $\mathbf{C}_{\mathrm{m_0 m_0}}$ and they minimize the objective function

$$s\left(\widehat{\mathbf{m}}^{(1)},\widehat{\mathbf{m}}^{(2)}\right) = \left\|\begin{matrix}\mathbf{d}_0^{(1)} - f^{(1)}\left(\widehat{\mathbf{m}}^{(1)}\right) \\ \mathbf{d}_0^{(2)} - f^{(2)}\left(\widehat{\mathbf{m}}^{(2)}\right)\end{matrix}\right\|^2_{\mathbf{C}_{dd}^{-1}} + \left\|\begin{matrix}\alpha^{(1)}\mathbf{D}\widehat{\mathbf{m}}^{(1)} \\ \alpha^{(2)}\mathbf{D}\widehat{\mathbf{m}}^{(2)}\end{matrix}\right\|^2 + \left\|\begin{matrix}\widehat{\mathbf{m}}^{(1)} - \widehat{\mathbf{m}}_0^{(1)} \\ \widehat{\mathbf{m}}^{(2)} - \widehat{\mathbf{m}}_0^{(2)}\end{matrix}\right\|^2_{\mathbf{C}_{m_0 m_0}^{-1}} \tag{4.15}$$

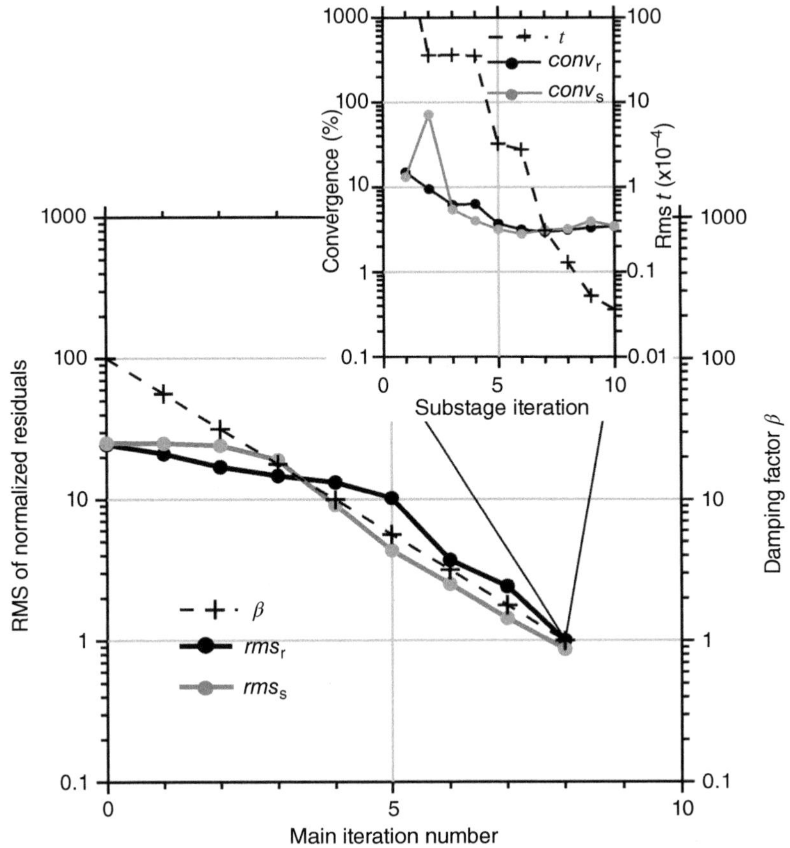

Figure 4.5 Example of RMS values of the normalized residuals and damping coefficient β per iteration for joint inversion of DC resistivity and seismic refraction data from Quorn, England. (After *Gallardo and Meju* [2004].) The inset shows the actual convergence measures and the trend of the RMS values of the cross-gradients function (i.e., substage minimization) for the last main iteration.

subject to $\{\vec{\tau}_i\} = \{\vec{0}\}$ in the discretized model volume. In this expression, **d** is the data vector, $f(\mathbf{m})$ is the forward operator, and **D** is a discrete version of a second-order Tikhonov regularization matrix and is weighted by a constant damping parameter $\alpha^{(i)}$ for each model. Since $f^{(1)}, f^{(2)}$, and $\vec{\tau}$ are nonlinear functions, the solution to the objective function (4.15) is accomplished by linearizing about a starting model ($\mathbf{m}_0$ in the first instance), and the two-stage minimization (one for the data misfit and another for the structural constraints) proceeds iteratively as detailed in *Gallardo and Meju* [2004] for practical exploration data from Quorn in England. For the Quorn multiple geophysical data, the convergence characteristics of the two-stage minimization scheme are shown in Figure 4.5 while the model evolution for each main iteration is shown in Figure 4.6. The solution at iteration $k + 1$ can be expressed as [*Gallardo and Meju*, 2003, 2011]

$$\hat{\mathbf{m}}_{k+1}^{(j)} = \hat{\mathbf{m}}_k^{(j)} + \mathbf{N}_k^{(j)^{-1}}\left(\mathbf{n}_k^{(j)} - \mathbf{B}_k^{(j)^T}\mathbf{q}_k\right), j = 1,2 \qquad (4.16)$$

where

$$\mathbf{N}_k^{(j)} = \mathbf{A}_k^{(j)^T}\mathbf{C}_{dd}^{(j)^{-1}}\mathbf{A}_k^{(j)} + \alpha^{(j)}\mathbf{D}^T\mathbf{D} + \left(\mathbf{C}_{m_0m_0}^{(j)}\right)^{-1}, \qquad (4.17)$$

$$\mathbf{n}_k^{(j)} = \mathbf{A}_k^{(j)^T}\mathbf{C}_{dd}^{(j)^{-1}}\left(\mathbf{d}_0^{(j)} - \mathbf{f}^{(j)}\left(\hat{\mathbf{m}}_k^{(j)}\right) + \mathbf{A}_k^{(j)}\hat{\mathbf{m}}_0^{(j)}\right) - \alpha^{(j)}\mathbf{D}^T\mathbf{D}\hat{\mathbf{m}}_k^{(j)} + \left(\mathbf{C}_{m_0m_0}^{(j)}\right)^{-1}\left(\hat{\mathbf{m}}_0^{(j)} - \mathbf{m}_k^{(j)}\right), \qquad (4.18)$$

$$\mathbf{q}_k = \left\{\mathbf{B}_k^{(1)}\mathbf{N}_k^{(1)^{-1}}\mathbf{B}_k^{(1)^T} + \mathbf{B}_k^{(2)}\mathbf{N}_k^{(2)^{-1}}\mathbf{B}_k^{(2)^T}\right\}^{-1} \left\{\mathbf{B}_k^{(1)}\mathbf{N}_k^{(1)^{-1}}\mathbf{n}_k^{(1)} + \mathbf{B}_k^{(2)}\mathbf{N}_k^{(2)^{-1}}\mathbf{n}_k^{(2)} + \Delta\phi\left(\hat{\mathbf{m}}_k^{(1)}, \hat{\mathbf{m}}_k^{(2)}\right)\right\}, \qquad (4.19)$$

$A = \dfrac{\partial f(m)}{\partial m_k}$ are the partial derivatives at the point $\mathbf{m} = \mathbf{m}_k$, and $\mathbf{B}_k$ is the Jacobian matrix associated with the cross-gradient function. The structural dissimilarity $\Delta\phi$ in Eq. (4.19) is measured by the cross-gradient function

$$\tau\left(m^{(1)}, m^{(2)}\right) = \nabla m^{(1)} \times \nabla m^{(2)} \qquad (4.20)$$

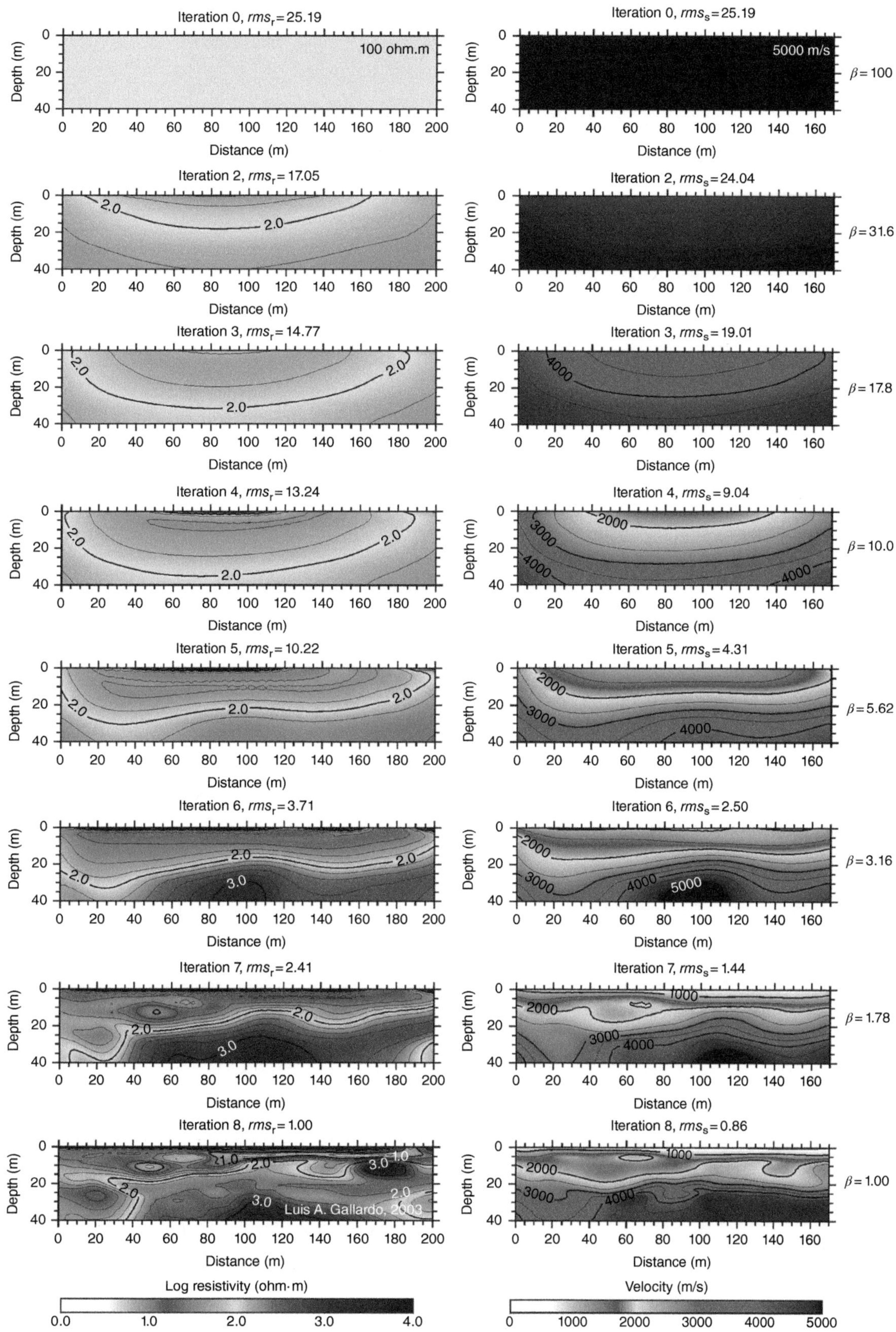

Figure 4.6 Model evolution in the joint cross-gradient inversion of Quorn field data whose convergence characteristics are summarized in Figure 4.5 [*Gallardo and Meju*, 2004]. Shown are the resultant resistivity and velocity models at each iteration. Note the gradual development of common structural features in both sets of models during the inversion process.

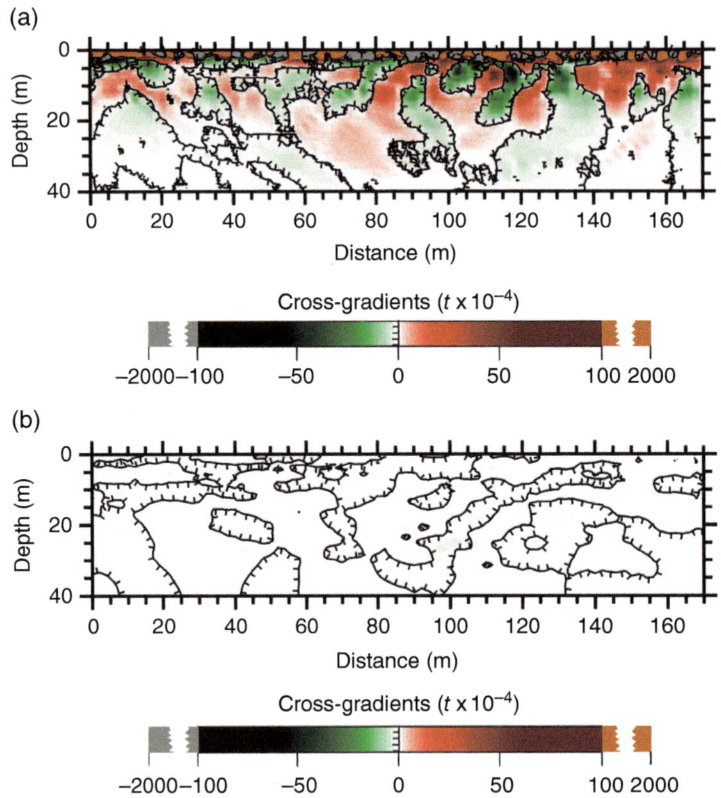

Figure 4.7 Comparison of the cross-gradients function computed for **(a)** the separate inversion and **(b)** joint inversion of the Quorn field data. (After *Gallardo and Meju* [2004].)

and the partial derivatives of the cross-gradient function are given by a discrete version of the operator

$$\frac{\partial \tau}{\partial m^{(j)}} = \left(\delta_{2,j} \nabla m^{(1)} - \delta_{1,j} \nabla m^{(2)}\right) \times \nabla(\cdot). \quad (4.21)$$

Using this derivative, a Taylor series expansion about $m_0^{(1)}$ and $m_0^{(2)}$ yields

$$\nabla m^{(1)} \times \nabla m^{(2)} \cong \nabla m^{(1)} \times \nabla m_k^{(2)} + \nabla m_k^{(1)} \times \nabla m^{(2)} - \nabla m_k^{(1)} \times \nabla m_k^{(2)}. \quad (4.22)$$

Therefore,

$$\tau\left(m^{(1)}, m^{(2)}\right) \cong \tau\left(m^{(1)}, m_k^{(2)}\right) + \tau\left(m_k^{(1)}, m^{(2)}\right) - \tau\left(m_k^{(1)}, m_k^{(2)}\right). \quad (4.23)$$

Equation (4.23) can be understood as suggesting that in Eqs. (4.16) to (4.18), the linearized cross-gradient constraint considers the structural dissimilarities of the previous models $\tau(m_k^{(1)}, m_k^{(2)})$ when updating both models $\mathbf{m}^{(1)}$ and $\mathbf{m}^{(2)}$. Following Eq. (4.23), it is easy to demonstrate that while the joint inversion algorithm converges, this linearized function will converge towards the exact cross-gradient constraint required to fully enforce the sought structural resemblance. Figure 4.7 shows the computed values of the cross-gradients function for the results of separate inversion and joint cross-gradients inversion of the Quorn field data. It is obvious that for this 2D field study, the cross-gradient function was successfully minimized by the implemented joint inversion process.

Note that for 3D inverse problems, the formulation given by Eq. (4.15) leads to an overconstrained problem since it results in three equality constraints on two properties per image cell. In this case, the direct solution of Eq. (4.19) leads to numerical inaccuracies and singularities. The Jacobian of the cross-gradients constraint $\mathbf{B}_k$ must have full row rank; otherwise the null space of $\mathbf{B}_k$ or its transpose could be activated, leading to numerical instabilities in iterative inversion. An effective strategy for solving this problem in the exact approach involves the removal of redundant constraints from $\mathbf{B}_k$ based on geometrical simplifications (such as two-dimensional formulation) or using the singular value decomposition (SVD) or other factorizations (see *Nocedal and Wright* [1999]). *Fregoso and Gallardo* [2009] and *Fregoso* [2010] solved this problem in a full 3D formulation by reducing the row rank of $\mathbf{B}_k$ using the SVD method.

What are the limitations of using this exact resemblance approach? The algorithm relies on the measure of structural resemblance for an accurate reconstruction of the shape of the sought object. There is also the possibility of solution equivalency since the information provided by the resemblance measure can lead to many equally suitable structural models. However, a good algorithm for enforcing similar structure should be able to find the most acceptable solution from among all the structurally equivalent data-consistent images of the sought object.

4.2.2.2. Inversion with Inexact Structural Resemblance Measures

A second approach adopts inexact measures of structural resemblance but have the added flexibility that such measures may be treated as statistical realizations, and the solution of the inverse problem can be tailored to meet specific requirements for large-scale systems. *Tryggvason and Linde* [2006], *Linde et al.* [2006, 2008], *Hu et al.* [2009], and *Doetsch et al.* [2010a, 2010b] adopt this approach and form objective functions that include the cross-gradient constraint. It is instructive to show the connection to the exact approach before exploring the various approaches adopted by these researchers. If the structural resemblance measure in Eq. (4.15) is assumed to be inaccurate and hence assigned an a priori value ($\Delta\phi_o$) with associated covariance matrix $\mathbf{C}_{\phi\phi}$, then one may state the joint multiphysics inverse problem for the inexact case as (cf. Eq. (4.15))

$$s\left(\hat{\mathbf{m}}^{(1)},\hat{\mathbf{m}}^{(2)}\right)=\left\|\begin{matrix}\mathbf{d}_0^{(1)}-f^{(1)}\left(\hat{\mathbf{m}}^{(1)}\right)\\ \mathbf{d}_0^{(2)}-f^{(2)}\left(\hat{\mathbf{m}}^{(2)}\right)\end{matrix}\right\|^2_{\mathbf{C}_{dd}^{-1}} +\left\|\Delta\phi\left(\hat{\mathbf{m}}^{(1)},\hat{\mathbf{m}}^{(2)}\right)-\Delta\phi_0\right\|^2_{\mathbf{C}_{\phi\phi}^{-1}}+\left\|\begin{matrix}\hat{\mathbf{m}}^{(1)}-\hat{\mathbf{m}}_0^{(1)}\\ \hat{\mathbf{m}}^{(2)}-\hat{\mathbf{m}}_0^{(2)}\end{matrix}\right\|^2_{\mathbf{C}_{m_0m_0}^{-1}}, \tag{4.24}$$

where, for simplicity, $\mathbf{C}_{\phi\phi}^{-1}=\alpha\mathbf{I}$ and α is a constant trade-off parameter. Adopting the notations in Eq. (4.16), an iterative formula for solving Eq. (4.24) is given by

$$\hat{\mathbf{m}}_{k+1}^{(j)}=\hat{\mathbf{m}}_k^{(j)}+\mathbf{H}_k^{-1}\mathbf{r}_k^{(j)},\ j=1,2, \tag{4.25}$$

where

$$\mathbf{H}_k=\begin{bmatrix}\mathbf{A}_k^{(1)^T}\mathbf{C}_{dd}^{(1)^{-1}}\mathbf{A}_k^{(1)}+\mathbf{B}_k^{(1)^T}\mathbf{C}_{\phi\phi}^{-1}\mathbf{B}_k^{(1)}+\left(\mathbf{C}_{m_0m_0}^{(1)}\right)^{-1} & -\mathbf{B}_k^{(1)^T}\mathbf{C}_{\phi\phi}^{-1}\mathbf{B}_k^{(2)}\\ -\mathbf{B}_k^{(2)^T}\mathbf{C}_{\phi\phi}^{-1}\mathbf{B}_k^{(1)} & \mathbf{A}_k^{(2)^T}\mathbf{C}_{dd}^{(2)^{-1}}\mathbf{A}_k^{(2)}+\mathbf{B}_k^{(2)^T}\mathbf{C}_{\phi\phi}^{-1}\mathbf{B}_k^{(2)}+\left(\mathbf{C}_{m_0m_0}^{(2)}\right)^{-1}\end{bmatrix}^{-1} \tag{4.26}$$

and

$$\begin{aligned}\mathbf{r}_k^{(j)}=&\ \mathbf{A}_k^{(j)^T}\mathbf{C}_{dd}^{(j)^{-1}}\left(\mathbf{d}_0^{(j)}-\mathbf{f}^{(j)}\left(\hat{\mathbf{m}}_k^{(j)}\right)\right)\\ &+\mathbf{B}_k^{(j)^T}\mathbf{C}_{\phi\phi}^{-1}\left(\Delta\phi_0-\Delta\phi\left(\hat{\mathbf{m}}_k^{(1)},\hat{\mathbf{m}}_k^{(2)}\right)\right)\\ &+\left(\mathbf{C}_{m_0m_0}^{(j)}\right)^{-1}\left(\hat{\mathbf{m}}_0^{(j)}-\mathbf{m}_k^{(j)}\right).\end{aligned} \tag{4.27}$$

Here, the derivation of the Hessian matrix $\mathbf{H}_k$ is based upon the Gauss–Newton inversion iteration. In this solution approach, the structural coupling terms are included in the off-diagonal elements of the Hessian matrix $\mathbf{H}_k$ (which is required to be full rank). Any redundancy or incompatibility in the structural measure is regarded as evidence of the inaccuracy of the function and should ideally be accounted for in the associated covariance matrix $\mathbf{C}_{\phi\phi}$. Conceptually, the objective function in Eq. (4.24) is identical to that given by Eq. (4.15) when $\mathbf{C}_{\phi\phi}^{-1}=\alpha\mathbf{I}$ and $\alpha\to\infty$, and the first term on the right-hand side containing the observables $f(\mathbf{m})$ are augmented with the regularization terms ($\mathbf{Dm}$) as in Eq. (4.15). However, computational differences can make Eq. (4.25) more suitable to inverse problems with special requirements. For example, in large-scale 3D multiphysics imaging problems, it may be desirable to use the nonlinear conjugate gradients method in the minimization process. This method is linear convergent and relies on gradient information on the objective function and not its Hessian. Hence, it can be readily applied to Eq. (4.25) but not to either Eq. (4.17) or (4.26).

Tryggvason and Linde [2006] minimize a quadratic objective function that determines model perturbations from a background model, and the minimization of the cross-gradient constraint seeks to encourage structural resemblance within the model perturbations rather than the actual physical property fields (seismic compressional-wave velocity, V_p and shear-wave velocity, V_s). Their objective function is similar to Eq. (4.24) but formulated in terms of model perturbations for V_p and V_s ($\Delta\mathbf{m}^p$ and $\Delta\mathbf{m}^s$). After linearization of the travel-time equation, their resulting system of equations to solve is

$$\begin{bmatrix}\mathbf{A}\\ \alpha\mathbf{D}\\ \lambda\mathbf{B}\end{bmatrix}\left[\Delta\mathbf{m}\right]=\begin{bmatrix}\Delta\mathbf{d}\\ 0\\ -\lambda\tau\end{bmatrix}, \tag{4.28}$$

where $\Delta\mathbf{d}$ and $\mathbf{A}$ are the travel times and their partial derivatives with respect to the slowness (reciprocal of velocity) parameters and $\Delta\mathbf{m}=[\Delta m^p:\Delta m^s]^T$. $\mathbf{D}$ is a

Laplacian operator used to control the roughness of the slowness perturbation field, and α is the regularization parameter. **B** is the matrix of partial derivatives of the cross-gradient function (τ) with respect to the model parameters, and λ is a constant weighting factor chosen heuristically. The system of equations is solved in a least squares sense using a nonlinear conjugate gradient method. This is an attractive approach for large-scale multiphysics inverse problems. *Linde et al.* [2006, 2008] adopt a similar approach and present their objective function at the $k + 1$ iteration as

$$s\left(\widehat{\mathbf{m}}_{k+1}^{(1)},\widehat{\mathbf{m}}_{k+1}^{(2)}\right)=\left\|\begin{array}{c}\mathbf{d}_0^{(1)}-\left[f^{(1)}\left(\widehat{\mathbf{m}}_k^{(1)}\right)+\mathbf{A}_k^{(1)}\left(\widehat{\mathbf{m}}_{k+1}^{(1)}-\widehat{\mathbf{m}}_k^{(1)}\right)\right]\\ \mathbf{d}_0^{(2)}-\left[f^{(2)}\left(\widehat{\mathbf{m}}_k^{(2)}\right)+\mathbf{A}_k^{(2)}\left(\widehat{\mathbf{m}}_{k+1}^{(2)}-\widehat{\mathbf{m}}_k^{(2)}\right)\right]\end{array}\right\|_{\mathbf{C}_{dd}}^2 +\alpha^2\left\|\begin{array}{c}\widehat{\mathbf{m}}_{k+1}^{(1)}-\widehat{\mathbf{m}}_0^{(1)}\\ \widehat{\mathbf{m}}_{k+1}^{(2)}-\widehat{\mathbf{m}}_0^{(2)}\end{array}\right\|_{\mathbf{C}_{mm}^{-1}}^2+\lambda^2\left\|\Delta\phi\left(\widehat{\mathbf{m}}_k^{(1)},\widehat{\mathbf{m}}_k^{(2)}\right)+\mathbf{B}_k^{(1)}\left(\widehat{\mathbf{m}}_{k+1}^{(1)}-\widehat{\mathbf{m}}_k^{(1)}\right)+\mathbf{B}_k^{(2)}\left(\widehat{\mathbf{m}}_{k+1}^{(2)}-\widehat{\mathbf{m}}_k^{(2)}\right)\right\|^2 \tag{4.29}$$

where $\Delta\phi_i(\widehat{\mathbf{m}}^{(1)},\widehat{\mathbf{m}}^{(2)})=\tau_i(\widehat{\mathbf{m}}^{(1)},\widehat{\mathbf{m}}^{(2)})$, $\mathbf{B}_k$ is the Jacobian matrix associated with the cross-gradient function, λ is a weighting factor, and α is a regularization weight which is varied within an Occam framework [*Constable et al.*, 1987] to find the smoothest model that is in agreement with the field data and the cross-gradients penalty. A particular choice of *Linde et al.* [2006, 2008] is to set λ large enough so as to mimic an exact structural resemblance approach. Note that their work is based on an actual model covariance matrix with a spatial support, but with no cross-coupling between model property types. A conjugate gradient method was used to solve the problem and they have shown successful results of joint 3D inversion of electrical resistivity and cross-hole georadar and seismic data. *Doetsch et al.* [2010a, 2010b] use the above-mentioned approach of *Linde et al.* [2006, 2008] and *Tryggvason and Linde* [2006].

Hu et al. [2009] propose an alternative joint multiphysics inversion approach with inexact structural constraints. Their objective function may be generally stated as finding the updated seismic and electromagnetic model models $(\widehat{\mathbf{m}}_{k+1}^{(1)},\widehat{\mathbf{m}}_{k+1}^{(2)})$ that minimize

$$s\left(\widehat{\mathbf{m}}_{k+1}^{(1)},\widehat{\mathbf{m}}_{k+1}^{(2)}\right)=\left\|\begin{array}{c}\mathbf{d}_0^{(1)}-f^{(1)}\left(\widehat{\mathbf{m}}_{k+1}^{(1)}\right)\\ \mathbf{d}_0^{(2)}-f^{(2)}\left(\widehat{\mathbf{m}}_{k+1}^{(2)}\right)\end{array}\right\|_{\mathbf{C}_{dd}^{-1}}^2+\left\|\begin{array}{c}\mathbf{D}\widehat{\mathbf{m}}_{k+1}^{(1)}\\ \mathbf{D}\widehat{\mathbf{m}}_{k+1}^{(2)}\end{array}\right\|_{\mathbf{C}_{DD}^{-1}}^2 +\left\|\begin{array}{c}\mathbf{B}_k^{(1)}\widehat{\mathbf{m}}_{k+1}^{(1)}\\ \mathbf{B}_k^{(2)}\widehat{\mathbf{m}}_{k+1}^{(2)}\end{array}\right\|_{\mathbf{C}_{\phi\phi}^{-1}}^2. \tag{4.30}$$

They employ block-diagonal weighting covariance matrices (i.e., no implicit cross-correlation). The structural constraint in Eq. (4.30) is notably different from that of Eq. (4.29). The least squares normal equations derived from Eq. (4.30) have two fundamental differences from those given in Eqs. (4.26) and (4.27). Firstly, the corresponding Hessian matrix $\mathbf{H}_k$ lacks the cross-correlating block term $\mathbf{B}_k^{(1)^T}\mathbf{C}_{\phi\phi}^{-1}\mathbf{B}_k^{(2)}$ present in Eq. (4.26). This fully decouples the solution of Eq. (4.30) into two separate minimization problems that make the problem less computationally expensive but at the cost of losing valuable cross-correlating terms. Secondly, the corresponding gradient vector [cf. Eq. (4.27)] lacks the structural resemblance term $\mathbf{B}_k^{(j)^T}\mathbf{C}_{\phi\phi}^{-1}(\Delta\phi(\widehat{\mathbf{m}}_k^{(1)},\widehat{\mathbf{m}}_k^{(2)}))$. This effectively makes the updated models at iteration $k + 1$ ($\mathbf{m}_{k+1}$) insensitive to the mutual structural dissimilarities $\Delta\phi(\widehat{\mathbf{m}}_k^{(1)},\widehat{\mathbf{m}}_k^{(2)})$ borne by their predecessors ($\mathbf{m}_k$). The combined effect is that the cross-gradient function may never attain values close enough to zero unless one of the property gradients becomes zero itself.

4.2.3. Joint Inversion with Direction-Independent Structural Constraints

There are a few joint inversion approaches that disregard the actual direction of the property changes. *Zhang and Morgan* [1996] and *Haber and Oldenburg* [1997] propose the use of normalized versions of Laplacian operators ($\phi = \kappa\nabla^2 m$) as measures of image morphology and specifically the edges of objects present in the images [Eqs. (4.10) and (4.11)]. *Haber and Oldenburg* [1997] define the objective function as minimizing

$$s\left(\widehat{\mathbf{m}}^{(1)},\widehat{\mathbf{m}}^{(2)}\right)=\left\|\begin{array}{c}\mathbf{d}_0^{(1)}-f^{(1)}\left(\widehat{\mathbf{m}}^{(1)}\right)\\ \mathbf{d}_0^{(2)}-f^{(2)}\left(\widehat{\mathbf{m}}^{(2)}\right)\end{array}\right\|_{\mu}^2+\left\|\phi\left(\widehat{\mathbf{m}}^{(1)}\right)-\phi\left(\widehat{\mathbf{m}}^{(2)}\right)\right\|^2 \tag{4.31}$$

where the structural measure ϕ is defined in Eq. (4.11) and μ is a weighting factor that controls the trade-off between structural difference and data misfit. The functions $f^{(j)}$ are linearized and the equation solved iteratively using Krylov space techniques. In contrast to the previously described approaches, this particular scheme cannot emulate an exact resemblance approach as the damping factor μ cannot be restricted to a large value and must be designed experimentally. The problems of convergence and local minima are partially alleviated by restricting the step size in the iteration by adding an extra term and damping factor, $\xi\|\delta\mathbf{m}\|^2$, in the objective function. In this case the selection of both μ and ξ is done using different approaches. The procedure of stepwise length control resembles the data weighting relaxation used by the other approaches. *Zhang and Morgan* [1996] propose an objective function that closely resembles Eq. (4.31). In their case, the corresponding structure curvatures are normalized by the average values of the Laplacian operator across each image. This normalizing factor must be chosen based on a previous estimate of the models in order to conserve the linearity of the structural measure. Note that Eq. (4.31) can at best be interpreted as containing only partial a priori information about the subsurface, that is, structure [also compare Eq. (15) in *Haber and Gazit* [2013]]. Unlike Eq. (4.24), it does not allow for the use of known physical properties of the two models involved so as to further reduce nonuniqueness in the inversion [*Jackson*, 1979]. For a two-property inverse problem, *Haber and Gazit* [2013, see their Eq. (20)] propose an optimization process composed of only the data misfit term and the joint total variation term. They state that if both forward models and the structural-coupling term [Eq. (4.13)] are convex, then the complete optimization process is convex and will lead to a unique solution independent of the starting model. This may be the case for linear or weakly nonlinear inverse problems (compare *Meju* [1994a, pp. 68–69]), but we caution that obtaining a geologically unique solution in nonlinear geophysical inversion would require the incorporation of realistic additional a priori information [*Jackson*, 1979; *Meju*, 1994b].

4.2.4. Can *n*-Property Fields Be Combined Using a Structural Approach?

There is no theoretical limitation in applying the principle of morphological invariance [*Serra*, 1982] of imaged objects to any two property representations, but it is also well known that a unique solution in nonlinear geophysical inversion can only be obtained with the use of accurate a priori information [*Jackson*, 1979]. For the general case of multiple geophysical methods, let us assume that we have at least *n*-types of field measurements that are associated to variations in *n*-different properties of the same object as described by the relation

$$\mathbf{d}_0^{(j)} = f^{(j)}\left(\widehat{\mathbf{m}}^{(j)}\right) + e^{(j)}, \quad j = 1,\ldots,n, \tag{4.32}$$

where **e** is the vector of additive random noise.

Following the principle of morphological invariance [*Serra*, 1982], we may posit that the morphology of each one of these model representations corresponds to that of the object itself (ϕ_{obj}) and that it can be quantified by a morphological measure ϕ given by

$$\phi\left(m^{(j)}\right) = \phi_{obj} + e_\phi^{(j)}, \quad j = 1,\ldots,n. \tag{4.33}$$

Then multiple joint inversion based on morphological invariance can be defined as finding the *n*-model representations that satisfy (4.32) and (4.33). As in the previous sections, assumptions regarding the accuracy of ϕ (i.e., the magnitude of $e_\phi^{(j)}$) permit exact and inexact solution approaches. While the expansion of the data vectors to accommodate *n*-data types is trivial, morphological resemblances $\Delta\phi(m^{(i)}, m^{(j)})$ should be applied with utmost care as the amount of combinations $\left(\dfrac{n!}{2(n-2)!}\right)$ surpasses *n* when more than three images are analyzed at a time and leads to overconstrained problems. *Gallardo* [2007] minimizes the objective function

$$s(\widehat{\mathbf{m}}^{(1)}, \widehat{\mathbf{m}}^{(2)}, \ldots, \widehat{\mathbf{m}}^{(n)}) = \sum_{j=1}^{n} \left\{ \left\| \mathbf{d}_0^{(j)} - f^{(j)}(\widehat{\mathbf{m}}^{(j)}) \right\|^2_{\mathbf{C}_{dd}^{j-1}} + \left\| \alpha^{(j)} \mathbf{D} \widehat{\mathbf{m}}^{(j)} \right\|^2 + \left\| \widehat{\mathbf{m}}^{(j)} - \widehat{\mathbf{m}}_0^{(j)} \right\|^2_{\mathbf{C}_{00}^{j1-1}} \right\} \tag{4.34}$$

subject to

$$\nabla \widehat{m}^{(g(x,y,z))} \times \nabla \widehat{m}^{(j)} = 0 \quad \forall\, j \neq g(x,y,z),\, j \leq n. \tag{4.35}$$

Equation (4.35) holds for all the possible cross-gradient combinations equal to zero as long as $\nabla \widehat{m}^{(g(x,y,z))} \neq 0$. This constraint does not produce redundant equations, and it will impose resemblance between all the images as long as the *g*-pivot image has no null gradient. Also, the pivot image varies for each position so as not to use the morphology of a single image property. The approach is schematized in the workflow presented in Figure 4.3 and was successfully applied in 2D studies at different spatial scales (see, e.g., *Gallardo* [2007], *Gallardo and Meju* [2011], and *Gallardo et al.* [2012]). *Doetsch et al.* [2010a] and *Moorkamp et al.* [2011] jointly invert three datasets and couple the different property fields using the cross-gradient function between each of the three models in 3D. Note that the Joint Total Variation (JTV) approach of *Haber and Gazit* [2013] can also be combined for more than two objective functions. Importantly, note also that the inverse problem formulation of *Gallardo and Meju* [2003, 2004] incorporate an additional term for a priori subsurface information (alias "model fusion" in their parlance), making it possible to include nonphysical lithological data in joint geophysical inversion (see, e.g., Figures 5 and 6 in *Gallardo and Meju* [2011]). Moreover, *Haber and Gazit* [2013] introduced a region-based smoothness such that the model is not smoothed in areas where structure is needed from the second model being jointly inverted. Thus *n*-property structural inversion has steadily evolved since the initial 2D work of *Gallardo* [2007].

It is noteworthy that *Zhdanov et al.* [2012a, 2012b] propose the use of Gramian constraints to invert jointly multimodal geophysical data by enforcing the correlations between the different model parameters or their attributes (e.g., spatial gradients). The method assumes that a correlation between the different model parameters or their attributes exists, but the specific forms are unknown. For the regularized joint inversion involving *n* geophysical methods, they minimize the parametric functional [*Zhdanov et al.*, 2012b, Eq. (4.1)]:

$$s\left(\widehat{\mathbf{m}}^{(1)}, \widehat{\mathbf{m}}^{(2)}, \ldots, \widehat{\mathbf{m}}^{(n)}\right) = \sum_{j=1}^{n} \left\| f^{(j)}\left(\widehat{\mathbf{m}}^{(j)}\right) - \mathbf{d}^{(j)} \right\|^2_{\mathbf{C}_{dd}^{-1}} + \alpha c_1 \sum_{j=1}^{n} R^{(j)} + \alpha c_2 S_{G_T} \tag{4.36}$$

where $\boldsymbol{R}^{(i)}$ are the stabilization functionals of the corresponding model parameters, S_{G_T} is the Gramian stabilizing functional for transformed model parameters given by [Eq. (2) in *Zhdanov et al.*, 2012b]

$$S_{G_T} = \|T\hat{\mathbf{m}}^{(n)}\|^2_{G_T^{(n)}} = G\left(T\hat{\mathbf{m}}^{(1)}, G\hat{\mathbf{m}}^{(2)}, \ldots, G\hat{\mathbf{m}}^{(n)}\right) \quad (4.37)$$

and implies that the model transform operator T may be any function, including the identity operator (for correlations between the model parameters) gradient operator (for structural correlations between the model parameter gradients), or any other operation (e.g., logarithms, Fourier transform, etc.). The premultiplier α is the regularization parameter, and the weighting coefficients c_1 and c_2 may be selected as unities at the initial stages of the inversion [*Zhdanov et al.*, 2012b]. Equation (4.36) is structurally similar to the standard regularized joint geophysical inversion schemes for n-correlated parameters with a priori information set equal to zero and T can be an identity or smoothness matrix depending on the inversion objectives (see, e.g., Eqs. (2)–(10) in *Sasaki and Meju* [2006]. It is yet to be demonstrated that it is effective in the general challenging situation involving seismic, electromagnetic, and potential field data, in which the model parameters are not correlated.

The foregoing discussions have focused on structure-coupled joint inversion using model-space methods. Multidimensional structural joint inversion is computationally intensive, and *Gallardo and Meju* [2011] suggested that for large-scale three-dimensional (3D) cross-gradients joint inversion of multiple geophysical data, there may be gains in computational efficiency using the data-space method of *Siripunvaraporn et al.* [2005]. *Pak et al.* [2015] implemented the cross-gradients joint inversion of magnetotelluric, magnetic, and gravity data in two-dimensions (2D) using the data-space method. The iterative inversion formula for conventional model-space cross-gradients inversion [Eqs. (4.16)–(4.19)] can be reexpressed as

$$\hat{\mathbf{m}}^{(j)}_{k+1} = \hat{\mathbf{m}}^{(j)}_k + \Delta\hat{\mathbf{m}}^{(j)}_k - \mathbf{N}^{(j)^{-1}}_k \mathbf{B}^{(j)^T}_k P^{(j)}_k, \quad j = 1, 2, \ldots, n, \quad (4.38)$$

where $\Delta\hat{\mathbf{m}}^{(j)}_k = \mathbf{N}^{(j)^{-1}}_k \mathbf{n}^{(j)}_k$ is the uncoupled least-squares model update, $P^{(j)}_k = \dfrac{(\tau^{(j)}_k + \mathbf{B}^{(j)}_k \Delta\hat{\mathbf{m}}^{(j)}_k)}{(\mathbf{B}^{(j)}_k \mathbf{N}^{(1)^{-1}}_k \mathbf{B}^{(j)^T}_k)}$ is the structural resemblance projector, and all the other variables are as previously defined. In the data-space cross-gradients formulation of *Pak et al.* [2015], the quantities $\Delta\mathbf{m}$ and $\mathbf{N}$ in Eq. (4.38) are replaced by

$$\Delta\hat{\mathbf{m}}^{(j)}_k = \mathbf{C}^{(j)}_{mm} \mathbf{A}^{(j)^T}_k \Lambda^{(j)}_k \hat{\mathbf{d}}^{(j)}_k \quad (4.39)$$

and

$$\mathbf{N}^{(j)}_k = \left(I - \mathbf{C}^{(j)}_{mm} \mathbf{A}^{(j)^T}_k \Lambda^{(j)}_k \mathbf{A}^{(j)}_k\right) \mathbf{C}^{(j)}_{mm}, \quad (4.40)$$

where $\Lambda^{(j)}_k = (\mathbf{A}^{(j)}_k \mathbf{C}^{(j)}_{mm} \mathbf{A}^{(j)^T}_k + \mathbf{C}^{(j)}_{dd})^{-1}$, $\hat{\mathbf{d}}^{(j)}_k = (\mathbf{d}^{(j)} - \mathbf{f}^{(j)}(\hat{\mathbf{m}}^{(j)}_k) + \mathbf{A}^{(j)}_k \hat{\mathbf{m}}^{(j)}_k)$, and $\boldsymbol{I}$ is the identity matrix. This leads to the iterative data-space inversion formula

$$\hat{\mathbf{m}}^{(j)}_{k+1} = \hat{\mathbf{m}}^{(j)}_k + \mathbf{C}^{(j)}_{mm} \mathbf{A}^{(j)^T}_k \left(\mathbf{A}^{(j)}_k \mathbf{C}^{(j)}_{mm} \mathbf{A}^{(j)^T}_k + \mathbf{C}^{(j)}_{dd}\right)^{-1} \hat{\mathbf{d}}^{(j)}_k - \left\{I - \mathbf{C}^{(j)}_{mm} \mathbf{A}^{(j)^T}_k \left(\mathbf{A}^{(j)}_k \mathbf{C}^{(j)}_{mm} \mathbf{A}^{(j)^T}_k + \mathbf{C}^{(j)}_{dd}\right)^{-1}\right\} \mathbf{C}^{(j)}_{mm} \mathbf{B}^{(j)^T}_k P^{(j)}_k. \quad (4.41)$$

Pak et al. [2015] showed that 2D cross-gradients joint inversion is slightly more computationally efficient in the data-space method than in the model-space method, but their tests only involved relatively small-size data sets from localized subsurface exploration and it remains to quantify this computational advantage in 3D. What is becoming obvious is that there are now other ways of solving the structural coupling problem and the implemented algorithms range from Gauss–Newton to conjugate gradient including block coordinate descent techniques.

4.3. DISCUSSION

Even when it is feasible to simultaneously invert the data from fundamentally different geophysical methods for common structure, there remains the challenge of fusing the structurally related images into one single model for decision making. This is an important image interpretation or visualization challenge. It is obvious that while the velocity and resistivity joint inversion models can be transformed into subsurface property distributions via appropriate rock-physics models, having a single interpretative model will make for easier understanding of subsurface structure by any nonspecialist user of geophysical models. It provides a basis for more realistic remote structural mapping of subsurface targets. Zonation and classification are steadily emerging as the way to combine different models and to define zones of similar characteristics within the model space (see, e.g., *Bosch* [1999], *Bosch et al.* [2002], *Meju et al.* [2003], *Gallardo and Meju* [2003, 2007], *Paasche and Tronicke* [2007], *Doetsch et al.* [2010a]). Details of zonation and classification are covered in a separate chapter of this book.

Another less commonly used technique for model blending is geospectral imaging. *Gallardo et al.* [2010, 2012] used the multiple cross-gradient inversion method for geospectral imaging [*Gallardo*, 2007] in their study. They applied the integrated imaging approach to 2D seismic reflection, magnetotelluric, gravity, and magnetic data from marine exploration in the Santos Basin, Brazil. The seismic reflection data are the two-way travel times (TWT) picked for five selected reflectors in the seismic-time section. The depths to these reflectors (floating

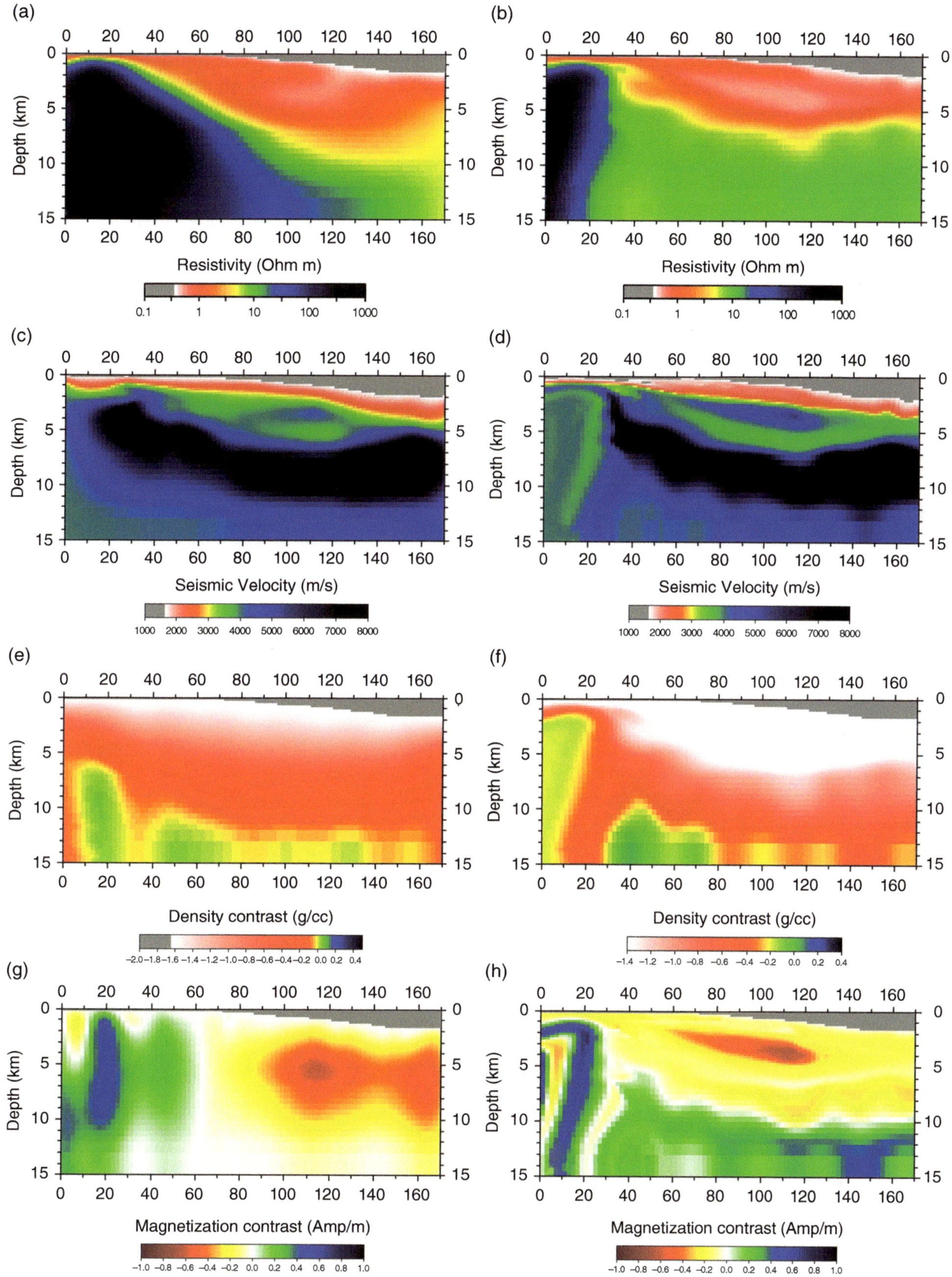

Figure 4.8 Example of regional-scale marine MT resistivity, seismic velocity, density, and magnetization models obtained by joint cross-gradient inversion of the corresponding datasets from Santos Basin, offshore Brazil. [After *Gallardo et al.*, 2012]. **(Left-hand panels)** Results obtained after separate inversion of individual datasets. **(Right-hand panels)** Results from joint inversion.

reflectors) were estimated using average interval velocities as in conventional practice. For the cross-gradient joint multiphysics inversion, a common inversion grid was selected for all the geophysical methods with dense sampling over the region of common data coverage. To allow for features of different scale length, the grid was resampled according to specific accuracy requirements for the seismic, magnetotelluric, gravity, and magnetic forward modeling and Jacobian computations. During the inversion process, the cross-gradient constrained physical property fields were estimated whereas the depths to the floating reflectors were held fixed as a priori information. The resulting joint inversion models (i.e., velocity, density, resistivity and magnetisation property fields) were then combined into a geospectral image, thus completing the data integration process (see Figure 11 in *Gallardo et al.* [2012]). In all cases, the results of joint structural inversion (Figure 4.8) showed a significant improvement over models derived from separate inversion of the individual datasets (see Figure 7 in *Gallardo et al.* [2012]).

4.4. CONCLUSION

The underlying philosophy in multiphysics structure-coupled imaging is that the datasets, though disparate, should have features in common because they sense or reflect the same geological structure. Structural similarity between the multiple physical property distributions is achieved by imposing the constraint that the cross-product, dot product, differences, or square root of sums of the gradients of the property fields should be zero at structural boundaries for a given geological frame of reference. This can be formulated mathematically through the model gradients by requiring the functions of the model gradients to approach zero everywhere.

It is now possible to perform joint inversion wherever two or more physically measured datasets are available, and the relevant parameters are controlled by the same structure or state variables of the physical system being investigated. Several approaches have been developed for joint multidimensional inversion. These include using (a) structural details furnished by seismic reflection image processing to constrain the inversion of the other methods and (b) the less subjective methods of physics-based coupling or structure-coupled simultaneous inversion of all the datasets. Although several workers have successfully applied joint inversion to seismic travel-time, EM, gravity, and magnetic datasets, joint three-dimensional full-waveform inversion of seismic data with other data types remains a formidable task. In all cases, the results of joint inversion showed a significant improvement over previous models derived from separate inversion of the individual datasets. However, while the joint inversion models can be transformed into reservoir property distributions via appropriate rock-physics models, having a single interpretative model will make for easier understanding of subsurface structure by any nonspecialist user of geophysical models. It provides a basis for more realistic remote structural mapping of subsurface targets and remains the main challenge for future research. It is slowly emerging that multispectral fusion of the models from joint inversion can lead to a single image that improves geological interpretation. However, geospectral images may not be easy to use in 3D, since only 2D planes of the full volume can be shown at any time. Zonation and classification work well in 2D and 3D cases [e.g., *Doetsch et al.*, 2010a] and have the advantage of allowing further use of the zoned model for reservoir modeling, say. Another important challenge in structural joint inversion is that not all physical property distributions in the subsurface will be structurally coincident and some flexibility in model reconstruction may be necessary where there is strong geological heterogeneity together with variations in salinity or water content as partly discussed in *Linde et al.* [2006]. Determining the limits of the structural approach in such geological environments is still an outstanding challenge. It is therefore suggested that joint multiphysics inversion with improved lithology classification coupled with fluid-flow modeling is a way forward in integrated imaging of the subsurface.

ACKNOWLEDGMENT

The insightful comments of two anonymous reviewers and Niklas Linde helped improve the clarity of this chapter and are gratefully acknowledged.

REFERENCES

Alpak, F.O., C. Torres-Verdin, and T. M. Habashy (2008), Estimation of in-situ petrophysical properties from wireline formation tester and induction logging measurements: A joint inversion approach, *J. Petroleum Sci. Eng.*, *63*, 1–17, doi:10.1016/j.petrol.2008.05.007.

Bai, D., and M. A. Meju (2003), Deep structure of the Longling–Ruili fault zone underneath Ruili basin near the Eastern Himalayan syntaxis: insights from magnetotelluric imaging, *Tectonophysics*, *364*,135–146.

Bai, D. B., M. A. Meju, and Z. Liao (2001), Magnetotelluric images of deep structure of the Rehai geothermal field near Tengchong, southern China, *Geophys. J. Int.*, *147*(3), 677–687.

Bedrosian, P. A., N. Maercklin, U. Weckmann, Y. Bartov, T. Ryberg, and O. Ritter (2007), Lithology-derived structure classification from the joint interpretation of magnetotelluric and seismic models, *Geophys. J. Int.*, *170*(2), 737–748.

Berge, P. A., J. G. Berryman, H. Bertete-Aguirre, P. Bonner, J. J. Roberts, and D. Wildenschild, Joint inversion of geophysical data for site characterization and restoration monitoring,

LLNL report number UCRL-ID-128343, Project 55411 U.S. D.O.E., 2000.

Binley A., P. Winship, L. J. West, M. Pokar, and R. Middleton (2002), Seasonal variation of moisture content in unsaturated sandstone inferred from borehole radar and resistivity profiles, *J. Hydrology*, *267*,160–172.

Binley, A., G. Cassiani, and P. Winship (2004), Characterization of heterogeneity in unsaturated sandstone using borehole logs and cross-borehole tomography, in *Aquifer Characterization*, Bridge and D. Hyndman, eds., SEPM, Society for Sedimentary Geology, Tulsa, OK, pp. 129–138.

Bosch, M. (1999), Lithologic tomography: From plural geophysical data to lithology estimation, *J. Geophys. Res.-Solid Earth*, *104*(B1), 749–766.

Bosch, M., M. Zamora, and W. Utama (2002), Lithology discrimination from physical rock properties, *Geophysics*, *67*(2), 573–581.

Cardiff, M., and P. K. Kitanidis (2009), Bayesian inversion for facies detection: An extensible level set framework, *Water Resources Res.*, *45*, W10416, doi:10.1029/2008WR007675.

Constable, S.C., R.L Parker and C.G. Constable (1987), Occam's inversion: a practical algorithm for generating smooth models from electromagnetic sounding data, *Geophysics*, *52*, 289–300.

Cumani, A. (1991), Edge-detection in multispectral images, *Cvgip-Graphical Models and Image Processing*, *53*(1), 40–51.

del Rio, J. A., and S. Whitaker (2001), Electrohydrodynamics in porous media, *Transp. Porous Media*, *44*(2), 385–405.

Doetsch, J., N. Linde, I. Coscia, S. A. Greenhalgh, and A. G. Green (2010a), Zonation for 3D aquifer characterization based on joint inversions of multimethod crosshole geophysical data, *Geophysics*, *75*, G53-G64, doi:10.1190/1.3496476.

Doetsch, J., N. Linde, and A. Binley (2010b), Structural joint inversion of time-lapse crosshole ERT and GPR traveltime data, *Geophys. Res. Lett.*, *37*, L24404, doi: 10.1029/2010gl045482.

Droske, M., and M. Rumpf (2003), A variational approach to nonrigid morphological image registration, *SIAM J. Appl. Math.*, *64*(2), 668–687.

Fregoso, E. (2010), 3D cross-gradient joint inversion of multiple geophysical data, PhD thesis, 243 pp, CICESE, Ensenada, Mexico.

Fregoso, E., and L. A. Gallardo (2009), Cross-gradients joint 3D inversion with applications to gravity and magnetic data, *Geophysics*, *74*(4), L31–L42. doi:10.1190/1.3119263.

Gallardo, L. A. (2004), Joint two-dimensional inversion of geoelectromagnetic and seismic refraction data with cross-gradients constraint, PhD thesis, Lancaster University, Lancaster, U.K.

Gallardo, L. A. (2007), Multiple cross-gradient joint inversion for geospectral imaging, *Geophys. Res. Lett.*, *34*(19), L19301, doi:19310.11029/12007GL030409.

Gallardo, L. A., and M. A. Meju (2004), Joint two-dimensional dc resistivity and seismic travel-time inversion with cross-gradients constraints, *J. Geophys. Res.*, *109*, B03311, doi: 10.1029/2003JB0022717.

Gallardo, L. A., and M. A. Meju (2003), Characterisation of heterogeneous near-surface materials by joint 2D inversion of dc resistivity and seismic data, *Geophys. Res. Lett.*, *30*(13), 1658–1661.

Gallardo, L. A., and M. A. Meju (2007), Joint two-dimensional cross-gradient imaging of magnetotelluric and seismic traveltime data for structural and lithological classification, *Geophys. J. Int.*, *169*, 1261–1272.

Gallardo, L.A. & Meju, M.A., 2011. Structure-coupled multiphysics imaging in geophysical sciences. *Reviews of Geophysics*, *49*, RG1003, doi:10.1029/2010RG000330.

Gallardo-Delgado, L. A., M. A. Perez-Flores, and E. Gomez-Trevino (2003), A versatile algorithm for joint 3D inversion of gravity and magnetic data, *Geophysics*, *68*, 949–959.

Gallardo, L. A., M. A. Meju, and M. A. Flores-Perez (2005), A quadratic programming approach for joint image reconstruction: Mathematical and geophysical examples, *Inverse Problems*, *21*, 435–452.

Gallardo, L. A., S. Fontes, M. A. Meju, P. de Lugao, and M. P. Buonora (2010), Joint cross-gradient inversion of offshore seismic reflection, MT, gravity and magnetic profiles over a petroliferous prospect in Santos basin, Brazil. Presented at 20th Workshop, IAGA WG 1.2 Electromagnetic Induction in the Earth, Giza, Egypt, September 18–24, 2010.

Gallardo, L. A., S. L. Fontes, M. A. Meju, M. P. Buonora, and P. de Lugao (2012), Robust geophysical integration through structure-coupled joint inversion and multispectral fusion of seismic reflection, magnetotelluric, magnetic and gravity images: Example from Santos Basin, offshore Brazil. *Geophysics*, *77*(5), B237-B251, doi:10.1190/GEO2011-0394.1.

Günther, T., and C. Rücker (2006), A new joint inversion approach applied to the combined tomography of DC resistivity and seismic refraction data. Presented at the 19th EEGS Symposium on the Application of Geophysics to Engineering and Environmental Problems.

Haber, E., and M. H. Gazit (2013), Model fusion and joint inversion, *Surv. Geophys.*, *34*, 675–695.

Haber E., and D. Oldenburg (1997), Joint inversion: A structural approach, *Inverse Problems*, *13*, 63–77.

Hu, W. Y., A. Abubakar, and T. M. Habashy (2009), Joint electromagnetic and seismic inversion using structural constraints, *Geophysics*, *74*(6), R99–R109, doi:10.1190/1.3246586.

Infante, V., L. A. Gallardo, J. C. Montalvo-Arrieta, and I. Navarro de Leon (2010), Lithological classification assisted by the joint inversion of electrical and seismic data at a control site in northeast Mexico, *J. Appl. Geophys.*, *70*, 93–102, doi:10.1016/j.jappgeo.2009.11.003.

Jackson, D. D. (1972), Interpretation of inaccurate, insufficient and inconsistent data, *Geophys. J. R. Astron. Soc.*, *28*, 97–110.

Jackson, D. D. (1979), The use of a priori data to resolve nonuniqueness in linear inversion, *Geophys. J. R. Astron. Soc.*, *57*, 137–157.

Jardani, A., A. Revil, and J. P Dupont (2013), Stochastic joint inversion of hydrological data for salt tracer test monitoring and hydraulic conductivity imaging, *Adv. Water Resources* *52*, 62–77.

Jegen, M. D., R. W. Hobbs, P. Tarits, and A. Chave (2009), Joint inversion of marine magnetotelluric and gravity data incorporating seismic constraints: Preliminary results of subbasalt imaging off the Faroe Shelf, *Earth and Planet. Sci. Lett.*, *282*, 47–55.

Jilinski P., S. Fontes, L. A. Gallardo (2010), Joint interpretation of maps using gradient directions, cross and dot-product values to determine correlations between bathymetric and gravity anomaly maps, *SEG Expanded Abstracts 29*, 1226–1229, doi:10.1190/1.3513066.

Jilinski, P., S. L. Fontes, and M. A. Meju (2013a), Estimating optimum density for regional Bouguer reduction by morphologic correlation of gravity and bathymetric maps: Examples from offshore south-eastern Brazil, *Geo-Marine Lett. 33*(1), 67–73, doi:10.1007/s00367-012-0312-0.

Jilinski, P., M. A. Meju, and S. L. Fontes (2013b), Demarcation of continental-oceanic transition zone using angular differences between gradients of geophysical fields, *Geophys. J. Int. 195*(1), 276–281; doi: 10.1093/gji/ggt216.

Lelievre, P. G. (2009), Integrating geologic and geophysical data through advanced constrained inversions, PhD thesis, University of British Columbia, Vancouver.

Lelievre, P. G., and D. W. Oldenburg (2009), A comprehensive study of including structural orientation information in geophysical inversions, *Geophys. J. Int. 178*, 623–637.

Linde, N., A. Binley, A. Tryggvason, L. B. Pedersen, and A. Revil (2006), Improved hydrogeophysical characterisation using joint inversion of cross-hole electrical resistance and ground penetrating radar traveltime data, *Water Resources Res. 42*, W12404, doi:10.1029/2006WR005131.

Linde, N., A. Tryggvason, J. E. Peterson, and S. S. Hubbard (2008), Joint inversion of crosshole radar and seismic traveltimes acquired at the South Oyster Bacterial Transport Site, *Geophysics*, *73*, G29–G37, doi: 10.1190/1.2937467.

Lines, L., A. Schultz, and S. Treitel (1988), Cooperative inversion of geophysical data, *Geophysics*, *53*, 8–20.

Lochbühler, T., J. Doetsch, R. Brauchler, and N. Linde (2013), Structure-coupled joint inversion of geophysical and hydrological data, *Geophysics*, *78*(3), ID1–ID14. doi:10.1190/GEO2012-0460.1.

Meju, M. A. (1994a), *Geophysical Data Analysis: Understanding Inverse Problem Theory and Practice*, Society of Exploration Geophysicists Course Notes Series, Vol. *6*, SEG Publishers, Tulsa, OK, 296 pp.

Meju, M. A. (1994b), Biased estimation: a simple framework for parameter estimation and uncertainty analysis with prior data, *Geophys. J. Int.*, *119*, 521–528.

Meju, M. A. (1998), A simple method of transient electromagnetic data analysis, *Geophysics*, *63*, 405–410.

Meju, M. A. (2002), Geoelectromagnetic exploration for natural resources: Models, case studies and challenges, *Surveys Geophys.*, *23*, 133–205.

Meju, M. A. (2005), Simple relative space-time scaling of electrical and electromagnetic depth sounding arrays: Implications for electrical static shift identification and joint DC-TEM data inversion, *Geophys. Prosp.*, *53*, 463–479.

Meju, M. A., and V. Sakkas (2007), Heterogeneous crust and upper mantle across Southern Kenya and the relationship to surface deformation as inferred from magnetotelluric imaging, *J. Geophys. Res.*, *112*, B04103, doi:10.1029/2005JB004028.

Meju, M. A., S. L. Fontes, M. F. B. Oliveira, J. P. R. Lima, E. U. Ulugergerli, and A. A. Carrasquilla (1999), Regional aquifer mapping using combined VES-TEM AMT/EMAP methods in the semi-arid eastern margin of Parnaiba Basin, Brazil, *Geophysics*, *64*, 337–356.

Meju, M. A., P. Denton and P. Fenning (2002), Surface NMR sounding and inversion to detect groundwater in key aquifers in England: comparisons with VES-TEM methods, *J. Appl. Geophys.*, *50*, 95–112.

Meju, M. A., L. A. Gallardo, and A. K. Mohamed (2003), Evidence for correlation of electrical resistivity and seismic velocity in heterogeneous near-surface materials, *Geophys. Res. Lett.*, *30*(7), 1373–1376.

Moorkamp, M., B. Heincke, M. Jegen, A. W. Roberts, R. W. Hobbs (2011), A framework for 3D joint inversion of MT, gravity and seismic refraction data, *Geophys. J. Int.*, *184*, 477–493.

Musil, M., H. R. Maurer and A. G. Green (2003), Discrete tomography and joint inversion for loosely connected or unconnected physical properties: Application to crosshole seismic and georadar data sets, *Geophys. J. Int.*, *153*, 389–402.

Newman, G. A., and M. Commer (2010), Joint electromagnetic-seismic inverse modelling for matched data resolution. Paper presented at International workshop on Electromagnetic, Gravity and Magnetic exploration methods (EGM2010), Capri, Italy, April 11–14.

Nocedal, J., and S. J. Wright (1999), *Numerical Optimization*, Springer, New York.

Oldenburg, D. W., Y. Li, and R. G. Ellis (1997), Inversion of geophysical data over a copper gold porphyry deposit—a case history for Mt. Milligan, *Geophysics*, *62*, 389–402.

Paasche, H., and J. Tronicke (2007), Cooperative inversion of 2D geophysical data sets: A zonal approach based on fuzzy c-means cluster analysis, *Geophysics*, *72*(3), A35–A39.

Pak, Y. C., T. L. Li, G. S. Kim, J. S. Kim, and B. M. Choe (2015), Data space multiple cross-gradient joint inversion of geophysical data: Two-dimensional analysis of MT, gravity and magnetic data, *J. Appl. Geophys.* (in review).

Pride, S. (1994), Governing equations for the coupled electromagnetics and acoustics of porous media, *Phys. Rev. B*, *50*(21), 15768–15696.

Revil, A. (2013), Effective conductivity and permittivity of unsaturated porous materials in the frequency range 1 mHz–1 GHz, *Water Resources Res.*, *49*(1), 306–327.

Revil, A., and H. Mahardika (2013), Coupled hydromechanical and electromagnetic disturbances in unsaturated porous materials, *Water Resources Res.*, *49*(1), 1–23.

Sakkas, V., Meju, M. A., Khan, M. A, Haak, V., and F. Simpson (2002), Magnetotelluric images of the crustal structure of Chyulu Hills volcanic field, Kenya, *Tectonophysics*, *346*, 169–185.

Sasaki, Y. (1989), Two-dimensional joint inversion of magnetotelluric and dipole–dipole resistivity data, *Geophysics*, *54*, 254–262.

Sasaki, Y., and M. A. Meju (2006), Three-dimensional joint inversion for magnetotelluric resistivity and static shift distributions in complex media, *J. Geophys. Res.*, *111*, B05101, doi:10.1029/2005JB004009.

Serra, J. (1982), *Image Analysis and Mathematical Morphology*, Academic Press, London, 610 pp.

Shin C., and Y. H. Cha (2008), Waveform inversion in the Laplace domain, *Geophys. J. Int.*, *173*(3), 922–931.

Shin C., and Y. H. Cha (2009), Waveform inversion in the Laplace–Fourier domain, *Geophys. J. Int.*, *177*(3), 1067–1079.

Shin, C., and W. Ha (2008), A comparison between the behaviour of objective functions for waveform inversion in the frequency and Laplace domains. *Geophysics*, *73*(5), VE119–VE133.

Siripunvaraporn, W., G. Egbert, Y. Lenbury, and M. Uyeshima (2005), Three-dimensional magnetotelluric inversion: Data-space method, *Phys. Earth Planet. Int.*, *150*, 3–14, doi:10.1016/j.pepi.2004.08.023.

Toivanen, P. J., J. Ansamaki, J. P. S. Parkkinen, and J. Mielikainen (2003), Edge detection in multispectral images using the self-organizing map, *Pattern Recognit. Lett.*, *24*(16), 2987–2994.

Tryggvason, A., and N. Linde (2006), Local earthquake (LE) tomography with joint inversion for P- and S-wave velocities using structural constraints, *Geophy. Res. Lett.*, *33*, L07303, doi:10.1029/2005GL025485.

Um, E. S., M. Commer, and G. A. Newman (2014), A strategy for coupled 3D imaging of large-scale seismic and electromagnetic data sets: Application to subsalt imaging, *Geophysics*, *79*(3), ID1–ID13, doi: 10.1190/geo2013-0053.1.

Vozoff, K., and D. L. B. Jupp (1975), Joint inversion of geophysical data, *Geophys. J. R. Astron. Soc.*, *42*, 977–991.

Zhang, J., and F. D. Morgan (1996), Joint seismic and electrical tomography, *Proceedings EEGS Symposium on Applications of Geophysics to Engineering and Environmental Problems*, Vol. *1*, Keystone, Colorado, pp.391–396.

Zhdanov, M. S., A.V. Gribenko, G. A. Wilson, and C. Funk (2012b), 3D joint inversion of geophysical data with Gramian constraints: A case study from the Carrapateena IOCG deposit, South Australia, *The Leading Edge*, November 2012, pp.1382–1388.

Zhdanov, M. S., A. Gribenko, and G. Wilson (2012a), Generalized joint inversion of multimodal geophysical data using Gramian constraints, *Geophys. Res. Lett.*, *39*, L09301, doi:10.1029/2012GL051233.

Zhou, L., A. Revil, M. Karaoulis, D. Hale, J. Doetsch, and S. Cutler (2014), Image-guided inversion of electrical resistivity data, *Geophys. J. Int. 197*, 292–309.

Zhu, T., and J. M. Harris (2011), Iterative joint inversion of P-wave and S-wave crosswell traveltime data, *SEG Expanded Abstracts*, *30*, 479–483.

5

Post-Inversion Integration of Disparate Tomographic Models by Model Structure Analyses

Hendrik Paasche

ABSTRACT

We review different approaches for post-inversion integration of multiple geophysical models according to model structure — for example, consistant shapes or color/chroma variations reflecting the physical model parameter variability. We provide the mathematical fundamentals of several approaches, ranging from simple edge detection to more advanced techniques, such as artificial neural networks and cluster analyses. We illustrate their application using a synthetic example comprising crosshole tomographic models. Most of the approaches are limited to the integration of fully co-located geophysical models and strive to ease the lithological interpretation of multiple geophysical models. However, some approaches discussed allow for the incorporation of geophysical model errors and the integration of partially co-located geophysical models in a numerical fashion. Finally, we illustrate the potential of structural model integration to support petrophysical parameter estimation without the need to formulate a deterministic transfer function describing the relationship between multiple geophysical models and additional sparse target parameters — for example, achieved from hydrological borehole logging. Thus, we extend our discussion clearly beyond the limit inherent to joint interpretations of disparate geophysical models, illustrating the potential of post-inversion model integration to substantially improve the information extraction from multiple geophysical models.

5.1. INTRODUCTION

Every human is well versed in the analysis of the structure of an image. For example, the human eye captures a spatial image in more than one dimension and the brain analyzes the structural variability of the information within this image with regard to different criteria, such as consistent shape, color and/or chroma reflecting the numerical (physical) values underlying the image, and texture [*Nixon and Aguado*, 2012]. Such abilities are fundamental for human survival — for example, for orientation in our environment, food quality assessment, or recognition of menaces. Identification of specific features in an image is often linked to a classification system in our brain which is learned under supervision of parental guidance and then later extended and refined by personal experience. For example, terms like *birds*, *buildings*, or *faces* serve as class descriptions for features or objects detected in an image. All objects assigned to one class have a certain similarity with regard to different detection criteria and differ substantially from objects assigned to other classes, albeit objects within one class are not necessarily identical.

When looking at a geophysical model obtained from inversion, we naturally start to analyze the structure of the

UFZ-Helmholtz Centre for Environmental Research, Department Monitoring and Exploration Technologies, Leipzig, Germany

Integrated Imaging of the Earth: Theory and Applications, Geophysical Monograph 218, First Edition.
Edited by Max Moorkamp, Peter G. Lelièvre, Niklas Linde, and Amir Khan.

model with regard to color/chroma, shapes/boundaries, or/and texture. Over time, we gain experience and are able to come up with an interpretation of the physical parameter variability imaged in the model. Experience may guide our structural model analysis more or less intentionally and lead the interpreter to individually weight the emphasis put on the different detection criteria. Problems arise, since geophysical model reconstruction by data inversion usually suffers ambiguity due to limited amount and accuracy of available observations (see, e.g., *Friedel* [2003] and *Aster et al.* [2005]; see also Chapter 2). When analyzing the structural variability of a geophysical model, it is difficult for a human brain to take the uncertainty resulting from the ambiguous model generation process into account. Our human vision and image analysis system is not used to ambiguity. In situations where we cannot classify what we have seen — for example, if we have never seen an aircraft before — we have to create a new class and name it. But even in this situation we quickly define detection criteria for the object put into this class, which will potentially be refined by every new detection. During this procedure of learning and detecting features suitable as detection criteria, we fundamentally assume that the object reveals its true structure to our visual detection system and is free of artifacts. For example, in our analyses we critically rely on the assumption that an object classified as an aircraft is truly an aircraft and not a bird obscuring its true nature by some artifacts.

To tackle the problem of ambiguity in geophysical model analysis, it became common practice to acquire several independent data sets capturing information about the same ground (see, e.g., *Butler* [2006]). By either joint or individual inversion of the disparate data sets, we obtain multiple fully or partially co-located geophysical models providing complementary information about the subsurface as well as artifacts, that is, deviations from ground reality in geophysical models with regard to the imaging of incorrect shapes, color/chroma, or textures, specific to each of the considered models. By analyzing the structural variability of more than one co-located geophysical model, we hope to improve the final interpretation and to reduce the risk of misinterpreting individual model artifacts.

Integrated post-inversion analyses of geophysical model structure is highly related to the research fields of feature extraction and image segmentation in computer vision (see, e.g., *Nixon and Aguado* [2012]), spatial data mining (see, e.g., *Shekar et al.* [2004]), multivariate statistical analyses (see, e.g., *Reyment and Savazzi* [1999]), machine learning in terms of pattern recognition (see, e.g., *Bishop* [2006]), and knowledge discovery in databases (KDD; see, e.g., *Fayyad et al.* [1996]), particularly when applied to remote sensing databases (see, e.g., *Aksoy et al.* [2010]). The integrative aspect substantially requires that two or more fully or partially co-located geophysical models are analyzed according to distinct structural detection criteria. This can be done by a pure visual inspection of multiple geophysical models. This process heavily relies on human image analysis capabilities and is known under the term joint interpretation of geophysical models. Such a joint interpretation approach is inherently subjective, since the detection criteria used and the weights assigned to the detection criteria are inherently related to the personal experience of the human interpreter. Joint interpretation is a qualitative model integration approach, since uncertainty appraisal of the resultant interpretation — for example, the shape of outlined lithological units — is not possible in an objective and quantitative manner.

Here, we will focus on quantitative post-inversion model integration by model structure analyses. Generally, a post-inversion integration of models achieved by separate inversion of the underlying data sets is considered less desirable than a simultaneous inversion of two or more data sets subject to a common structural constraint (see Chapter 4). However, results of joint inversion approaches rigorously enforcing common model structure — for example, by defining common layer boundaries for the geophysical models to be determined — do not need post-inversion integration by model structure analysis. Individual inversion of two or more data sets rarely results in models with equal or co-linear structure with regard to shapes, color and/or chroma, and texture (see the examples in later chapters of this book). This holds also for structurally coupled simultaneous joint inversion approaches (see Chapter 4) honoring a common structural objective — for example, as implemented in the popular cross-gradient approach of *Gallardo and Meju* [2003]. In this work, we focus on the integration of models achieved by deterministic inversion; that is, one model is achieved per data set inverted. However, most of the techniques discussed in the following chapters can be applied to stochastic inversion results comprising ensembles of plausible models for each data set (see Chapters 2 and 6).

Quantitative integration of geophysical models should result in one image showing relevant structural information of all geophysical models considered in the integrated structural analysis. The resultant integrated image should be reproducible by other human interpreters. This means that transparent rules for the model integration have to be defined and implemented in a numerical algorithmic framework. These rules should not include elements of a qualitative integration — for example, human (subjective) visual analyses results. Resultant images should ease interpretation and ideally allow for quantitative uncertainty appraisal of the detected structural units.

Here, we review different quantitative approaches for structural model integration that have been previously applied to 2D and 3D geophysical models. Starting with simple edge detection techniques for visual integration, we come to numerically more complex approaches, such as self-organizing feature maps and cluster analyses. Using synthetic crosshole tomographic models, we will explain the mathematical setup of the methods and illustrate their strengths and weaknesses. We limit ourselves to unsupervised integration techniques reported in geophysical literature and applicable to the integrated analyses of at least two disparate geophysical models. A starting point for all our analyses is the assumption that the models are equally discretized; that is, the co-located model area must be described by equal number of model cells with identical spatial arrangement for every model considered. The shape of the grid cells (e.g., triangular or rectangular), their dimensionality (e.g., 2D or 3D), and variability in size within a single model is no restriction for the applicability of the methods discussed in the following. All that is required is that appropriate operators for those discretizations can be generated. If necessary, the considered models must be interpolated on an equal grid prior to integration (see, e.g., *Bedrosian et al.* [2007]), and interpolation method and settings should be chosen such that interpolation errors are kept to a minimum. For example, a geophysical model **M** should be defined as matrix stretching along two or three orthogonal spatial dimensions $\hat{x}$, $\hat{y}$, $\hat{z}$. The m elements of **M** are referred to as model cells. These model cells can be identical to the elements of the model parameter vector **m** determined during inversion, if the same grid-based parameterization of the model reconstruction area had been chosen for all considered geophysical models.

5.2. THE SYNTHETIC DATABASE — AN ILLUSTRATIVE EXAMPLE OF CROSSHOLE TOMOGRAPHY

Throughout this work, we will use a synthetic crosshole tomographic study to illustrate the employed post-inversion model integration approaches. This database has been used before by *Paasche and Tronicke* [2007, 2014] as test data set for newly developed joint inversion approaches. The input model for our synthetic study is shown in Figure 5.1a. It consists of seven lithologic units labeled A–G. We superimpose stochastic noise with Gaussian probability density functions generated using a von Kármán auto-covariance function (Figure 5.1b; see, e.g., *Tronicke and Holliger* [2005]). Mean values and standard deviations of physical properties have been assigned to every lithological unit and the corresponding noise model, respectively. The resultant geophysical models in Figures 5.1c and 5.1d illustrate the seismic P-wave and radar wave velocity distributions, respectively. We refer to these models as the original models and consider them to represent ground truth in our synthetic experiment. Note that several pairs of lithological units are characterized by very similar radar and P-wave velocities: A and D, B and C, and F and G (see Figures 5.1c and 5.1d). Albeit using seven lithological layers, we can only expect to find four physical zones that are clearly distinct when analyzing the original models. These four physical zones are outlined by the spatial extent of lithological layers A and D, B and C, F and G, and E. Note that the physical zone defined by lithological layers A and D is spatially disrupted by the physical zone defined by lithological layers B and C.

To simulate seismic and georadar crosshole tomographic experiments, we place sources and receivers with a vertical spacing of 0.25 m at the left and right model edges, respectively. Travel-time data sets are created from full-waveform synthetic data modeled using finite difference techniques (*Paasche et al.* [2006]; commercial software ReflexW, Sandmeier Scientific Software, Germany). Typical radar and seismic source waveforms are used with center frequencies of 100 MHz and 1 kHz, respectively, and bandwidths of 2–3 octaves. We contaminate the seismic and radar data sets with random noise of 0.8% and 1.5%, respectively.

Both data sets are separately inverted using a regularized least squares inversion scheme (see, e.g., *Aster et al.* [2005]) based on a finite difference solution of the Eikonal equation [*Schneider et al.*, 1992]. The ill-posed problems have been regularized using first-order Tikhonov regularization (FTR; see, e.g., *Aster et al.* [2005]). The resultant velocity models shown in Figures 5.1e and 5.1f can be regarded as blurred reconstructions of the original models, which we will refer to as FTR models.

Additionally, both data sets have been inverted using the zonal cooperative inversion (ZCI) methodology, which employs fuzzy c-means cluster analysis to realize a structural link between both inversion problems [*Paasche and Tronicke*, 2007]. The results of the cooperative inversion are shown in Figures 5.1g and 5.1h. These models are less blurred and characterized by larger regions of almost constant seismic and radar velocities. We will refer to these models as ZCI models. All models in Figures 5.1e–5.1h explain the underlying data sets within the global noise level of the data. In the following, we will use the FTR and the ZCI models as two different pairs of fully co-located geophysical models, each serving as individual input information for post-inversion model integration by model structure analyses.

5.3. VISUAL INTEGRATION

5.3.1. Boundary Information in the Model Domain

In image analysis, various techniques have been developed to highlight potential edges in images (see, e.g., *Nixon and Aguado* [2012]). When utilizing them in the context of geophysical model structure analysis, edge operators can be used to visualize lineament-like features outlining boundaries between potential lithological units — for example, characterized by high gradients between adjacent model cells. Geophysical models are often blurred due to the utilization of smoothness when regularizing

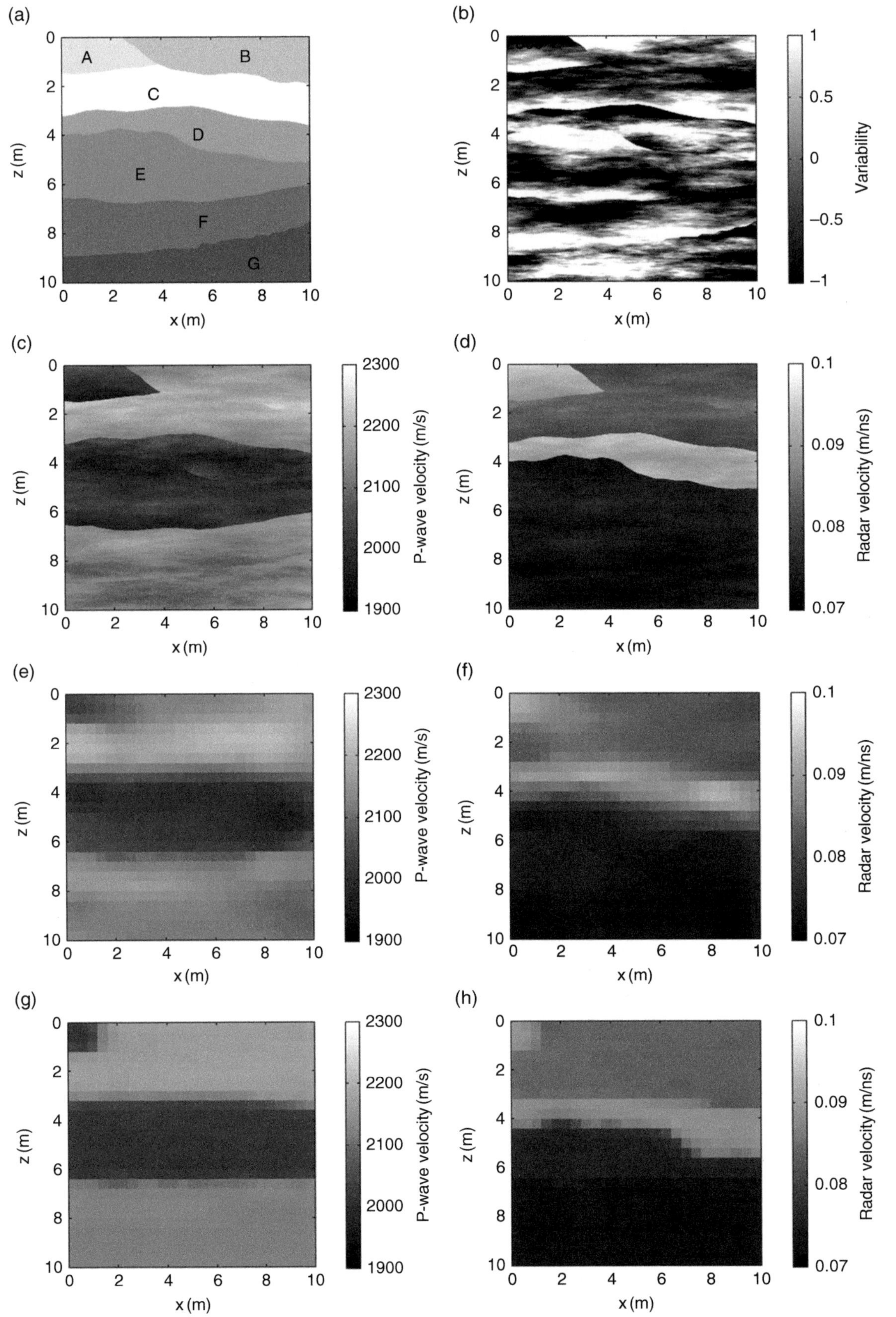

Figure 5.1 **(a)** Lithological model and **(b)** corresponding noise information employed to generate the original **(c)** P-wave and **(d)** radar velocity models considered to represent ground truth. Smooth FTR **(e)** P-wave and **(f)** radar velocity models achieved by separate inversion of underlying crosshole tomographic data sets. **(g)** P-wave and **(h)** radar velocity models achieved by zonal cooperative inversion (ZCI).

the inverse problem. Hence high-frequency spatial noise is usually not present in geophysical models, guaranteeing that even simple edge detection algorithms building on local analyses of gradients in an image matrix can be applied successfully to highlight potential boundaries present in a geophysical model.

In Figures 5.2a and 5.2b the normalized total gradient derived from the FTR models in Figures 5.1e and 5.1f is shown. When calculating the total gradient of a geophysical image, we convolve the discretized model $\mathbf{M}$ with a kernel matrix [–1 0 1] applied along each spatial dimension $\hat{x}$, $\hat{y}$, $\hat{z}$. This results in matrices $\mathbf{M}_{\hat{x}}$, $\mathbf{M}_{\hat{y}}$, $\mathbf{M}_{\hat{z}}$ of the same size as $\mathbf{M}$ containing the directional model gradients. The total gradient at the location of the ith model cell is then achieved by

$$\mathbf{M}_t(i)=\sqrt{\left(\mathbf{M}_{\hat{x}}(i)\right)^2+\left(\mathbf{M}_{\hat{y}}(i)\right)^2+\left(\mathbf{M}_{\hat{z}}(i)\right)^2}. \quad (5.1)$$

In our two-dimensional example, only $\mathbf{M}_{\hat{x}}$ and $\mathbf{M}_{\hat{y}}$ are calculated. Choosing a threshold allows for converting $\mathbf{M}_t$ into a binary matrix with nonzero entries at local maximal values of the total gradient. After applying an image morphological erosion technique (see, e.g., *Nixon and Aguado*, 2012]) to the binarized $\mathbf{M}_t$, we achieve edge-indicating information not exceeding one model cell in width. The eroded (edge-thinned) binarized $\mathbf{M}_t$ matrix can be further analyzed and converted into boundary information by simply connecting the center positions of all adjacent nonzero elements. This has been done for both maps shown in Figures 5.2a and 5.2b. The resultant boundaries are plotted in Figure 5.2c. Figure 5.2c is a visual composite integrating the detected edge attribute positions numerically derived from the separately inverted FTR models shown in Figures 5.1e and 5.1f. When comparing the edges detected in the seismic and radar model, it is obvious that they are not always coincident. However, the integrated information allows for an improved outline of potential lithological boundaries that match well with the boundaries of the lithological units in the original model (Figure 5.1a). The visual composite needs further interpretation since the integrated image only provides information about potential boundary shapes, but not about the physical meaning of structures separated by detected boundaries.

We repeat this analysis using the ZCI models (Figures 5.1g and 5.1h) as input information. The total gradients as well as the detected boundaries are shown in Figures 5.2d–5.2f.

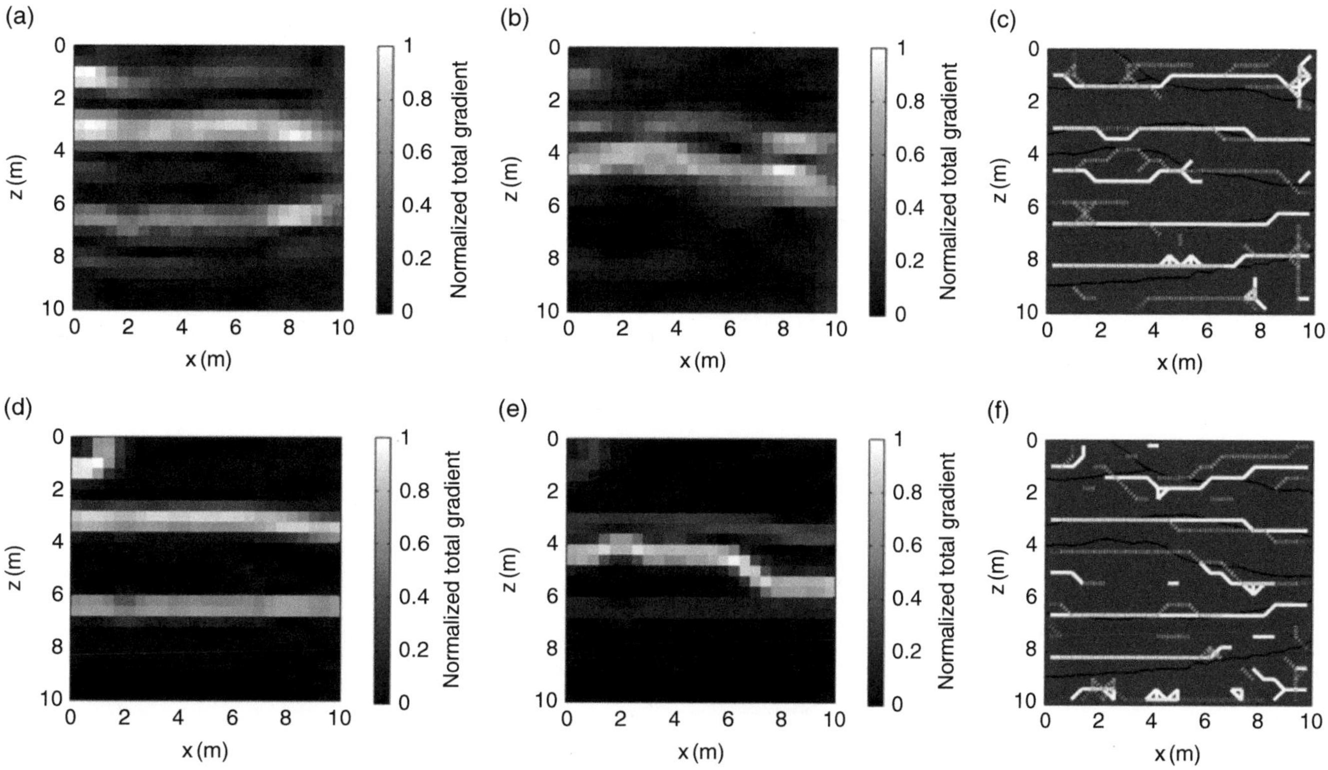

Figure 5.2 Total gradients calculated for the FTR **(a)** P-wave and **(b)** radar velocity models. **(c)** Visual composite of edge information present in the FTR models in Figures 5.1e and 5.1f achieved by tracking the maxima of the total gradient information in part a (*solid white lines*) and part b (*dashed gray lines*). **(d–f)** the same as in **(a–c)** but for the ZCI models (Figures 5.1g and 5.1h).The black lines in parts c and f outline the true lithological layer boundaries (see Figure 5.1a).

The gradients are more sharply determined, but the final composite of the visually integrated boundaries is of similar quality to that achieved for the FTR models. The integrated information allows for more holistic detection and easier identification of potential lithological boundaries which match the true boundaries of our original model well. Again, artifacts are present complicating the correct interpretation. Such visual composites gain increasing complexity with growing number of considered models. However, there exists no upper limit restricting the number of models to be integrated in a visual composite. Many authors have used such visual composites of boundary information (see, e.g., *Bedrosian* [2007] and *Gallardo et al.* [2012]).

Gradient-based analysis techniques can also be employed beyond a simple visual integration. *Gallardo and Meju* [2011] review different direction dependent and direction independent measures of gradient-based attributes derived from geophysical models. They show how these attributes can be applied to measure the difference between two disparate geophysical models with regard to their boundary locations, which brings even more rigor into the integration of boundary information derived from geophysical models than a visual composite. One example that gained popularity for joint inversion of disparate geophysical data sets by model structure coupling using gradient-based measures of image or model similarity are cross-gradients (see, e.g., *Gallardo and Meju* [2003, 2011], Chapter 4).

5.3.2. Color/Chroma Information in the Model Domain

Another concept of visual integration is popular in the analysis of radiometric maps. Here, ternary maps (see, e.g., *IAEA-TECDOC-1363* [2003]) are generated by scaling the gamma radiation intensity specific to the energy windows of potassium, thorium, and uranium on the red, green, and blue color intensity of an image in RGB color space (see, e.g., *Nixon and Aguado* [2012]), respectively. *Gallardo* [2007], *Infante et al.* [2010], and *Gallardo et al.* [2012] adopt this concept to the integration of fully co-located geophysical models and coin the term geospectral image for the resultant ternary RGB composite model.

We illustrate this concept using again our FTR and ZCI models as input examples. The resultant geospectral images are shown in Figure 5.3. Following *Infante et al.* [2010], in our two-model integration example, the three-element color vector $[r(i)\ g(i)\ b(i)]$ of a model cell i in a geospectral image is achieved by

$$[r(i)\ g(i)\ b(i)] = 255 \cdot \left[\frac{\mathbf{M}_1(i) - \min(\mathbf{M}_1)}{\max(\mathbf{M}_1) - \min(\mathbf{M}_1)} \quad \frac{\mathbf{M}_2(i) - \min(\mathbf{M}_2)}{\max(\mathbf{M}_2) - \min(\mathbf{M}_2)} \quad 0 \right], \tag{5.2}$$

with $\mathbf{M}_1$ and $\mathbf{M}_2$ being the seismic and radar velocity model of the considered model pair, respectively. If a third model would have been available, it could be scaled on the blue intensity channel b(i). The vector elements achieved by Eq. (5.2) have to be rounded to integer values matching the 24-bit color definition of RGB computer graphics prior to visualization. This approach is fundamentally limited to the visual integration of three models. A fourth model could be scaled to color intensity (brightness). Alternatively, ratios of two models could be employed for color scaling. *Gallardo et al.* [2012] visually integrate a geospectral image composed of three geophysical input models with isoline information illustrating boundaries in the structural variability of a fourth model. Care must be taken when interpreting such geospectral images, since the human color reception is not equally sensitive to red, green, and blue color components.

Geospectral images analyze the color/chroma information of the individual models instead of potential boundaries. When integrating blurred geophysical models, the geospectral image will also be of blurred character (see Figure 5.3a)

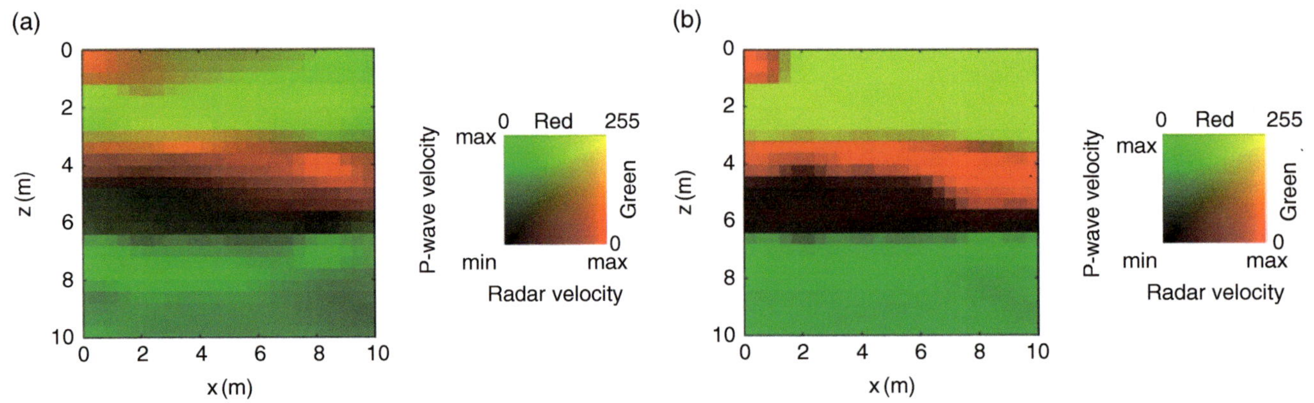

Figure 5.3 Geospectral images of **(a)** the FTR models and **(b)** the ZCI models.

and boundary detection may still be a challenging task. Nevertheless, color variations can be simply analyzed by a human interpreter to segment the model into regions of piecewise homogeneous color information. When comparing the models in Figure 5.3 with the original lithological model in Figure 5.1a, we see that all major physical zones can be clearly identified in the geospectral images, albeit the fact that identification of the contacts separating lithological layers B and C as well as F and G is not particularly clear.

5.3.3. Color/Chroma Information in the Parameter Space

Up to now we analyzed model structure information with regard to boundaries or color/chroma in the model domain itself. However, multiple geophysical models can be integrated in other domains as well, which may eventually offer advantages. For example, two equally discretized models $\mathbf{M}_1$ and $\mathbf{M}_2$ with m co-located model cells can be scaled along two orthogonal axes spanning a t-dimensional parameter space common to both models with $t = 2$. In such a parameter space, m samples $\mathbf{s}_i$ with $i = 1,\ldots,m$ will be determined by the corresponding information in the ith model cells of both input models $\mathbf{s}_i = [\mathbf{M}_1(i)\ \mathbf{M}_2(i)]$. Thus, the visualization of samples in the parameter space represents an integration of both models in one image. This concept holds also for more than two models resulting in a parameter space with its dimensionality equal to the number of considered models. The samples' arrangement in the parameter space is solely determined by the model cell values — that is, the color/chroma in a visualization of the models in the model domain. Information about the spatial neighborhood of model cells in the model domain is not present in the spanned parameter space. Here, samples closely located in the parameter space may correspond to model cells far away from each other in the model domain and vice versa.

In Figure 5.4a–5.4c we show scatter plots illustrating the parameter space spanned by the original models, the FTR models, and the ZCI models, respectively. Every symbol in the scatter plot marks a sample position. In Figures 5.4d–5.4f we illustrate the same parameter spaces by 2D histogram plots illustrating the sample density in the parameter space. For this purpose, we rasterized the parameter space and binned the samples with 10 m/s and 0.001 m/ns side lengths. Relative frequencies were calculated for every class by dividing the number of samples in a bin by m.

When comparing the arrangements of samples in Figure 5.4, we clearly see that they are significantly different for all three model pairs revealing that none of the model pairs, achieved by inversion, was able to resemble the physical relation of the original models. The reason lies in the utilization of Tikhonov regularization in the model generation procedure acting on the actual velocity entries of spatially neighbored model cells (see, e.g., *Moorkamp et al.* [2013]) — that is, enforcing a nondiagonal model covariance matrix. While integrated analyses of the inverted models with regard to boundary and color/chroma information in the map domain revealed quite acceptable reconstruction

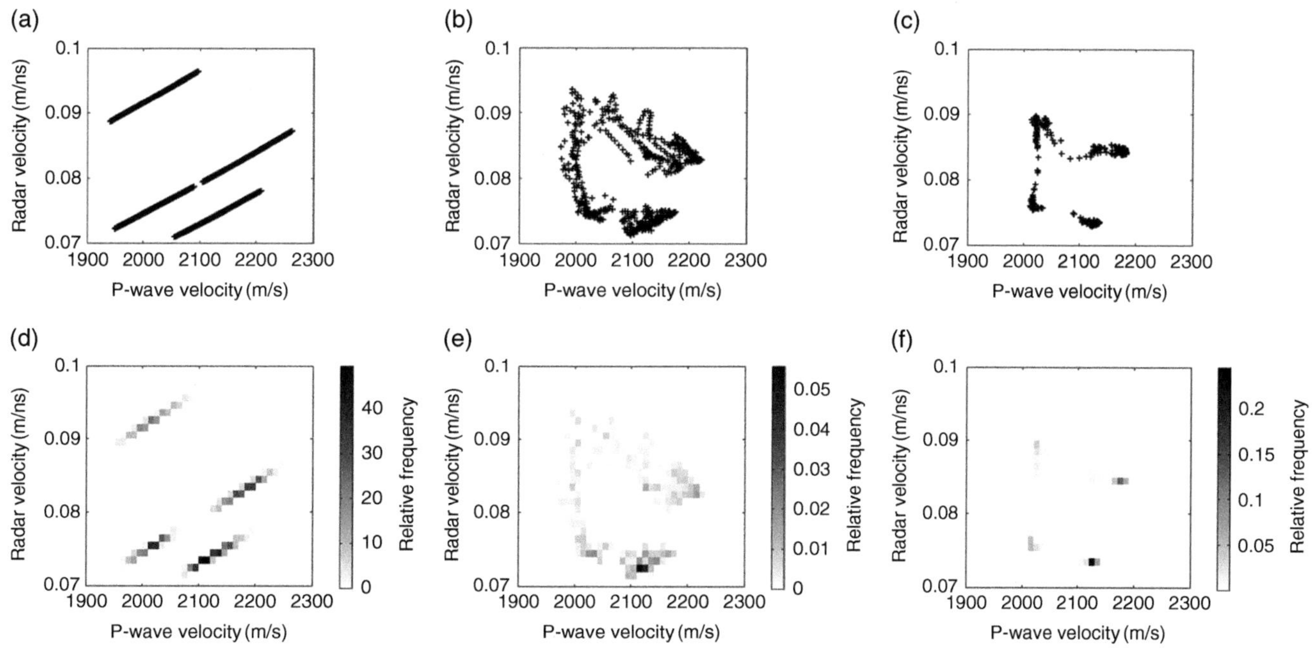

Figure 5.4 Scatter plots illustrating the sample distribution in the parameter spaces spanned by **(a)** the original, **(b)** FTR, and **(c)** the ZCI models. **(d–f)** The same as in parts a–c but illustrated by sample density in the rasterized parameter spaces.

of the structural information of the layered original model (see Figures 5.2 and 5.3; cf. Figure 5.1a), the transformation of the models into the parameter space reveals severe mismatch in the physical properties.

5.3.3.1. Analyzing the Parameter Space — Visual Approaches

However, many authors found it helpful to analyze the parameter space with regard to the arrangement of the samples. For example, motivation may come from the objective to identify significant parameter states in both models helping to identify lithological units or to derive or validate deterministic transfer functions linking the physical quantities imaged in the geophysical models achieved by smoothness constrained inversion. A number of publications are available where human interpreters manually analyze the sample distributions in the parameter space according to accumulations of samples and/or distinct slopes and shapes of sample arrangements in the parameter space (see, e.g., *Gallardo and Meju* [2003], *Linde et al.* [2006], *Jousset et al.* [2011], and *Carter-McAuslan et al.* [2015]). However, care should be taken not to overinterpret the sample arrangement in the parameter space, since regularization employed in the model generation procedure may critically effect the sample arrangement (Figure 5.4).

When analyzing the sample density in the parameter space spanned by the ZCI models, one may recognize that four regions of maximal sample density coincide largely with the maximal sample density in the parameter space spanned by the original models (cf. Figures 5.4f and 5.4d). Here, the simultaneous inversion outperformed the individual inversion (Figure 5.4e), which was less obvious in the previous analyses (Figures 5.2 and 5.3) but coincides with the finding of *Moorkamp et al.* [2013].

In Figures 5.5a and 5.5b we color the samples in the parameter space according to Eq. (5.2). In doing so, we

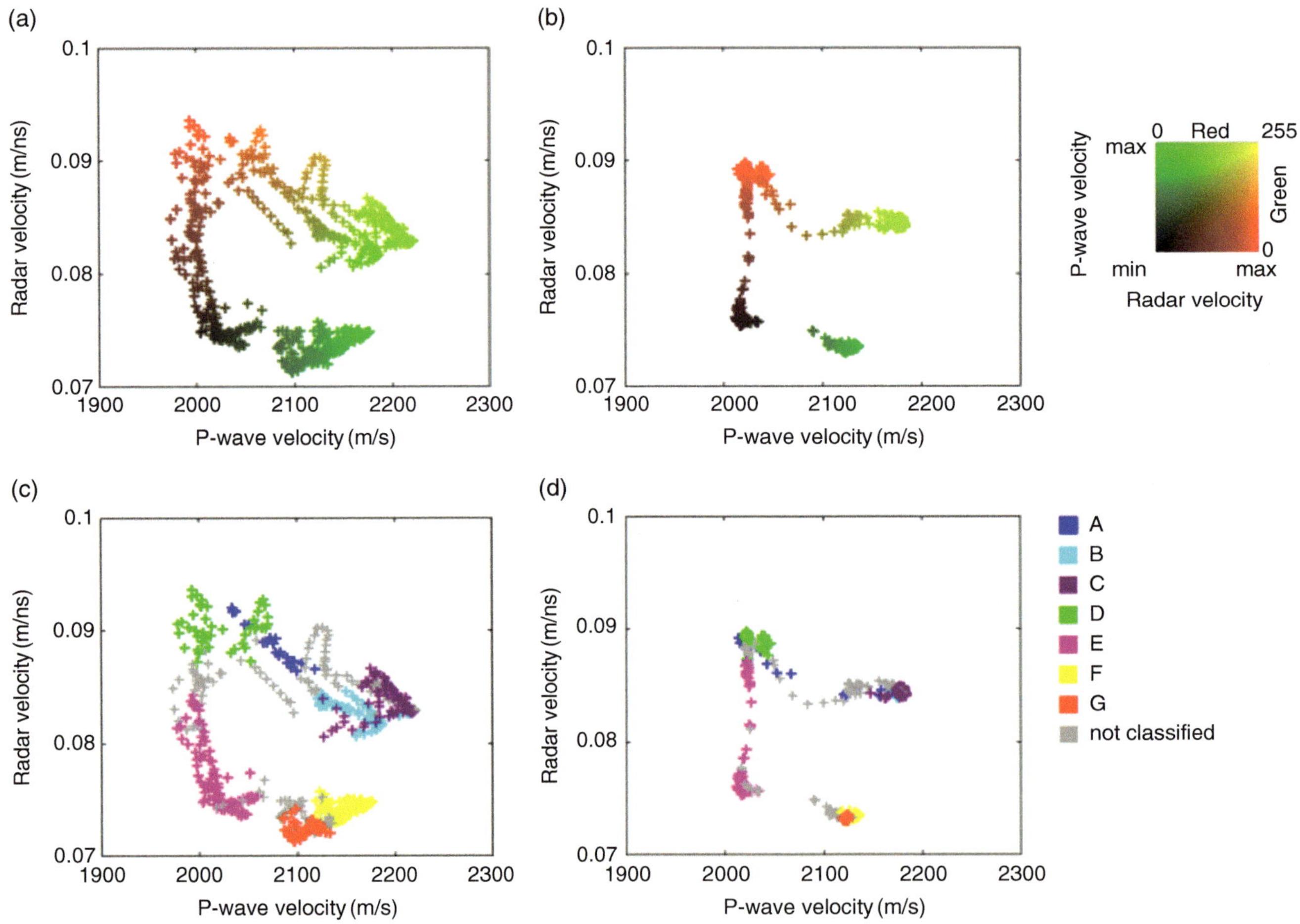

Figure 5.5 Scatter plots of the parameter spaces spanned by the **(a)** FTR and **(b)** ZCI models with color information according to the spectral models (Figure 5.3). **(c, d)** The same as in part a and b but with color information according to the lithological zonation of the original models (see Figure 5.1a). Samples in gray color are not assigned to a distinct lithology, since they fall in more than one lithology. Note that we do not show similarly colored versions for the original models because of plotting resolution issues.

follow *Infante et al.* [2010] and *Gallardo et al.* [2012]. Identification of distinct lithological units appears very difficult, particularly in the parameter space spanned by the FTR models (Figure 5.5a). In the model domain (Figure 5.3a) it appeared easier to detect distinct lithological units. The interpretation appears to be slightly easier in the parameter space spanned by the ZCI models. For example, two closely neighbored groups of samples with slightly different yellow-green color can be identified. Separation of these groups could have been easily overseen in the map domain because of the low color variation (see Figure 5.3b).

In Figures 5.5c and 5.5d we color the samples according to the lithological units in the original model they belong to (see Figure 5.1a). Samples that are falling into more than one lithological unit due to the coarse discretization of the inverted models are not classified. One can see that neither the FTR nor the ZCI models allow for a clear identification of lithological units in the parameter space. Sometimes, samples belonging to a distinct lithological unit scatter over a large area, while for other lithological units the samples form compact clusters. Significant overlap between samples belonging to different lithological units exists. Instead of visually analyzing slopes or sample accumulations in the parameter space, which will likely not result in optimal results, *Haberland et al.* [2003] define thresholds separating different classes in a 2D parameter space using reference models valid for similar geologies and with known physical parameter ranges for distinct subsurface materials.

The quantitative integration of models in a common parameter space requires a consecutive step of interpretation, which has been done manually up to now in our discussions in this chapter, particularly in view of the classification or grouping of samples. Such visual analysis of sample distribution in parameter space is only feasible up to $t = 3$. Parameter spaces with more than three dimensions cannot be visualized in one image. Instead, only projections on two or three dimensions could be analyzed by a human interpreter not allowing for objective assessment of sample distributions. In the following sections, we discuss now numerical analysis techniques allowing the classification of samples present in a parameter space. Some of them may even be able to take model resolution information into account. All techniques discussed hereafter have been tested on the FTR models as well as on the ZCI models. Most techniques perform well on both databases and provide similar results. In such cases, we limit our discussions and illustrations to the FTR models which form the database offering a more smeared and less grouped arrangement of samples in the parameter space.

5.4. INCORPORATION OF MODEL RESOLUTION IN THE PARAMETER SPACE

Bauer et al. [2003] developed a methodology to add model resolution information to the parameter space spanned by co-located models. We apply it to the FTR models (Figures 5.1e and 5.1f). Instead of true model parameter errors, we employ matrices $\mathbf{C}_1$ and $\mathbf{C}_2$, which are achieved by summing the columns of the absolute sensitivity matrices (see, e.g., *Aster et al.* [2005]) corresponding to the finally achieved velocity models (Figures 5.6a and 5.6b). Elements of low values in $\mathbf{C}_1$ and $\mathbf{C}_2$ indicate poorly determined regions of the model reconstruction area. In the following we refer to $\mathbf{C}_1$ and $\mathbf{C}_2$ as coverage matrices. Instead of just calculating sample density information (Figure 5.4e), we follow *Bauer et al.* [2003] and calculate a probability density function $\mathbf{pdf}(i)$ for every sample i:

$$\mathbf{pdf}(i) = \frac{1}{\sqrt{2\pi\,\delta\mathbf{M}_1(i)\delta\mathbf{M}_2(i)}} .\exp\left\{-\frac{1}{2}\left[\left(\frac{(\mathbf{B}_1-\mathbf{M}_1(i))^2}{(\delta\mathbf{M}_1(i))^2}\right)+\left(\frac{(\mathbf{B}_2-\mathbf{M}_2(i))^2}{(\delta\mathbf{M}_2(i))^2}\right)\right]\right\}. \tag{5.3}$$

$\mathbf{M}_1(i)$ and $\mathbf{M}_2(i)$ refer to the ith model cell of the seismic and radar velocity model, respectively. Model velocity errors $\delta\mathbf{M}_q(i)$ are determined for each model $q = 1, 2$:

$$\delta\mathbf{M}_q(i) = e_q \frac{\left(\dfrac{\sum_{g=1}^{m}\log_{10}(\mathbf{C}_q(g))}{m}\right)}{\log_{10}(\mathbf{C}_q(i))}. \tag{5.4}$$

The mean error e_q in a model q has to be determined and is considered constant. For simplicity, we estimate $e_1 = 40$ m/s and $e_2 = 0.003$ m/ns for seismic and radar velocities, respectively. *Bauer et al.* [2003] and *Bedrosian et al.* [2007] determine e_q by more quantitative analyses. Note that utilization of a mean error assumes a Gaussian error function, which may sometimes be a rough approximation for a model cell error $\delta\mathbf{M}_q(i)$ (see, e.g., *Tronicke et al.* [2012]). Additionally, spatially variable covariance between model parameters and the utilization of spatially nonuniform regularization could require the definition of more realistic and spatially variable model velocity errors (see, e.g., *Day-Lewis and Lane* [2004]). $\mathbf{B}_q$ are matrices describing the rasterized parameter space. In our case, $\mathbf{B}_1$ is a 30×40 matrix with every row vector equal to [1905 1915 1925 … 2285 2295] m/s. $\mathbf{B}_2$ is a 30×40 matrix with every column vector equal to $[0.0705\ 0.0715\ 0.0725\ \ldots\ 0.0985\ 0.0995]^T$ m/ns. The elements in $\mathbf{B}_1$ and $\mathbf{B}_2$ specify the

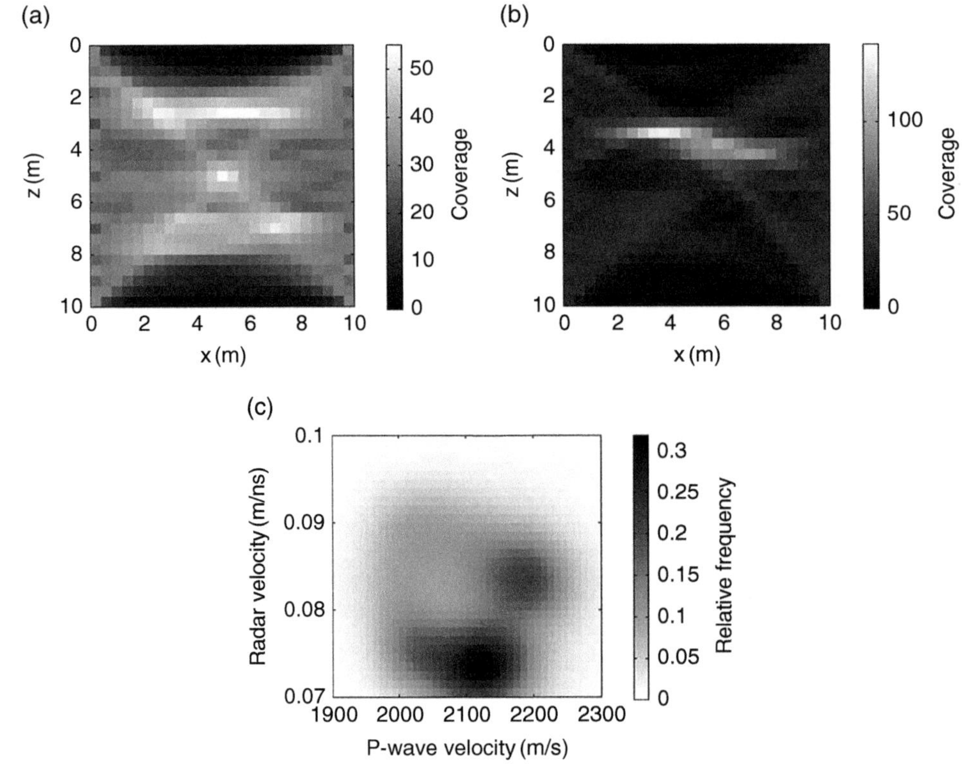

Figure 5.6 Coverage matrices of the FTR **(a)** P-wave and **(b)** radar velocity model. **(c)** Probability density function illustrating the parameter space after consideration of model errors.

center locations of every bin in the rasterized parameter space. The resultant probability density function for every sample **pdf**(i) achieved by Eq. (5.3) is of the same size as $\mathbf{B}_1$ and $\mathbf{B}_2$. Finally, all probability density functions are stacked to achieve the global probability density function **pdf** of all considered models

$$\mathbf{pdf} = \frac{1}{m}\sum_{i=1}^{m}\mathbf{pdf}(i). \tag{5.5}$$

The probability density function achieved for the FTR models and the corresponding coverage matrices is shown in Figure 5.6c. When comparing it to the 2D histogram illustration of the parameter space spanned by both models (Figure 5.4e), we see that incorporation of model resolution information resulted in significant smearing of the relative frequency information. In Figure 5.6c, we can identify four regions of increased relative frequency roughly coinciding with the positions of maximal sample density in the parameter space spanned by the original models (Figure 5.4d) and the ZCI models (Figure 5.4f). *Bauer et al.* [2003] consider the local maxima of the achieved probability density function as center locations of distinct classes in the parameter space assigned to different lithologies in the model domain. For the example in Figure 5.6, this suggests the utilization of four classes.

5.5. ARTIFICIAL NEURAL NETWORKS OF KOHONEN ARCHITECTURE

Artificial neural networks (ANN) attracted geophysicists for a long time (see, e.g., *Poulton* [2002]). ANNs are simplified mathematical models using calculation units (neurons) connecting informational input and output layers. The information transfer between input and output layer by the neurons is learned using examples presented to the network. Particularly ANNs of Kohonen architecture have been used with a self-organizing map (SOM) learning method [*Kohonen*, 2001], which enables an unsupervised learning of information transfer rules between input and output layer. Surprisingly, only very few studies have employed SOM for the integration of geophysical models. None consider model resolution information. To our knowledge, *Bauer et al.* [2008] were the first to employ SOM to integrate and classify crosshole tomographic images. Later, *Bauer et al.* [2012] used SOMs to integrate and classify fully collocated seismic and magnetotelluric models. Generally, the SOM method topologically maps the information of any number of input models on a two-dimensional model map, which allows for clear separation of different features in the input models. Neighboring regions in the model map are related to similar physical input properties.

In the following, we illustrate this approach and apply it to the integration of the ZCI models (Figures 5.1g and 5.1h). Following *Bauer et al.* [2008], we do not consider model resolution information. We begin by normalizing both models to a mean value of zero and standard deviation of unity. The m samples defined in the common normalized parameter space of both models serve as input information to the Kohonen layer, which is a two-dimensional arrangement of neurons (see, e.g., *Poulton* [2002]). The size and dimensionality of the Kohonen layer has no physical meaning and the optimal number of neurons can only be guessed according to rather vague rules (see, e.g., *Kohonen* [2001]). Usually, the number of neurons is significantly smaller than the number of model parameters. Here we choose 100 neurons arranged in a square matrix, but other numbers of neurons and matrix dimensions have been tested as well and gave consistent results as long as the number of neurons was not too low and the neuron arrangement was closer to a square matrix than a 1D vector arrangement. When starting the training phase, each neuron is randomly associated with a vector comprising elements taken from the range of each input model. In our example, each neuron is associated with a two-element vector $\mathbf{l}_\eta$ comprising a normalized seismic velocity and a normalized radar velocity. A sample $\hat{\mathbf{s}}_i$ present in the normalized parameter space is then compared to all neurons. The neuron $\mathbf{l}_w$ fulfilling the condition

$$\forall \eta, \|\hat{\mathbf{s}}_i - \mathbf{l}_w\| \leq \|\hat{\mathbf{s}}_i - \mathbf{l}_\eta\| \qquad (5.6)$$

is called the winning neuron. Iteratively, a learning rule is then applied to the neurons, which acts strongest on the winning neuron and its spatial neighbors. The neuron vectors $\mathbf{l}_\eta$ are changed towards the direction of the sample $\hat{\mathbf{s}}_i$ in the next iteration $d + 1$:

$$\mathbf{l}_\eta^{d+1} = \mathbf{l}_\eta^d + \beta^d \cdot \exp\left[-\left(\frac{r_w^d}{2\sigma^d}\right)^2\right] \cdot \left(\hat{\mathbf{s}}_i^d - \mathbf{l}_\eta^d\right). \qquad (5.7)$$

The exponential term describes a Gaussian distance function with distance r_w between the winning neuron and the updated neuron. The width of the neighborhood function σ decreases with increasing d. The weighting parameter β is called the learning rate and also decreases with increasing d. During the unsupervised iterative training phase, the ordering in the Kohonen layer increases. The elements (weights) of all $\mathbf{l}_\eta$ after 40,000 iterations of learning are illustrated in Figures 5.7a and 5.7b. Next, we calculate the total gradient of these maps (see Eq. (5.1) and Figure 5.7c). In Figure 5.7d, the histogram of all total gradients is shown. Following *Bauer et al.* [2012], we split the histogram into a "lowland" and "mountainous" part defining a threshold of 0.12. When applying this threshold to the Kohonen layer, we achieve four lowland segments labeled 1, 2, 3, and 4 in our example that are completely distinct from one another and separated by neurons assigned to the mountainous region. A water shed algorithm (see, e.g., *Vincent and Soille* [1991]) is then used to iteratively assign mountainous neurons to lowland segments using the total gradients. Overlap of segments is not allowed, and a number of mountainous neurons separating different lowland segments are not assigned to any segment (Figure 5.7e). Samples in the parameter space are now classified into four groups according to the neuron most close to them. Some of the samples remain unclassified since they are associated to remaining mountainous neurons outlining the boundaries between segments in the Kohonen layer. Backtransformation of the classified samples into the model domain results in a zonal model integrating the information of both input models (Figure 5.7f). The zonal model outlines the major lithology generally well but uses fewer classes than the number of lithologies present in the original models (see Figure 5.1). This is consistent with the original velocity models shown in Figures 5.1c and 5.1d. Note that this approach may bear some subjectivity in defining the optimal number of classes if difficulties exist in fixing the threshold between "lowland" and "mountainous" neurons in the histogram. However, lithological units characterized by very similar P-wave and radar velocity values are summarized in one class, which is acceptable from a physical point of view due to limited resolution capabilities of the inversion procedure (see Figure 5.6c). Classes do not overlap. Misclassification occurs only in regions where at least one input model has significant gradients (see Figure 5.2).

We have also applied this approach to the integration and zonation of the FTR models. The results are not shown here, since we found the approach lacking sufficient stability when applied to the FTR models. The resultant classification was critically dependent on the number of neurons. We tested all possible sizes of the 2D Kohonen layer by combining side lengths from 8 to 20 neurons. Classification of neurons resulted in highly different zonal models not allowing for clear identification of the four relevant physical zones for many tested Kohonen layers. The reason is not quite clear. However, the FTR models span a parameter space that is poorly clustered and rather offers a relatively even sample density (Figure 5.4e). The highly nonlinear mapping of the models on the Kohonen layer may result in stability problems in spotting the relevant center regions of each physical zone in the spanned parameter space.

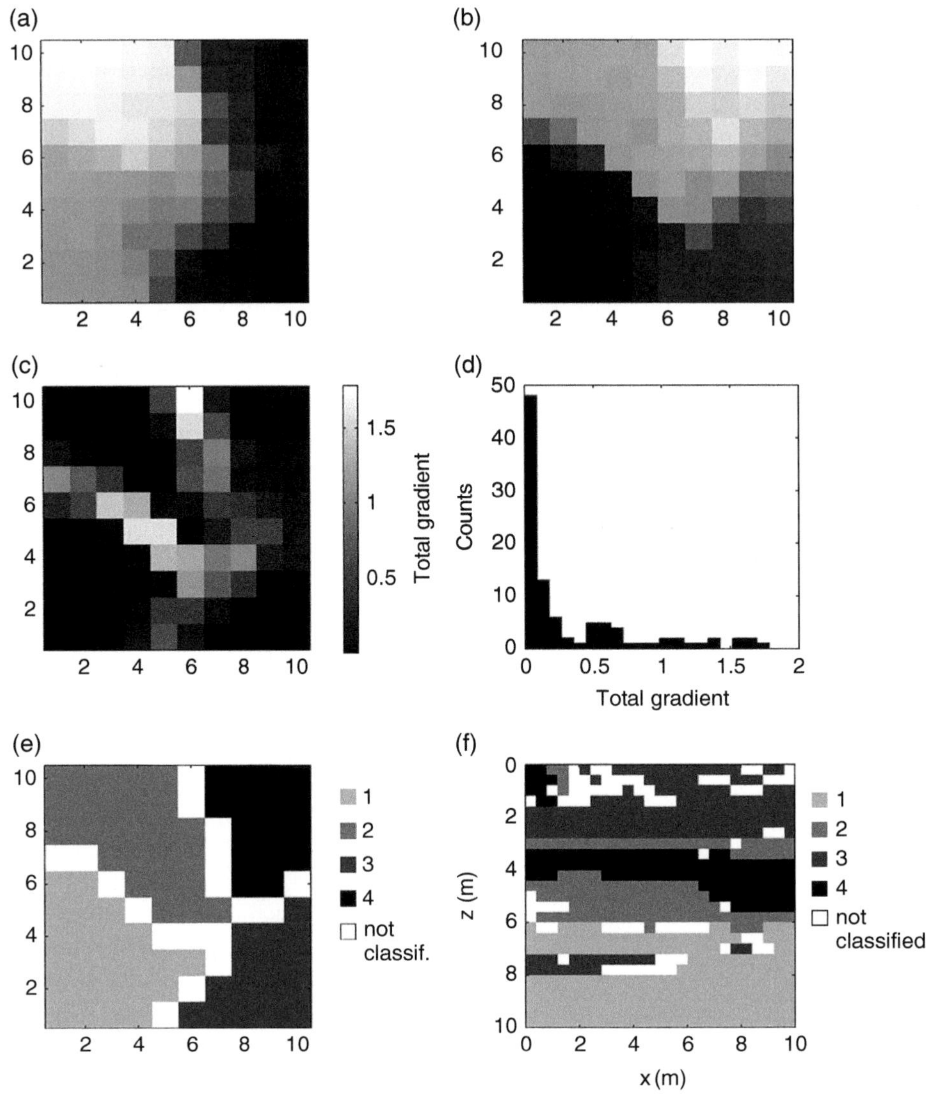

Figure 5.7 Normalized **(a)** P-wave and **(b)** radar velocity weights assigned to the topological map defined by 100 neurons. **(c)** Total gradient of the maps in parts a and b. **(d)** Histogram of the total gradient information in part c. **(e)** Segmentation of the topological map achieved by a watershed algorithm sorting the neurons into four different groups plus another for unclassified neurons outlining the remaining mountainous regions (see text for explanation). **(f)** Zonal model integrating the information of the ZCI models after transforming the neuron segmentation in part e back to the model domain.

5.6. CLUSTER ANALYSES

Probably the most popular family of algorithms for post-inversion integration of geophysical models is cluster analysis. Available algorithms can be grouped into distinct algorithmic families referred to as hierarchical, model based, partitioning, and spectral cluster analysis. Hierarchical cluster analysis has rarely been considered for the integration of disparate geophysical sets of information (see, e.g., *Martelet et al.* [2006]). They result in geophysical taxonomy trees of increasing refinement level when using a divisive approach and decreasing refinement level when employing agglomerative hierarchical clustering. In the latter, each sample forms an individual cluster (class) when initializing the algorithm. At each iteration, two clusters are merged according to some distance or similarity decision criteria — for example, shortest distance between clusters. Albeit very natural to human thinking, problems may arise in finding the optimal number of classes suiting all regions in the model reconstruction area.

A fundamental distinction between cluster algorithms should be made with regard to the resultant information provided by a cluster algorithm. A crisp cluster algorithm assigns each sample in the parameter space to a distinct cluster (class). No information is provided about the statistical degree of reliability — for example, in the form of classification probability or class membership. A fuzzy cluster algorithm assigns a sample partially to all available clusters. The degree of membership differs for every class. Fuzzy class membership information can be reduced to crisp classification information employing a defuzzification technique (see, e.g., *van Leekwijck and Kerre* [1999]).

5.6.1. Model-Based Cluster Analysis

Model-based cluster analyses have been used for the integration of multimethod crosshole tomographic models by *Doetsch et al.* [2010]. They use a classification scheme employing a maximum likelihood estimation technique to determine the parameters of a Gaussian mixture model that defines zonal geometries. *Doetsch et al.* [2010] consider three co-located models in their integration and classification but do not account for model resolution information. The employed expectation maximization (EM) cluster algorithm is essentially equivalent to the fuzzy maximum likelihood estimation (FMLE) algorithm (see, e.g., *Hilger* [2001] and *Canty and Nielsen* [2006]) also known as Gath–Geva (GG) cluster algorithm (*Gath and Geva* [1989]), which is usually considered a partitioning cluster algorithm. This makes the distinction between model-based and partitioning cluster analysis somewhat fuzzy.

Inspired by *Bedrosian et al.* [2007], we use an EM algorithm to find optimal parameters for bivariate Gaussian probability distributions allowing us to reconstruct the probability density function **pdf** shown in Figure 5.6c. A number of authors (see, e.g., *Bedrosian et al.* [2007], *Bedrosian* [2007], *Zhang et al.* [2009], *Muñoz et al.* [2010], *Stankiewicz et al.* [2010, 2011]) set up c bivariate Gaussian distributions manually and use Levenberg–Marquardt optimization to optimize their shape, volume, and position honoring an objective function measuring the misfit between the **pdf** and the probability distribution of the c superimposed Gaussian distributions. The Gaussian mixture model probability density is described by

$$f(\mathbf{s}) = \sum_{j=1}^{c} \frac{a_j}{2\pi\left|\mathbf{F}_j\right|^{1/2}} \exp\left\{-\frac{1}{2}\left[\left(\mathbf{s}_i - \mathbf{v}_j\right)^T \mathbf{F}_j^{-1}\left(\mathbf{s}_i - \mathbf{v}_j\right)\right]\right\}, \quad (5.8)$$

with amplitude a_j and covariance $\mathbf{F}_j$. Center positions of each Gaussian distribution j in the parameter space are defined by the position vector $\mathbf{v}_j$.

Here, we follow a slightly different approach, which could be easily adapted to any other classification algorithm acting in a parameter space — for example, SOM or partitioning clustering if model resolution information should be taken into account. Instead of analyzing the samples $\mathbf{s}_i$ or the **pdf** defined by all $\mathbf{s}_i$ and the model errors, we generate a new set of samples $\mathbf{s}^*$ approximating the **pdf** in Figure 5.6c. For this purpose, we generate 50,000 samples $\mathbf{s}^*$ and distribute them in the parameter space spanned by the FTR models. Their distribution in the parameter space is proportional to the **pdf** defined by all $\mathbf{s}_i$ and the model errors [Eq. (5.5)]. Within the range of each bin defined by its side lengths in terms of seismic and radar velocities, the samples are distributed randomly. Thus, the 2D histogram (i.e., probability density function) of the generated set of samples $\mathbf{s}^*$ closely approximates the **pdf** of the FTR models and their model errors.

Next, we subject all $\mathbf{s}^*$ to a cluster analysis selecting a predefined number of clusters. According to *Bauer et al.* [2003], it is recommended that the number of clusters equals the number of local maxima of the **pdf**. We subject the data set $\mathbf{s}^*$ to an EM cluster algorithm which strives to find an optimal mixture of multivariate normal distributions to represent the probability density function of the samples $\mathbf{s}^*$. Following *Bedrosian et al.* [2007], we constrain cluster center locations to regions of local maximal values in the **pdf**. In Figure 5.8a the probability density function of a solution with four bivariate Gaussian distributions is shown, which resembles the **pdf** in Figure 5.6c closely. We repeat the clustering of all $\mathbf{s}^*$ for different number of clusters and calculate the absolute difference between the **pdf** and the clustering results (Figure 5.8b). According to *Bedrosian et al.* [2007], the point of largest curvature indicates optimal number of classes, which would be three in this case. Alternatively, increasing number of clusters allows for better matching of the **pdf**. In our synthetic example we select the solution with four clusters as the optimal solution, since we know that this is the most reasonable choice with regard to the original models (cf. Figures 5.1c and 5.1d). Following *Bedrosian et al.* [2007], we binarize the information in Figure 5.8a. We set all bins with probability density values less than 0.66 of the maximal amplitude of its class to zero. The remaining bins of each class are set to unity (Figure 5.8c). Four non-overlapping classes are achieved labeled 1, 2, 3, and 4, and the model cells in the model domain are classified according to the binary information in the parameter space. The resultant crisp zonal model is shown in Figure 5.8d. It outlines the major lithology zonation based on both input models and their error estimates and is similar to the model achieved by the SOM (Figure 5.7), albeit not identical. Again, some model cells remain unclassified, which is acceptable because of resolution considerations.

5.6.2. Partitioning Cluster Analysis

Partitioning cluster analysis is based on iterative relocation of m samples in a t-dimensional parameter space

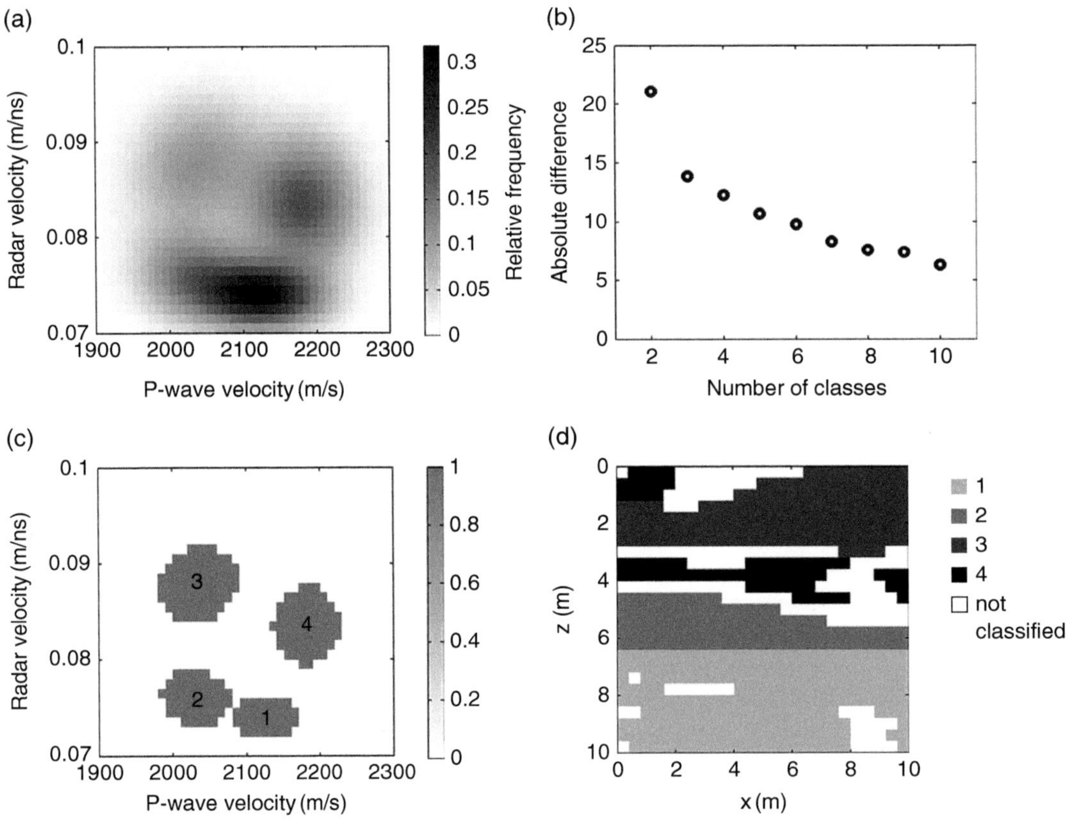

Figure 5.8 **(a)** Mixture of four bivariate Gaussian distributions achieved by an expectation maximization algorithm (compare to Figure 5.6c). **(b)** Absolute difference of the probability density information in Figure 5.6c and mixtures of bivariate Gaussian distributions for different number of distributions (classes). **(c)** Binarized image of part a after setting all values <66% of the maximum value of a distribution to zero. Different nonoverlapping classes are labeled 1–4. **(d)** Zonal map after backtransformation of (c) to the model domain.

by moving samples from one class to another until no further improvement of the sample classification is achieved. Various crisp and fuzzy cluster algorithms are available (see, e.g., *Höppner et al.* [1999]). A distinct algorithm should be chosen with care since the suitability of its classification rules for the integration of geophysical models may critically influence the quality of the achieved results. In the following we present two simple crisp and fuzzy cluster algorithms that are particularly popular among geophysicists.

5.6.2.1. Crisp k-Means Cluster Analysis

Crisp k-means cluster analysis has been used extensively in the past to integrate multiple crosshole tomographic geophysical or hydrological models (see, e.g., *Dietrich et al.* [1998], *Tronicke et al.* [2002], *Tronicke et al.* [2004], *Dietrich and Tronicke* [2009], *Brauchler et al.* [2012]). Here, we employ this algorithm to integrate the FTR models of our synthetic example (Figures 5.1e and 5.1f). The algorithm acts in the parameter space spanned by the two models after scaling them to a mean value of zero and a standard deviation of unity. Samples will be assigned to the cluster with the nearest mean by minimizing the following objective function

$$J_{kmeans}(\mathbf{V}) = \sum_{j=1}^{k}\sum_{i=1}^{m} \| \hat{\mathbf{s}}_i - \mathbf{v}_j \|^2 . \tag{5.9}$$

The term $\| \hat{\mathbf{s}}_i - \mathbf{v}_j \|$ is a chosen distance measure between a sample $\hat{\mathbf{s}}_i$ and a cluster center $\mathbf{v}_j$. The number of clusters k is prescribed a priori. Cluster validity measures — for example, the variance ratio criterion (*Calinski and Harabasz* [1974] — exist for finding the optimal k_{opt} by posterior analysis of results for different k. Cluster centers $\mathbf{v}_j$ are randomly initialized and iteratively updated to minimize J_{kmeans}. After minimization of J_{kmeans}, an index vector $\mathbf{h}$ of length m is achieved by

$$\mathbf{h} = \arg\min \sum_{j=1}^{k}\sum_{i=1}^{m} \| \hat{\mathbf{s}}_i - \mathbf{v}_j \|^2, \tag{5.10}$$

assigning each sample in the parameter space to a cluster. In Figures 5.9a and 5.9b the results of the k-means cluster analysis are shown when partitioning the samples

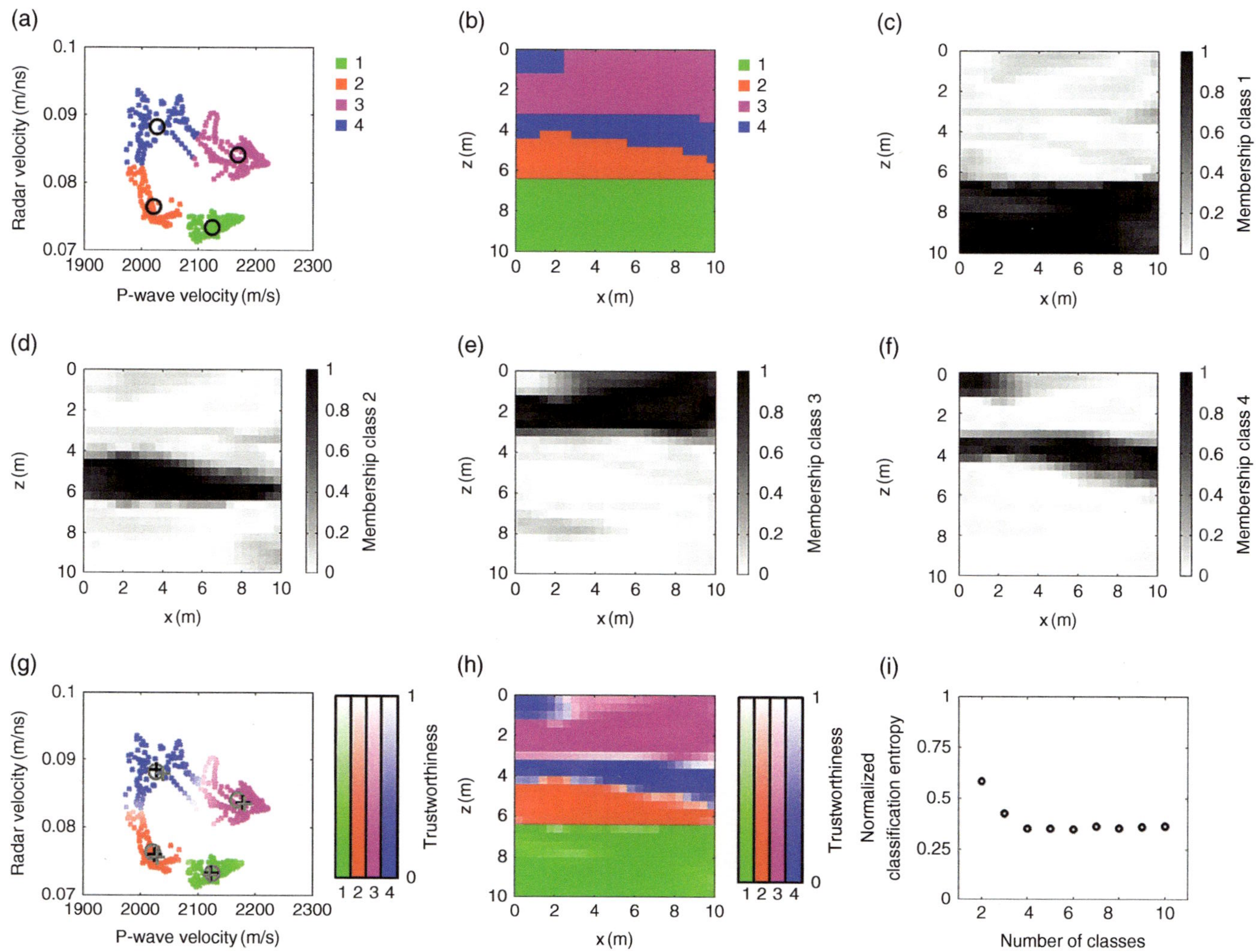

Figure 5.9 **(a)** Scatter plot of the FTR models and **(b)** corresponding zonal model illustrating the result of a *k*-means cluster analysis using four clusters (classes). **(c–f)** Membership information displayed in the model domain achieved by fuzzy *c*-means cluster analysis of the FTR models using four clusters (classes). **(g)** Scatter plot of the FTR models and **(h)** corresponding zonal model achieved by defuzzification of the membership information in parts c–f. Color saturation is scaled according to trustworthiness information [see Eq. (5.16)]. **(i)** Normalized classification entropy for different numbers of clusters employed by fuzzy *c*-means cluster analysis.

in four clusters. Noteworthy, all samples are assigned to a distinct cluster and no sample remains unclassified. The integrated zonal model is close to those achieved by the previously employed structural integration techniques. If desired, posterior classification probabilities for every sample can be calculated by performing a discriminant analysis trained with the full index vector **h** (see, e.g., *Eberle* [1993]). Generally the algorithm is considered robust and can be easily extended to different, more complex distance measures than the Euclidian distance standardly used [*Späth*, 1985]. If model errors are to be considered in a partitioning cluster analysis, samples $\hat{\mathbf{s}}_i$ should be replaced by $\hat{\mathbf{s}}_i^*$, which is the generated set of samples $\mathbf{s}^*$ after normalization. Samples $\hat{\mathbf{s}}_i$ can be assigned to a specific cluster according to the classification of the sample $\hat{\mathbf{s}}_i^*$ with minimum distance from $\hat{\mathbf{s}}_i$ in the parameter space.

5.6.2.2. Fuzzy c-Means Cluster Analysis

The fuzzy *c*-means (FCM) cluster analysis is a fuzzy partitioning cluster algorithm using the same distance measure as the *k*-means algorithm. Its fuzzy character offers some advantages over crisp algorithms. However, it was not before 2006 that the FCM algorithm was used to integrate co-located geophysical tomograms [*Paasche et al.*, 2006].

We apply this algorithm to integrate and classify the FTR models after normalizing them to a mean value of zero and standard deviation of unity. We minimize the objective function

$$J_{fcm}(\mathbf{U},\mathbf{V}) = \sum_{j=1}^{k}\sum_{i=1}^{m}\left(u_{ji}\right)^f \|\hat{\mathbf{s}}_i - \mathbf{v}_j\|^2, \qquad (5.11)$$

where u_{ji} denotes the degree of membership of sample $\hat{\mathbf{s}}_i$ to cluster j defined by its center $\mathbf{v}_j$. The elements u_{ji} of the membership matrix **U** are constrained to be positive and accumulate to column sums of unity. The weighting exponent f, also referred to as the fuzzification parameter, controls the degree of fuzziness and lies in the range of $1 \le f < \infty$. A selection between 1.5 and 2 is regarded as suitable choice for most databases [*Hathaway and Bezdek*, 2001]. For f approaching unity, the FCM algorithm approximates the *k*-means algorithm. Note that for high f the cluster separation becomes very fuzzy; that is, the samples $\hat{\mathbf{s}}_i$ belong to every cluster with almost the same degree of membership. The Euclidian distance from a sample to a cluster center is calculated by

$$\|\hat{\mathbf{s}}_i - \mathbf{v}_j\| = \sqrt{\sum_{a=1}^{t}\left(\hat{\mathbf{s}}_{ia} - v_{ja}\right)^2}. \qquad (5.12)$$

Iterative alternating optimization is used to update the membership values

$$u_{ji} = \sum_{b=1}^{k} \|\frac{\hat{\mathbf{s}}_i - \mathbf{v}_j}{\hat{\mathbf{s}}_i - \mathbf{v}_b}\|^{\frac{-2}{f-1}} \qquad (5.13)$$

and the cluster centers

$$\mathbf{v}_j = \frac{\sum_{i=1}^{m}\left(u_{ji}\right)^f \hat{\mathbf{s}}_i}{\sum_{i=1}^{m}\left(u_{ji}\right)^f}. \qquad (5.14)$$

We set $f = 2$ and $k = 4$ and apply this approach to the FTR models. The resultant membership information is illustrated in Figures 5.9c–5.9f. We use core selection defuzzification [*van Leeckwijck and Kerre*, 1999] to achieve an index vector **h** containing the crisp cluster numbers the m samples are assigned to for every element i of **h**

$$h_i = \arg\max\left(\mathbf{u}_{1,\ldots,k,i}\right), \qquad (5.15)$$

with $h_i \in \{1, \ldots, k\}$. In Figures 5.9g and 5.9h colors denote the cluster a sample has been assigned to. Membership information in Figures 5.9c–5.9f can be used to derive a classification trustworthiness for every sample:

$$\phi_i = \left[\frac{\max\left(\mathbf{u}_{1,\ldots,k,i}\right) - k^{-1}}{1 - k^{-1}}\right]^y. \qquad (5.16)$$

Throughout this manuscript we set $y = 1$, which ensures a linear link between cluster membership of a sample and trustworthiness. Color saturation superimposed on the class assignment is weighted by the trustworthiness. This allows one to easily spot regions of reduced zonation reliability in the integrated model in Figure 5.9h. For example, in cluster 1, a region with low color saturation can be identified between $z = 7$ and $z = 8$ m and $x < 5$ m. It coincides with differences between the classification results achieved by the approach of model-based cluster analysis (Figure 5.8d) and the SOM (Figure 5.7f). Incorporation of trustworthiness information derived from the fuzziness of the clustering result in the visualization of the integrated model can help to improve the interpretation and to even identify regions of minor parameter variations within a zone.

As for all cluster algorithms, the selection of an optimal number of classes is critical. We perform a posterior statistical analysis of solutions achieved for different k by calculating the normalized classification entropy *NCE*:

$$NCE = \frac{-\frac{1}{n}\sum_{j=1}^{k}\sum_{i=1}^{m} u_{ji} \cdot \log_{10}(u_{ji})}{\log_{10}(k)} \quad (5.17)$$

for each solution with $k > 1$ and $k < 11$ (Figure 5.9i). Solutions with minimal *NCE* are considered optimal. Here, we find a rather constant *NCE* for solutions with $k \geq 4$. Following Occam's razor, we consider $k = 4$ as the solution with optimal structural complexity.

5.6.3. Integration of Partially Co-located Models by Cluster Analysis

All numerical integration techniques reviewed up to now require full co-location of the considered geophysical models, since they employ distance measures in the spanned parameter space. However, samples referring to model cells not covered by all models cannot be uniquely positioned in the parameter space. Distances of such samples from other samples or cluster centers cannot be computed uniquely anymore. Statistical imputation techniques (see, e.g., *Little and Rubin* [1987]) exist allowing the estimation of the missing information (i.e., model areas not covered by the data set), based on available information — for example, by means of likelihood techniques [*Little and Rubin*, 1987; *Schafer*, 1997]. Such information completion techniques are purely mathematic and cannot honor any physical relations underlying geophysical model information in the areas not covered by a geophysical data set. We do not recommend using such techniques in the context of geophysical model integration. Instead, *Paasche et al.* [2010] developed a modified fuzzy partitioning cluster algorithm allowing the integration of partially co-located models without prior imputation of the missing information. Since model regions not covered by all models suffer a reduced amount of information available for integration and classification, they are particularly vulnerable to misclassification. Hence we additionally regularized the cluster analysis by a penalty term ensuring that samples, which are spatial neighbors in the model domain, will be characterized by similar class membership. The objective function in Eq. (5.11) changes to

$$J^*_{fcm}(\mathbf{U},\mathbf{V}) = \sum_{j=1}^{k}\sum_{i=1}^{m}(u_{ji})^f \, \|\hat{\mathbf{s}}_i - \mathbf{v}_j\|^2 + \frac{\beta}{2}\sum_{j=1}^{k}\sum_{i=1}^{m}(u_{ji})^f \sum_{\alpha\in A_j}\sum_{\gamma\in G_i}(u_{\alpha\gamma})^f . \quad (5.18)$$

G_i is the set of spatial neighbors of $\hat{\mathbf{s}}_i$ and $A_j = \{1,\ldots,k\}\backslash\{j\}$. β controls the regularization strength. The distance between sample $\hat{\mathbf{s}}_i$ and the cluster center $\mathbf{v}_j$ is calculated by

$$\|\hat{\mathbf{s}}_i - \mathbf{v}_j\| = \sqrt{\sum_{a=1}^{t} w_{ia}(\hat{\mathbf{s}}_{ia} - v_{ja})^2} . \quad (5.19)$$

w_{ia} is the Boolean ath element of the vector $\mathbf{w}_i$ with $w_{ia} = 1$ if the ath element of sample vector $\hat{\mathbf{s}}_i$ is observed and $w_{ia} = 0$ otherwise. The cluster center positions are updated by

$$v_{ja} = \frac{\sum_{i=1}^{m}(u_{ji})^f \, \hat{\mathbf{s}}_{ia} \cdot w_{ia} \cdot \omega_{ia}}{\sum_{i=1}^{m}(u_{ji})^f \cdot w_{ia} \cdot \omega_{ia}}, \quad (5.20)$$

where

$$\omega_{ia} = \frac{\sum_{a=1}^{t} w_{ia}}{t} . \quad (5.21)$$

The weighting factors ω_{ia} compensate for the effect of the underestimated distances for samples with incomplete position vectors. The membership values are updated in all iterations by

$$u_{ji} = \sum_{b=1}^{k}\left[\frac{\|\hat{\mathbf{s}}_i - \mathbf{v}_j\|^2 + \beta\sum_{\alpha\in A_j}\sum_{\gamma\in G_i}(u_{\alpha\gamma})^f}{\|\hat{\mathbf{s}}_i - \mathbf{v}_b\|^2 + \beta\sum_{\alpha\in A_j}\sum_{\gamma\in G_i}(u_{\alpha\gamma})^f}\right]^{\frac{-1}{f-1}} . \quad (5.22)$$

We demonstrate the potential of this regularized missing value fuzzy *c*-means (RMVFCM) cluster algorithm by applying it to the FTR models after normalizing them to a mean value of zero and a standard deviation of unity. We reduce the region of the model area that is covered by both models (Figures 5.10a and 5.10b) to 28% of the total model area. Thus most of the samples present in the two-dimensional parameter space spanned by both models have an incomplete position vector. When integrating and clustering these models using the RMVFCM cluster algorithm, the zonal model in Figure 5.10c is achieved. For consistency we used four classes. The model is fully coincident with the zonal model achieved when applying the fcm algorithm to the fully co-located FTR models (Figure 5.9h). In areas covered by only one model the zonation is different from those in Figure 5.9h. However, it is fully compliant with the information provided by the two reduced FTR models in Figures 5.10a and 5.10b. Hence interpretation of an integrated model relying on a spatially variable informational base should be done with care. To illustrate the reduced informational base, we screen the regions that are not covered by both models (Figure 5.10c).

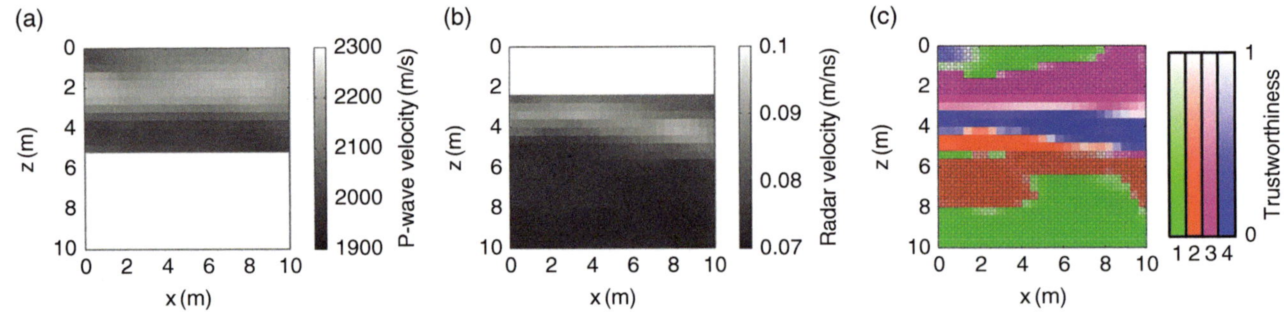

Figure 5.10 **(a, b)** The FTR models shown in Figures 5.1e and 5.1f, respectively, after reduction of model area coverage. The fully co-located model area is reduced to 28%. **(c)** The zonal model achieved after employing the regularized missing value fuzzy *c*-means cluster analysis to integrate the partially co-located models in parts a and b. Regions in the zonal model not covered by both input models are screened.

5.7. UTILIZING NUMERICAL INTEGRATION BEYOND CLASSIFICATION AND LITHOLOGY ASSESSMENT

If model structure information can be described in sufficient complexity and completely separated from the physical attributes of the available models (e.g., radar and seismic velocity), model structure information achieved by post-inversion integration of complementary geophysical models can even support the optimal sampling of sparse additional parameters. For example, such knowledge can guide the identification of optimal locations for the acquisition of sparse hydrological or geotechnical properties of the ground and the consecutive interpolation of this sparse information. This issue is related to the field of unconditional and conditional simulations of spatially continuous distributions of sparse logging data by multiple disparate geophysical tomograms simultaneously considered to guide the spatial simulation rules (see, e.g., *Chen et al.* [2001] and *Hubbard et al.* [2001]).

Paasche et al. [2006] and *Hachmöller and Paasche* [2013] integrate multiple geophysical models using fuzzy cluster analysis. The structural heterogeneity of all input models is captured in the membership information, whereas the physical attributes of the geophysical input models are described by the cluster center values and are separated from the model structure information. With increasing number of clusters the information loss will be reduced when transforming the heterogeneity of all input models into the fuzzy membership matrix. We apply the approach of *Paasche et al.* [2006] to achieve a two-dimensional estimation of the porosity model underling the generation of the original seismic and radar data sets. The original porosity model is shown in Figure 5.11a. Figure 5.11b shows a decimated version of this model employing the discretization used in the geophysical inversions (see Figures 5.1e–5.1h). From the models in Figures 5.11a and 5.11b we extract two logs at $x = 0$ m and $x = 10$ m, respectively (Figure 5.11c). Note the reduced small-scale variability of the decimated logs (black lines). Following *Paasche et al.* [2006] we split the porosity logs according to the fuzzy membership information achieved by integrating the FTR models (Figure 5.11d). By doing so, we are able to assign a mean porosity value $\overline{p}_j$ with $j = 1,\ldots, k$ to every cluster. Using a simple linear mixture model with the fuzzy memberships of every model cell i as weighting parameters

$$p_i = \sum_{j=1}^{k} \overline{p}_j \cdot u_{ji}, \tag{5.23}$$

we can achieve a 2D porosity distribution $\mathbf{P}$ with elements p_i with $i = 1,\ldots, m$ guided by the structural heterogeneity of both geophysical models. We do not need to specify an expectation about the relationship between the geophysical models and the target parameter (e.g., porosity). When using at least two different geophysical models and three clusters, this approach is capable of accommodating nonlinear and even nonunique relations between geophysical models and the target parameter. In Figure 5.11e the reconstructed porosity model is shown. The decimated borehole logging information and the reconstructed porosity information are compared in Figure 5.11f. We see that the reconstructed information is slightly reduced in amplitude. If no spatial agreement between the logging data and the geophysical models had been present, the reconstruction would have resulted in a homogeneous (structureless) porosity model. We rescale the amplitude of the reconstructed porosity model according to the borehole logging information extracted from the decimated models (the dotted lines in Figure 5.11f). The resultant porosity model in Figure 5.11g is considered the best possible reconstruction of porosity information and shows good agreement with the original decimated model (Figure 5.11h). It has been achieved by the structural

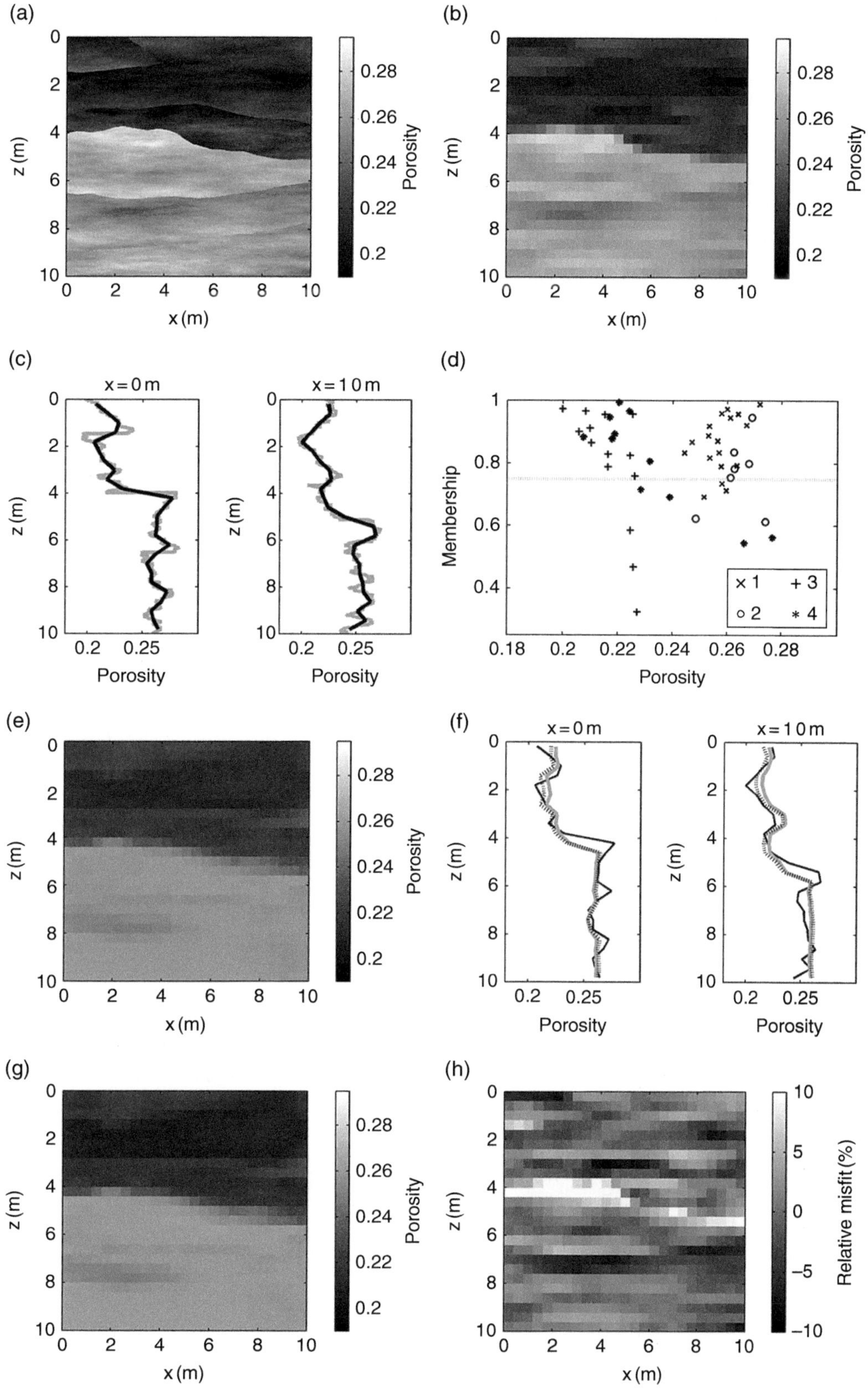

Figure 5.11 **(a)** Original and **(b)** decimated porosity models underlying the geophysical models in Figure 5.1. **(c)** Synthetic borehole porosity logs extracted from part a (*gray lines*) and part b (*black lines*). **(d)** Sorting the decimated borehole logging information according to the results of a four cluster solution achieved by fuzzy *c*-means cluster analysis of the FTR models (Figures 5.9c–5.9f). **(e)** Reconstruction of the two-dimensional porosity information based on the decimated porosity logs (part c) and the membership information in Figures 5.9c–5.9f. **(f)** Comparison of the decimated (*black lines*), reconstructed (*gray lines*), and scaled reconstructed (*dotted line*) porosity information at the locations of the synthetic boreholes. **(g)** Reconstructed porosity after scaling (corresponds to the dotted lines in part f). **(h)** Deviation between the porosity distributions shown in parts b and g.

integration of two disparate geophysical models and sparse porosity data. Its spatial resolution is controlled by the geophysical models.

5.8. FUTURE RESEARCH CHALLENGES IN POST-INVERSION MODEL INTEGRATION

5.8.1. Combining Different Features and Different Analysis Domains

All numerical integration techniques discussed above analyze model structure either in the model domain or the parameter space. The only exception is the RMVFCM algorithm which combines sample classification in the parameter space with information about spatial neighborhoods of model cells (samples) in the model domain. None of the discussed algorithms analyze multiple features — for example, boundary and color/chroma and texture information. The latter may be of lower importance when integrating geophysical models, since regularized inversion procedures may largely govern model texture. However, spectral cluster algorithms (see, e.g., *von Luxburg* [2007]) potentially allow for a more flexible definition of rules for classifying samples defined by co-located models than any other family of cluster algorithms. Instead of scaling models directly on the axes of a parameter space, a graph (see, e.g., West [2007]) is defined by its weighted adjacency matrix and cut into segments. The definition of similarity measures used as weights connecting the nodes of the graph may offer potential for integrating and classifying multiple geophysical models according to a number of different decision rules at the same time. Thus, it appears feasible to integrate different concepts of edge detection and color/chroma analyses. *Hachmöller and Paasche* [2013] were the first to employ a spectral cluster algorithm for the post-inversion integration of geophysical models. However, they used a Gaussian similarity measure to populate the weighted adjacency matrix, which makes their analysis still similar to a fuzzy c-means cluster algorithm and limits the model structure analysis to color/chroma information analysis.

5.8.2. Integration Across Scales and Different Types of Information

Another topic of interest is the additional consideration of borehole logging data in the post-inversion integration and classification of geophysical models. Such information could help to overcome problems in geophysical model integration resulting from ambiguity of the model generation process. Challenges to be tackled are based in the different scales of spatial resolution of logging data and geophysical models and in the partial co-location of information since logging data usually do only cover a very small fraction of the geophysical model area. In principle, the RMVFCM algorithm may offer the potential to consider down-scaled sparse borehole logging data in the integration of multiple geophysical models. Some tests with this approach have shown problematic results, and this method requires further research before it can be generally applied to any data scenario. The experienced problems may be due to insufficient capability to consider measurement errors of the logging data in the cluster analysis. Furthermore, a rather empirical adjustment of the neighborhood G_i in Eq. (5.18) may be required to ensure a sufficient spatial outreach of the borehole log information in the integrated zonal model to avoid artifacts in the zonal model at the positions of the boreholes.

In some cases, geological models may be available that could form an additional class of information to be considered in geophysical model integration. However, geological models are usually preclassified in the sense that a number of homogeneous layers are separated by sharp boundaries. Such geological models may partly rely on subjective insights of a human expert. The resultant geological model could be fully correct at some regions of the model area, but it may also bear significant inaccuracies — for example, due to limited information. Numerical assessment of the accuracy of the geological model is usually not possible, which contrasts with geophysical models (e.g., see Section 5.4). Considering such geological models as just one additional set of information in a post-inversion model integration approach without considering the special information characteristics of geological models in the integration procedure will likely fail to produce reliable results. No sophisticated approaches have been developed to date that incorporate the specific error characteristics of numerically derived and subjectively sampled models.

5.9. CONCLUSIONS

Numerical integration of disparate geophysical models achieved by inversion of underlying geophysical data sets allows for improved interpretation of geophysical models. Numerical integration should result in one image comprising information from multiple geophysical models. The integration procedure should follow defined decision rules allowing for repetition of the procedure. Various integration techniques are available ranging from simple edge detection techniques integrated in visual composites to mathematically more complex techniques like artificial neural networks or cluster analyses. Currently, none of the available techniques allows for model integration in different domains — for example, model domain and parameter space.

The approaches in use nowadays usually limit themselves to either edge analysis or color/chroma analysis — that is, analyzing model structure according to the values assigned to model grid cells. Differences may exist in the approaches regarding their suitability to handle fully as well as partially co-located models. The latter is more challenging and has rarely been done. Many approaches offer at least the potential to consider geophysical model error information. However, in most applications, model errors are ignored for simplicity or due to lacking ability to provide realistic error estimates.

While most post-inversion integration approaches focus on the identification of potential lithological units, some approaches may offer potential beyond lithology discrimination — for example, with regard to petrophysical parameter estimation or geoscientifically constrained interpolation of sparse hydrological or geotechnical parameters. Such numerical analyses building on integrated model structure information make this field of research a promising field with significant potential to advance information extraction from geophysical models. However, many model integration approaches are inspired by computer vision analysis and data mining — two subjects that are often not part of the standard training of the current generation of young geoscientists.

For those readers missing a summative section comparing and eventually ranking the different integration approaches discussed, this fundamental conclusion may be of particular importance: All approaches have their strengths as well as their limitations and their own benefit/cost ratio, which critically depends on the objectives to be addressed by the user. The latter cannot all be foreseen, which does not allow for a representative summation section, since particularly the mathematically more complex integration approaches have often only been applied to a limited number of case studies by the same group of authors. However, if the interest is merely directed towards a quick assessment of model structure similarity as, for example, defined by gradients/edges, a visual integration as described in Section 5.3 would be fully sufficient and efficient. However, edges in the visual composite may not always be continuous and coincident and leave the user sometimes with a challenging interpretation. Artificial neural networks or cluster analyses allow for more rigorous integration and provide a single integrated image always clearly outlining distinct regions, but with higher computational costs. If trustworthiness assessment of the detected structures is desired, the computational effort increases further. To conclude, the suitability of any of the discussed model integration approaches depends critically on a clear antecedent and prudential reflection on the motivation leading to their utilization.

REFERENCES

Aksoy, S., N. H. Younan, and L. Bruzzone (2010), Pattern recognition in remote sensing, *Pattern Recogn. Lett.*, *31*, 1069–1070.

Aster, R. C., B. Borchers, and C. H. Thurber (2005), *Parameter Estimation and Inverse Problems*, Academic Press, New York.

Bauer, K., A. Schulze, T. Ryberg, S. V. Sobolev, and M. H. Weber (2003), Classification of lithology from seismic tomography: A case study from the Messum igneous complex, Namibia, *J. Geophys. Res.*, *108*, doi: 10.1029/2001JB001073.

Bauer, K., R. G. Pratt, C. Haberland, and M. Weber (2008), Neural network analysis of crosshole tomographic images: The seismic signature of gas hydrate bearing sediments in the Mackenzie Delta (NW Canada), *Geophys. Res. Lett.*, *35*, L19306.

Bauer, K., G. Muñoz, and I. Moeck (2012), Pattern recognition and lithological interpretation of collocated seismic and magnetotelluric models using self-organizing maps. *Geophys. J. Int.*, *189*, 984–998.

Bedrosian, P. A. (2007), MT+, integrating magnetotellurics to determine earth structure, physical state, and process. *Surv. Geophys.*, *28*, 121–167.

Bedrosian, P. A., N. Maercklin, U. Weckmann, Y. Bartov, T. Ryberg, and O. Ritter (2007), Lithology-derived structure classification from the joint interpretation of magnetotelluric and seismic models, *Geophys. J. Int.*, *170*, 737–748.

Bishop, C. M. (2006), *Pattern Recognition and Machine Learning*, Springer, New York.

Brauchler, R., J. Doetsch, P. Dietrich, and M. Sauter (2012), Derivation of site-specific relationships between hydraulic parameters and P-wave velocities based on hydraulic and seismic tomography. *Water Resources Res.*, *48*, W03531.

Butler, D. K. (2006), *Near-Surface Geophysics*, Society of Exploration Geophysicists, Tulsa, OK.

Calinski, T., Harabasz, J., 1974, A dendrite method for cluster analysis, *Commun. Stat.*, *3*, 1–27.

Canty, M. J., and A. A. Nielsen (2006), Visualization and unsupervised classification of changes in multispectral satellite imagery, *Int. J. Remote Sens.*, *27*, 3961–3975.

Carter-McAuslan, A., P. G. Lelièvre, and C. G. Farquharson (2015), A study of fuzzy *c*-means coupling for joint inversion, using seismic tomography and gravity data test scenarios, *Geophysics*, *80*, W1–W15.

Chen, J., S. Hubbard, and Y. Rubin (2001), Estimating the hydraulic conductivity at the South Oyster Site from geophysical tomographic data using Bayesian techniques based on the normal linear regression model. *Water Resources Res.*, *37*, 1603–1613.

Day-Lewis, F. D., and W. Lane (2004), Assessing the resolution-dependent utility of tomograms for geostatistics, *Geophys. Res. Lett.*, *31*, L07503, doi:10.1029/2004GL019617.

Dietrich, P., and J. Tronicke (2009), Integrated analysis and interpretation of cross-hole P- and S-wave tomograms: A case study, *Near Surface Geophys.*, *7*, 101–109.

Dietrich, P., T. Fechner, J. Whittacker, and G. Teutsch (1998), An integrated hydrogeophysical approach to subsurface characterization, in M. Herbert, and K. Kovar, eds.,

Groundwater quality: Remediation and protection. *Int. Assoc. Hydrol. Sci.*, *250*, 513–519.

Doetsch, J., N. Linde, I. Coscia, S. A. Greenhalgh, and A.G. Green (2010), Zonation of 3D aquifer characterization based on joint inversions of multimethod crosshole geophysical data, *Geophys.*, *75*, G53–G64.

Eberle, D. (1993), Geologic mapping based upon multivariate statistical analysis of airborne geophysical data. *Inst. Aerospace Survey Earth Sci. (ITC) J.*, *1993-2*, 173–178.

Fayyad, U., G. Piatetsky-Shapiro, and P. Smyth (1996), From data mining to knowledge discovery in databases, *AI Magazine*, *17*, 37–54.

Friedel, S. (2003), Resolution, stability and efficiency of resistivity tomography estimated from a generalized inverse approach, *Geophys. J. Int.*, *153*, 305–316.

Gallardo, L. (2007), Multiple cross-gradient joint inversion for geospectral imaging, *Geophys. Res. Lett.*, *34*, L19301.

Gallardo, L. A., and M. Meju (2003), Characterization of heterogeneous near-surface materials by joint 2D inversion of dc resistivity and seismic data. *Geophys. Res. Lett.*, *30*, 1658. Doi:10.1029/2003GL017370.

Gallardo, L. A., and M.A. Meju (2011), Structure-coupled multiphysics imaging in geophysical sciences, *Rev. Geophys.*, *49*, RG1003.

Gallardo, L. A., S. L. Fontes, M. A. Meju, M. P. Buonora, and P. P. de Lugao (2012), Robust geophysical integration through structure-coupled joint inversion and multispectral fusion of seismic reflection, magnetotelluric, magnetic, and gravity images: Example from Santos Basin, offshore Brazil, *Geophysics*, *77*, B237–B251.

Gath, I., and A. B. Geva (1989), Unsupervised optimal fuzzy clustering. *IEEE Trans. Pattern Analysis Machine Intell.*, *11*, 773–780.

Haberland, C., A. Rietbrock, B. Schurr, and H. Brasse (2003), Coincident anomalies of seismic attenuation and electrical resistivity beneath the southern Bolivian Altiplano plateau, *Geophys. Res. Lett.*, *30*, 1923.

Hachmöller, B., and H. Paasche (2013), Integration of surface-based tomographic models for zonation and multimodel guided extrapolation of sparsely known petrophysical parameters, *Geophysics*, *78*, EN43–EN53.

Hathaway, R. J., and J. C. Bezdek (2001), Fuzzy c-means clustering with incomplete data. *IEEE Trans. Syst. Man, Cybern., Part B*, *31*, 735–744.

Hilger, K. B. (2001), Exploratory analysis of Multivariate data, Ph.D. thesis, Technical University of Denmark.

Höppner, F., F. Klawonn, R. Kruse, and T. Runkler (1999), *Fuzzy Cluster Analysis: Methods for Classification, Data Analysis and Image Recognition*, John Wiley & Sons, New York.

Hubbard, S. S., J. Chen, J. Peterson, E. L. Majer, K. H. Williams, D. J. Swift, B. Mailloux, and Y. Rubin (2001), Hydrogeological characterization of the South Oyster bacterial transport site using geophysical data, *Water Resources res.*, *37*, 2431–2456.

IAEA-TECDOC-1363 (2003), *Guidelines for Radioelement Mapping Using Gamma Ray Spectrometry Data*, IAEA, Vienna.

Infante, V., L. A. Gallardo, J. C. Montalvo-Arrieta, and I. Navarro de Leon (2010), Lithological classification assisted by the joint inversion of electrical and seismic data at a control site in northeast Mexico, *J. appl. Geophys.*, *70*, 93–102.

Jousset, P., C. Haberland, K. Bauer, and K. Arnason (2011), Hengill geothermal volcanic complex (Iceland) characterized by integrated geophysical observations, *Geothermics*, *40*, 1–24.

Kohonen, T. (2001), *Self-Organizing Maps*, Springer Series in Information Sciences, Vol. *30*, Springer, New York.

Linde, N., A. Binley, A. Tryggvason, L. B. Pedersen, and A. Revil (2006), Improved hydrogeophysical characterization using joint inversion of cross-hole electrical resistance and ground-penetrating radar traveltime data,*Water Resources Res.*, *42*, W12404.

Little, R. J. A., D. B. Rubin (1987), *Statistical Analysis with Missing Data*, John Wiley & Sons, New York.

Martelet, G., C. Truffert, B. Tourlière, P. Ledru, and J. Perrin (2006), Classifying airborne radiometry data with agglomerative hierarchical clustering: A tool for geological mapping in the context of rainforest (French Guiana), *Int. J. App. Earth Obs. Geoin.*, *8*, 208–223.

Moorkamp, M., A. W. Roberts, M. Jegen, B. Heincke, and R. W. Hobbs (2013), Verification of velocity–resistivity relationships derived from structural joint inversion with borehole data, *Geophys. Res. Lett.*, *40*, 3596–3601.

Muñoz, G., K. Bauer, I. Moeck, A. Schulze, and O. Ritter (2010), Exploring the Groß Schönebeck (Germany) geothermal site using a statistical joint interpretation of magnetotelluric and seismic tomography models, *Geothermics*, *39*, 35–45.

Nixon, M. S., and A. S. Aguado (2012), *Feature Extraction & Image Processing for Computer Vision*, Academic Press, New York.

Paasche, H., and J. Tronicke (2007), Cooperative inversion of 2D geophysical data sets: A zonal approach based on fuzzy *c*-means cluster analysis. *Geophysics*, *72*, A35–A39.

Paasche, H., and J. Tronicke (2014), Non-linear joint inversion of tomographic data using swarm intelligence,*Geophysics*, *79*, R133–R149.

Paasche, H., J. Tronicke, K. Holliger, A. G. Green, and H. R. Maurer (2006), Integration of diverse physical-property models: Subsurface zonation and petrophysical parameter estimation based on fuzzy *c*-means cluster analyses, *Geophysics*, *71*, H33–H44.

Paasche, H., J. Tronicke, and P. Dietrich (2010), Automated integration of partially colocated models: Subsurface zonation using a modified fuzzy *c*-means cluster analysis algorithm, *Geophysics*, *75*, P11–P22.

Poulton, M. M. (2002), Neural networks as an intelligence amplification tool: A review of applications, *Geophysics*, *67*, 979–993.

Reyment, R. A., and E. Savazzi (1999), *Aspects of Multivariate Statistical Analysis in Geology*, Elsevier, Amsterdam.

Schafer, J. L. (1997), *Analysis of Incomplete Multivariate Data*, Chapman and Hall, London.

Schneider, W. A., K. A. Ranzinger, A. H. Balch, and C. Kruse (1992), A dynamic programming approach to first arrival traveltime computation in media with arbitrarily distributed velocities, *Geophysics*, *57*, 39–50.

Shekar, S., P. Zhang, Y. Huang, and R. Vatsavai (2004), Trends in spatial data mining, in H. Kargupta, A. Joshi,

K. Sivakumar, and Y. Yesha, eds., *Data Mining: Next Generation Challenges and Future Directions*, American Association for Artificial Intelligence, Cambridge, MA.

Späth, H. (1985), *Cluster Dissection and Analysis*, Horwood, Little Rock, AR.

Stankiewicz, J., K. Bauer, and T. Ryberg (2010), Lithology classification from seismic tomography: Additional constraints from surface waves, *J. African Earth Sci.*, *58*, 547–552.

Stankiewicz, J., G. Muñoz, O. Ritter, P. A. Bedrosian, T. Ryberg, U. Weckmann, and M. Weber (2011), Shallow lithological structure across the Dead Sea transform derived from geophysical experiments, *Geochem. Geophys. Geosyst.*, *12*, Q07019.

Tronicke, J., and K. Holliger (2005), Quantitative integration of hydrogeophysical data: Conditional geostatistical simulation for characterizing heterogeneous alluvial aquifers, *Geophysics*, *70*, H1–H10.

Tronicke, J., P. Dietrich, U. Wahlig, and E. Appel (2002), Integrating surface georadar and crosshole radar tomography: A validation experiment in braided stream deposits, *Geophysics*, *67*, 1516–1523.

Tronicke, J., K. Holliger, W. Barrash, and M. D. Knoll (2004), Multivariate analysis of crosshole georadar velocity and attenuation tomograms for aquifer zonation, *Water Resources Res.*, *40*, 159–178.

Tronicke, J., H. Paasche, and U. Böniger (2012), Crosshole traveltime tomography using particle swarm optimization: A near-surface field example, *Geophysics*, *77*, R19–R32.

van Leekwijck, W., and E. E. Kerre (1999), Defuzzification: Criteria and classification, *Fuzzy Sets Syst.*, *108*, 159–178.

Vincent, L., and P. Soille (1991), Watersheds in digital spaces: An efficient algorithm based on immersion simulation, *IEEE Trans. Pattern Analysis Machine Intell.*, *13*, 583–598.

von Luxburg, U. (2007), A tutorial on spectral clustering, *Statis. Comput.*, *17*, 395–416.

West, D. (2007), *Introduction to Graph Theory*, Prentice Hall, Upper Saddle River, NJ.

Zhang, H., C. Thurber, and P. Bedrosian (2009), Joint inversion for Vp, Vs, and Vp/Vs at SAFOD, Parkfield, California, *Geochem. Geophys. Geosyst.*, *10*, Q11002.

6

Probabilistic Integration of Geo-Information

Thomas Mejer Hansen, Knud Skou Cordua, Andrea Zunino, and Klaus Mosegaard

ABSTRACT

The problem of inferring information about the Earth can be described as a data integration problem, where the solution is a probability distribution that combines all available information. The theory is conceptually simple, but application in practice can be challenging. Probabilistic data integration requires that the information at hand can be quantified in the form of a probability distribution, either (a) directly through specification of an analytical description of a probability distribution or (b) indirectly through algorithms that can sample an often unknown probability distribution. Once all information has been quantified, efficient numerical algorithms are needed for inferring information from the combined probability distribution. In the following, methods for probabilistic characterization of different kinds of geo-information are presented. Then a number of methods that allow inferring information from the probability distribution that combines all available information will be discussed. Straight forward application of classic sampling algorithms such as the rejection sampler and the Metropolis algorithm will in most cases lead to computationally intractable problems. However, a number of methods exist that can turn an otherwise intractable data integration problem into a manageable one.

6.1. INTRODUCTION

A fundamental problem in Earth sciences is how to combine available information about the Earth (geo-information) into one consistent model of the subsurface. One difficulty is that the available information is of very different nature. Examples of geo-information are geophysical measurements, well logs, remote sensing, knowledge about geological processes, and so on. One commonly used approach to solve this problem is to make use of inverse problem theory. An inverse problem is typically defined as a problem where information about unknown parameters of a physical system are inferred from indirect physical measurements (see, e.g., *Tarantola* [2005]; *Mosegaard and Hansen* [2015]). The term "inverse problem" implies that one seeks to invert a process. For example, in a forward process some physical response from the Earth is measured in the form of some data. In the inverse process, an Earth model (or a collection of Earth models) explaining the data is sought. Alternatively, inferring information about the Earth can be considered as a problem of integration of information, where information from indirect information may, or may not, be available.

Let $I_1, I_2, \ldots, I_N$ represent N different types of sources of information available about the Earth. Let the Earth be described by a set of M model parameters $\mathbf{m} = [m_1, m_2, \ldots, m_M]$. In a probabilistic formulation, the information about $\mathbf{m}$, from a specific type of information I_i, can be quantified by a probability distribution $f(\mathbf{m} \mid I_i)$, and hence N probability distributions $f(\mathbf{m} \mid I_1), f(\mathbf{m} \mid I_2), \ldots, f(\mathbf{m} \mid I_N)$ describe all the information available about $\mathbf{m}$.

If the information is independent—that is, if $f(\mathbf{I}) = f(I_1, I_2, \ldots, I_N) = f(I_1)\, f(I_2), \ldots, f(I_N)$—then $f(\mathbf{m}|I_1)$,

Solid Earth Physics, Niels Bohr Institute, University of Copenhagen, Copenhagen, Denmark

Integrated Imaging of the Earth: Theory and Applications, Geophysical Monograph 218, First Edition.
Edited by Max Moorkamp, Peter G. Lelièvre, Niklas Linde, and Amir Khan.

$f(\mathbf{m},|I_2,)$, ..., $f(\mathbf{m}|I_N)$ can be considered statistically independent, and then the combined information from all sources of information is given by the probability distribution

$$\begin{aligned} f(\mathbf{m} \mid \mathbf{I}) &= f(\mathbf{m} \mid I_1, I_2, \ldots, I_N) \\ &= f(\mathbf{m} \mid I_1) f(\mathbf{m} \mid I_2) \ldots f(\mathbf{m} \mid I_N) \\ &= \prod_{i=1}^{N} f(\mathbf{m} \mid I_i). \end{aligned} \quad (6.1)$$

The probabilistic formulation in Eq. (6.1) is similar to the concept of "conjunction of states of information" proposed as an approach to solve inverse problems [*Tarantola and Valette*, 1982; *Tarantola*, 2005]. $f(\mathbf{m}|I_i)$ represents one state of information. *Tarantola*, [2005] considers the conjunction of two states of information: the *a priori* probability and the theoretical probability density (given by a *likelihood* function). The conjunction of these two states of information is referred to as the a posteriori probability distribution. Here we intentionally avoid using the terms a priori and *likelihood* for several reasons. First, we argue that the two states of information (the prior and likelihood) simply represent different types of information about the model parameters, as described by, in this case, two probability distributions $f(\mathbf{m}|I_1)$ and $f(\mathbf{m}|I_2)$, respectively. Second, traditionally most focus in inverse problems has been on the information available from indirect information, related to geophysical data, while the use of a priori information has historically been debated. Some support the argument that the a priori probability distribution should be chosen as noninformative as possible [*Scales and Sneider*, 1997; *Buland and Omre,* 2003]. Others have argued that direct information about the model parameters may be of more value than geophysical data [*Journel*, 1994]. Here, in line with *Jaynes* [1984], we suggest to make use of whatever information, $f(\mathbf{m}|I_i)$, that is available (be it more or less informative) about the model parameters. The importance of each type of information is independent of the source of the information and is solely quantified by $f(\mathbf{m}|I_i)$.

Probabilistic data integration using Eq. (6.1) is conceptually very simple, namely an application of statistical independence. In practice, however, inferring information about $f(\mathbf{m}|\mathbf{I})$ may not be trivial. First, the information available has to be quantified probabilistically. This can be either in the form of an analytical description of $f(\mathbf{m}|I_i)$ (e.g., a normal distribution) or in the form of an algorithm that samples $f(\mathbf{m}|I_i)$ (e.g., a pruned partially ordered Markov mixture model as sampled by the SNESIM algorithm [*Strebelle*, 2002; *Cordua et al.*, 2015]). Next, it may be a challenge to computationally efficiently infer information from $f(\mathbf{m}|\mathbf{I})$, even in cases where a mathematical expression for $f(\mathbf{m}|\mathbf{I})$ exist. The computational complexity is highly linked to the method that is applied for inferring such information.

In the following, we will discuss methods and algorithms that allow probabilistic integration of geo-information, as given in Eq. (6.1), such that inference from $f(\mathbf{m}|\mathbf{I})$ is possible.

First, methods for quantifying different types of geo-information (information about the Earth) through probability distributions will be reviewed. Geo-information differs in the form in which it is available and we argue that it can, crudely, be divided into two categories: "direct" and "indirect" information about **m**. Direct information allows characterizing the model parameters directly, which can be done using a variety of methods based on, for example, geostatistics, Markov models, and parsimonious model assumptions. We show examples on how to infer direct information from a sample model, related to a variety of types of statistical models (Section 6.2). We also recall how indirect information (e.g., where geophysical data provides information about some property related to the model parameters) can be quantified by data, measuring uncertainty and modeling errors (Section 6.3).

Then we discuss and compare a number of widely used sampling methods for inference of information from the probability distribution representing the combined information $f(\mathbf{m}|\mathbf{I})$. Specifically, we discuss how the numerical efficiency of such methods is strongly related to the type and amount of information available (Sections 6.4 and 6.5), and demonstrate this in a case study (Section 6.6). Finally, we discuss how the entropy related to different types of information affect the complexity of the data integration problem (Section 6.7).

Any knowledge, direct or indirect, about the model parameters **m** is conditional to a specific type of information I_i. Hence, the use of the notation $f(\mathbf{m}|I_i)$. However, for brevity, we will occasionally make use of the shorter notation $f_{I_i}(\mathbf{m}) = f(\mathbf{m} \mid I_i)$, and $f_{\mathbf{I}}(\mathbf{m}) = f(\mathbf{m}|\mathbf{I})$ in parts of the remainder of the text.

6.2. QUANTIFYING DIRECT GEO-INFORMATION USING PROBABILITY DISTRIBUTIONS

Working with geo-data, model parameters **m** typically describe an earth model, where each model parameter m_i refers to a physical property, or geological unit, of a point or volume located somewhere in a three-dimensional space. When information about the model parameters **m** is available, it will be referred to as "direct" information about the model parameters (as opposed to indirect information that provide information related to the model parameters through some function $g(\mathbf{m})$). This type of information must be quantified through the probability distribution $f(\mathbf{m}|I_{direct})$.

Direct information can, for example, refer to knowledge about the value a model parameter can take. Physical laws may impose restrictions on the values that specific types of model parameters can attain. For example, a velocity cannot be negative and cannot exceed the speed of light. Other types of direct information are rooted in geological knowledge. For example, knowledge about how the Earth has evolved in time can lead to some information about what type of structures that can and cannot be expected in the Earth. It can also give rise to information about what kind of geology that can be expected and, hence, information about **m**. Such information can be rooted in both observations, theoretical studies, and numerical simulation. Sometimes, a "sample model" may be available. A sample model is an example of (perhaps a part of) a realization from the unknown probability distribution $f(\mathbf{m}|I_{direct})$. This is the case when, for example, outcrops available at one location can be considered representative at another location; that is, the same probability distribution is expected to represent the same subsurface variability at the location of the sample model and at the unknown location. See, for example, *Holliger and Levander* [1994] for an example of using a sample model.

In the following a wide variety of types of probability distributions will be considered that allow characterizing $f(\mathbf{m}|I_{direct})$. They differ in the type of assumptions that is made regarding the statistical properties of $f(\mathbf{m}|I_{direct})$. Each choice of type of probability distribution requires a specific set of statistical properties in order to define the probability distribution (such as, for example, the mean and covariance for a Gaussian probability distribution). Also, different methods exist for generating realizations for each type of probability distribution.

Whether one makes use of a simple, high-entropy probability distribution, or a more complex, low-entropy type of probability distribution, the workflow of quantifying the available information is the same: (1) Select a type of probability distribution and (2) infer the properties, for example, from a sample model that defines this probability distribution. In that sense the only difference between probability distributions based on one-point, two-point, and multiple-point statistics is related to what type of statistics is taken into account.

6.2.1. Quantifying $f(\mathbf{m} \mid I_{direct})$

In the following, a number of methods for characterization of direct information, $f(\mathbf{m}|I_{direct})$, will be given, both in case an analytical expression of $f(\mathbf{m}|I_{direct})$ is assumed and in case it is unknown, but numerical algorithms exist that allow sampling from $f(\mathbf{m}|I_{direct})$.

6.2.1.1. Probability Distributions Based on One-Point Statistics

If we assume that the information for each individual model parameters m_i is independent on other model parameters, then

$$\begin{aligned} f_I(\mathbf{m}) &= f_I(m_1, m_2, \ldots, m_M) \\ &= f_I(m_1) f_I(m_2) \ldots f_I(m_M) \\ &= \prod_i^M f_I(m_i). \end{aligned} \tag{6.2}$$

When **m** describes an Earth model, where each model parameter is related to a location in space, we say that the model parameters are spatially uncorrelated.

Uniform Distribution. The most simple model for direct characterization of **m** is the uncorrelated uniform model. Assuming that all model parameters are independent and are uniformly distributed between m_{min} and m_{max}, then $f_I(\mathbf{m})$ can be described using Eq. (6.2), where each 1D marginal distribution is given by

$$f_{I,\mathcal{U}}(m_i) = \begin{cases} \dfrac{1}{m_{max} - m_{min}} & \text{for } m_{min} \le m_i \le m_{max} \\ 0 & \text{else.} \end{cases} \tag{6.3}$$

The spatially uncorrelated uniform distribution is the distribution that provides least information and maximum entropy (i.e., maximum disorder) given only an upper and lower limit (which in the limit may tend to $-\infty$ to ∞) [*Shannon*, 1948]. If one wants to assume as little as possible about **m**, then the spatially uncorrelated uniform distribution $f_{I,\mathcal{U}}(m_i)$ is often suggested [*Scales and Sneider*, 1997; *Sambridge and Mosegaard*, 2002].

Univariate Normal Distribution. Another type of maximum entropy model (given a mean and a variance) is the uncorrelated Gaussian model. $f_I(\mathbf{m})$ is then described by Eq. (6.2), where each 1D marginal distribution is given by

$$f_{I,\mathcal{N}}(m_i) = \left(\sigma\sqrt{2\pi}\right)^{-1} \exp\left(-\frac{1}{2}\frac{(m_i - \mu)^2}{\sigma^2}\right), \tag{6.4}$$

where μ and σ represent the mean and the standard deviation of the univariate normal distribution.

The assumption of spatial independence may be convenient in that using any of $f_{I,\mathcal{U}}(m_i)$ or $f_{I,\mathcal{N}}(m_i)$ leads to a probability distribution for which the probability distribution value can be easily evaluated. However, such simple, spatially uncorrelated model parameters may not allow realistic characterization of actual available information. The assumption of spatial independence implies that two model parameters located infinitely close together is assumed to be independent—an assumption that in general may not be consistent with most natural phenomena, as these may display highly correlated features. Fortunately, a number of probability distributions and methods exist that allow describing spatially dependent model parameters, along with characterization of more geologically realistic structures.

6.2.1.2. Probability Distributions Based on Two-Point (Gaussian) Statistics

In the special case where $f(\mathrm{m}|I_{direct})$ can be described fully by the mean and covariance between pairs of model parameters m_i and m_j and where m_i is normally distributed, then $f_I(\mathbf{m})$ is a Gaussian probability distribution with mean m_0 and covariance $\mathbf{C}_\mathrm{m}$ ($\mathcal{N}(\mathrm{m}_0, \mathbf{C}_\mathrm{m})$), which is given analytically by

$$f_I(\mathbf{m} \mid \mathbf{m}_0, \mathbf{C}_\mathrm{m}) = \left((2\pi)^M |\mathbf{C}_\mathrm{m}|\right)^{-.5} \times \exp\left(-\frac{1}{2}(\mathbf{m}-\mathbf{m}_0)^\top \mathbf{C}_\mathrm{m}^{-1}(\mathbf{m}-\mathbf{m}_0)\right). \quad (6.5)$$

The Gaussian description of $f_I(\mathbf{m})$ given in Eq. (6.5) is mathematically convenient. However, the Gaussian probability distribution is also the probability distribution with maximum entropy of all probability distributions with a given mean $\mathbf{m}_0$ and covariance $\mathbf{C}_\mathrm{m}$. The multivariate Gaussian distribution maximizes spatial disorder such that the Gaussian choice of probability distribution is not able to describe more structured features such as, for example, channel structures [*Journel and Deutsch*, 1993].

A rather rich family of probability distributions, reflecting quite different spatial structures, can be obtained from simple operations on realizations of a Gaussian probability distribution [*Emery*, 2007; *Armstrong et al.*, 2011]. In addition, numerical algorithms exist that, based on Gaussian statistics, can generate realizations that expose non-Gaussian spatial features, such as indicator simulation [*Journel and Isaaks*, 1984] and direct sequential simulation [*Soares*, 2001]. Note that in these cases no analytical description of the underlying probability distribution $f(\mathbf{m}|I_{direct})$ may exist, but numerical methods exist that allow sampling from $f(\mathbf{m}|I_{direct})$.

6.2.1.3. Probability Distributions Based on Multiple-Point Statistics

An alternative to using the Gaussian framework is to consider a probability distribution over **m** based on statistics that describes the (co)relation between more than two model parameters at a time. This is known as probability distributions based on multiple-point statistics. In this case, $f(\mathbf{m}|I_{direct})$ cannot simply be described by the statistical variation between pairs of model parameters (as given in covariance-based probability distributions described above). Instead, the variation between multiple model parameters needs to be quantified. Usually, no parametric description exists to quantify such distributions. Instead, nonparametric distributions based on multiple-point statistics are obtained from sample models. For examples of methods that utilize multiple-point statistics see *Guardiano and Srivastava* [1993], *Tjelmeland and Besag* [1998], *Strebelle* [2002], *Mariethoz et al.* [2010], *Dimitrakopoulos et al.* [2010], *Lange et al.* [2012], *Mariethoz and Caers* [2014], and *Cordua et al.* [2015]

When $f_I(\mathbf{m})$ is based on multiple-point statistics, it is often a type of partially ordered Markov model (POMM) [*Cressie and Davidson*, 1998; *Cordua et al.*, 2015]:

$$f_{I,POMM}(\mathbf{m}) = \prod_{i=1}^{M} p(m_i \mid pa(m_i)). \quad (6.6)$$

$p(m_i|pa(m_i))$ is the conditional probability of m_i given the so-called parents $pa(m_i)$ of the model parameter m_i, which are the model parameters that m_i is conditional dependent on.

In practice, a realization of $f_{I,POMM}(\mathbf{m})$, Eq. (6.6), can be generated by sequentially visiting all model parameters (optionally in random order), while at each step generating a realization of $p(m_i|pa(m_i))$. Note, however, that in practice, $f_{I,POMM}(\mathbf{m})$ will change for different choices of simulation path [*Cordua et al.*, 2015]. Moreover, the distribution is also dependent on the individual outcome realizations because the algorithm that samples from this distribution will prune the number of parents [*Strebelle*, 2002]. If the random path used for simulation from a POMM Is chosen from a uniform distribution, the probability distribution being sampled is a so-called pruned mixture model (PMM) of partially ordered Markov models [*Cordua et al.*, 2015; *Daly*, 2005]:

$$f_{I,PMM} = \sum_{path} w^{path} f_{I,POMM}^{path}(\mathbf{m}), \quad (6.7)$$

where the weights are given as $w_{path} = \frac{1}{M!}$ for all paths and the sum is taken over all possible simulation paths ($M!$). $f_{I,PMM}(\mathbf{m})$ in Eq. (6.7) is, however, computationally intractable to obtain because the individual partial ordered Markov models depend on the pruning of the algorithm. This demands that the pruning related to all possible outcomes for all possible simulation paths have to be known in order to obtain an actual explicit mathematical expression of this probability distribution [*Cordua et al.*, 2015].

6.2.1.4. Parsimonious/Trans-Dimensional Models

For the probability distributions considered previously, it has been assumed that the parameterization of **m** (i.e., the location and density of model parameters) has been chosen densely enough, as part of parameterization, to allow a realistic representation of spatial features [*Mosegaard and Hansen*, 2015].

However, one can choose to treat the number of model parameters as an unknown model parameter itself, which is referred to as using a trans-dimensional or parsimonious parameterization [*Constable et al.*, 1987; *Malinverno*, 2002; *Bodin et al.*, 2009].

Malinverno [2002] suggested a type of Monte Carlo-based inversion that allows, in their presented 1D case, the number of subsurface layers to vary. *Bodin et al.* [2009] explored this further and suggested a "self-parameterizing partition model" (trans-dimensional) approach that allows defining the subsurface using a number of basis functions, in this case exemplified using a number of Voronoi cells. In both studies, the number of layers/cells control the complexity of the subsurface. And, in both cases, algorithms are presented that allow randomly perturbing a subsurface model to update the number, location, and value of layers/cells.

A general formulation of a transdimensional description of the model parameters space, in form of N_b basis functions can be given by (*Bodin et al.* [2009])

$$\mathbf{m} = \sum_{i=1}^{N_b} a_i B_i(\mathbf{x}), \tag{6.8}$$

where B_i is a specific choice of kernel function, $\mathbf{x}$ is a location in space, and a_i is its associated amplitude.

In the case where the basis function defines 2D Voronoi cells, then $\mathbf{m}$ can be completely characterized by the $3N_b$ parameters, N_b values of each Voronoi cell, $\mathbf{v}_c$, as well as N_b values for the x- and y-location for the center of each Voronoi cell, $\mathbf{x}_c$ and $\mathbf{y}_c$. A statistical model over these parameters can be given by $f_{I,V}(N_b, \mathbf{v}_c, \mathbf{x}_c, \mathbf{y}_c)$. For each realization of $f_{I,V}$, the value of any corresponding model parameter, m_i, regardless of the sampling density of the model parameters, can then be computed using Eq. (6.8).

Here we will simply consider the trans-dimensional model as a specific type of information about $\mathbf{m}$, for which no explicit description of $f(\mathbf{m}|I)$ may be given, but where algorithms exist to allow sampling $f(\mathbf{m}|I)$.

Note that in practice one will almost always implicitly make use of basis functions as part of parameterizing the model parameters. For example, when illustrating a set of model parameters, parameterized over a 2D grid, one tends to show this as an image of pixels, where each pixel reflect the value of one model parameter. This implies that each model parameter is assumed to reflect an average value within an area (as spanned by the pixel size) and not the value of a point. For a more detailed discussion on the implicit use of basis functions as part of parameterizing inverse problem, see *Mosegaard and Hansen* [2015].

6.2.2. Sampling from $f(\mathbf{m} \mid I_{direct})$

Many different methods exist that allow sampling (i.e., generating a sample of Earth models) from $f(\mathbf{m}|I_{direct})$. Here we pay special attention to sampling methods based on *sequential simulation*, as we shall later exploit some features of the sequential simulation approach that allow efficient sampling from $f(\mathbf{m} \mid I_1, I_2) \propto f(\mathbf{m} \mid I_1) f(\mathbf{m} \mid I_2)$ when either $f(\mathbf{m}|I_1)$ or $f(\mathbf{m}|I_2)$ can be sampled using sequential simulation.

6.2.2.1. Sequential Simulation

Sequential simulation is a method that can be used to generate a realization of a joint probability distribution $f(\mathbf{m}) = f(m_1, m_2, \ldots, m_M)$ in the case where the *conditional* distribution

$$f(m_i \mid \mathbf{m}_c) = f(m_i \mid m_1, m_2, \ldots, m_{i-1}) \tag{6.9}$$

can be evaluated for all sets of conditional model parameters $\mathbf{m}_c$. It is based on the product rule

$$f(\mathbf{m}) = f(m_1) f(m_2 \mid m_1) \prod_{k=3}^{M} f(m_k \mid m_1, m_2, \ldots, m_{k-1}). \tag{6.10}$$

A realization of $f(\mathbf{m})$ can be generated as $\mathbf{m}^*$ using the sequential simulation algorithm as follows:

SEQUENTIAL SIMULATION

Visit model parameter 1, m_1. Generate a realization m_1^* of $f(m_1)$.

Visit model parameter 2, m_2. Generate a realization m_2^* of $f(m_2 \mid m_1^*)$,

Visit model parameter 3, m_3. Generate a realization m_3^* of $f(m_3 \mid m_1^*, m_2^*)$.

⋮

Visit model parameter M, m_M. Generate a realization m_M^* of $f(m_M \mid m_1^*, m_2^*, \ldots, m_{M-1}^*)$.

Then $\mathbf{m}^* = [m_1^*, m_2^*, \ldots, m_M^*]$ will be a realization of $f(\mathbf{m})$. The model parameters can be visited in any order, as long as all model parameters are eventually visited [*Gomez-Hernandez and Journel*, 1993].

At each step in the sequential simulation algorithm, one will typically compute the conditional distribution, $f(m_i|\mathbf{m}_c) = f(m_i|m_1, m_2, \ldots, m_{i-1})$, and then draw a realization of this distribution. Note, though, that in order to use sequential simulation, $f(m_i|\mathbf{m}_c)$ does not need to be explicitly computed. It is sufficient that a realization from $f(m_i|\mathbf{m}_c)$ can be generated.

In most practical applications of sequential simulation, it can be computationally difficult or impossible to describe the full conditional distribution, Eq. (6.9). Instead, one can retain only a limited number of conditional model parameters—for example, based on proximity to the model parameter being simulated. This is referred to as using a "neighborhood" where the size of the neighborhood reflects the number of conditional model parameters. Such an application of sequential simulation will not sample the joint distribution exactly, but instead an approximation of it that can be described by a partially ordered Markov model (see *Cressie and Davidson* [1998]; *Cordua et al.* [2015]).

An early example of what can be seen as an application of sequential simulation using a neighborhood, is presented by *Shannon* [1948]. Here, sequential simulation is applied in order to simulate a sequence of English text character by character (based on a nonparametric probability distribution describing the occurrence of sets of characters inferred form on an English textbook used as a sample model). For each new character location a new character is simulated by generating a realization from a conditional distribution, conditional to a fixed number of preceding characters. The full conditional distribution is not computed. Instead, the first match, from a random starting point in the book (sample model) to the conditioning data, is chosen as a realization of the conditional distribution. This is equivalent to inferring the full conditional distribution from the sample model, followed by drawing a realization from the conditional distribution. More detailed descriptions of the theory and application of sequential simulation developed in geostatistical community can be found in, for example, *Gomez-Hernandez and Journel* [1993] and *Deutsch and Journel* [1998].

6.2.2.2. Sequential Simulation of Probability Distributions Based on Two-Point (Gaussian) Statistics

If $f(\mathbf{m})$ is distributed according to a multivariate Gaussian model [see Eq. (6.5)], $f(\mathbf{m}) \sim \mathcal{N}(\mathbf{m}_0, \mathbf{C}_\mathrm{m})$, then the conditional distribution, $f(m_i|\mathbf{m}_c)$, will be a 1D Gaussian distribution

$$f\left(m_i \mid m_1, m_2, \ldots, m_{i-1}\right) \sim \mathcal{N}\left(\mathbf{m}_0^*, \sigma^{2*}\right). \qquad (6.11)$$

The mean and the variance can be found by solving a simlpe kriging system, *Journel and Huijbregts* [1978], or equivalently by solving linear least squares system, *Hansen and Mosegaard* [2008]. Sequential simulation, based on Eq. (6.11) is also known as sequential Gaussian simulation, a widely used two-point statistical simulation algorithm, *Deutsch and Journel* [1998]; *Remy et al.* [2008].

Other variants of sequential Gaussian simulation are direct sequential simulation [*Soares*, 2001; *Oz et al.*, 2003; *Hansen and Mosegaard*, 2008], sequential indicator simulation [*Caers*, 2000] and plurigaussian simulation [*Armstrong et al.*, 2011].

6.2.2.3. Sequential Simulation of Probability Distributions Based on Multiple-Point Statistics

Sequential simulation from a probability distribution based on multiple-point statistics such as the pruned mixture model based on partially ordered Markov models, Eq. (6.7) can be obtained through sequential simulation. The conditional distribution $f(m_i|\mathbf{m}_c)$ needed for sequential simulation is the term $p(m_i|pa(m_i))$ in Eq. (6.6). As noted previously, when $f(\mathbf{m})$ is based on multiple-point statistics, a parametric analytical description of both $f(\mathbf{m})$ and the conditional distribution $f(m_i|\mathbf{m}_c)$ are typically not provided and a nonparametric description of the joint distribution is computationally intractable to obtain [*Cordua et al.*, 2015].

However a nonparametric formulation of the individual conditional distributions, $f(m_i|\mathbf{m}_c)$, needed for sequential simulation can be obtained directly from a sample model, most often in the form of a training image. A training image is a specific type of sample model (which in 2D is given by an image of pixels and in 3D by a cube of voxels) that represents realistic spatial variability. Such an image can, for example, be provided by a geological expert or from outcrops.

In the case where $f(\mathbf{m})$ represents a discrete probability distribution, *Guardiano and Srivastava* [1993] propose to scan the training image for a specific data event, as defined by the conditioning data, from which $f(m_i|\mathbf{m}_c)$ can be constructed. This is done at each step in the sequential simulation approach and, therefore, is computationally expensive.

Strebelle [2002] proposes to scan the training image only once for a large collection of data events and then store the result in a search tree. The conditional distribution $f(m_i|\mathbf{m}_c)$ can then be relatively efficiently obtained from the search tree during sequential simulation.

The direct sampling method [*Mariethoz et al.*, 2010] essentially makes use of the approach proposed by *Shannon* [1948] described above. Here, the conditional distribution $f(m_i|\mathbf{m}_c)$ is never explicitly computed. Instead, a realization of $f(m_i|\mathbf{m}_c)$ is found by scanning the training image, from a random starting location, until the first matching data event is found (or within some tolerance).

These three methods represent different ways to generate a realization from $f(m_i|\mathbf{m}_c)$, and differ mostly in computational CPU and memory requirements. For an overview of related mulitple-point based sequential simulation sampling algorithms, see, for example, *Mariethoz and Caers* [2014].

6.2.2.4. Sampling Methods not Based on Sequential Simulation

Many other types of methods, not based on sequential simulation, exist to generate realizations from probability distributions based on two-point or multiple-point statistics.

The fast Fourier transform moving average (FFT-MA) method is especially efficient for generating independent unconditional realizations of a stationary Gaussian distribution [*Le Ravalec et al.*, 2000]. Realizations can also be generated using LU decomposition of the covariance matrix. While this allows for a nonstationary covariance model, it is also computational inefficient for anything but very small models (see, e.g, *Deutsch and Journel* [1998]).

Realizations from probability distributions based on two-point or multiple-point statistics can also be obtained by locating models whose frequency distribution of patterns match the frequency distribution obtained from a sample model [*Peredo and Ortiz,* 2011; *Lange et al.*, 2012; *Cordua et al.*, 2015].

6.2.3. Quantifying $f(\mathbf{m}|I_{direct})$ from a Sample Model

A probability distribution describing direct information $f(\mathbf{m}|I_{direct})$ is almost never available directly. Instead the information may be available in form of a sample model, from which information about $f(\mathbf{m}|I_{direct})$ can be inferred.

Figure 6.1 shows an image of meandering sand channels (from *Strebelle* [2000]) that we will consider as an example of a sample model $\mathbf{m}_{sm}$. $\mathbf{m}_{sm}$ represents a 2D regular grid of electromagnetic wave velocity values, consisting of 125 × 125 cells with a cell distance of 0.15 m (the physical model is 18.75 m wide and deep). The sample model in Figure 6.1 only takes two values (0.11 m/ns and 0.13 m/ns). By assuming stationarity, the mean and standard deviation of all pixel values can be determined as $m_0 = 0.1155$ m/ns and $\sigma = 0.0089$ m/ns, respectively.

In the following, we will demonstrate how information from the sample model can be inferred and also used to characterize $f(\mathbf{m}|I_{direct})$ using the different type of probability distributions defined in the previous sections. For all considered cases, $f(\mathbf{m}|I_{direct})$ will describe a distribution over the model model parameters $\mathbf{m}$ which are spatially ordered in a 2D grid defined over 40 × 84 model parameters organized in a 2D grid, with a cell distance 0.15 m (5.85 m wide and 12.45 m deep). We shall later combine these different types of information with indirect information.

$f(\mathbf{m}|I_{d1})$, Uncorrelated Gaussian. A stationary probability distribution that describes uncorrelated Gaussian distributed model parameters is completely described by a mean and a variance, as given above. Figure 6.2a shows five realizations from such a Gaussian model where $f(m_i\,|\,I_{d1}) = \mathcal{N}(m_0,\sigma^2)$.

$f(\mathbf{m}|I_{d2})$, Uncorrelated Binary Distribution. The para meters of an uncorrelated probability distribution with a binary 1D marginal distribution can be inferred assuming stationarity, and considering the 1D marginal distribution of the training image as representative for all model parameters. This leads to $f(m_i = 0\,|\,\mathbf{m}_{sm}) = 0.72$ and $f(m_i = 1\,|\,\mathbf{m}_{sm}) = 0.28$. Figure 6.2b shows five realizations from such a skewed binary distribution, $f(\mathbf{m}\,|\,I_{d2})$.

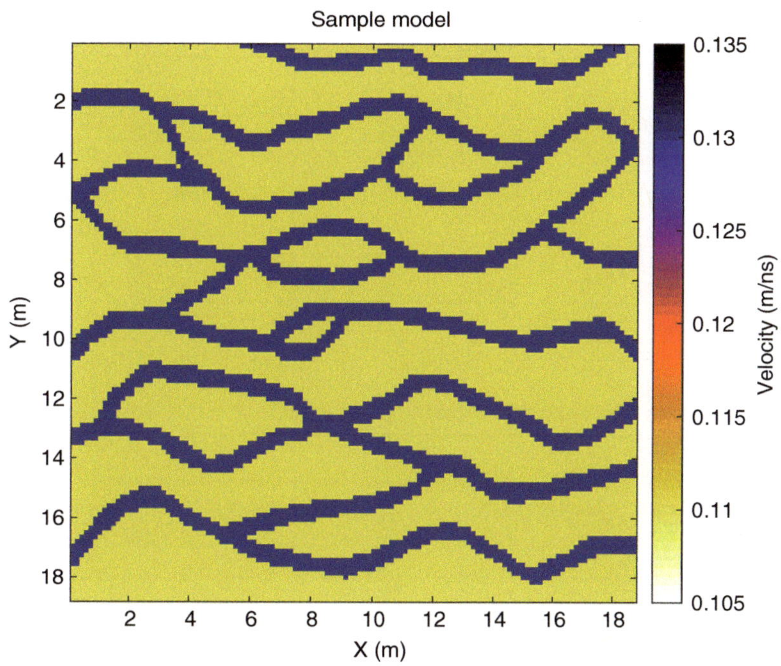

Figure 6.1 Example of a sample model, $\mathbf{m}_{sm}$. From *Strebelle* [2000].

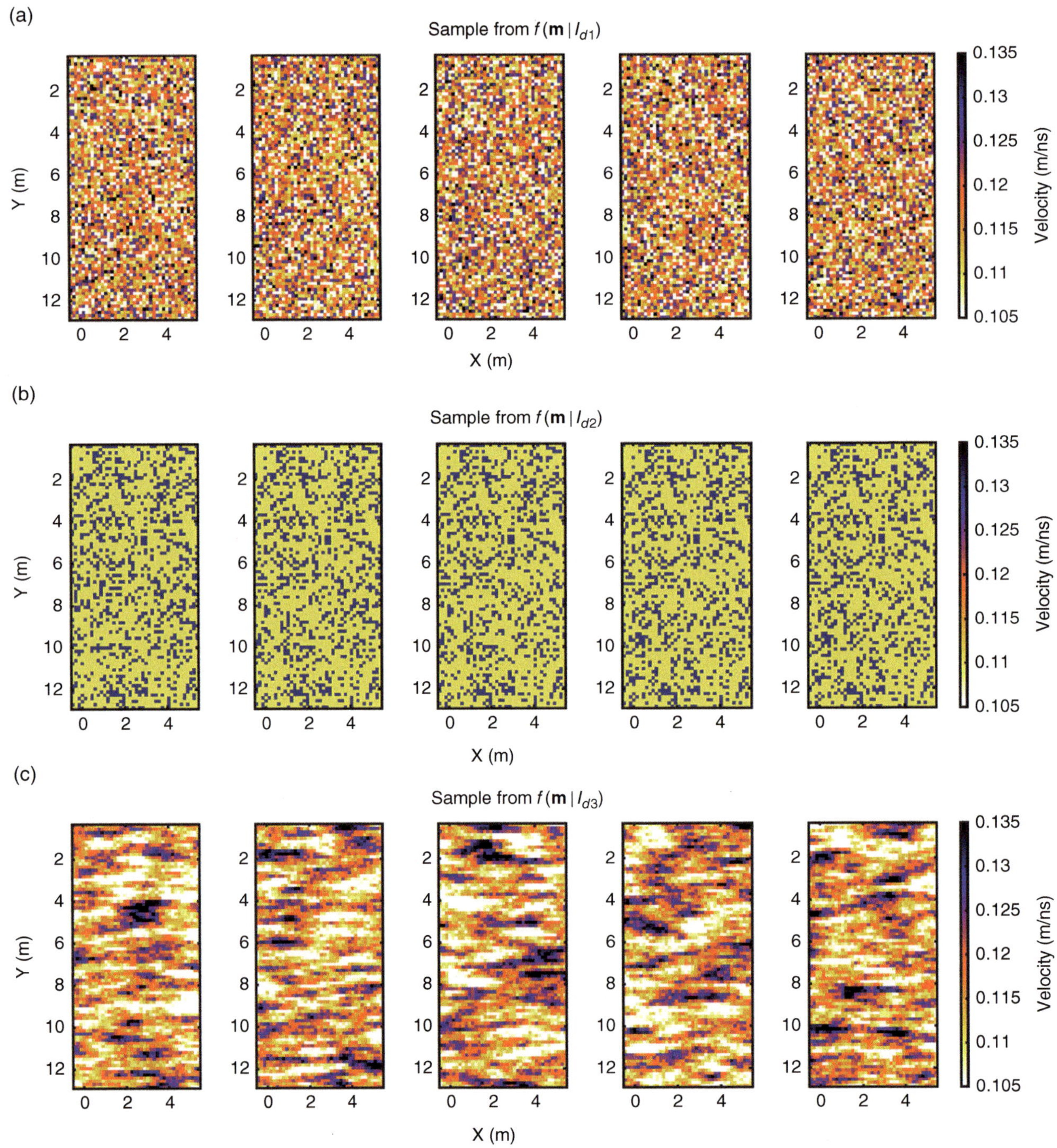

Figure 6.2 Five realizations from **(a)** $f(\mathbf{m} \mid I_{d1})$, **(b)** $f(\mathbf{m} \mid I_{d2})$, and **(c)** $f(\mathbf{m} \mid I_{d3})$. See text for details.

$f(\mathbf{m} \mid I_{d3})$, **Correlated Gaussian Distribution.** A Gaussian probability distribution is completely described by a mean vector and a covariance matrix. Both the mean and the covariance (equivalent to a semi-variogram model) can be inferred from a sample model $\mathbf{m}_{sm}$. Figure 6.3 shows the experimental semi-variogram found from the sample model along the x- and y-axis, compared to the parametric semi-variogram model used to describe the experimental sem-variogram. From this model, a covariance matrix is constructed as $\mathbf{C}_m$, such that $f(\mathbf{m} \mid I_{d3})$ can be described by the Gaussian distribution $\mathcal{N}(m_0, \mathbf{C}_m)$. Figure 6.2c shows five realizations from such a correlated Gaussian distribution.

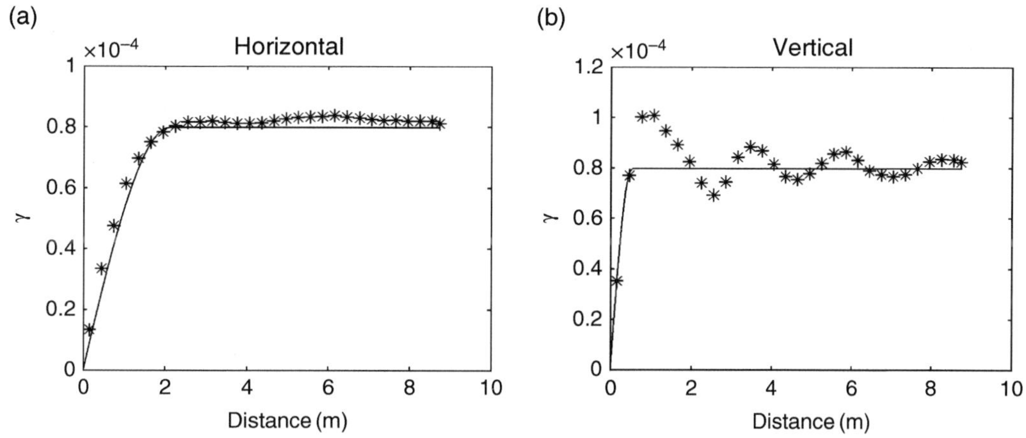

Figure 6.3 Experimental semivariogram model inferred from the sample model in Figure 6.1 (*black asterisks*) compared to the semivariogram model chosen to represent the covariance model (*solid line*) of $f(\mathbf{m} \mid I_{d3})$ along the **(a)** horizontal axis and **(b)** vertical axis.

$f(\mathbf{m} \mid I_{d4})$, **Correlated Transformed Gaussian Distribution.** A simple way to simulate spatially correlated model parameters with an arbitrary non-Gaussian 1D marginal distribution is to apply an inverse normal score transformation to a realization from a Gaussian probability distribution. In the extreme case, a binary distribution, such as that of the distribution of the values in the sample model, can be assumed. This can also be obtained by truncating realizations from a Gaussian model. Figure 6.4a shows five realizations of such a probability distribution, $f(\mathbf{m} \mid I_{d4})$, reflecting the mean, covariance, and 1D marginal distribution obtained from $\mathbf{m}_{sm}$. The 2D covariance models used to describe the Gaussian distribution in the normal score space is chosen such that the experimental semi-variogram of the back-transformed realizations, along the x- and y-axis, reflects that of the sample model, as shown in Figure 6.5.

$f(\mathbf{m} \mid I_{d5})$, **Probability Distribution Based on Multiple-Point Statistics.** Fig. 6.4b shows five realizations generated using the SNESIM algorithm with $\mathbf{m}_{sm}$ as a training image [*Remy et al.*, 2008]. In this case, part of the multiple-point statistics of $\mathbf{m}_{sm}$ is consistent with the shown realizations. Strictly speaking, the SNESIM algorithm only samples from the same probability distribution if the same path is used and in the case where the neighborhood is kept constant [*Cordua et al.*, 2015]. But we will refer to these realizations as realizations from $f(\mathbf{m} \mid I_{d5})$.

$f(\mathbf{m} \mid I_{d6})$, **Voronoi Cells.** One could choose to describe information about $\mathbf{m}$ using the parsimonious approach and 2D Voronoi cells. Figure 6.4c shows a realization from, $f(\mathbf{m} \mid I_{d6})$, where the number of 2D Voronois cells N_b is assumed uniformly distributed between 3 and 200. The x- and y-locations of the center of each cell is assumed to be located at a random location on the model parameter grid. The value of each Voronoi cell is assumed to be either 0.11 m/ns or 0.13 m/ns, with the same 1D marginal distribution as found in the sample model. Note that each realization in Figure 6.4c represents one set of parameters describing the Voronoi cells mapped into the exact same 2D 40 × 84 model parameter grid as for the other considered models of direct information. While the number of parameters that describe $f(\mathbf{m} \mid I_{d6})$ varies, the actual number of model parameters in $\mathbf{m}$ is fixed.

The probability distributions $f(\mathbf{m} \mid I_{d1}), \ldots, f(\mathbf{m} \mid I_{d5})$ are all consistent with the statistics from the sample model, in that $f(\mathbf{m}_{sm} \mid I) > 0$. In other words, part of the sample model the same size as $\mathbf{m}$ is possible as a realization of $f(\mathbf{m} \mid I_{d1}), \ldots, f(\mathbf{m} \mid I_{d5})$. This may not be the case for $f(\mathbf{m} \mid I_{d6})$. Further, $f(\mathbf{m} \mid I_{d1}) - f(\mathbf{m} \mid I_{d4})$ represent models with maximum disorder (maximum entropy) for the statistical correlations not specifically accounted for.

The main goal, when quantifying $f(\mathbf{m} \mid I_{direct})$, should be to define a probability distribution that has outcome realizations with the spatial (one-, two-, or multiple-point) statistics as obtained from the known sample model. Further, such a statistical model should represents the spatial structures as observed from an outcrop, or as known from geological expert knowledge. A simple way to validate the choice of probability distribution describing $\mathbf{m}_{sm}$ is to generate a set of independent realizations of $f(\mathbf{m} \mid I)$, as shown in Figures 6.2 and 6.4, and visually compare the realization to the sample model, Figure 6.1.

If the connectivity of the channel structures of the sample model, Figure 6.1, is an essential feature when characterizing the subsurface, then it may not be very useful to make use of the spatially uncorrelated models

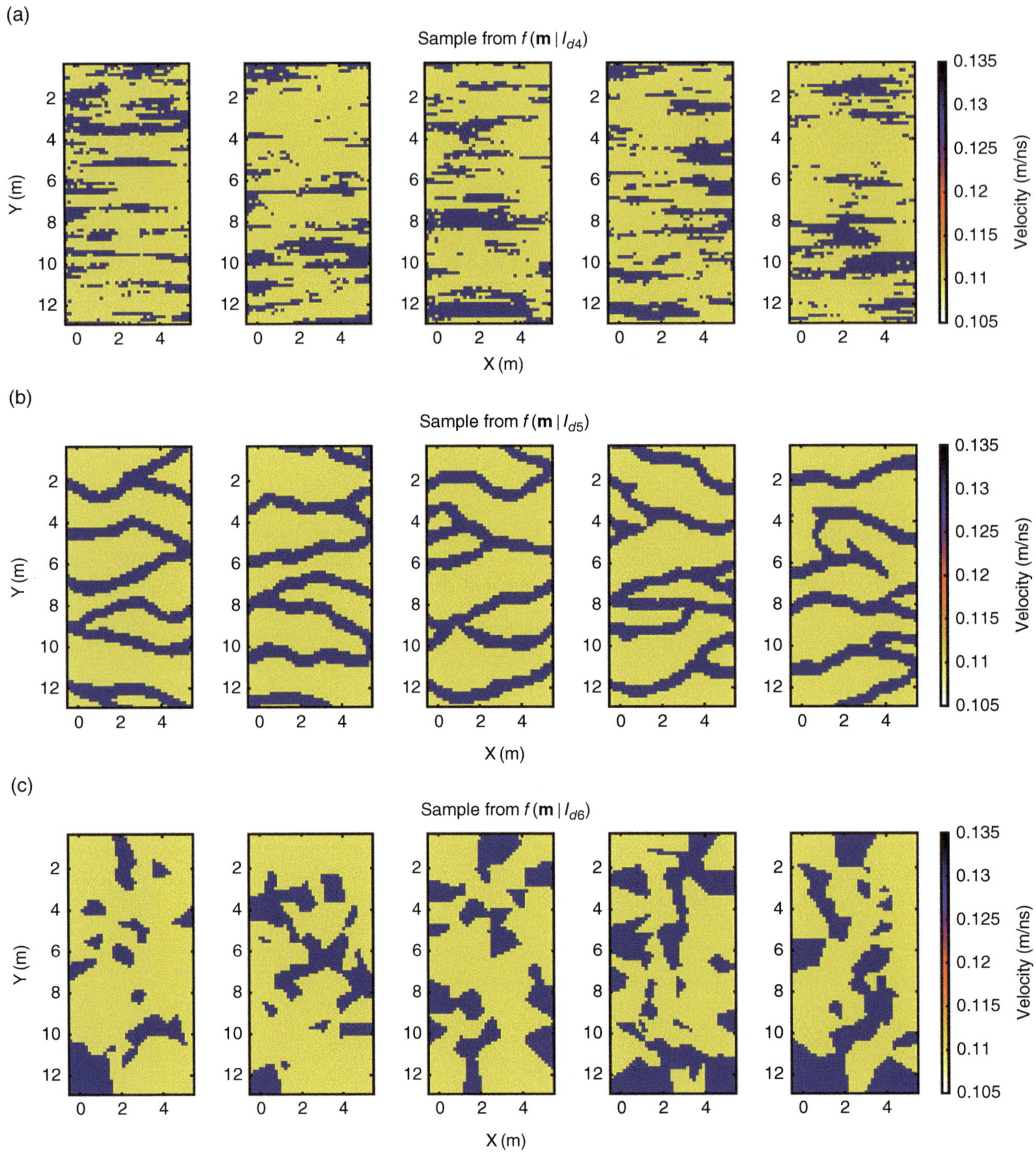

Figure 6.4 Five realizations from **(a)** $f(\mathbf{m}|I_{d4})$, **(b)** $f(\mathbf{m}|I_{d5})$, and **(c)** $f(\mathbf{m}|I_{d6})$. See text for details.

$f(\mathbf{m} \mid I_{d1})$ and $f(\mathbf{m} \mid I_{d2})$ even if they may be consistent with some statistical properties of the sample model, as discussed by *Journel and Deutsch* [1993]. While the model based on Voronoi cells, $f(\mathbf{m} \mid I_{d6})$, may possess some features that can be mathematically useful, it is also evident from Figure 6.4c that such a model does not seem particularly useful to describe natural geological variability. Note that when the number of Voronoi cells becomes very high, $f(\mathbf{m} \mid I_{d6})$ may reflect the same kind of information as $f(\mathbf{m} \mid I_{d2})$.

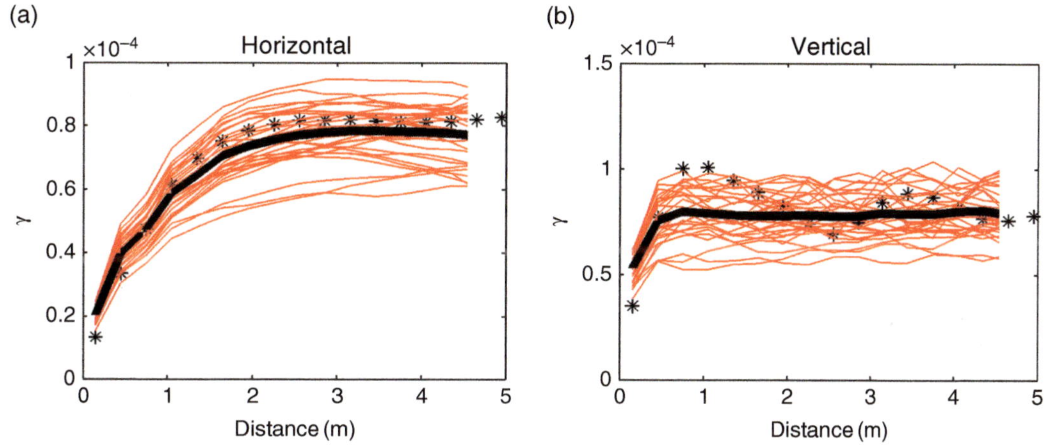

Figure 6.5 Experimental semivariogram model inferred from the sample model in Figure 6.1 (*black asterisks*) compared to the experimental semivariogram of 10 realizations of $f(\mathbf{m} \mid I_{d4})$ (*red lines*) along the **(a)** horizontal axis and **(b)** vertical axis. *Black line* indicates the mean of the 10 semivariograms.

6.3. QUANTIFYING INDIRECT GEO-INFORMATION USING PROBABILITY DISTRIBUTIONS

As opposed to direct information, indirect information, $I_{indirect}$, is available in the form of data, **d**, that is related to the model parameters **m** through a function g as

$$\mathbf{d} = g(\mathbf{m}). \tag{6.12}$$

Evaluating Eq. (6.12) is often referred to as solving the forward problem. Examples of such indirect information are geophysical or remote sensing data.

A general probabilistic description of the relative probability of a certain model **m** given such indirectly observed data is the likelihood function [*Tarantola,* 2005]:

$$f(\mathbf{m} \mid I_{indirect}) \propto L(\mathbf{m}) = \int_{\mathcal{D}} \frac{\rho_D(g(\mathbf{m}))\theta(\mathbf{d} \mid \mathbf{m})}{\mu_D(\mathbf{d})}. \tag{6.13}$$

$\rho_D(\mathbf{d})$ describes *measurement uncertainties*, typically related to the instrument recording the data. $\theta(\mathbf{d} \mid \mathbf{m})$ is a probabilistic formulation of the forward modeling that describes the probability of a set of calculated data given a model **m**. $\mu_D(\mathbf{d})$ is the homogeneous probability distribution (see *Tarantola* [2005] for more details).

The uncertainty related to the forward modeling may be significant and higher than the measurement uncertainty [*Hansen et al.*, 2014]. However, in many cases the modeling error is ignored (i.e., described by a delta function) in which case Eq. (6.13) reduces to

$$L(\mathbf{m}) = \rho_D(g(\mathbf{m})) \tag{6.14}$$

In this case, evaluation of $f(\mathbf{m} \mid I)$ can be achieved as long as a probability distribution describing the measurement uncertainty can be evaluated. Very often the measurement errors are considered zero mean Gaussian distributed $\mathcal{N}(0, \mathbf{C}_d)$, in which case

$$\rho_D(g(\mathbf{m})) = \left((2\pi)^2 |\mathbf{C}_d|\right)^{-.5} \times \exp\left(-\frac{1}{2}\left(\mathbf{d}_{obs} - g(\mathbf{m})^{\top}\right)\mathbf{C}_d^{-1}\left(\mathbf{d}_{obs} - g(\mathbf{m})\right)\right). \tag{6.15}$$

If the modeling error is Gaussian, it can be described simply as an addition to the Gaussian measurement uncertainty and can then be accounted for through Eq. (6.15). More details on this topic can be found in *Hansen et al.* [2014].

Thus, in the latter simple case, the conditional probability $f(\mathbf{m} \mid I_{indirect})$ can be evaluated through Eq. (6.15) by solving the forward problem, Eq. (6.12), and evaluating the resulting data residual, $\mathbf{d}_{obs} - g(\mathbf{m})$.

The forward relation, Eq. (6.12), may be quite complex and involve mapping of the model parameters **m** into secondary parameters from which data can be computed. For example, seismic inversion can be formulated such that the primary model parameters reflect rock physical parameters. These must be transformed, for example, to elastic parameters in order to solve the forward problem in order to compute a seismic response. For a detailed discussion on complex forward models see *Bosch* [2015].

Note that the likelihood function, Eq. (6.13), is not strictly a probability distribution, as $\int L(\mathbf{m})\, d\mathbf{m}$ in general will not be 1. However, if the goal is to sample from $f(\mathbf{m} \mid I_1, I_2) = f(\mathbf{m} \mid I_1) f(\mathbf{m} \mid I_2)$, then a relative measure proportional to $f(\mathbf{m} \mid I_{indirect})$, such as the likelihood, will suffice, and hence the normalization of the likelihood is not needed.

6.4. SAMPLING FROM A PROBABILITY DISTRIBUTION, $f(\mathbf{m} \mid \mathbf{I})$

In the ideal case, $f(\mathbf{m} \mid \mathbf{I})$—which describes all available information about **m**—can be described analytically. In practice, however, this may not be possible, unless restrictions, such as Gaussian assumptions, are imposed on $f(\mathbf{m} \mid I_i)$.

A general approach for characterizing $f(\mathbf{m} \mid \mathbf{I})$ is by *sampling* it, which is done by generating a (representative) sample from $f(\mathbf{m} \mid \mathbf{I})$ that consists of a number of realizations that are distributed according to $f(\mathbf{m} \mid \mathbf{I})$. If this sample is large enough, any statistical measure or question related to $f(\mathbf{m} \mid \mathbf{I})$ can be probabilistically evaluated and answered.

In this section, a number of widely used methods for sampling from $f(\mathbf{m} \mid \mathbf{I})$ when a measure proportional to $f(\mathbf{m} \mid \mathbf{I})$ can be evaluated will be described. This implies that a measure proportional to the probability distribution value related to any of the independent types of information $f(\mathbf{m} \mid I_i)$ can be computed. This means that the previous defined models $f(\mathbf{m} \mid I_{d1})$, $f(\mathbf{m} \mid I_{d2})$, and $f(\mathbf{m} \mid I_{d3})$ can all readily be used. On the other hand, information available and quantified through numerical simulation algorithms, such as $f(\mathbf{m} \mid I_{d4})$, $f(\mathbf{m} \mid I_{d5})$, and $f(\mathbf{m} \mid I_{d6})$ where $f(\mathbf{m} \mid I_i)$ cannot readily be computed, cannot be considered by the algorithms discussed in this section.

6.4.1. Rejection Sampling

Any probability distribution for which $f(\mathbf{m})$ can be evaluated can in principle be sampled using rejection sampling. If $h(\mathbf{m})$ is a *proposal* distribution from which a realization can be generated (preferably in a computationally efficient manner) and for which $h(\mathbf{m}) \geq f(\mathbf{m})\ \forall \mathbf{m}$, then $f(\mathbf{m})$ can be sampled using the rejection sampling algorithm as follows:

REJECTION SAMPLING ALGORITHM
1. **Propose** a model $\mathbf{m}_{propose}$, as a realization of $h(\mathbf{m})$.
2. **Accept** this model with probability P_{acc}

$$P_{acc} = \frac{f(\mathbf{m}_{propose})}{h(\mathbf{m}_{propose}) \max(f(\mathbf{m}))}, \tag{6.16}$$

where $\max(f(\mathbf{m}))$ is the maximum value of $f(\mathbf{m})$.

Each accepted model will be a realization from $f(\mathbf{m})$, and the series of models accepted when the algorithm is run iteratively will be a representative sample from $f(\mathbf{m})$.

$h(\mathbf{m})$ is often chosen as the uniform distribution, in which case the acceptance probability becomes

$$P_{acc} = \frac{f(\mathbf{m}_{propose})}{\max(f(\mathbf{m}))}, \tag{6.17}$$

In many cases it may not be possible to estimate $\max(f(\mathbf{m}))$. Further, even in cases where $\max(f(\mathbf{m}))$ can be evaluated, the acceptance probability of the rejection sampler may be extremely low. Consider, for example, the Gaussian model [as in Eq. (6.5)] where the value

$$-2\log(f(\mathbf{m}^*)) \propto (\mathbf{m}^* - \mathbf{m}_0)^T\ \mathbf{C}_m^{-1}\ (\mathbf{m}^* - \mathbf{m}_0), \tag{6.18}$$

related to a realization $\mathbf{m}^*$, is distributed according to the χ^2 distribution with M degrees of freedom (where M is the number of parameters of **m**) [*Tarantola*, 2005]. For high values of M the χ^2 distribution will tend to be Gaussian distributed as $\mathcal{N}(M, 2M)$. This means that for high values of M, $\log(f(\mathbf{m}^*))$ will tend to be normally distributed as $\mathcal{N}(-M/2, \sqrt{M/2})$. In other words, the most frequent probability value of a realization $\mathbf{m}^*$ of $f(\mathbf{m})$ will be $f(\mathbf{m}^*) \approx \exp(-M/2)$.

Considering $M = 10$, model parameters will lead to $f(\mathbf{m}^*) \approx \exp(-10/2) = 0.0067$. This means that in order to accept a typical realization $\mathbf{m}^*$ of $f(\mathbf{m})$ using the rejection sampler, it has to be proposed on average 1/0.0067 = 148 times. Considering $M = 20$ model parameters will lead to $f(\mathbf{m}^*) \approx \exp(-20/2) = 0.000045$, which means that in order to accept a typical realization $\mathbf{m}^*$ from $f(\mathbf{m})$ using the rejection sampler, it has to be proposed on average 22,026 times.

Thus, the rejection sampler, with a uniform proposal distribution, is extremely inefficient except for very-low (less than about five)-dimensional problems.

6.4.2. Metropolis–Hastings Algorithm

The Metropolis–Hastings algorithm is a Monte Carlo–based method for sampling a probability distribution $f(\mathbf{m})$ [*Metropolis et al.*, 1953]. At each step in a random walk the algorithm goes through two phases. In the "exploration" phase a new model is proposed in the vicinity of a current model. Then, in an "exploitation" phase the new model is either accepted or rejected as a realization from $f(\mathbf{m})$ as follows:

THE METROPOLIS–HASTINGS ALGORITHM
0. Generate a starting model, $\mathbf{m}_{current}$.
1. **Exploration.** Propose a new realization from $\mathbf{m}_{propose}$, in the vicinity of $\mathbf{m}_{current}$ by generating a realization from a transition probability $h(\mathbf{m}_{proposed} \mid \mathbf{m}_{current})$.

2. **Exploitation.** Accept the move to $\mathbf{m}_{propose}$ with the acceptance probability P_{acc}:

$$P_{acc} = \min\left(1, \frac{f(\mathbf{m}_{propose})\, h(\mathbf{m}_{current} \mid \mathbf{m}_{propose})}{f(\mathbf{m}_{current})\, h(\mathbf{m}_{propose} \mid \mathbf{m}_{current})}\right). \quad (6.19)$$

For simplicity it is often assumed that the proposal distribution is symmetrical such that $h(\mathbf{m}_{propose} \mid \mathbf{m}_{current}) = h(\mathbf{m}_{current} \mid \mathbf{m}_{propose})$. In this case, Eq. (6.19) reduces to

$$P_{acc} = \min\left(1, \frac{f(\mathbf{m}_{propose})}{f(\mathbf{m}_{current})}\right). \quad (6.20)$$

If the move is accepted, $\mathbf{m}_{propose}$ becomes $\mathbf{m}_{current}$. Otherwise the random walk stays at the $\mathbf{m}_{current}$ location.

3. Goto 1.

It can be shown that $f(\mathbf{m})$ will be asymptotically sampled by running the Metropolis–Hasting algorithm.

Often a new model is proposed using a uniform or Gaussian transition distribution centered on $\mathbf{m}_{current}$. In such a case the exploration step simply consists of adding a realization of a uniform or Gaussian model, $\mathbf{m}_{\delta}$, with mean zero, to the current model, such that $\mathbf{m}_{proposed} = \mathbf{m}_{current} + \mathbf{m}_{\delta}$. The amplitude of $\mathbf{m}_{\delta}$ is referred to as the step-length.

One major advantage using the Metropolis–Hastings algorithm—as opposed to, for example, rejection sampling—is that this algorithm only relies on the relative change in probability value between the current and the proposed model for computing the acceptance probability, Eq. (6.19). Therefore the value of $\max(f(\mathbf{m}))$ does not need to be known as is the case using the rejection sampler.

A disadvantage is that the series of realizations generated by the Metropolis–Hastings algorithm are not independent. Thus, in order to obtain a statistical independent realization from $f(\mathbf{m})$, a number of iterations of the algorithm must be run. It may not be trivial to estimate how many iterations are needed in order to obtain an independent realization.

In addition, when the Metropolis–Hastings algorithm is started, it will, most often, not sample $f(\mathbf{m})$ immediately. Initially the algorithm will be in what is referred as the "burn-in" phase, in which state the algorithm searching for models that are consistent with $f(\mathbf{m})$. When the algorithm starts to sample $f(\mathbf{m})$, it is said to have reached burn-in.

The average distance between $\mathbf{m}_{propose}$ and $\mathbf{m}_{current}$ is called the exploration step-length. A large exploration step results in a more exploratory algorithm spanning relatively large volumes of probability at the expense of increasing computational demands. It is nontrivial to choose an exploration step-length that leads to maximum efficiency of the Metropolis sampling algorithm. It has been suggested that an exploration step-length leading to an accepted move in every third to fourth iteration provides a good compromise between exploration and computational efficiency [*Geman and Geman*, 1984]. In practice an optimal choice of exploration step-length is closely linked to the shape of the probability distribution being sampled.

The Metropolis–Hasting algorithm is guaranteed to asymptotically sample $f(\mathbf{m})$ in finite time. In practice, however, the Metropolis–Hastings algorithm can have difficulties sampling multimodal problems in high dimensions (i.e., problems where local areas of high probability exist, which are disconnected by areas of zero probability). In such cases, it may end up sampling a local area of high probability. There are no trivial tests to ensure that the full probability distribution is being sampled. A simple approach is to start more sampling algorithms (sometimes called chains) in parallel and then test whether they end up sampling the same distribution. A more formal approach is to make use of parallel tempering, where multiple chains run in parallel, where jumps between chains are allowed. Each chain is run with a different temperature, as known from simulated annealing. Parallel tempering is promising for lower-dimensional problems [*Sambridge*, 2013].

For all its shortcomings, the Metropolis–Hastings algorithm is computationally superior to rejection sampling, for sampling anything but very-low-dimensional probability distributions.

The computational efficiency of the Metropolis-Hastings algorithm is closely related to the choice of transition probability. The efficiency of the rejection sampler is linked to the choice of proposal distribution. Ideally such transition probabilities and proposal distributions should be chosen such that the acceptance rate is maximized. However, this is often not a trivial task. For example, a straightforward application of the Metropolis algorithm to sample from a multivariate Gaussian probability distribution with a Gaussian-type covariance using a symmetric proposal distribution will in practice be computationally extremely inefficient. This is due to the fact that any proposed model will lead to a discontinuity in the proposed model, which is inconsistent with the (spatial) smoothness implied by the Gaussian-type covariance model. The Hamiltonian Monte Carlo approach suggests to make use of the local gradient of the probability distribution being sampled, in order to allow faster mixing and higher acceptance probability of the Monte Carlo Chain [*Duane et al.*, 1987]. However, the Hamiltonian Monte Carlo requires that the gradient of the probability distribution being sampled can be evaluated. In the following, we will consider to sample probability distributions where the probability distribution value, and hence the gradient, may not be available.

We will, therefore, not consider the use of the Hamiltonian Monte Carlo any further.

The rejection sampler and Metropolis–Hasting algorithm as described above will, in the following, be referred to as the "classic" rejection sampler, and the "classic" Metropolis algorithm.

6.5. SAMPLING OF $f(\mathbf{m}\,|\,I_1,I_2) \propto f(\mathbf{m}\,|\,I_1)\, f(\mathbf{m}\,|\,I_2)$

Consider the case where $f(\mathbf{m}\,|\,\mathrm{I})$ is proportional to the product of two probability densities $f(\mathbf{m}\,|\,I_1)$ and $f(\mathbf{m}\,|\,I_2)$:

$$f(\mathbf{m}\,|\,\mathrm{I}) \propto f(\mathbf{m}\,|\,I_1)\; f(\mathbf{m}\,|\,I_2), \tag{6.21}$$

that is, a case identical to the information integration problem in Eq. (6.1), where information is available from two independent sources. One can choose to sample directly from $f(\mathbf{m}\,|\,\mathbf{I})$, using the methods described in the previous section, in case $f(\mathbf{m}\,|\,\mathbf{I})$ can be evaluated.

But, in many data integration problems one may not be able to evaluate $f(\mathbf{m}\,|\,\mathbf{I})$, as not all $f(\mathbf{m}\,|\,I_i)$ can be evaluated. For example, if $f(\mathbf{m}\,|\,I_i)$ describes the statistical information inferred from a training image, then, in most cases, the evaluation of $f(\mathbf{m}\,|\,I_i)$ is, until now, not possible.

It turns out that when $f(\mathbf{m}\,|\,I_1)$ and $f(\mathbf{m}\,|\,I_2)$ have certain properties, $f(\mathbf{m}\,|\,I_1,I_2)$ may be sampled even when either $f(\mathbf{m}\,|\,I_1)$ or $f(\mathbf{m}\,|\,I_2)$ cannot be evaluated.

Further, even when both $f(\mathbf{m}\,|\,I_1)$ or $f(\mathbf{m}\,|\,I_2)$ can be evaluated, simple alterations of the classical rejection sampler and Metropolis–Hastings, algorithm can lead to computationally much more efficient sampling methods.

6.5.1. Extended Rejection Sampling

Say that an algorithm exists that allows generation of independent realizations from $f(\mathbf{m}\,|\,I_1)$. Using $f(\mathbf{m}\,|\,I_1)$ as a proposal distribution for the rejection sampler results in a more efficient rejection sampler, specifically for the case of sampling the product $f(\mathbf{m}\,|\,\mathbf{I}) \propto f(\mathbf{m}\,|\,I_1)\; f(\mathbf{m}\,|\,I_2)$:

EXTENDED REJECTION SAMPLING ALGORITHM OF $f(\mathbf{m}\,|\,\mathbf{I}) \propto f(\mathbf{m}\,|\,I_1)\, f(\mathbf{m}\,|\,I_2)$

1. **Propose** a model $\mathbf{m}_{propose}$, as a realization of from $f(\mathbf{m}\,|\,I_1)$.
2. **Accept** this model with probability P_{acc}:

$$\begin{aligned} P_{acc} &= \frac{f(\mathbf{m}_{propose}\,|\,I_1,I_2)}{f(\mathbf{m}_{propose}\,|\,I_1)\max(f(\mathbf{m}\,|\,\mathbf{I}))} \\ &= \frac{f(\mathbf{m}_{propose}\,|\,I_1)\; f(\mathbf{m}_{propose}\,|\,I_2)}{f(\mathbf{m}_{propose}\,|\,I_1)\max(f(\mathbf{m}\,|\,\mathbf{I}))} \end{aligned} \tag{6.22}$$

$$\propto \frac{f(\mathbf{m}_{propose}\,|\,I_2)}{\max(f(\mathbf{m}\,|\,I_2))}, \tag{6.23}$$

where $\max(f(\mathbf{m}\,|\,I_2))$ is the maximum probability distribution value of $f(\mathbf{m}\,|\,I_2)$.

Note that in this case the actual probability distribution value of $f(\mathbf{m}\,|\,I_1)$ or $f(\mathbf{m}\,|\,I_1,I_2)$ need never be evaluated, as long as an algorithm exists that generates realizations of $f(\mathbf{m}\,|\,I_1)$. If the algorithm that samples $f(\mathbf{m}\,|\,I_1)$ is reasonably efficient, the extended rejection sampling algorithm may be computationally much more efficient than the classic rejection sampler.

As demonstrated previously (Sections 6.2.1.2 and 6.2.1.3), a large collection of algorithms have been developed in recent decades, which are able to generate realizations of a (possibly unknown) probability distribution, such as, for example $f(\mathbf{m}\,|\,I_1)$, which can therefore be used as part of a rejection sampler to sample from $f(\mathbf{m}\,|\,I_1,I_2) \propto f(\mathbf{m}\,|\,I_1)\; f(\mathbf{m}\,|\,I_2)$, if only $f(\mathbf{m}\,|\,I_2)$ can be evaluated.

6.5.2. The Extended Metropolis Algorithm

The extended Metropolis algorithm [*Mosegaard and Tarantola*, 1995] is a modified version of the classic Metropolis–Hastings algorithm designed to sample the product of two probability distributions, $f(\mathbf{m}\,|\,I_1,I_2) = k\, f(\mathbf{m}\,|\,I_1) f(\mathbf{m}\,|\,I_2)$, in the specific case where an algorithm exists to iteratively sample $f(\mathbf{m}\,|\,I_1)$. It can be applied as follows:

0. **Init** Generate a starting model, $\mathbf{m}_{current}$, as a realization of $f(\mathbf{m}\,|\,I_1)$.
1. **Exploration.** Propose a new realization of $f(\mathbf{m}\,|\,I_1)$, $\mathbf{m}_{propose}$, in the vicinity of $\mathbf{m}_{current}$.
2. **Exploitation.** Accept the move to $\mathbf{m}_{propose}$ with the acceptance probability P_{acc}:

$$P_{acc} = \min\left(1, \frac{f(\mathbf{m}_{proposed}\,|\,I_2)}{f(\mathbf{m}_{current}\,|\,I_2)}\right). \tag{6.24}$$

If the move is accepted, $\mathbf{m}_{propose}$ becomes $\mathbf{m}_{current}$. Otherwise the random walk stays at the $\mathbf{m}_{current}$ location. Goto 1.

The exploration must be implemented in such way that when iterating, only the exploration step (i.e., accepting all model proposals) should lead to an algorithm sampling $f(\mathbf{m}\,|\,I_1)$.

To apply the extended Metropolis algorithm, one must (a) be able to compute a value proportional to $f(\mathbf{m}\,|\,I_2)$ for any proposed model $\mathbf{m}_{proposed}$ and (b) be able to perform a random walk that will sample $f(\mathbf{m}\,|\,I_1)$. There is no requirement to be able to evaluate neither $f(\mathbf{m}\,|\,I_1)$

nor the product $f(\mathbf{m} \mid I_1, I_2) \propto f(\mathbf{m} \mid I_1)\, f(\mathbf{m} \mid I_2)$. A "black box" algorithm that can perform a random walk which samples $f(\mathbf{m} \mid I_1)$ is sufficient [*Mosegaard and Tarantola*, 1995].

All the advantages and disadvantages of using the classic Metropolis–Hastings algorithm listed above also applies when using the extended Metropolis algorithm. However, if an algorithm exists that allows performing a random walk, such that $f(\mathbf{m} \mid I_1)$ is sampled, then the extended Metropolis algorithm may be orders of magnitude more efficient than using the Metropolis–Hastings algorithm.

6.5.2.1. Sequential Gibbs Sampling

The sequential simulation algorithm, Section 6.2.2, was originally developed to allow efficient simulation of independent realizations from probability distributions $f(\mathbf{m} \mid I_1)$ based on two- and multiple-point statistics, as demonstrated in Sections 6.2.1.1 and 6.2.1.3.

Therefore, any sampling algorithm based on sequential simulation can be used to perform a random walk, where each visited model is independent of its neighbors. This corresponds to a random walk with maximum exploration and hence maximum step-length. However, a crucial part of applying the extended Metropolis algorithm is the ability to control the step-length—that is, controlling the exploratory nature of the algorithm performing the random walk sampling $f(\mathbf{m} \mid I_1)$.

Sampling of $f(\mathbf{m} \mid I_1)$, using an arbitrary step-length, can be accomplished using sequential Gibbs sampling [*Hansen et al.*, 2008, 2012], for any probability distribution that can be sampled using sequential simulation. See also *Fu* and *Gómez-Hernández* [2008]; *Irving et al.* [2010] for related methods specific for Gaussian based models, and *Mariethoz et al.* [2010] for a method similar to *Hansen et al.* [2008].

Assume that $\mathbf{m}_1$ is a "current" model, which is a realization of $f(\mathbf{m} \mid I_1)$. Then one step of the sequential Gibbs sampling algorithm will generate a new realization $\mathbf{m}_2$ of $f(\mathbf{m} \mid I_1)$ in the vicinity of $\mathbf{m}_1$ using the following steps:

1. Select a subset U of all M model parameters, $\mathbf{m}_{1,i\in U}$.
2. Use sequential simulation to generate a realization $\mathbf{m}^*_{i\in U}$ of $f(\mathbf{m}_{i\in U} \mid \mathbf{m}_{i\notin U})$; that is, re-simulate the model parameters in U conditional to the model parameters not in U.
3. Update the next model, $\mathbf{m}_2$, as $\mathbf{m}_{2,i\notin U} = \mathbf{m}_{1,i\notin U}$ and $\mathbf{m}_{2,i\in U} = \mathbf{m}^*_{i\in U}$.

Performing these steps iteratively will generate a series of models that will represent a random walk sampling $f(\mathbf{m} \mid I_1)$. This is exactly the requirements of the "black" box algorithm needed by the extended Metropolis algorithm to sample $f(\mathbf{m} \mid I_1)$.

The number of model parameters in the subset U reflects the step-length. The longest step-length is when U contains all the model parameters in which case $\mathbf{m}_2$ will be independent of $\mathbf{m}_1$.

The sequential Gibbs sampler can in principle be used to sample any of the probability distributions described in Sections 6.2.1.1–6.2.1.3 through a random walk with an arbitrary step size.

Note that a perfect application of sequential Gibbs sampling requires sampling from the full conditional distribution, Eq. (6.9), at each iteration. Most of the simulation algorithms based on sequential simulation described previously make use of a data neighborhood in which case the conditional distribution will only be approximately correct. If the neighborhood is chosen sufficiently large for probability distributions based on two-point statistics, this approximation does in practice provide the same results as when using a full neighborhood. *Cordua et al.* [2015] observed that using sequential Gibbs sampling with the multiple-point based SNESIM algorithm, *Strebelle* [2002], will render a sampling algorithm, where the sampled probability distribution depends on the step perturbation size of the sequential Gibbs perturbation. A correction using frequency matching *Lange et al.* [2012] is suggested to remedy the unwanted effect of perturbations size and in this way remain to sample from a probability distribution that satisfies the multiple-point statistics from the training image.

6.5.2.2. Independent Extended Metropolis Algorithm

A simple variant of the extended Metropolis algorithm is when the step-length is set to its maximum; that is, a new independent realization of the $f(\mathbf{m} \mid I_1)$ is proposed in the exploration step, similar to the metropolized independence sampler proposed by *Liu* [1996]. In this case, any probability distribution from which independent realizations can be generated can be used for probabilistic data integration. Thus, there is no need to use the sequential Gibbs sampler. This means that, in principle, most developed geostatistical algorithms can be used to describe information that can be used for data integration problems. Application of the independent extended Metropolis algorithm avoids the problem of estimating the normalization constant in the acceptance ratio, as is needed when applying the rejection sampler, which may lead to a computationally much faster algorithm. This algorithm is as simple to implement as the rejection sampler, which in practice render the rejection sampler obsolete. Compared to the extended Metropolis algorithm, the independent extended Metropolis algorithm is easier to implement, but also much less computationally efficient.

6.6. EXAMPLE OF SAMPLING $f(\mathrm{m} \mid I_1, I_2)$

To demonstrate different aspects of some of the presented algorithms, consider a crosshole tomographic inverse problem. 40 × 84 model parameters represent a 2D electromagnetic velocity field of size 5.85 m × 12.45 m. This is exactly the same model size as considered in Figures 6.2 and 6.4.

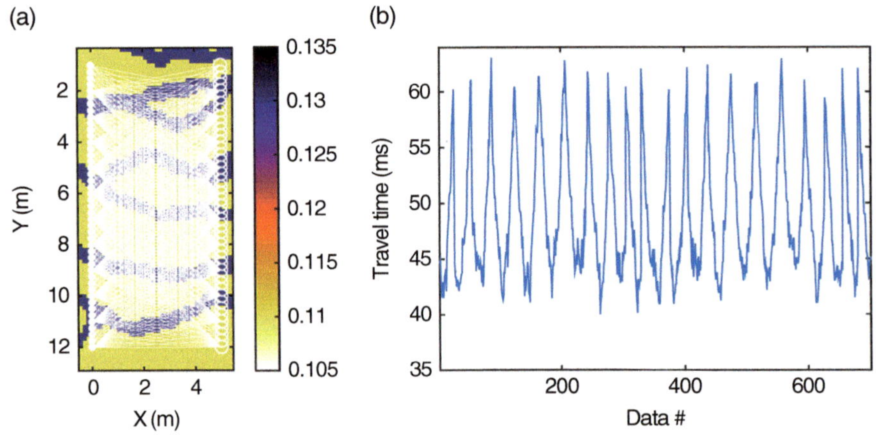

Figure 6.6 **(a)** Reference velocity model (*white lines* connect source and receiver locations). **(b)** Reference travel-time dataset.

Figure 6.6 shows a reference model generated as a realization from $f(\mathbf{m} \mid I_{d5})$, which is based on the statistics inferred from the sample model in Figure 6.1.

Mimicking a cross borehole tomographic experiment, travel times of electromagnetic waves from 702 source locations to 702 receiver locations, as indicated in Figure 6.6a, are computed using finite frequency theory using *Hansen et al.* [2013a]. Then a realization of zero mean uncorrelated Gaussian noise with standard deviation of 0.8 ns, $\mathbf{C}_d = 0.8^2\mathbf{I}$, is added to the travel-time data, which are then considered as "observed" data.

Thus one type of indirect information, which we will refer to as $I_{indirect}$, related to geophysical travel time measurement, is available. $I_{indirect}$ then specifies not only the travel-time data and the measurement uncertainty, $\mathcal{N}(0,\mathbf{C}_d)$, but also knowledge about how to solve the forward problem. This means that we are able to use the likelihood function in Eq. (6.13) to evaluate $f(\mathbf{m} \mid I_{indirect})$.

Any of the previously defined probability distributions describing different types of direct information $f(\mathbf{m} \mid I_{d1}),\ldots,f(\mathbf{m} \mid I_{d6})$ is also, in turn, considered as information available about the model parameters. Recall that Figures 6.2 and 6.4 show realizations from these probability distributions.

The problem is now to solve the data integration problem by generating a sample from $f(\mathbf{m} \mid I_{di}, I_{indirect}) = k\, f(\mathbf{m} \mid I_{di})\, f(\mathbf{m} \mid I_{indirect})$, where $i = 1,\ldots,6$.

6.6.1. Sampling $f(\mathbf{m} \mid I_{di}, I_{indirect})$ Using Rejection Sampling

The extended rejection sampler presented previously, making use of sequential simulation to generate realizations of the direct information, can in principle be used to sample the joint distribution $f(\mathbf{m} \mid I_{di}, I_{indirect})$. However, in practice the rejection sampler is only applicable to very-low-dimensional sampling problems and could not be applied for the current case.

6.6.2. Sampling $f(\mathbf{m} \mid I_{di}, I_{indirect})$ Using the Extended Metropolis Algorithm

For all the considered probability distributions based on direct information, $f(\mathbf{m} \mid I_{d1}),\ldots, f(\mathbf{m} \mid I_{d6})$, a random walk that samples the probability distribution, with arbitrary step-length, can be performed using sequential Gibbs sampling. Hence, the combined information $f(\mathbf{m} \mid I_{di}, I_{indirect})$ can be sampled using the extended Metropolis algorithm, without ever evaluating $f(\mathbf{m} \mid I_{di})$.

The extended Metropolis algorithm has been run for 100,000 iterations drawing realizations from $f(\mathbf{m} \mid I_{di}, I_{Indirect})$, for each of the six types of direct information. In all runs, the step-length is selected such that the acceptance rate of the algorithm is around 30%. For details about running the extended Metropolis algorithm, see, for example, *Cordua et al.* [2012] and *Hansen et al.* [2013a].

The extended Metropolis sampler was especially prone to be caught in local minima sampling $f(\mathbf{m} \mid I_{d6}, I_{indirect})$, and therefore the parallel tempering algorithm was used in this case [*Sambridge*, 2013].

Figures 6.7 and 6.8 show five realizations from the probability distribution describing the combined information of $f(\mathbf{m} \mid I_{d1}, I_{indirect}),\ldots,f(\mathbf{m} \mid I_{d6}, I_{indirect})$. These realizations should be compared to the realizations from the probability distribution based on direct information in Figures 6.2 and 6.4.

Comparing Figures 6.7 and 6.8 to Figures 6.2 and 6.4, it is obvious that the spatial variability from the direct information is preserved in the realizations from the combined probability distributions. If the direct information defines the subsurface as a set of Voronoi cells, as for

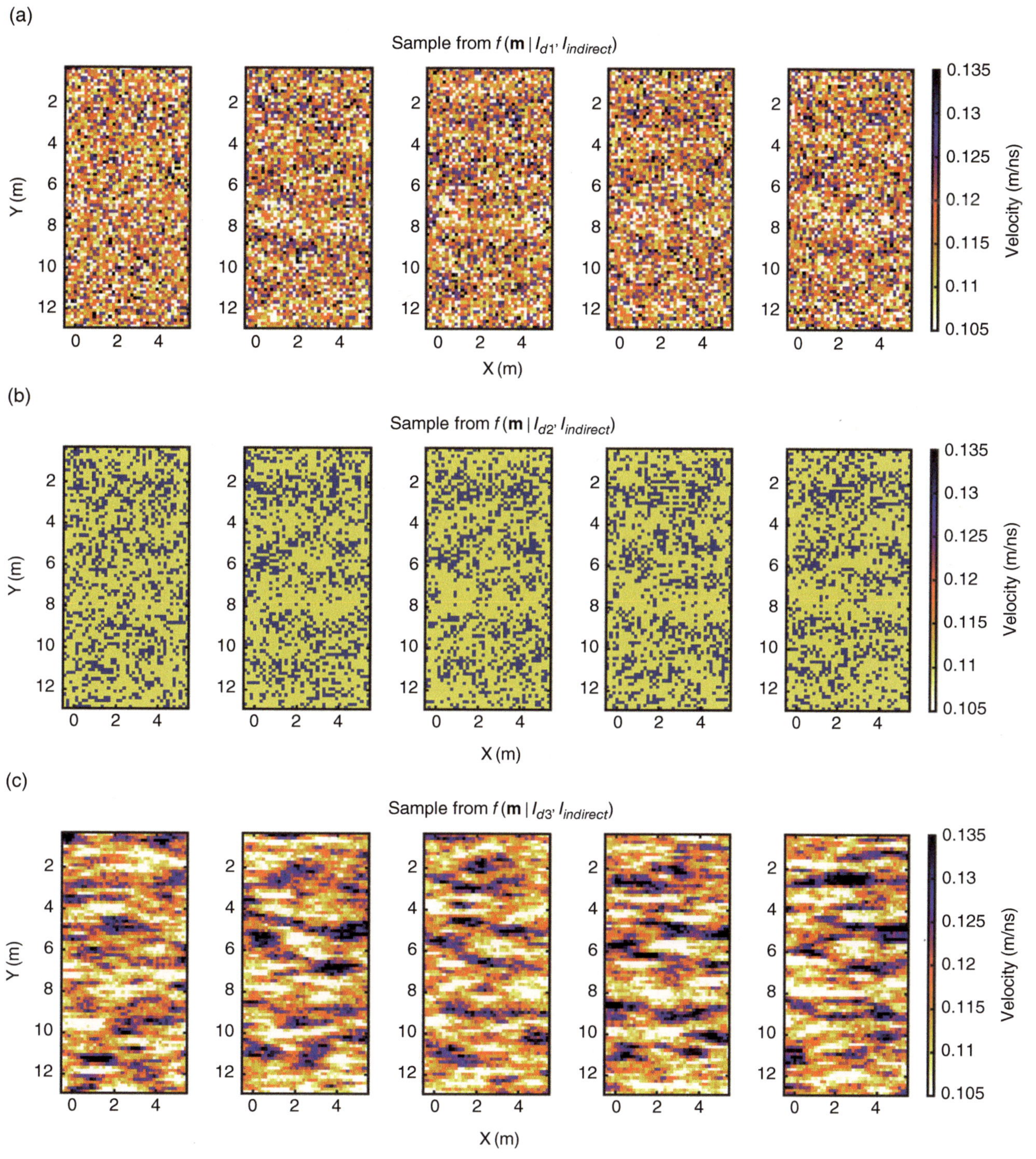

Figure 6.7 Five realizations from **(a)** $f(\mathbf{m} \mid I_{d1}, I_{indirect})$, **(b)** $f(\mathbf{m} \mid I_{d2}, I_{indirect})$, and **(c)** $f(\mathbf{m} \mid I_{d3}, I_{indirect})$. See text for details.

$f(\mathbf{m} \mid I_{d6})$, then realizations from the combined probability distribution will consist of Voronoi cells (Figure 6.8c). Then one should off course consider whether a set of Voronoi cells provide a geologically reasonable description of Earth structures. In this case, the realizations of $f(\mathbf{m} \mid I_{d6}, I_{indirect})$ does not seem to resemble realistic geological variability.

The choice of a spatially uncorrelated probability distribution to describe direct information, such as $f(\mathbf{m} \mid I_{d1})$ and $f(\mathbf{m} \mid I_{d2})$, will also affect the combined information content of $f(\mathbf{m} \mid I_{d1}, I_{indirect})$ and $f(\mathbf{m} \mid I_{d1}, I_{indirect})$, which will also exhibit maximum spatial disorder in the outcome realizations. If more indirect information is available (e.g., less noise or more data),

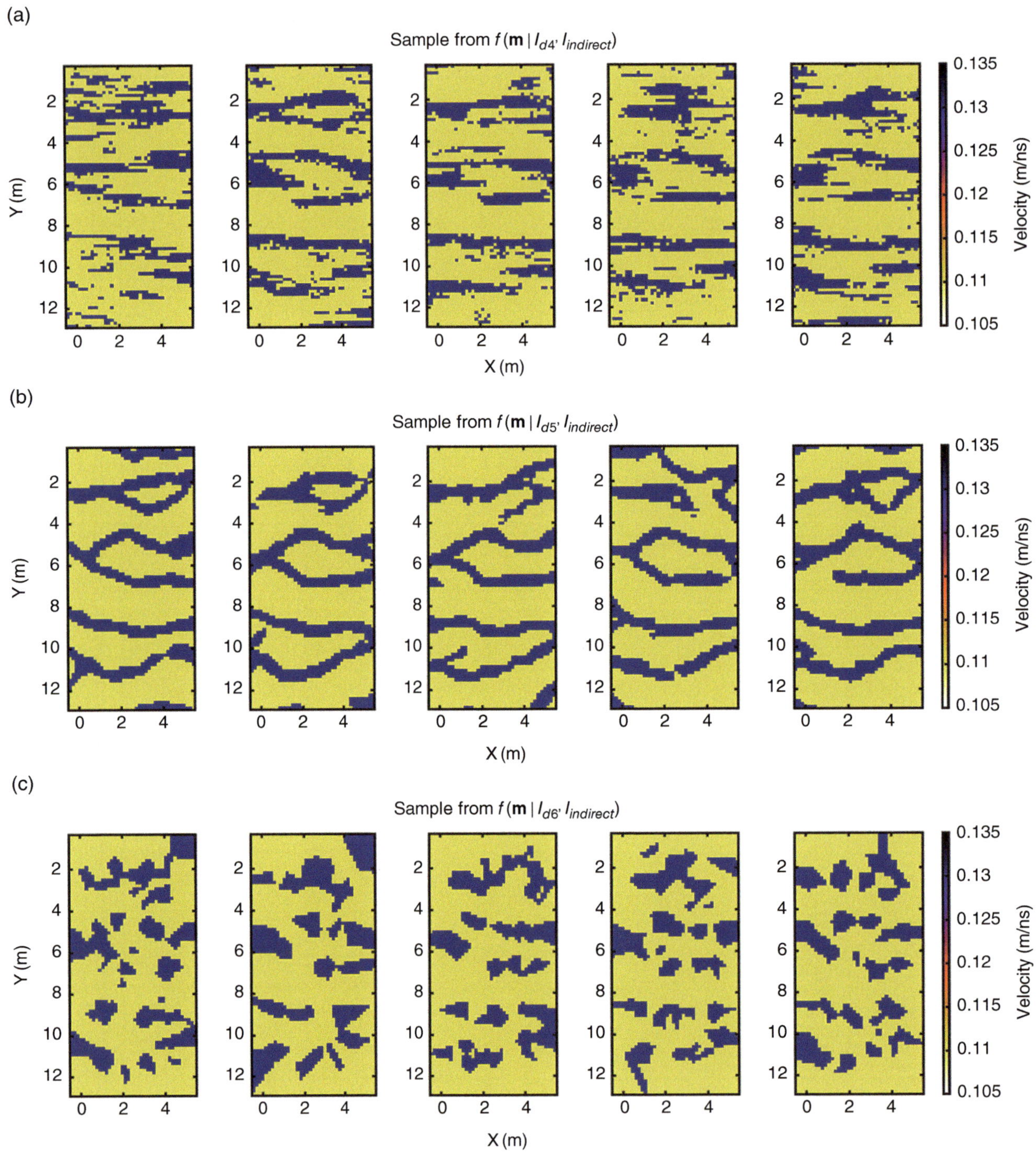

Figure 6.8 Five realizations from **(a)** $f(\mathbf{m} \mid I_{d4}, I_{indirect})$, **(b)** $f(\mathbf{m} \mid I_{d5}, I_{indirect})$, and **(c)** $f(\mathbf{m} \mid I_{d6}, I_{indirect})$. See text for details.

such that $f(\mathbf{m} \mid I_{indirect})$ will be more informed, then realizations of the combined probability distribution $f(\mathbf{m} \mid I_{d1}, I_{Indirect})$ may expose more correlated features, corresponding the actual reference model. However, the information that cannot be resolved by the indirect information will stem from the probability distribution of direct information.

Figure 6.9 shows the pointwise mean (sometime called the etype mean) computed from all realizations. This indicates that, on average, the correct location of the

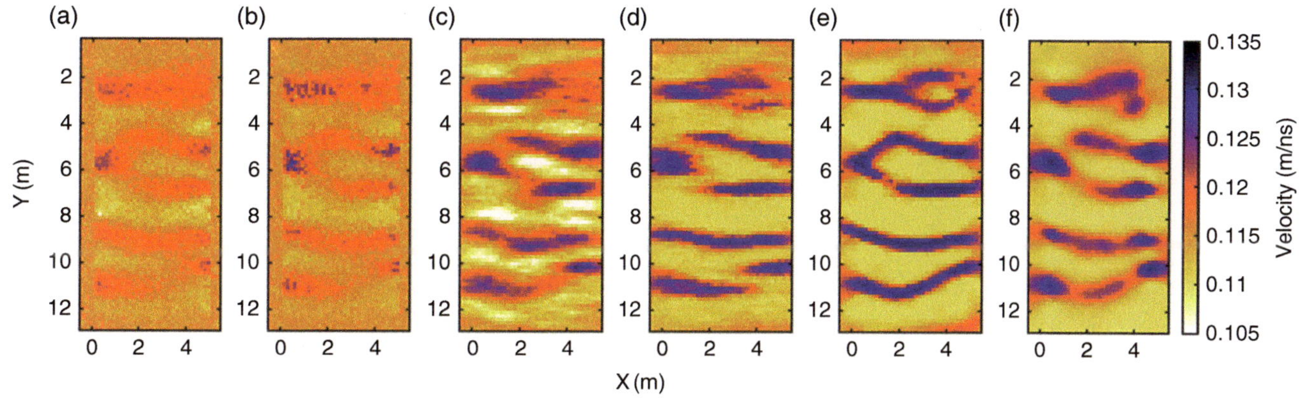

Figure 6.9 Pixelwise mean model obtained from a sample of **(a)** $f(\mathbf{m}\,|\,I_{d1}, I_{indirect})$, **(b)** $f(\mathbf{m}\,|\,I_{d2}, I_{indirect})$, **(c)** $f(\mathbf{m}\,|\,I_{d3}, I_{indirect})$, **(d)** $f(\mathbf{m}\,|\,I_{d4}, I_{indirect})$, **(e)** $f(\mathbf{m}\,|\,I_{d5}, I_{indirect})$, and **(f)** $f(\mathbf{m}\,|\,I_{d6}, I_{indirect})$.

Table 6.1 Correlation Coefficient Between Independent Realizations of $f(\mathbf{m}\,|\,I_{di}, I_{indirect})$

	I_{d1}	I_{d2}	I_{d3}	I_{d4}	I_{d5}	I_{d6}
CC	0.08	0.08	0.35	0.44	0.55	0.05

channel structures can be identified, even if they cannot be identified on the individual realizations. Specifically, using direct information probability distribution based on Voronoi cells results in individual realizations from $f(\mathbf{m}\,|\,I_{d6}, I_{Indirect})$ that are clearly geologically unrealistic (compared to the sample model) (Figure 6.8c), while realizations from $f(\mathbf{m}\,|\,I_{d5}, I_{Indirect})$ results in geologically highly realistic realizations (Figure 6.8b). On average, though, the pointwise mean is remarkable similar (Figures 6.9e and 6.9f). Note that such average models are, in general, not solutions to the data integration problem, as they may be inconsistent with both the direct and indirect information.

If the goal is to simulate geologically realistic features, then Figures 6.7 and 6.8 clearly show that, for this case, direct information describing geological realistic features are essential.

Table 6.1 provides the correlation coefficient between independent realizations of $f(\mathbf{m}\,|\,I_{di}, I_{indirect})$. A high number indicates that independent realizations are very similar and, hence, that the model parameters are well-resolved. Relying on the spatially uncorrelated models, $f(\mathbf{m}\,|\,I_{d1}, I_{indirect})$ and $f(\mathbf{m}\,|\,I_{d1}, I_{indirect})$ provides a very low correlation coefficient, which may suggest a poor resolution. The correlation coefficient increases as information about the model parameters, consistent with the reference model, increases. This indicates that as more information is available, consistent with the actual unknown subsurface, the resolution will increase.

6.6.3. Sampling $f(\mathbf{m}\,|\,I_{d3}, I_{indirect})$ Using the Classic Metropolis Algorithm

$f(\mathbf{m}\,|\,I_{d3})$ represents a Gaussian probability distribution and can be evaluated directly using Eq. (6.5). Therefore $f(\mathbf{m}\,|\,I_{d3}, I_{indirect})$ can be evaluated and, hence, sampled using the classic Metropolis algorithm.

Using a spatially uncorrelated uniform proposal distribution, with velocity values between 0.0755 m/ns and 0.1555 m/s, the classic Metropolis algorithm has been run for 4 million iterations in order to sample $f(\mathbf{m}\,|\,I_{d3}, I_{indirect})$.

Figure 6.10 shows three independent realizations from $f(\mathbf{m}\,|\,I_{d3}, I_{indirect})$ as well as the corresponding pointwise mean model. These results are comparable to the results obtained using the extended Metropolis sampler (Figures 6.7b and 6.9c).

Figure 6.11 shows the logarithm of the probability distribution values for $f(\mathbf{m}\,|\,I_{d3})$, $f(\mathbf{m}\,|\,I_{indirect})$, and $f(\mathbf{m}\,|\,I_{d3}, I_{indirect})$ as a function of iteration number using classic Metropolis algorithm, and $f(\mathbf{m}\,|\,I_{indirect})$ using the extended Metropolis algorithm. Both algorithms tend to sample models with comparable values for $f(\mathbf{m}\,|\,I_{indirect})$—that is, suggesting, as Figure 6.10, that the same probability distribution has been sampled.

However, it also highlights that the number of iterations needed to achieve burn-in—that is, where the algorithm starts to generate realizations of $f(\mathbf{m}\,|\,I_{d3}, I_{indirect})$—is very different. Using the extended Metropolis algorithm burn-in is reached after around 10^3 iterations, whereas it takes about 10^6 iterations to reach burn-in using the classic Metropolis algorithm with a uniform proposal distribution. Further, the number of iterations between independent realizations is about 4×10^3 using the extended Metropolis algorithm but about 1.5×10^6 using the classic Metropolis algorithm. Hence, the difference in computational requirements for sampling $f(\mathbf{m}\,|\,I_{d3}, I_{indirect})$

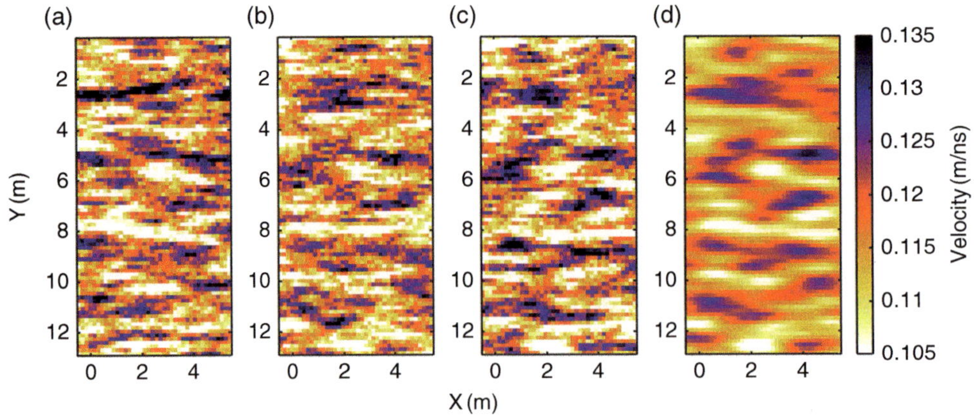

Figure 6.10 Three realizations from $f(\mathbf{m} \mid I_{d3}, I_{indirect})$, **(a–c)** and the pointwise average **(d)** obtained using 4,000,000 iterations of the classic Metropolis alogrithm.

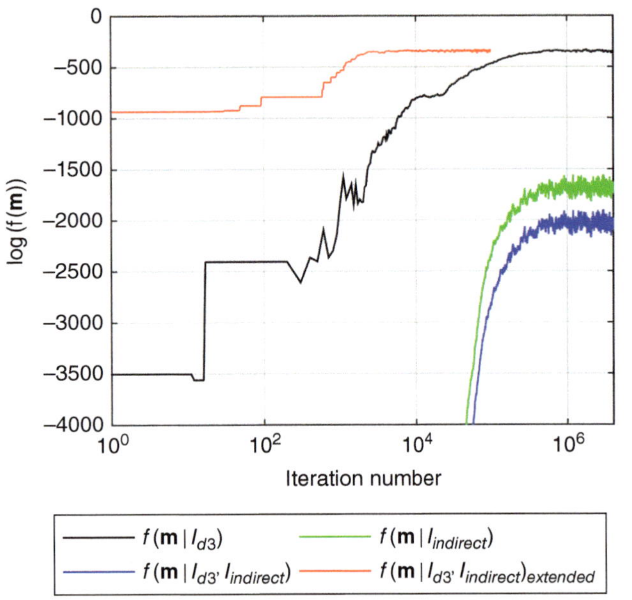

Figure 6.11 log(f(**m**)) as a function of iteration number for $f(\mathbf{m}|I_{indirect})$ (*green*), $f(\mathbf{m}|I_{d3})$ (*black*), and $f(\mathbf{m}|I_{d3}, I_{indirect})$ (*blue*) using the classic Metropolis algorithm and for $f(\mathbf{m}|I_{d3}, I_{indirect})_{extended}$ using the extended Metropolis algorithm (*red*).

using the two types of Metropolis algorithms is close to a factor of 1000.

The main reason for this huge difference in computational efficiency is related to the fact that using the classic Metropolis algorithm, one must sample $f(\mathbf{m} \mid I_{d3})$ as part of sampling $f(\mathbf{m} \mid I_{d3}, I_{indirect})$. On the other hand, using the extended Metropolis algorithm, the use of sequential Gibbs sampling ensures that all proposed models are realizations of $f(\mathbf{m} \mid I_{d3})$, and hence the computational requirements are mostly related to evaluating $f(\mathbf{m} \mid I_{d3}, I_{indirect})$.

6.7. DISCUSSION

The example in the previous section demonstrates the benefits of being able to use information about for example geologically plausible structures. It also demonstrates a case where the information quantified by $f(\mathbf{m} \mid I_{d1})$, ..., $f(\mathbf{m} \mid I_{d5})$ is consistent with the actual "Earth" as shown in Figure 6.6a. However, in practice the inference of statistical properties from a sample model, such as the one shown in Figure 6.1, may be associated with varying degrees of subjectivity. Also, an inferred statistical model may not be able to describe the actual spatial properties of the subsurface.

Consider the sample model in Figure 6.1. The width of the channels in this sample model is consistently around 0.6 m. The same is the case for the width of the channels in the realizations of $f(\mathbf{m} \mid I_{d5})$ shown in Figure 6.4b. In fact, the probability of locating a channel (that is, not intersecting other channels) with width $w > 1$ m or $w < 0.45$ m is zero. Further, in this sample model each model parameter can only take two values. This means that any other value will have a probability of zero of occurring. For any real case, the information exemplified in the sample model in Figure 6.1 will most likely exhibit too little variability. Hence, $f(\mathbf{m} \mid I_{d5})$ may be low in entropy and, in fact, inconsistent with the true Earth. Therefore it may be difficult, if not impossible, to integrate this information with other types of data. For an example on the use of inconsistent direct information, see, for example, *Hansen et al.* [2008]. The difficulty in quantifying direct information is that one should try to quantify as much direct information as possible, while at the same time allow realistic uncertainty [*Jaynes*, 1984; *Journel and Deutsch*, 1993].

Extreme High-Entropy Uniform Model. An extreme choice of an uninformed statistical model is the uniform model. Consider the integration of two types of independent information, $f(\mathbf{m} \mid I_1)$ and $f(\mathbf{m} \mid I_2)$, where $f(\mathbf{m} \mid I_1)$ represents a uniform distribution $\mathcal{U}(-\infty,\infty)$. Then $f(\mathbf{m} \mid I_1, I_2)$ is given by

$$f(\mathbf{m} \mid I_1, I_2) = f(\mathbf{m} \mid I_1) f(\mathbf{m} \mid I_2) \tag{6.25}$$

$$\propto f(\mathbf{m} \mid I_2) \tag{6.26}$$

In other words, when $f(\mathbf{m} \mid I_1)$ is a uniform distribution, it adds no information about the model parameters, as $f(\mathbf{m} \mid I_1, I_2) \propto f(\mathbf{m} \mid I_2)$. In principle, any 40 × 84 pixel random cutout of the reference model in Figure 6.6a is as probable an outcome of the uniform model $f(\mathbf{m} \mid I_{d2})$ as any of the single realizations shown in Figures 6.2b, 6.4a, 6.4b, and 6.4c. In reality, though, the uniform model as a choice for a description of the distribution of **m** involves a rather extreme assumption about maximum entropy, or maximum disorder. Any typical realization of $f(\mathbf{m} \mid I_1)$ will expose high disorder. In other words, the probability of realizing a model with a high degree of disorder is very high. The probability of realizing a highly ordered model, such as the reference model with ordered channel like structures, is extremely low. This is exactly what is exemplified by the realizations from $f(\mathbf{m} \mid I_{d2})$ in Figure 6.2b. The high entropy assumptions of $f(\mathbf{m} \mid I_{d2})$ will also be associated to $f(\mathbf{m} \mid I_{d2}, I_{indirect})$ as shown in Figure 6.7b.

This poses a problem, not only related to visual plausibility. In real life, end users may not be interested in the model parameters **m** themselves, but in a variable **k** linked to the model parameters through some transfer function h as $\mathbf{k} = h(\mathbf{m})$. **k** may be very sensitive to the type of spatial variability. Consider, for example, flow modeling of groundwater reservoir or hydrocarbon reservoirs. Say the channel structures (blue in Figure 6.6a) represents highly permeable structures embedded in low permeability material. Then, flow modeling results will provide radically different results depending on which model is chosen to describe spatial variability. For illustrative examples see, for example, *Journal and Deutsch* [1993]; *Journel and Zhang* [2006].

Another property of spatially uncorrelated models, such as $f(\mathbf{m} \mid I_{d1})$ and $f(\mathbf{m} \mid I_{d2})$, is that the number of effective "free" model parameters M_f is the same as the number of model parameters M, $M_f = M$. The number of "free" model parameters is the minimum number of model parameters needed to represent **m** [*Hansen et al.*, 2009]. When the number of model parameters increases, the data integration problem may become increasingly more difficult in terms of sampling from the distribution of combined information.

Extreme-Low Entropy Models. Other types of models represent cases of extreme low entropy. Consider, for example, a checkerboard model in a regular grid, where each model parameter (pixel) takes the value "black" or "white". The neighbor pixel up or down, left or right to one centered pixel has the opposite value as the center pixel. This also means that for such a model, the number of free parameters is $M_f = 1$, independent of the actual number of parameters. If such a checkerboard model is used to describe direct information, then an exhaustive search of all possible models can be undertaken simply by evaluating two models, one with a white pixel centered at a reference parameter and one with a black pixel centered at a reference parameter.

Another extreme type of low-entropy model is the multivariate Gaussian model, where all the model parameters are completely correlated. Again, this would indicate that one only needs to know the value of one model parameter in order to know the value of all model parameters ($M_f = 1$) independent of the number of model parameters.

Intermediate Entropy Models. In general, the number of free model parameters will depend on the chosen a priori model. For multivariate Gaussian models, *Hansen et al.* [2009] demonstrate that in general the number of effective free parameters is related to the correlation length. The longer the correlation length, the smaller the value of M_f. When the correlation length is zero, the model parameters become independent, and hence $M_f = M$.

Low Entropy as the Source of Inconsistencies. In order to avoid inconsistencies in data integration, careful consideration should be used when quantifying different types of information $f(\mathbf{m} \mid I_i)$. If only known information is quantified and all uncertainties are taken into account, inconsistencies should not arise.

However, sometimes data integration, in the form of sampling from $f(\mathbf{m} \mid I_1, I_2)$, can become unsolvable if there is inconsistency between the available information [*Hansen et al.*, 2008]. There are at least four explanations: (1) The direct information is specified such that the data cannot be matched within their uncertainty, (2) the modeling uncertainty related to the forward model is underestimated, (3) the measurement uncertainty is underestimated, and (4) the parameterization has been chosen too sparse to allow realistic representation of Earth structures [*Mosegaard and Hansen*, 2015]. In any case inconsistencies may arise when some of the information has been described with too little uncertainty.

Sampling from $f(\mathbf{m} \mid I_{d5}, I_{indirect})$ Using Different Neighborhood. The entropy, and the degree of spatial variability, is affected when the size of the neighborhood is changed (i.e., when changing the number of conditional data), which is used to compute/evaluate the conditional distribution as part of running sequential simulation. The smaller the amount of conditional data, the smaller the amount of information that is assumed (the entropy increases).

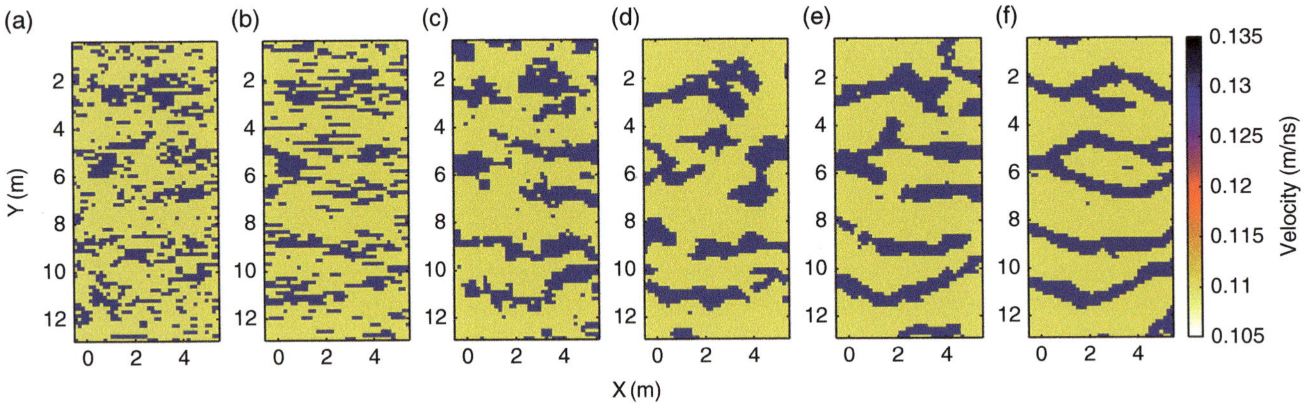

Figure 6.12 One realization from $f(\mathrm{m}|I_{d5}, I_{indirect})$ using different number for conditioning data, N_c (i.e., different size data neighborhood). **(a)** $N_c = 1$, **(b)** $N_c = 2$, **(c)** $N_c = 4$, **(d)** $N_c = 8$, **(e)** $N_c = 15$, **(f)** $N_c = 30$.

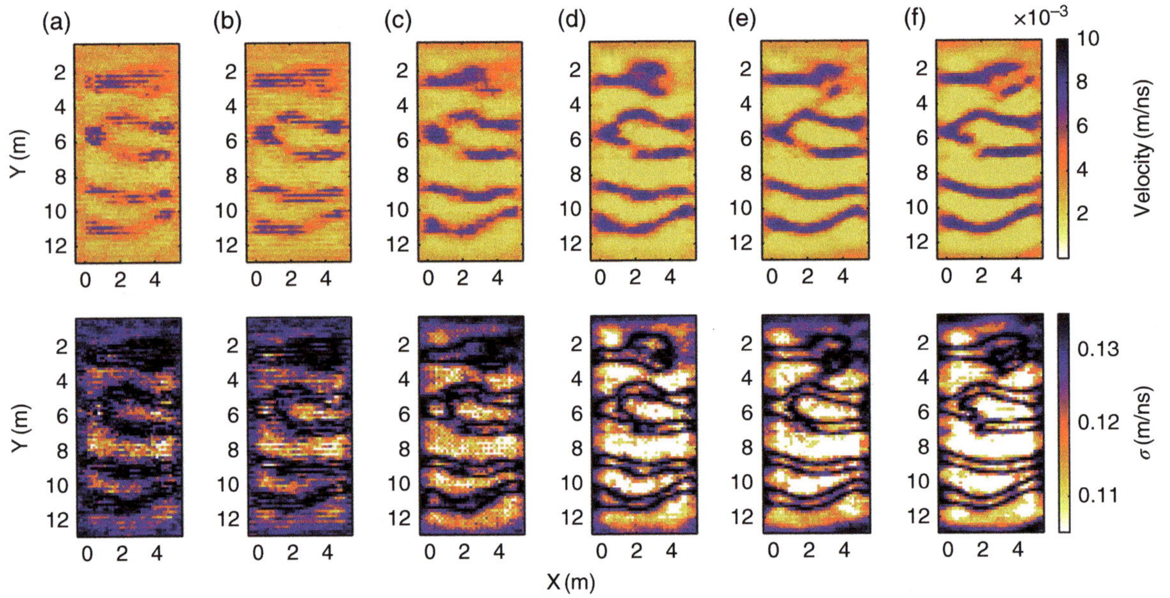

Figure 6.13 The pointwise mean (*top*) and standard deviation (*bottom*) from a sample of $f(\mathbf{m}|I_{d5}, I_{indirect})$ using different number for conditioning data, N_c (i.e., different size data neighborhood). **(a)** $N_c = 1$, **(b)** $N_c = 2$, **(c)** $N_c = 4$, **(d)** $N_c = 8$, **(e)** $N_c = 15$, **(f)** $N_c = 30$.

For the realizations generated from $f(\mathbf{m} \mid I_{d5})$ and $f(\mathbf{m} \mid I_{d5}, I_{indirect})$, shown in Figures 6.4b and 6.8b, the number of conditional points for the sequential simulation algorithm used is $N_c = 60$. Figure 6.12 shows one realization obtained from sampling $f(\mathbf{m} \mid I_{d5}, I_{indirect})$ using $N_c = [1, 2, 4, 8, 15, 30]$. The same type of extended Metropolis algorithm as described earlier is used. Figure 6.13 shows the corresponding pointwise mean (top row) and point wise standard deviation (bottom row) obtained from all generated realizations. Note how the variability is increasingly associated with the location of the channel edges as the number of conditional data increases. Note that $f(\mathbf{m} \mid I_{d5})$ corresponds to $f(\mathbf{m} \mid I_{d2})$ when no conditioning points data are used (i.e., assuming no spatial dependency). The spatial disorder is clearly seen to decrease as the number of conditioning points increase.

6.8. CONCLUSIONS

The goal of probabilistic data integration is to (1) integrate all available information $\mathbf{I} = [I_1, I_2, \ldots, I_N]$ related to model parameters $\mathbf{m}$ into one probability distribution $f(\mathbf{m} \mid \mathbf{I})$ and (2) generate a large sample from $f(\mathbf{m} \mid \mathrm{I})$ allowing detailed uncertainty analysis and propagation of uncertainty into other types of parameters (such as, for example, related to flow simulations).

In some rare cases, $f(\mathbf{m} \mid \mathbf{I})$ can be evaluated, in which case the "classic" Metropolis algorithm (or in principle the rejection sampler) can be used to sample $f(\mathbf{m} \mid \mathbf{I})$ directly. However, the type of (usually simplistic) information that can be quantified and allows evaluation of $f(\mathbf{m} \mid \mathbf{I})$ is often not adequate to describe information at hand. Further, even when this is the case, such a sampling problem can become prohibitively computationally demanding, even for the relatively small 2D models considered here, which will be, in practice, intractable. Direct sampling of $f(\mathbf{m} \mid \mathbf{I})$ using the rejection sampler or the classic Metropolis algorithm will, in general, lead to a computationally intractable problem.

On the other hand, complex models of direct information can be quantified in a way that allows efficient sampling, based on sequential simulation, from these models, without the need to evaluate $f(\mathbf{m} \mid I_1)$ and, hence, $f(\mathbf{m} \mid \mathbf{I})$. When such information is available, together with other types of information such as indirect information from, for example, geophysical data $I_{indirect}$, where $f(\mathbf{m} \mid I_{indirect})$ can be evaluated, then $f(\mathbf{m} \mid I_1, I_{indirect})$ can be sampled efficiently using the extended Metropolis algorithm utilizing the sequential Gibbs sampler to sample $f(\mathbf{m} \mid I_1)$.

Compared to using direct sampling of $f(\mathbf{m} \mid \mathbf{I})$ using the classic Metropolis algorithm with a uniform proposal distributions, the use of extended Metropolis can lead to a sampling problem that is orders of magnitude more tractable.

A wide range of statistical methods, providing varying degrees of information content, are currently available that can be used with the extended Metropolis algorithm and that allow characterization of probability distributions describing quite complex and geologically realistic spatial features. These methods allow building statistical models that assume, in principle, a lot more than is typically known. Therefore, care should be taken when quantifying direct information, to avoid subjective information such that only information that is actually known is quantified and taken into account and such that all uncertainties are taken into account. If this is not the case, then the data integration problem may become either inconsistent and unsolvable, or solvable but providing biased results with too little associated uncertainty.

On the other hand, realistic description of direct information has several advantages: (1) Realizations from the probability distribution describing the combined information will be consistent with structural geological information. (2) Sampling from $f(\mathrm{m} \mid \mathrm{I})$ will be computationally more efficient. (3) The complexity of the inverse problem can be dramatically reduced due to the reduced number of effective free model parameters.

ACKNOWLEDGMENTS

All computations have been performed using the SIPPI Matlab package [*Hansen et al.*, 2013b], and codes for reproducing the results can be found at http://sippi.sourceforge.net/. We thank Mats Lundh Gulbrandsen for discussions and editing. We thank two reviewers for their constructive and useful critique.

REFERENCES

Armstrong, M., A. Galli, H. Beucher, G. Loc'h, D. Renard, B. Doligez, R. Eschard, and F. Geffroy (2011), *Plurigaussian Simulations in Geosciences*, Springer Science & Business Media, New York.

Bodin, T., M. Sambridge, and K. Gallagher (2009), A self-parametrizing partition model approach to tomographic inverse problems, *Inverse Problems*, *25*(5), 055,009.

Bosch, M. (2015), Inference networks in earth models with multiple components and data, in *Integrated Imaging in Earth Science*, M. Moorkamp, N. Linde, P. Lelievre, and A. Khan, eds., AGU, Washington, DC.

Buland, A., and H. Omre (2003), Bayesian linearized avo inversion, *Geophysics*, *68*(1), 185–198.

Caers, J. (2000), Direct sequential indicator simulation, in *Proceedings of the 6th International Geostatistics Congress, Cape Town, South Africa, April 10–14, 2000*, W. Kleingeld and D. Krige, eds., 12 pp.

Constable, S. C., R. L. Parker, and C. G. Constable (1987), Occam's inversion: A practical algorithm for generating smooth models from electromagnetic sounding data, *Geophysics*, *52*(3), 289–300.

Cordua, K. S., T. M. Hansen, and K. Mosegaard (2012), Monte Carlo full waveform inversion of crosshole GPR data using multiple-point geostatistical a priori information, *Geophysics*, *77*, H19–H31, doi:10.1190/geo2011-0170.1.

Cordua, K. S., T. M. Hansen, and K. Mosegaard (2015), Improving the pattern reproducibility of multiple-point-based prior models using frequency matching, *Mathe. Geosci.*, *47*, 317–343.

Cressie, N., and J. L. Davidson (1998), Image analysis with partially ordered Markov models, *Comput. Stat. Data Anal.* *29*(1), 1–26.

Daly, C. (2005), Higher order models using entropy, Markov random fields and sequential simulation, Geostatistics Banff 2004, Springer Netherlands, pp. 215–224.

Deutsch, C. V., and A. G. Journel (1998), *GSLIB, Geostatistical Software Library and User's Guide, Applied Geostatistics*, 2nd ed., Oxford University Press, New York, 384 pp.

Dimitrakopoulos, R., H. Mustapha, and E. Gloaguen (2010), High-order statistics of spatial random fields: Exploring spatial cumulants for modeling complex non-Gaussian and non-linear phenomena, *Mathematical Geosciences*, *42*(1), 65–99.

Duane, S., A. D. Kennedy, B. J. Pendleton, and D. Roweth (1987), Hybrid monte carlo, *Physics Lett. B*, *195*(2), 216–222.

Emery, X. (2007), Using the gibbs sampler for conditional simulation of Gaussian-based random fields, *Comput. Geosci.*, *33*(4), 522–537.

Fu, J., and J. J. Gómez-Hernández (2008), Preserving spatial structure for inverse stochastic simulation using blocking Markov chain Monte Carlo method, *Inverse Probl. Sci. Eng.*, *16*(7), 865–884.

Geman, S., and D. Geman (1984), Stochastic relaxation, Gibbs distributions, and the Bayesian restoration of images, *IEEE Trans. Pattern Anal. Machine Intell.*, *6*, 721–741.

Gomez-Hernandez, J., and A. Journel (1993), Joint sequential simulation of multi-Gaussian fields, *Geostatistics Troia, 92*, 85–94.

Guardiano, F., and R. Srivastava (1993), Multivariate geostatistics: Beyond bivariate moments, *Geostatistics-Troia, 1*, 133–144.

Hansen, T. M., and K. Mosegaard (2008), VISIM: Sequential simulation for linear inverse problems, *Comput. Geosci., 34*(1), 53–76.

Hansen, T. M., K. Mosegaard, and K. C. Cordua (2008), Using geostatistics to describe complex a priori information for inverse problems, in *VIII International Geostatistics Congress*, Vol. 1, J. M. Ortiz and X. Emery, eds., Mining Engineering Department, University of Chile, pp. 329–338.

Hansen, T. M., K. S. Cordua, and K. Mosegaard (2009), Reducing complexity of inverse problems using geostatistical priors, in *Proceeding from IAMG 09, August 23–28, 2009, Stanford, CA*.

Hansen, T. M., K. C. Cordua, and K. Mosegaard (2012), Inverse problems with nontrivial priors—Efficient solution through sequential Gibbs sampling, *Computational Geosciences, 16*(3), 593–611, doi:10.1007/s10596-011-9271-1.

Hansen, T., K. Cordua, M. Looms, and K. Mosegaard (2013a), SIPPI: A Matlab toolbox for sampling the solution to inverse problems with complex prior information: Part 2, Application to cross hole GPR tomography, *Comput. Geosci., 52*, 481–492, doi:10.1016/j.cageo.2012.10.001.

Hansen, T., K. Cordua, M. Looms, and K. Mosegaard (2013b), SIPPI: A Matlab toolbox for sampling the solution to inverse problems with complex prior information: Part 1, methodology, *Comput. Geosci., 52*, 470–480, doi:10.1016/j.cageo.2012.09.004.

Hansen, T. M., K. S. Cordua, B. H. Jacobsen, and K. Mosegaard (2014), Accounting for imperfect forward modeling in geophysical inverse problems exemplified for crosshole tomography, *Geophysics, 79*(3), H1–H21.

Holliger, K., and A. Levander (1994), Lower crustal reflectivity modeled by rheological controls on mafic intrusions, *Geology, 22*(4), 367–370.

Irving, J., and K. Singha (2010), Stochastic inversion of tracer test and electrical geophysical data to estimate hydraulic conductivities, *Water Resour Res., 46*, W11514.

Jaynes, E. T. (1984), Prior information and ambiguity in inverse problems, *Inverse Problems, 14*, 151–166.

Journel, A., and E. Isaaks (1984), Conditional indicator simulation: Application to a saskatchewan uranium deposit, *J. Int. Associ. Math. Geol., 16*(7), 685–718.

Journel, A., and T. Zhang (2006), The necessity of a multiple-point prior model, *Math. Geol., 38*(5), 591–610.

Journel, A. G. (1994), Modeling uncertainty: Some conceptual thoughts, in *Geostatistics for the next Century*, Springer, New York, pp. 30–43.

Journel, A. G., and C. V. Deutsch (1993), Entropy and spatial disorder, *Math. Geol., 25*(3), 329–355.

Journel, A. G., and C. J. Huijbregts (1978), *Mining Geostatistics*, Academic Press, New York, 600 pp.

Lange, K., J. Frydendall, K. S. Cordua, T. M. Hansen, Y. Melnikova, and K. Mosegaard (2012), A frequency matching method: Solving inverse problems by use of geologically realistic prior information, *Math. Geosci., 44*(7), 783–803.

Le Ravalec, M., B. Noetinger, and L. Y. Hu (2000), The FFT moving average (FFT-MA) generator: An efficient numerical method for generating and conditioning Gaussian simulations, *Math. Geol., 32*(6), 701–723.

Liu, J. S. (1996), Metropolized independent sampling with comparisons to rejection sampling and importance sampling, *Stat. Comput., 6*(2), 113–119.

Malinverno, A. (2002), Parsimonious bayesian markov chain Monte Carlo inversion in a nonlinear geophysical problem, *Geophysical Journal International, 151*(3), 675–688.

Mariethoz, G., and J. Caers (2014), *Multiple-Point Geostatistics: Stochastic Modeling with Training Images*, Wiley-Blackwell, Hoboken, NJ, 376 pp.

Mariethoz, G., P. Renard, and J. Straubhaar (2010), The direct sampling method to perform multiple-point geostatistical simulations, *Water Resources Res., 46*(11).

Metropolis, N., M. Rosenbluth, A. Rosenbluth, A. Teller, and E. Teller (1953), Equation of state calculations by fast computing machines, *J. Chem. Phys., 21*, 1087–1092.

Mariethoz G., P. Renard, and J. Caers (2010), Bayesian inverse problem and optimization with iterative spatial resampling, *Water Resour. Res., 46*, W11530, http://dx.doi.org/10.1029/2010WR009274.

Mosegaard, K., and T. M. Hansen (2015), Inverse methods: Problem formulation and probabilistic solutions, in *Integrated Imaging in Earth Science*, M. Moorkamp, N. Linde, P. Lelievre, and A. Khan, eds., AGU, Washington, DC.

Mosegaard, K., and A. Tarantola (1995), Monte Carlo sampling of solutions to inverse problems, *J. Geophy. Res., 100*(B7), 12,431–12,447.

Oz, B., C. V. Deutsch, T. T. Tran, and Y. Xie (2003), DSSIM-HR: a FORTRAN 90 program for direct sequential simulation with histogram reproduction, *Comput. Geosci., 29*(1), 39–51, doi:http://dx.doi.org/10.1016/S0098-3004(02)00071-7.

Peredo, O., and J. M. Ortiz (2011), Parallel implementation of simulated annealing to reproduce multiple-point statistics, *Comput. Geosci., 37*(8), 1110–1121.

Remy, N., A. Boucher, and J. Wu (2008), *Applied Geostatistics with SGeMS: A User's Guide*, Cambridge University Press, New York.

Sambridge, M. (2013), A parallel tempering algorithm for probabilistic sampling and multimodal optimization, *Geophys. J. Int.*, p. ggt342.

Sambridge, M., and K. Mosegaard (2002), Monte Carlo methods in geophysical inverse problems, *Rev. Geophy., 40*(3), 3–1.

Scales, J. A., and R. Sneider (1997), To Bayes or not to Bayes?, *Geophysics, 62*(4), 1045–1046.

Shannon, C. E. (1948), A mathematical theory of communication, *ACM SIGMOBILE Mobile Comput Commun. Rev—reprint 2001, 5*(1), 3–55.

Soares, A. (2001), Direct sequential simulation and cosimulation, *Math. Geol., 33*(8), 911–926.

Strebelle, S. (2000), Sequential simulation drawing structures from training images, Ph.D. thesis, Stanford University.

Strebelle, S. (2002), Conditional simulation of complex geological structures using multiple-point statistics, *Math. Geol., 34*(1), 1–20.

Tarantola, A. (2005), *Inverse Problem Theory and Methods for Model Parameter Estimation*, SIAM, Philadelphia.

Tarantola, A., and B. Valette (1982), Inverse problems—Quest for information, *J. Geophys. 50*(3), 150–170.

Tjelmeland, H., and J. Besag (1998), Markov random fields with higher-order interactions, *Scand. J. Stat., 25*(3), 415–433.

Part II
Applications

7

Joint Inversion in Hydrogeophysics and Near-Surface Geophysics

Niklas Linde[1] and Joseph Doetsch[2]

ABSTRACT

The near-surface environment is often too complex to enable inference of hydrological and environmental variables using one geophysical data type alone. Joint inversion and coupled inverse modeling involving numerical flow and transport simulators have, in the last decade, played important roles in pushing applications towards increasingly challenging targets. Joint inversion of geophysical data that is based on structural constraints is often favored over model coupling based on explicit petrophysical relationships. More specifically, cross-gradient joint inversion has been applied to a wide range of near-surface applications and geophysical data types. To infer hydrological subsurface properties, the most appropriate approach is often to use temporal changes in geophysical data that can be related to hydrological state variables. This allows using geophysical data as indirect hydrological observables, while the coupling with a flow and transport simulator ensures physical consistency. Future research avenues include investigating the validity of different coupling strategies at various scales, the spatial statistics of near-surface petrophysical relationships, the influence of the model conceptualization, fully probabilistic joint inversions, and how to include complex prior information in the joint inversion.

7.1. INTRODUCTION

The increasing pace of unsustainable man-induced activities and associated threats [*Rockström et al.*, 2009] will, for the foreseeable future, continue to push researchers and practitioners towards environmental problems of growing complexity. Remote sensing and geophysics are playing ever-increasing roles in describing near-surface environments at scales spanning the root zone to major aquifers [*NRC*, 2012]. For instance, critical zone research focusing on "*the heterogeneous, near-surface environment in which complex interactions involving rock, soil, water, and living organisms regulate the natural habitat and determine the availability of life-sustaining resources*" [*NRC*, 2001] requires spatially distributed data. In particular, process understanding in the deep critical zone can be enhanced by geophysical data, as it is largely out of reach for classic methods based on coring and trial pits [*Parsekian et al.*, 2015]. To illustrate the need for multiple data and joint inversion, let us consider the discipline of hydrogeophysics that relies on geophysical data to gain information about hydrological processes or controlling subsurface structures (see, e.g., *Rubin and Hubbard* [2005] and *Hubbard and Linde* [2011]). Early hydrogeophysical research often relied on the assumption that geophysical tomograms could be treated as spatially distributed and exhaustive "data". Petrophysical relationships (sometimes estimated through inversion; see, e.g., *Hyndman et al.* [2000] and *Linde et al.* [2006b]) were then used to

[1] *Applied and Environmental Geophysics Group, Institute of Earth Sciences, University of Lausanne, Lausanne, Switzerland*

[2] *Swiss Competence Center for Energy Research, Supply of Electricity (SCCER-SoE), ETH Zurich, Zurich, Switzerland*

Integrated Imaging of the Earth: Theory and Applications, Geophysical Monograph 218, First Edition.
Edited by Max Moorkamp, Peter G. Lelièvre, Niklas Linde, and Amir Khan.

convert these "data" into hydrological properties (see, e.g., *Rubin et al.* [1992], *Copty et al.* [1993], *Cassiani and Medina* [1997], *Cassiani et al.* [1998]). Such approaches have been criticized because they often neglect the resolution limitations of geophysical tomograms [*Day-Lewis and Lane*, 2004; *Day-Lewis et al.*, 2005] and because of the strong assumptions made about the relationship between the geophysical tomogram and the hydrogeological system of interest [*Linde et al.*, 2006c]. Indeed, using one data type alone is often insufficient to adequately constrain key target variables (see, e.g., *Linde et al.* [2006a]). Joint inversion provides formal approaches to integrate multiple data such that the resulting subsurface models and interpretations are more consistent and reliable than those obtained by comparing the results obtained by individual inversion of different data types (see, e.g., *Doetsch et al.* [2010b]). *Linde et al.* [2006c] conclude their review on hydrogeophysical parameter estimation approaches by stating that joint inversion is the way forward, but that is still in its infancy. The situation has changed radically in the last 10 years, and joint inversion is nowadays widely used.

Most of the joint inversion methodologies in near-surface environments have been developed and applied for hydrogeophysical applications, and it is hoped that this review can stimulate applications in other related fields such as archeological geophysics and civil engineering geophysics. Archeological prospecting would be an ideal target for joint inversion, because archeological remains cause signatures in multiple geophysical methods (e.g., magnetometry, ground penetrating radar (GPR), electrical resistance tomography (ERT)).Many archeological sites have been surveyed with complementary geophysical methods, which offer the possibility of joint interpretation and data integration (see, e.g., *Böniger and Tronicke* [2010] and *Keay et al.* [2009]). To date, integrated analysis has been performed on the 2D and 3D models that result from the individual geophysical surveys using visualization tools, advanced image processing, and statistical analysis [*Kvamme*, 2006; *Watters*, 2006]. Joint inversion could here help to shift the integration to the data level and ensure that resultant models honor all available data. In engineering and public safety applications, joint inversion can improve the reliability of feature detection (see, e.g., unexploded ordnance (UXO) characterization; *Pasion et al.* [2003]), which could also be used for safety assessment of critical infrastructure such as roads and bridges.

This review focuses on joint inversion of geophysical data, as well as joint inversion of geophysical and hydrological data, in near-surface environments. To enable a comprehensive review, we primarily consider contributions that fulfill the following conditions:

1. At least one type of near-surface geophysical data is considered. We will not consider the vast literature on joint inversion of different types of hydrological data, such as, hydraulic head and tracer test data [*Nowak and Cirpka*, 2006]);

2. At least two different physical or geological subsurface properties are considered. For example, this implies that we will not discuss joint inversion of electrical and electromagnetic (EM) induction methods, as they are both sensitive to electrical resistivity (see, e.g., *Kalscheuer et al.* [2010] and *Rosas-Carbajal et al.* [2014]). These type of applications can be handled within the same framework as individual inversions.

3. There is one common objective function that considers the data misfit of all data types simultaneously. We will not discuss inversion of one geophysical data type that uses prior information about property variations (see, e.g., *Saunders et al.* [2005]) or lithological interfaces [*Doetsch et al.*, 2012] inferred from previously processed or inverted geophysical data. These are very useful approaches, but they do not qualify as joint inversion.

4. We consider joint inversion of dependent data only (e.g., electrical resistances, seismic travel times, tracer breakthrough curves). That is, we do not consider geophysical inversion of one data type conditional to interpreted borehole logging data (e.g., fracture zonation by combining permeability estimates based on flowmeter data with crosshole seismic travel times [*Chen et al.*, 2006]). Again, such approaches can be very useful, but they do not qualify as joint inversion.

5. The primary depth range is from the centimeter to the kilometer scale and we exclude application areas treated elsewhere in this book. Hence, we will not consider mineral exploration (see Chapter 8) or hydrocarbon exploration (see Chapter 9).

This review is organized as follows. Section 7.2 describes petrophysical approaches and Section 7.3 describes structural approaches to joint inversion in near-surface environments. Section 7.4 focuses on hydrogeophysical applications in which hydrological flow and transport simulations are combined with petrophysical relationships that link hydrological state variables and geophysical properties. Section 7.5 discusses outstanding challenges and Section 7.6 provides concluding remarks.

7.2. PETROPHYSICAL JOINT INVERSION IN NEAR-SURFACE ENVIRONMENTS

Petrophysical joint inversion relies on a statistical or theoretically based relationship between different model parameter types (e.g., P-wave velocity and density; electrical conductivity and permeability). One attractive feature of joint inversion by petrophysical coupling is that it allows formulating the inverse problem in terms of the target properties of primary interest (e.g., porosity, permeability,

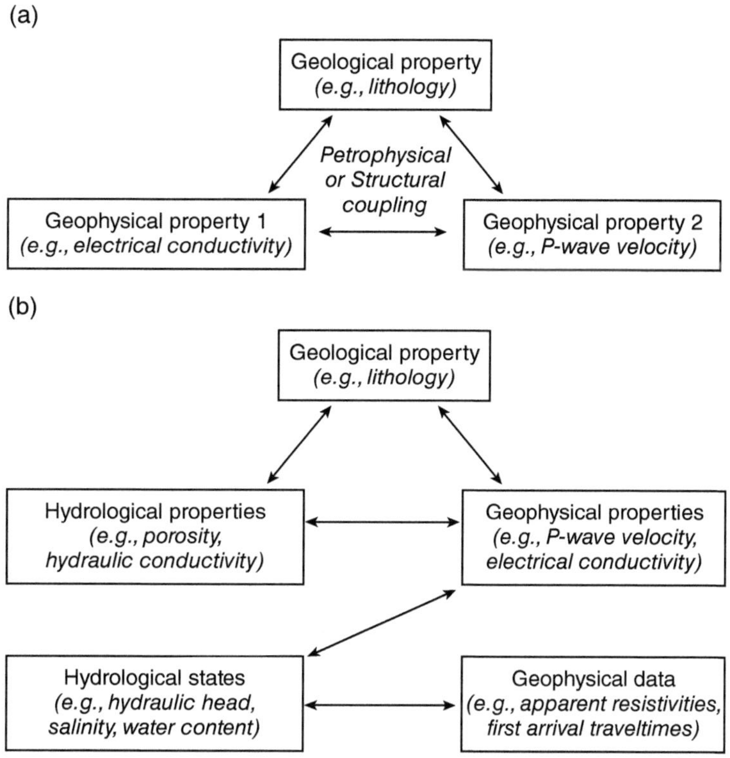

Figure 7.1 **(a)** Common coupling strategies for the joint inversion of near-surface geophysical data. **(b)** Joint inversion of hydrological and geophysical data opens up additional possibilities for model coupling. These schematic figures should not be interpreted as flowcharts of joint inversion methods as they simply highlight the possible information exchange between multiple properties and data.

lithology). In this approach that is referred to as lithological tomography, the petrophysical relationships are used to transform the primary property fields into geophysical property fields that the observed geophysical data are sensitive to (see the upper arrows in Figure 7.1). Comparisons of the simulated forward responses and the data are then used to guide further model updates. For more details, we refer to the work by *Bosch* [1999, Chapter 3 in this volume], Another option is to impose a direct petrophysical relationship between different geophysical properties (e.g., electrical conductivity, dielectric permittivity; see lower arrow in Figure 7.1a). If the petrophysical relationships include primary hydrological properties (e.g., porosity or permeability) and not hydrological state variables (e.g., pressure, water content, salinity), then there is no methodological difference in including hydrological data in the joint inversion (see the upper three rectangles in Figure 7.1b). In the following, hydrological data should be understood as measurements of hydrological state variables (e.g., hydraulic pressure or salinity).

There were until recently only few examples of petrophysics-based joint inversion in near-surface environments. The likely reason for this is that the approach is mainly feasible when working with sampling or global optimization methods [*Sen and Stoffa*, 2013] that can easily consider nonlinear, complex, and uncertain petrophysical relationships. Local inverse formulations (e.g., Gauss–Newton) are very sensitive to errors in the petrophysical model and it is likely that model artifacts will be introduced to compensate for these errors. It is only recently that the near-surface community has adapted global methods, and computational constraints will continue to limit their applicability for many applications. For example, *Hertrich and Yaramanci* [2002] used simulated annealing to jointly invert both synthetic and field-based surface nuclear magnetic resonance (SNMR) and vertical electrical sounding (VES) data under the assumption of a common layered 1-D earth structure. Each layer was parameterized in terms of mobile (sensed by SNMR and VES data) and immobile (sensed by the VES data only) water content, fluid resistivity, and the cementation exponent in Archie's law. *Jardani et al.* [2010] used an adaptive Metropolis algorithm to jointly invert synthetic seismic and seismoelectric data to infer, in a 2-D setting, permeability, porosity, electrical conductivity, and four different moduli for three zones of known geometry.

We are not aware of work on joint inversions of geophysical and hydrological data that use a petrophysical relationship to relate primary hydrological properties with geophysical properties. Most joint inversions of geophysical and hydrological data rely on petrophysical relationships between geophysical properties and hydrological state variables. Such approaches are discussed in Section 7.4.

7.3. STRUCTURALLY COUPLED JOINT INVERSION IN NEAR-SURFACE ENVIRONMENTS

Structurally coupled joint inversion seek multiple distributed models that share common interfaces or have similar model gradients (see Figure 7.1a). They can be grouped into three subcategories: common layers, common lithological units, and gradient-based joint inversion.

7.3.1. Common Layer Structure

It is possible to conceptualize the subsurface by layers of constant (1-D) or variable (2-D and 3-D) thickness and to ignore vertical property variations within layers. The model parameters to infer are thus related to the interface depths between layers and lateral property variations with each layer. This is often a suitable model parameterization for large-scale aquifer characterization in sedimentary settings, and it facilitates the integration of unit boundaries identified in borehole logs. The coupling for the joint inversion is here achieved by imposing common layer boundaries [*Auken and Christiansen*, 2004].

Hering et al. [1995] jointly inverted VES data together with Rayleigh and Love wave dispersion curves for a layered 1-D model. In a synthetic example it was found that the resulting model parameters were better defined than for individual inversions. *Misiek et al.* [1997] applied this joint inversion approach to two field datasets and reported improved results compared with individual inversions. *Kis* [2002] proposed a joint inversion of VES and seismic refraction travel times that she applied to both synthetic and field-based data. The seismic and electrical properties were assumed constant within each layer and they shared common interfaces in 2-D by using a common parameterization in terms of Chebyshev polynomials. *Wisén and Christiansen* [2005] derived 2-D layered models based on 1-D modeling for both synthetic and field-based VES and surface wave seismic data. The models were made laterally continuous by penalizing lateral gradients of layer properties and interface locations. The electrical resistivity and shear wave velocity models were coupled by assuming similar layer interfaces. This coupling led to a better-constrained shear wave velocity model and a good agreement of interfaces identified with lithological drill logs. *Juhojuntti and Kamm* [2010] further extended this approach to 2-D joint inversion of both synthetic and field-based seismic refraction and ERT data. The resulting interfaces agreed well with independent information (cone-penetration tests, drillings, a reflection seismic section). *Santos et al.* [2006] jointly inverted VES and gravity data for a 2-D layered model (the VES data were simulated using a local 1-D model). The density of each layer was assumed known, and simulated annealing was used to determine the layer resistivities and thicknesses. Synthetic test models indicated that the joint inversion provided a more stable and reliable parameter estimation than individual inversions. A field example related to water prospection provided results in agreement with local geology. *Jardani et al.* [2007] used simulated annealing to jointly invert field-based self-potential and apparent electrical conductivity data (EM-34). Their model consisted of two layers with an irregular interface. Each layer was characterized by an electrical resistivity and an apparent voltage-coupling coefficient. The modeling was 1-D, but lateral constraints were imposed to obtain a map view of the depth to the interface, which was used to identify sinkholes. *Moghadas et al.* [2010] jointly inverted full-waveform data from synthetic off-ground zero-offset GPR and EM data to determine the electrical conductivity and permittivity of two soil layers, as well as the thickness of the upper layer. A Bayesian formulation to the joint inversion provided the best results and they found that sequential inversion alternatives could lead to biased estimates. *Günther and Müller-Petke* [2012] jointly inverted SNMR and VES data using a Marquardt-type Gauss–Newton inversion. In addition to the thickness of each layer, the algorithm provided the porosity, decay time, and electrical resistivity of each layer. In a field application, they found that the inclusion of SNMR data resulted in an improved lithological description of the subsurface compared with inversion of VES data alone.

7.3.2. Common Lithological Units

The layered model parameterization allows for variable layer thicknesses, but the number of layers is the same throughout the model domain and each layer has an infinite lateral extent. This is a questionable assumption in geological settings characterized by finite-size bodies, for example, channel forms or clay lenses. In such cases, it is better to parameterize the subsurface in terms of lithological units and to assume that the internal heterogeneity within each unit is negligible compared to the contrast with neighboring units. The main challenge with this formulation is the need for an adaptive model parameterization.

Musil et al. [2003] used discrete tomography based on mixed-integer linear programming to jointly invert synthetic crosshole seismic and radar data to locate air and water-filled cavities. Inequality constraints forced model properties to the expected values (rock-, air- and water-filled cavities) with discrete jumps at the interfaces. *Paasche and Tronicke* [2007] proposed a cooperative inversion, in which the results of individual inversions (synthetic crosshole GPR and P-wave travel times in their examples) underwent a fuzzy *c*-means clustering step after each iteration step. The resulting zonal model was subsequently used as the starting model for a conventional individual inversion, and so on. This allows information sharing from different data types, even if it is strictly speaking not a joint inversion. *Linder et al.* [2010] extended this approach to three data types (field-based crosshole GPR, P-wave and S-wave travel times).

Considering a synthetic test case, *Hyndman et al.* [1994] inverted for lithological models by simultaneously minimizing the misfit between simulated and observed seismic travel times and tracer test data. A standard seismic travel-time inversion was first carried out to obtain an initial seismic velocity model. A separation velocity between homogeneous high- and low-velocity zones was then sought that minimized the travel-time residuals. Under the assumption that this zonal model was also representative of permeability, the tracer test data were inverted for a constant permeability in each zone. In the next step, the seismic velocity model was updated by performing a new seismic inversion with the zonal model as starting model. The separation velocity was this time determined by simultaneously minimizing the misfit in seismic travel times and the tracer test data. These steps were repeated multiple times. In a field application, *Hyndman and Gorelick* [1996] extended this approach from 2-D to 3-D, into three different lithology types, and including hydraulic data as well. The resulting 3-D lithological model provides one of the most convincing hydrogeophysical case studies to date. *Cardiff and Kitanidis* [2009] developed an extended level set method to obtain zonal models with geometries (shape and location) being dependent on all data types. A synthetic 2-D example was used to demonstrate joint inversion of seismic travel times and drawdown data from pumping tests at steady state. The final zonal model was vastly improved compared with individual inversions (see Figure 7.2), in terms of both shape and parameter values of the sand and clay inclusions in

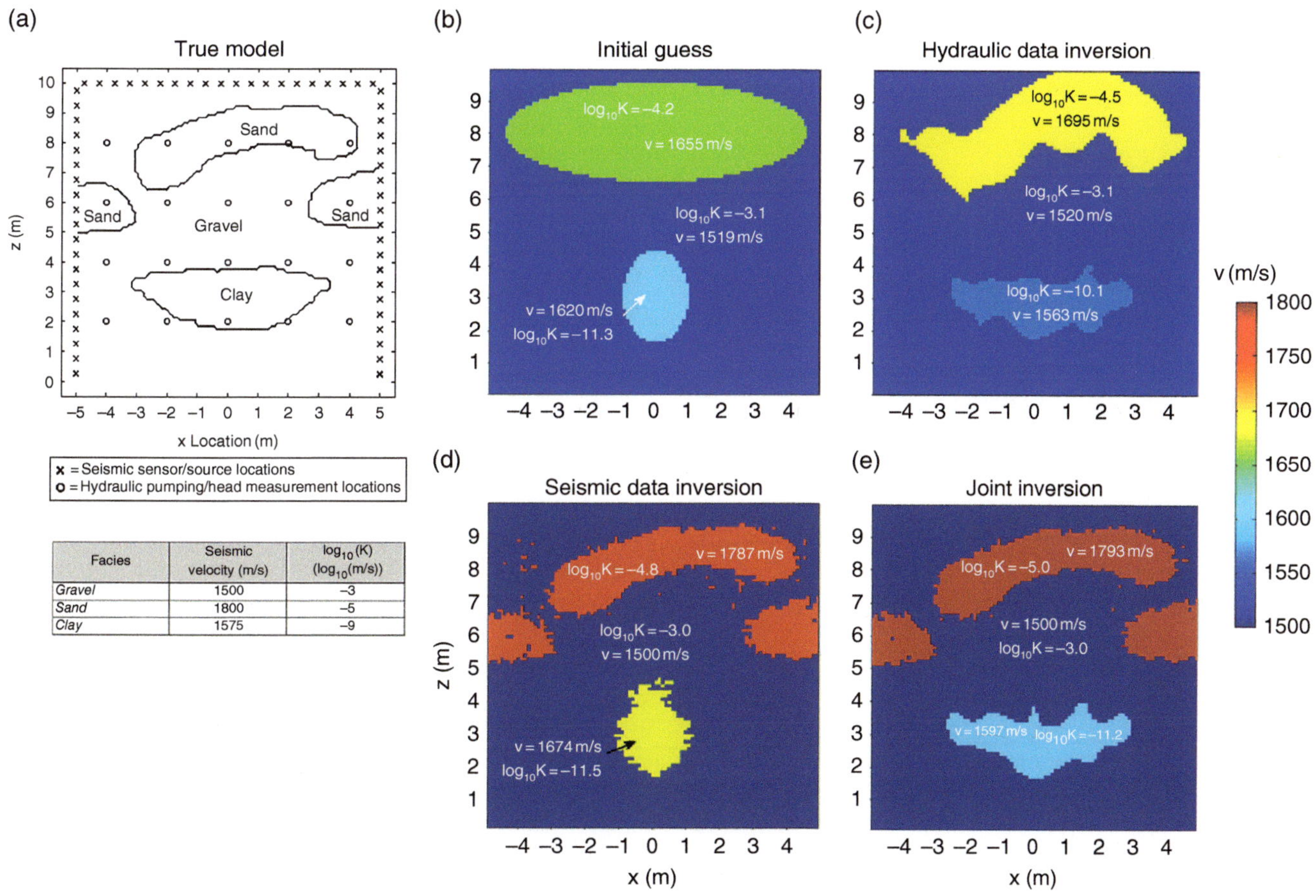

Facies	Seismic velocity (m/s)	$\log_{10}(K)$ ($\log_{10}$(m/s))
Gravel	1500	−3
Sand	1800	−5
Clay	1575	−9

Figure 7.2 Results from zonal inversion using level sets, with the true model in **(a)** and the initial guess in **(b)**. The individual inversion results of hydraulic and seismic data **(c,d)** are clearly improved in the joint inversion results **(e)**. Modified from *Cardiff and Kitanidis* [2009].

the gravel background. For a synthetic text case, *Aghasi et al.* [2013] presented a level set method to infer a contaminant's source zone architecture in 3-D by joint inversion of ERT and contaminant concentration. Level sets have the distinct advantage that the interfaces between different geological units are discontinuous. In many settings, this is necessary to reproduce hydrogeological properties and responses.

Conceptually similar to the case of common lithological units is the detection of man-made structures and objects in the subsurface. Examples could be archeological remains hidden in sediments or buried metallic objects, such as UXO. *Pasion et al.* [2003] developed a methodology for jointly inverting magnetic and time-domain electromagnetic (TDEM) data for characterization of UXO. Due to the known size and property range for UXO and the strong magnetic and electrical contrast to the background, the inversion could be formulated to invert directly for UXO characteristics such as position, orientation, and magnetic and electrical properties. Magnetic and TDEM data jointly contributed to resolving position and orientation, while the other parameters were not shared between the methods. In the synthetic example of *Pasion et al.* [2003], joint inversion were found to improve size and shape estimates of the buried target.

7.3.3. Gradient Constraints

The most popular model parameterization strategy in geophysical inversion is to use a very fine model discretization that remains fixed during the inversion process. The inverse problem is formulated to maximize the weight assigned to model regularization constraints that quantify model structure, provided that the data are fitted to the expected noise level. This Occam-style inversion can easily be solved using a least-squares formulation and it leads to a minimum-structure model, in which all resolved features are necessary to explain the observed data [*Constable et al.*, 1987]. The resulting images are smoothly varying fields that are visually very different from the discontinuous property fields that are discussed in Sections 7.3.1 and 7.3.2. Gradient-based joint inversion has played an important role in popularizing joint inversion as they are easily implemented in existing Occam-style inversion algorithms. Most published papers rely on the cross-gradients function introduced by *Gallardo and Meju* [2003]. This approach is valid when changes in the physical properties of interest are aligned (i.e., gradients are parallel or anti-parallel, or the gradient is zero for one property). This is a reasonable assumption when a single lithological property dominates the subsurface response (e.g., porosity) or when changes in state variables are also large across lithological units. The cross-gradient constraints are invalid (in its standard formulation) when there are important uncorrelated changes in lithological and state variables [e.g., *Linde et al.*, 2006a].

The original cross-gradients joint inversion formulation [*Gallardo and Meju*, 2003] was first applied to 2-D surface-based near-surface geophysical field data [*Gallardo and Meju*, 2003, 2004, 2007]. A slightly modified formulation was introduced by *Linde et al.* [2006a] and applied to joint 3-D inversion of field-based crosshole GPR and ERT data. This approach was later adapted to joint inversion of synthetic and field-based crosshole seismic and GPR data [*Linde et al.*, 2008], to synthetic and field-based time-lapse crosshole ERT and GPR data [*Doetsch et al.*, 2010a], and to synthetic and field-based three-method joint inversion in combination with classification, clustering, and zonal joint inversion [*Doetsch et al.*, 2010b]. *Bouchedda et al.* [2012], *Karaoulis et al.* [2012], *Hamdan and Vafidis* [2013], and *Revil et al.* [2013] presented related applications. This suite of real-world case studies based on different data types and geological settings suggest that cross-gradient joint inversion is presently one of the most robust approaches to joint inversion of near-surface geophysical field data. The resulting models have higher resolution and cross-property plots are more focused than for individual inversions (see Figure 7.3). The focused scatter plots enable visual and automatic clustering that facilitate geological interpretations and petrophysical inference (see, e.g., *Gallardo and Meju* [2004] and *Doetsch et al.* [2010b]).

Günther and Rücker [2006] introduced an alternative gradient-based joint inversion approach that they applied to synthetic 2-D seismic refraction and ERT data. They used the model gradients in one model to locally scale the regularization constraints in the other model, and vice versa. If one property displays large spatial changes at one location, this approach helps to introduce larger changes in the other property field by locally decreasing the regularization weights. This approach has not been used extensively, and any added value with respect to cross-gradient constraints remains to be demonstrated.

Lochbühler et al. [2013] presented the first joint inversion of synthetic and field-based geophysical and hydrological data based on gradient-constraints. The cross-gradient function was used to impose structural similarities between radar slowness and the logarithm of the hydraulic diffusivity (or the permeability field) when inverting crosshole GPR and hydraulic tomography (or tracer) data. Similar to previous applications, the joint inversion provided models with higher resolution and cross-property plots were less scattered than for individual inversions.

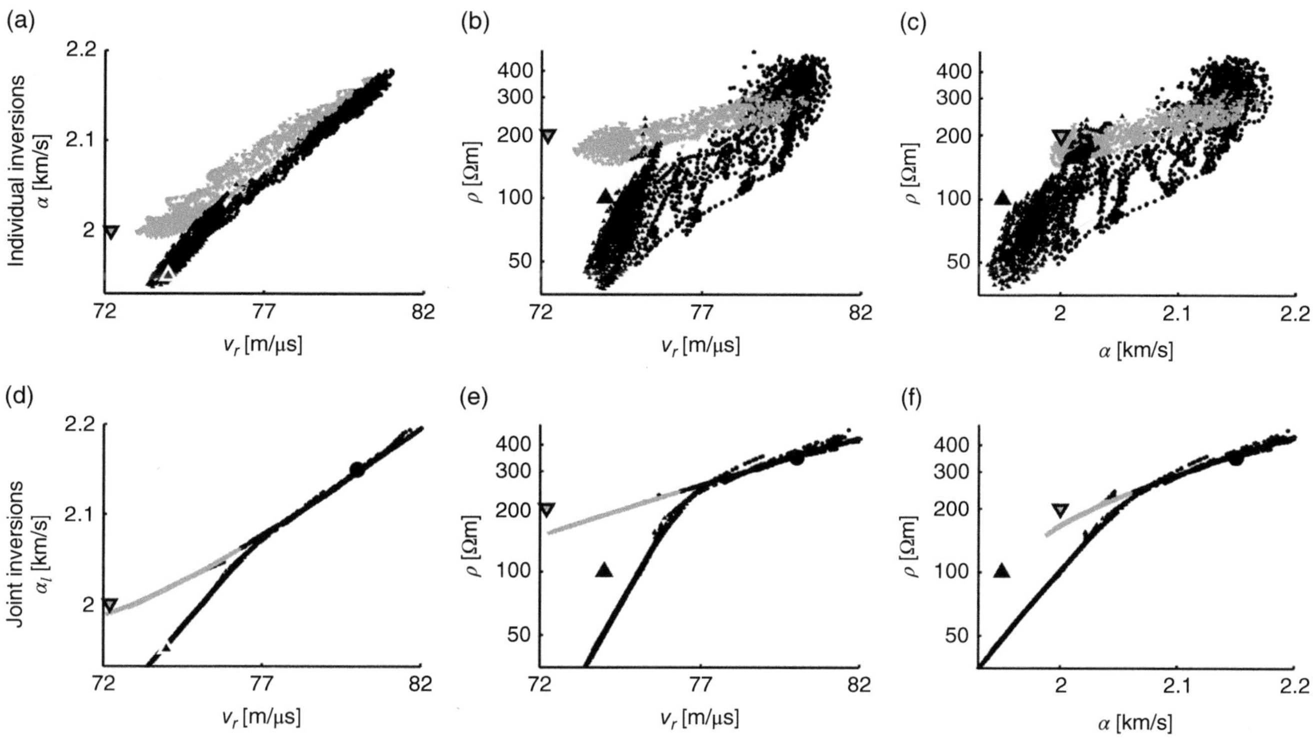

Figure 7.3 Cross-property plots of seismic velocity α, GPR velocity v_r, and resistivity ρ. The diffuse scatter clouds from individual inversions (*top*) focus to clear linear features in the joint inversions (*bottom*). The joint inversion results are also closer to the zonal properties of the underlying synthetic lithological model (*larger symbols*). Modified from *Doetsch et al.* [2010b].

7.4. COUPLED HYDROGEOPHYSICAL INVERSION

One popular approach to jointly invert geophysical and hydrological data is to link hydrological state variables to geophysical properties. This approach has primarily been applied to transient hydrological phenomena using time-lapse geophysical data. Typical examples include water flow in the unsaturated (vadose) zone and salt tracer movement in saturated aquifers. Classical time-lapse geophysical inversion suffers from resolution limitations. In tracer experiments, this might lead to the inferred plumes being unphysical (see, e.g., loss of mass [*Day-Lewis et al.*, 2007]). When considering geophysical data within a hydrological inversion context, all results are, by construction, in agreement with mass conservation laws and other constraints imposed by the hydrological model. Another advantage is that petrophysical relationships related to a perturbation in a hydrological state variable (e.g., salinity, water content) are much better constrained than for primary properties. In time-lapse inversions it is expected that only state variables change with time, whereas primary properties and petrophysical parameter values remain unchanged. This reduces the number of unknown petrophysical parameters that need to be assigned or inverted for.

For coupled hydrogeophysical inversions, it is challenging to provide a general implementation and parameterization for a wide range of applications, hydrological settings, and geophysical data, and most published works are specific to certain settings (e.g., water flow in unsaturated soil) and geophysical data types. It is also often necessary to assume that certain properties are constant (e.g., porosity) when inverting for the spatial variations of others (e.g., permeability). Furthermore, petrophysical parameters (e.g., the cementation exponent in Archie's law) are often assumed constant throughout the study area, and the consequences of such assumptions are seldom addressed in detail.

7.4.1. Petrophysical Coupling Applied to Vadose Zone Hydrology

Variations in water content have a clear geophysical signature—for example, in terms of electrical permittivity and resistivity. This implies that time-lapse geophysical data can be used to monitor vadose zone processes, but also to constrain subsurface architecture and hydrological properties within a coupled hydrogeophysical inversion. The most commonly used geophysical data have been GPR and ERT data acquired with crosshole or surface acquisition geometries.

Kowalsky et al. [2004] pioneered coupled hydrogeophysical inversion by linking an unsaturated flow simulator and a GPR forward solver. In their approach, the soil hydraulic properties were described by the Mualem-van Genuchten parameterization [*van Genuchten*, 1980] that relates water content to permeability in partially saturated media. Water content was related to permittivity and GPR velocities using the complex refractive index model (CRIM [*Roth et al.*, 1990]). *Kowalsky et al.* [2004] used a pilot point parameterization to generate multi-Gaussian fields with a reduced set of model parameters. In a synthetic experiment, the crosshole GPR measurements were jointly inverted with water content values that were available along the boreholes. The coupled inversion improved the estimates of permeability and its spatial variation compared with considering the water content data only. It was also found that permeability could be very well resolved, when all other soil hydraulic properties and the petrophysical parameters were perfectly known. In addition, they retrieved soil hydraulic and petrophysical properties under the assumption of homogeneous distributions. *Kowalsky et al.* [2005] inverted multi-offset crosshole GPR field data with conditioning to point information of water content (estimated from neutron probe measurements). They simultaneously estimated homogeneous petrophysical parameters and the heterogeneous distribution of permeability. Including GPR data improved permeability estimates in a synthetic example (see Figure 7.4), which led to a better prediction of water content. *Kowalsky et al.* [2005] concluded that it was crucial to infer petrophysical parameter values within the inversion procedure, as incorrect assumptions might otherwise compromise the permeability estimates. Nevertheless, it is important to acknowledge that assuming homogeneous petrophysical parameters (as done herein) could also bias the permeability estimates. *Finsterle and Kowalsky* [2008]inverted for geostatistical parameters of unsaturated soil. Using filtration rates, water content, and GPR travel times, they were able to simultaneously invert synthetic data for three homogeneous hydraulic parameters, two petrophysical parameters, and three geostatistical parameters. The permeability distribution parameterized by pilot points was also part of the inversion.

Cassiani and Binley [2005] inverted for soil hydraulic properties (Mualem–van Genuchten parameters) of different layers by coupling a Richards' equation solver to a Monte Carlo sampler and linked water content to zero-offset GPR data through a known petrophysical relationship. In a field example, *Cassiani and Binley* [2005] found that unsaturated flow parameters were not

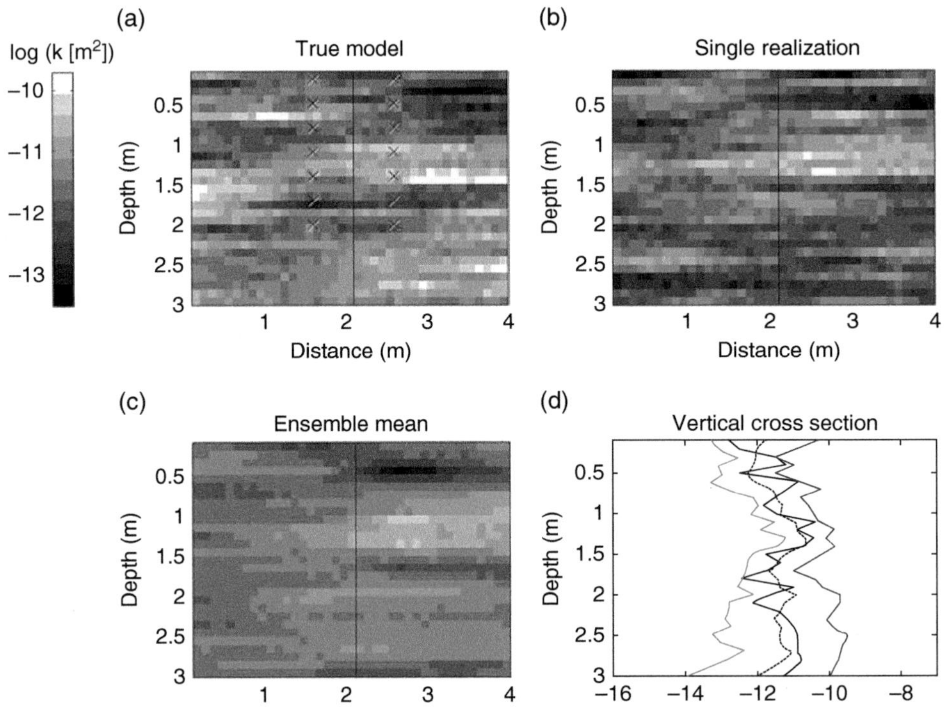

Figure 7.4 Input permeability distribution **(a)** and inversion results **(b,c)** from a synthetic tracer experiment. Both the single realization **(b)** and the ensemble mean **(c)** capture the main features, but the single realization shows more realistic variability. The vertical cross section **(d)** shows the true (*solid line*), mean (*dashed line*), and uncertainty bounds (*gray lines*) of permeability. From *Kowalsky et al.* [2005].

individually constrained by their data and that measurements under dynamic conditions would have been needed. They also stressed the importance of acquiring independent geological information. *Scholer et al.* [2011, 2012] used a Markov chain Monte Carlo (MCMC) inversion approach to study the influence of prior information on estimated soil hydraulic properties. When using both synthetic and field-based test cases, they found that the geophysical data alone contained valuable information but that significantly better results were obtained when using informative prior distributions that include parameter correlations.

All of the approaches presented above relied on coupled hydrogeophysical inversion based on crosshole GPR traveltime data to provide information in between the boreholes. A number of papers have addressed the coupling of unsaturated flow modeling (e.g., based on Richards' equation) with surface GPR data. *Lambot et al.* [2006] and *Jadoon et al.* [2008] presented numerical experiments with off-ground GPR monitoring data to constrain the hydraulic properties of the topsoil, and *Busch et al.* [2013] inferred such soil hydraulic properties from both synthetic and field-based surface GPR measurements. These methods could potentially cover larger areas than crosshole applications, although frequent repeat measurements would be needed to retrieve soil hydraulic properties.

The above-mentioned examples highlight the utility of GPR for retrieving soil hydraulic properties. *Looms et al.* [2008] combined 1-D flow simulations with ERT and GPR data to invert for permeability. The ERT and GPR forward simulators were both linked to the hydraulic simulator through a known and homogeneous petrophysical relationships. *Looms et al.* [2008] showed that five layers were needed to explain their field data but that only permeability and one additional fitting parameter could be retrieved for each layer.

Hinnell et al. [2010] and *Mboh et al.* [2012a] investigated how surface-based ERT monitoring of infiltration tests could improve soil hydraulic property estimation. *Hinnell et al.* [2010] showed in a synthetic example that petrophysical coupling can reduce parameter errors, but only if the conceptual hydraulic model is correct. Considering an undisturbed soil core, *Mboh et al.* [2012a] also improved their parameter estimates when combining inflow measurements with ERT data. They also highlighted the challenge of assigning appropriate weights to each dataset in the objective function.

Huisman et al. [2010] combined soil moisture measurements with ERT monitoring data acquired over a dike built of uniform sand. They constructed a 2-D hydrological model and inverted for the homogeneous soil parameters of the dike using MCMC. They included the ERT data through a petrophysical relationship and inverted for unknown petrophysical parameters. They found that the permeability of the dike was the best-resolved parameter and that the combination of ERT and water content measurements reduced uncertainty.

Mboh et al. [2012b] used self-potential monitoring data acquired during infiltration and drainage experiments in a sand-filled column to invert for key parameters that describe the soil water retention and relative permeability functions. The hydrogeophysical coupling was obtained by linking the predicted streaming potential to the simulated water flow and water content distribution. While promising in the laboratory, the weak signal makes applications to field studies challenging [*Linde et al.*, 2011]. The petrophysical relationship needed to calculate the streaming potential at partial saturation is currently being debated (see, e.g., *Jougnot et al.* [2012]), and associated uncertainties are generally larger than for GPR and ERT.

7.4.2. Petrophysical Coupling Applied to Transient Groundwater Processes

Another popular use of coupled hydrogeophysical inversion is for salt tracer tests in saturated aquifers. Dissolved salt increases the electrical conductivity of water, thereby decreasing electrical bulk resistivity. The changes in bulk resistivity can be sensed by ERT or EM methods and related to the fluid conductivity, which in turn can be linked to salinity. For moderate salinity, there is a linear relationship between salt concentration and fluid conductivity (see, e.g., *Keller and Frischknecht* [1966]), which can be calibrated for specific conditions.

Considering synthetic test cases, *Irving and Singha* [2010] and *Jardani et al.* [2013] both linked a flow and transport simulator at saturated conditions with an ERT forward solver to infer the permeability distribution. Both approaches used MCMC, but the parameterization was different. *Irving and Singha* [2010] used a fine spatial discretization, fixed the permeabilities of two facies, and inverted for the probability of each cell to belong to one of the two facies. ERT data were inverted jointly with concentration data measured in boreholes and were found to mainly improve the estimates of the spatial correlation length. *Jardani et al.* [2013] used a pilot point parameterization to decrease the number of model parameters and inverted for the permeability at each pilot point. They included ERT, self-potential, and tracer concentrations in the inversion and found that all three datasets contained valuable information on permeability.

Pollock and Cirpka [2010] considered a synthetic laboratory salt tracer experiment with ERT and hydraulic head measurements. They inverted for the permeability distribution using the mean arrival times of electrical potential perturbations and hydraulic head measurements. Their approach assumed a linear relationship between salt concentration and bulk electrical conductivity but avoided the actual conversion between concentration

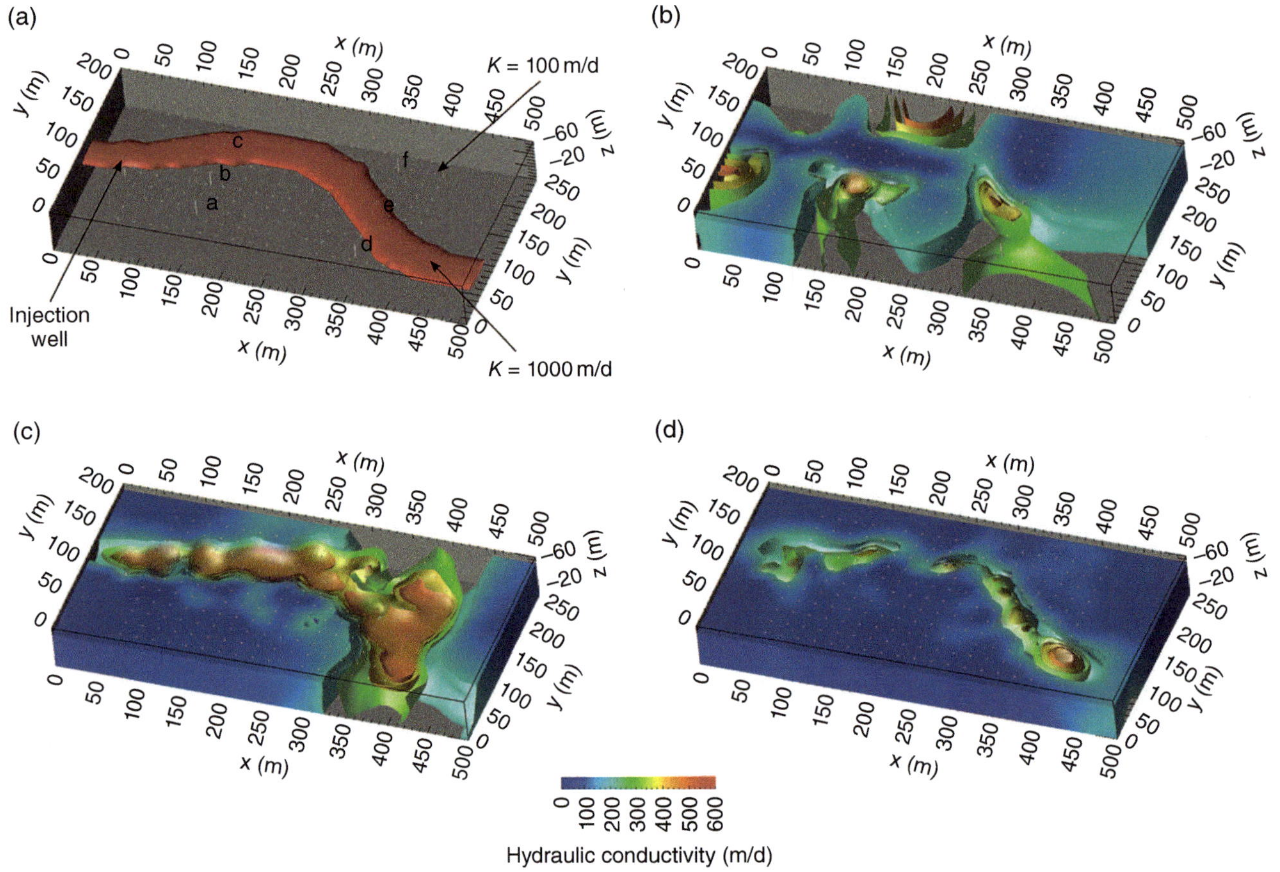

Figure 7.5 **(a)** True hydraulic conductivity distribution, **(b)** inversion of head and fluid conductivity data only, **(c)** ERT inversion results, and **(d)** inversion of head, fluid conductivity, and ERT data. From *Johnson et al.* [2009].

and electrical conductivity. In a real sandbox experiment, *Pollock and Cirpka* [2012] recovered the detailed permeability structure. The full transient behavior was predicted very well even though only mean arrival times were used in the inversion. The inversion methodology developed by *Pollock and Cirpka* [2010, 2012] is very efficient as the forward problem that relates electrical potential difference perturbations to the saline tracer distribution and, hence, to the permeability field is formulated in terms of temporal moment-generating equations [*Harvey and Gorelick*, 1995]. Temporal moment-generating equations are widely used in hydrogeology and allow, for example, calculating the mean arrival time of a tracer by solving a steady-state equation instead of performing a full transient simulation.

Johnson et al. [2009] presented a data correlation approach where no specific petrophysical relationship was assumed, only the type of relationship (e.g., linear) between changes in fluid conductivity and changes in bulk electrical conductivity had to be chosen. They parameterized a 3-D subsurface model using pilot points and inverted for the permeability distribution using hydraulic head, fluid conductivity in six wells and surface ERT data. In their synthetic example (Figure 7.5a), inversion of head and fluid conductivity data were unable to constrain the high-permeability zone (Figure 7.5b). ERT data alone and especially joint inversion of hydraulic and ERT data greatly improved permeability estimates (see Figures 7.5c and 7.5d). Due to the loose assumption of a correlation rather than a fixed petrophysical relationship between the hydrological state variable and the primary geophysical property, the approach of *Johnson et al.* [2009] could potentially be applied to a wide variety of applications.

Kowalsky et al. [2011] performed a coupled inversion of ERT and hydrogeochemical data to better understand the factors that influence flow and contaminant transport in a complex geological setting. They adapted their parameterization of the hydrogeological model at the field site to fit all available geophysical and geochemical data and to estimate permeability of the different model units, along with petrophysical parameters needed to use the ERT data. *Kowalsky et al.* [2011] were the first to link unsaturated and saturated flow and transport in a coupled hydrogeophysical inversion, and their application highlights the complexity of real-world problems. A parallel computing implementation of this approach [*Commer et al.*, 2014] will enable applications to even more complex environments and larger datasets.

Dorn et al. [2013] conditioned discrete fracture network realizations such that they were in agreement with field-based single-hole GPR reflection, tracer, hydraulic, and televiewer data. The tracer test data were used to identify active fractures that intersected the borehole (inferred from electrical conductivity logs) and those within the formation (inferred from time-lapse GPR images that were sensitive to the tracer distribution). No petrophysical link was needed here, as it was the fracture geometry of the active fractures that was constrained by the GPR data. *Dorn et al.* [2013] stochastically generated three-dimensional discrete networks of active fractures and used a hierarchical rejection sampling method to create large sets of conditional realizations. Constraints offered by the GPR data made the stochastic scheme computationally feasible as they strongly reduced the set of possible prior models.

Christiansen et al. [2011] and *Herckenrath et al.* [2012] used field-based time-lapse gravity data to estimate porosity and permeability in shallow unconfined aquifers. They inverted hydraulic head measurements along with changes in the gravity response of the mass change associated with the fluctuating water table. *Christiansen et al.* [2011] inverted for the homogeneous porosity, permeability, evapotranspiration, and riverbed conductance. They found that including gravity measurements significantly reduced parameter correlation between porosity and permeability and that especially porosity was better constrained when including gravity data. For a synthetic test case, *Herckenrath et al.* [2012] additionally included SNMR data in their coupled inversion for porosity and permeability.

7.4.3. Petrophysical Coupling Applied to Steady-State Groundwater Systems

Calibration of groundwater models to data acquired under steady-state conditions is more challenging, due to the lack of dynamic forcing terms. Nevertheless, regional groundwater models are often built using hydraulic head distributions along with borehole descriptions and knowledge of the local geology. *Herckenrath et al.* [2013b] developed a joint hydrological inversion approach that included synthetic and field-based ERT and TDEM data to obtain a calibrated groundwater model. They used a combination of common interfaces and petrophysical relationships to link the geophysical response to the groundwater model. Using the same methodology, *Herckenrath et al.* [2013a] calibrated a saltwater intrusion model to field-based TDEM and hydraulic head data.

Jardani and Revil [2009] linked hydraulic, thermal, and self-potential simulators to jointly invert downhole temperature and self-potential data measured at the surface. In a geothermal field application, they built a model with 10 geological units. Using a stochastic inversion scheme, they inferred horizontal and vertical permeability of each model unit using borehole temperature data and self-potentials measured along a surface profile. The derived permeability values agreed well with those from hydrological studies.

Straface et al. [2011] jointly inverted field-based hydraulic head and self-potential measurements for the 3-D distribution of permeability in a shallow aquifer. To do so, they rely on the sensitivity of self-potential data to the water table level. Each model cell was first classified using a geostatistical multi-continuum approach, and hydraulic inversion was performed on data collected during dipole pumping tests.

7.5. OUTSTANDING CHALLENGES

Joint inversions of near-surface geophysical data and coupled hydrogeophysical inversions have advanced subsurface characterization, but important outstanding research challenges remain to be solved before their full potential can be reached.

7.5.1. Petrophysical Relationships

When using petrophysical relationships within the inversion, it is necessary to first decide upon its functional form (e.g., is surface conductivity to be ignored when predicting electrical conductivity; is the petrophysical relationship based on volume-averaging or differential effective medium theory). The next step is to decide if petrophysical parameters are to be estimated within the inversion, assigned based on literature data, or estimated from independent laboratory or logging data. It is also essential to consider spatial variations of petrophysical parameters. For example, is it reasonable to assume that the cementation exponent in Archie-type relations is the same throughout a region consisting of different lithological units? This might be tempting as it simplifies the solution of the inverse problem, but it is often unlikely to be representative of reality.

Furthermore, is the petrophysical relationship assumed to be perfect (a one-to-one relation) or is it assumed to have uncertainty (almost always modeled with a Gaussian error model)? If it accounts for uncertainty, what is the spatial correlation structure of this error term? Should we assume that the error associated with the petrophysical relationship is the same throughout the model domain or that it is different at every grid cell? These questions are rarely addressed in near-surface geophysical or hydrogeophysical studies. Primary references concerning the most commonly used petrophysical relationships provide no or only very limited information about the spatial arrangement of the sample locations (the results are often presented in scatter plots).

To complicate things even further, most petrophysical relationships are originally defined at the scale of a representative elementary volume (REV), while model coupling is imposed at a much larger scale that is determined by the model parameterization and resolution of the geophysical or hydrological models. There are few reasons to believe that the same parameter values apply at both scales. This important topic has only been partially addressed in the hydrogeophysics and near-surface geophysics communities [*Day-Lewis and Lane*, 2004; *Moysey et al.*, 2005; *Singha and Moysey*, 2006].

The challenges discussed above are not specific to geophysics and arise in any application in which one state variable or property is linked to another one—for example, the relation between the water content and relative permeability or the relation between water content and capillary pressure, and so on. Nevertheless, ignoring the uncertainty related to the petrophysical relationship and its spatial dependence will often lead to overly optimistic uncertainty estimates that might overstate the information content in the geophysical data. How to properly account for these effects is an important challenge for future research.

7.5.2. Structural Constraints

Structural approaches to joint inversion are popular as they avoid introducing explicit petrophysical relationships. The resulting inversion results are similar to those obtained from joint inversion through a known petrophysical relationship, but the structural approach is more robust when the petrophysical relationship is uncertain or spatially variable [*Moorkamp et al.*, 2011]. The cross-gradient constraints are used to obtain smooth models at a resolution that is much larger than the model discretization, and the gradients are calculated with respect to neighboring cells and are thus strongly mesh-dependent. How do we assess if the underlying assumption of the cross-gradient function being zero is valid, or rather, for what applications and at what scale is the assumption reasonable? Can we formulate the cross-gradient constraints to operate at that scale? As we push the resolution limits of deterministic inversion, for example, by introducing full waveform inversion, it appears important to carefully assess if this type of constraint is meaningful at those finer scales. Furthermore, comparison with individual inversions, site-specific knowledge, and petrophysical relationships are crucial to assess the validity of cross-gradients constraints for a given application.

7.5.3. Probabilistic Versus Deterministic Inversion

Probabilistic inversions, for example, based on MCMC are well suited to integrate multiple geophysical datasets and arbitrary petrophysical relationships [*Bosch*, 1999]. Uncertainty estimation is straightforward (but may be strongly affected by minor assumptions; e.g., *Linde and Vrugt* [2013]) and it is possible to perform the inversion within a modern geostatistical simulation context [*Hansen et al.*, 2012]. But, MCMC inversion is still out of reach for most applications with large datasets (millions of data points), high model dimensions (thousands of parameters), and advanced forward solvers (e.g., 3-D solvers). To apply MCMC, it is often necessary to reduce the number of unknowns, the number of data, and the accuracy of the forward solver. This implies that theoretically solid approaches to uncertainty assessments may become strongly biased and questionable (see, e.g., *Linde and Revil* [2007]).

The smoothness of Occam-style cross-gradient deterministic inversions helps to avoid excessive overinterpretation as it is clear from the outset that only an upscaled minimum-structure representation of the subsurface is sought. Such deterministic inversion results are often rather robust as the regularization term can be tweaked (by the inversion code or by hand) to get models that appear reasonable to the modeler. No such tweaking term should exist in probabilistic inversions, even if there is often a tendency to achieve a similar effect by manipulating the prior probability density function [*Scales and Sneider*, 1997].

Both the deterministic and probabilistic approaches have their merits. It is seldom possible to accurately state the actual prior information on model parameters and petrophysical relationships. It is thus often useful to consider multiple inversions with different underlying assumptions. It is a mistake to ignore uncertainty, but it might also be as dangerous to blindly believe in the uncertainty estimates provided by linear deterministic inverse theory or advanced probabilistic inversions. Current error models that incorporate the effects of the model parameterization, simplifications of the physics, the prior property fields, and petrophysical relationships are still far too simplistic. Improving these aspects and applying them to the joint inverse problem is an important future challenge for the near-surface geophysics and hydrogeophysics research communities.

7.6. CONCLUDING REMARKS

Joint inversion can enhance model resolution and decrease interpretational ambiguities. Its application to near-surface geophysical data has become a robust and almost standard approach to integrate multiple geophysical datasets into consistent subsurface models at the highest possible resolution. The most robust coupling strategy is often structural (common layers, common lithological units or aligned gradients of property fields). Nevertheless, it is essential to carefully motivate a particular structural coupling approach for a given application to avoid the incorrect use of structural constraints and to identify when model coupling by petrophysical relationships is more suitable.

Joint inversion of geophysical and hydrological data has become very popular in recent years, and the vast majority of publications rely on a coupled hydrogeophysical inversion framework. Hydrological models provide by simulation a predicted hydrological state that is transformed into a geophysical model through a petrophysical relationship. The misfit between the predicted geophysical forward response and the observed geophysical data is then used to guide the update of the hydrological model. This approach is often favored among hydrologists as the focus is on calibrating the hydrological model.

In field-based studies, it is useful to first identify incoherencies in data, geometry, and modeling results by performing individual inversions and comparing the results that are obtained with different methods. Blind application of joint inversion algorithms without careful assessments of data quality and coverage, model parameterization, the error model, geological setting, and imposed prior or regularization constraints rarely leads to useful results. Future research avenues of interest include fully probabilistic joint inversions combined with complex prior models and improved statistical descriptions of petrophysical relationships and their scaling properties. These studies could also be most useful to better understand the fidelity of joint inversion results obtained within a deterministic framework for cases when a fully probabilistic treatment is computationally infeasible (e.g., most 3-D applications) and to understand which steps in the data acquisition, modeling, and inversion process are the most critical to ensure reliable results.

ACKNOWLEDGMENTS

We are grateful for the constructive comments and suggestions provided by associate editor Peter Lelièvre, Lee Slater, and three anonymous reviewers.

REFERENCES

Aghasi, A., I. Mendoza-Sanchez, E. L. Miller, C. A. Ramsburg, and L. M. Abriola (2013), A geometric approach to joint inversion with applications to contaminant source zone characterization, *Inverse Problems*, *29*(11), 115014, doi:10.1088/0266-5611/29/11/115014.

Auken, E., and A. V. Christiansen (2004), Layered and laterally constrained 2D inversion of resistivity data, *Geophysics*, *69*(3), 752–761, doi:10.1190/1.1759461.

Böniger, U., and J. Tronicke (2010), Integrated data analysis at an archaeological site: A case study using 3D GPR, magnetic, and high-resolution topographic data, *Geophysics, 75*(4), B169–B176, doi:10.1190/1.3460432.

Bosch, M. (1999), Lithologic tomography: From plural geophysical data to lithology estimation, *J. Geophys. Res.*, *104*(B1), 749–766, doi:10.1029/1998JB900014.

Bouchedda, A., M. Chouteau, A. Binley, and B. Giroux (2012), 2-D joint structural inversion of cross-hole electrical resistance and ground penetrating radar data, *J. Appl. Geophys.*, *78*, 52–67, doi:10.1016/j.jappgeo.2011.10.009.

Busch, S., L. Weihermüller, J. A. Huisman, C. M. Steelman, A. L. Endres, H. Vereecken, and J. van der Kruk (2013), Coupled hydrogeophysical inversion of time-lapse surface GPR data to estimate hydraulic properties of a layered subsurface, *Water Resour. Res.*, *49*(12), 8480–8494, doi:10.1002/2013WR013992.

Cardiff, M., and P. K. Kitanidis (2009), Bayesian inversion for facies detection: An extensible level set framework, *Water Resour. Res.*, *45*(10), W10416, doi:10.1029/2008WR007675.

Cassiani, G., and A. Binley (2005), Modeling unsaturated flow in a layered formation under quasi-steady state conditions using geophysical data constraints, *Adv. Water Resour.*, *28*(5), 467–477, doi:10.1016/j.advwatres.2004.12.007.

Cassiani, G., and M. A. Medina (1997), Incorporating auxiliary geophysical data into ground-water flow parameter estimation, *Ground Water*, *35*(1), 79–91, doi:10.1111/j.1745-6584.1997.tb00063.x.

Cassiani, G., G. Böhm, A. Vesnaver, and R. Nicolich (1998), A geostatistical framework for incorporating seismic tomography auxiliary data into hydraulic conductivity estimation, *J. Hydrol.*, *206*(1–2), 58–74, doi:10.1016/S0022-1694(98)00084-5.

Chen, J., S. Hubbard, J. Peterson, K. Williams, M. Fienen, P. Jardine, and D. Watson (2006), Development of a joint hydrogeophysical inversion approach and application to a contaminated fractured aquifer, *Water Resour. Res.*, *42*(6), W06425, doi:10.1029/2005WR004694.

Christiansen, L., P. J. Binning, D. Rosbjerg, O. B. Andersen, and P. Bauer-Gottwein (2011), Using time-lapse gravity for groundwater model calibration: An application to alluvial aquifer storage, *Water Resour. Res.*, *47*(6), W06503, doi:10.1029/2010WR009859.

Commer, M., M. B. Kowalsky, J. Doetsch, G. A. Newman, and S. Finsterle (2014), MPiTOUGH2: A parallel parameter estimation framework for hydrological and hydrogeophysical applications, *Comput. Geosci.*, *65*, 127–135, doi:10.1016/j.cageo.2013.06.011.

Constable, S. C., and R. L. Parker, and C. G. Constable (1987), Occam's inversion: A practical algorithm for generating smooth models from electromagnetic sounding data, *Geophysics*, *52*, 289–300.

Copty, N., Y. Rubin, and G. Mavko (1993), Geophysical-hydrological identification of field permeabilities through Bayesian updating, *Water Resour. Res.*, *29*(8), 2813–2825, doi:10.1029/93WR00745.

Day-Lewis, F. D., and J. W. Lane (2004), Assessing the resolution-dependent utility of tomograms for geostatistics, *Geophys. Res. Lett.*, *31*(7), L07503, doi:10.1029/2004GL019617.

Day-Lewis, F. D., K. Singha, and A. M. Binley (2005), Applying petrophysical models to radar travel time and electrical resistivity tomograms: Resolution-dependent limitations, *J. Geophys. Res.*, *110*(B8), B08206, doi:10.1029/2004JB003569.

Day-Lewis, F. D., Y. Chen, and K. Singha (2007), Moment inference from tomograms, *Geophys. Res. Lett.*, *34*(22), L22404, doi:10.1029/2007GL031621.

Doetsch, J., N. Linde, and A. Binley (2010a), Structural joint inversion of time-lapse crosshole ERT and GPR traveltime data, *Geophys. Res. Lett.*, *37*(24), L24404, doi:10.1029/2010GL045482.

Doetsch, J., N. Linde, I. Coscia, S. A. Greenhalgh, and A. G. Green (2010b), Zonation for 3D aquifer characterization based on joint inversions of multimethod crosshole geophysical data, *Geophysics*, *75*(6), G53–G64, doi:10.1190/1.3496476.

Doetsch, J., N. Linde, M. Pessognelli, A. G. Green, and T. Günther (2012), Constraining 3-D electrical resistance tomography with GPR reflection data for improved aquifer characterization, *J. Appl. Geophys.*, *78*, 68–76, doi:10.1016/j.jappgeo.2011.04.008.

Dorn, C., N. Linde, T. L. Borgne, O. Bour, and J.-R. de Dreuzy (2013), Conditioning of stochastic 3-D fracture networks to hydrological and geophysical data, *Adv. Water Resour.*, *62, Part A*, 79–89, doi:10.1016/j.advwatres.2013.10.005.

Finsterle, S., and M. B. Kowalsky (2008), Joint hydrological–geophysical inversion for soil structure identification, *Vadose Zone J.*, *7*(1), 287, doi:10.2136/vzj2006.0078.

Gallardo, L. A., and M. A. Meju (2003), Characterization of heterogeneous near-surface materials by joint 2D inversion of dc resistivity and seismic data, *Geophys. Res. Lett.*, *30*(13), 1658, doi:10.1029/2003GL017370.

Gallardo, L. A., and M. A. Meju (2004), Joint two-dimensional DC resistivity and seismic travel time inversion with cross-gradients constraints, *J. Geophys. Res.*, *109*(B3), B03311, doi:10.1029/2003JB002716.

Gallardo, L. A., and M. A. Meju (2007), Joint two-dimensional cross-gradient imaging of magnetotelluric and seismic traveltime data for structural and lithological classification, *Geophys. J. Int*, *169*(3), 1261–1272, doi:10.1111/j.1365-246X.2007.03366.x.

Günther, T., and M. Müller-Petke (2012), Hydraulic properties at the North Sea island of Borkum derived from joint inversion of magnetic resonance and electrical resistivity soundings, *Hydrol. Earth Syst. Sci.*, *16*(9), 3279–3291, doi:10.5194/hess-16-3279-2012.

Günther, T., and C. Rücker (2006), A new joint inversion approach applied to the combined tomography of DC resistivity and seismic refraction data.

Hamdan, H. A., and A. Vafidis (2013), Joint inversion of 2D resistivity and seismic travel time data to image saltwater intrusion over karstic areas, *Environ. Earth Sci.*, *68*(7), 1877–1885, doi:10.1007/s12665-012-1875-9.

Hansen, T. M., K. S. Cordua, and K. Mosegaard (2012), Inverse problems with nontrivial priors: Efficient solution through Gibbs sampling, *Comput. Geosci.*, *16*(3), 593–611, doi:10.1007/s10596-011-9271-1.

Harvey, C. F., and S. M. Gorelick (1995), Temporal moment-generating equations: modeling transport and mass transfer in heterogeneous aquifers, *Water Resour. Res.*, *31*, 1895–1911, doi:10.1029/95WR01231.

Herckenrath, D., E. Auken, L. Christiansen, A. A. Behroozmand, and P. Bauer-Gottwein (2012), Coupled hydrogeophysical inversion using time-lapse magnetic resonance sounding and time-lapse gravity data for hydraulic aquifer testing: Will it work in practice?, *Water Resour. Res.*, *48*, W01539, doi:10.1029/2011WR010411.

Herckenrath, D., N. Odlum, V. Nenna, R. Knight, E. Auken, and P. Bauer-Gottwein (2013a), Calibrating a salt water intrusion model with time-domain electromagnetic data, *Ground Water*, *51*(3), 385–397, doi:10.1111/j.1745-6584.2012.00974.x.

Herckenrath, D., G. Fiandaca, E. Auken, and P. Bauer-Gottwein (2013b), Sequential and joint hydrogeophysical inversion using a field-scale groundwater model with ERT and TDEM data, *Hydrol. Earth Syst. Sci.*, *17*(10), 4043–4060, doi:10.5194/hess-17-4043-2013.

Hering, A., R. Misiek, A. Gyulai, T. Ormos, M. Dobroka, and L. Dresen (1995), A joint inversion algorithm to process geoelectric and surface wave seismic data. Part I: basic ideas, *Geophys. Prospect.*, *43*(2), 135–156, doi:10.1111/j.1365-2478.1995.tb00128.x.

Hertrich, M., and U. Yaramanci (2002), Joint inversion of Surface Nuclear Magnetic Resonance and Vertical Electrical Sounding, *J. Appl. Geophys.*, *50*(1–2), 179–191, doi:10.1016/S0926-9851(02)00138-6.

Hinnell, A. C., T. P. A. Ferré, J. A. Vrugt, J. A. Huisman, S. Moysey, J. Rings, and M. B. Kowalsky (2010), Improved extraction of hydrologic information from geophysical data through coupled hydrogeophysical inversion, *Water Resour. Res.*, *46*(4), W00D40, doi:10.1029/2008WR007060.

Hubbard, S. S., and N. Linde (2011), Hydrogeophysics, in *Treatise on Water Science*, Vol. 2, S. Uhlenbrook, eds., Elsevier, Amsterdam, pp. 401–434.

Huisman, J. A., J. Rings, J. A. Vrugt, J. Sorg, and H. Vereecken (2010), Hydraulic properties of a model dike from coupled Bayesian and multi-criteria hydrogeophysical inversion, *J. Hydrol.*, *380*(1–2), 62–73, doi:10.1016/j.jhydrol.2009.10.023.

Hyndman, D. W., and S. M. Gorelick (1996), Estimating lithologic and transport properties in three dimensions using seismic and tracer data: The Kesterson aquifer, *Water Resour. Res.*, *32*(9), 2659–2670, doi:10.1029/96WR01269.

Hyndman, D. W., J. M. Harris, and S. M. Gorelick (1994), Coupled seismic and tracer test inversion for aquifer property characterization, *Water Resour. Res.*, *30*(7), 1965–1977, doi:10.1029/94WR00950.

Hyndman, D. W., J. M. Harris, and S. M. Gorelick (2000), Inferring the relation between seismic slowness and hydraulic conductivity in heterogeneous aquifers, *Water Resour. Res.*, *36*(8), 2121–2132, doi:10.1029/2000WR900112.

Irving, J., and K. Singha (2010), Stochastic inversion of tracer test and electrical geophysical data to estimate hydraulic conductivities, *Water Resour. Res.*, *46*(11), W11514, doi:10.1029/2009WR008340.

Jadoon, K. Z., E. Slob, M. Vanclooster, H. Vereecken, and S. Lambot (2008), Uniqueness and stability analysis of hydrogeophysical inversion for time-lapse ground-penetrating radar estimates of shallow soil hydraulic properties, *Water Resour. Res.*, *44*(9), doi:10.1029/2007wr006639.

Jardani, A., and A. Revil (2009), Stochastic joint inversion of temperature and self-potential data, *Geophys. J. Int.*, *179*(1), 640–654, doi:10.1111/j.1365-246X.2009.04295.x.

Jardani, A., A. Revil, F. Santos, C. Fauchard, and J. P. Dupont (2007), Detection of preferential infiltration pathways in sinkholes using joint inversion of self-potential and EM-34 conductivity data, *Geophys. Prospect.*, *55*(5), 749–760, doi:10.1111/j.1365-2478.2007.00638.x.

Jardani, A., A. Revil, E. Slob, and W. Söllner (2010), Stochastic joint inversion of 2D seismic and seismoelectric signals in linear poroelastic materials: A numerical investigation, *Geophysics*, *75*(1), N19–N31, doi:10.1190/1.3279833.

Jardani, A., A. Revil, and J. P. Dupont (2013), Stochastic joint inversion of hydrogeophysical data for salt tracer test monitoring and hydraulic conductivity imaging, *Adv. Water Resour.*, *52*, 62–77, doi:10.1016/j.advwatres.2012.08.005.

Johnson, T. C., R. J. Versteeg, H. Huang, and P. S. Routh (2009), Data-domain correlation approach for joint hydrogeologic inversion of time-lapse hydrogeologic and geophysical data, *Geophysics*, *74*(6), F127–F140, doi:10.1190/1.3237087.

Jougnot, D., N. Linde, A. Revil, and C. Doussan (2012), Derivation of soil-specific streaming potential electrical parameters from hydrodynamic characteristics of partially saturated soils, *Vadose Zone Journal*, *11*(1), doi:10.2136/vzj2011.0086.

Juhojuntti, N. G., and J. Kamm (2010), Joint inversion of seismic refraction and resistivity data using layered models—Applications to hydrogeology, *AGU Fall Meeting Abstracts*, *11*, 0850.

Kalscheuer, T., M. D. los Á. G. Juanatey, N. Meqbel, and L. B. Pedersen (2010), Non-linear model error and resolution properties from two-dimensional single and joint inversions of direct current resistivity and radiomagnetotelluric data, *Geophys. J. Int.*, *182*(3), 1174–1188, doi:10.1111/j.1365-246X.2010.04686.x.

Karaoulis, M., A. Revil, J. Zhang, and D. D. Werkema (2012), Time-lapse joint inversion of crosswell DC resistivity and seismic data: A numerical investigation, *Geophysics*, *77*(4), D141–D157, doi:10.1190/GEO2012-0011.1.

Keay, S., G. Earl, S. Hay, S. Kay, J. Ogden, and K. D. Strutt (2009), The role of integrated geophysical survey methods in the assessment of archaeological landscapes: The case of Portus, *Archaeol. Prospect.*, *16*(3), 154–166, doi:10.1002/arp.358.

Keller, G. V., and F. C. Frischknecht (1966), *Electrical Methods in Geophysical Prospecting*, Pergamon Press, London.

Kis, M. (2002), Generalised Series Expansion (GSE) used in DC geoelectric–seismic joint inversion, *J. Appl. Geophys.*, *50*(4), 401–416, doi:10.1016/S0926-9851(02)00167-2.

Kowalsky, M. B., S. Finsterle, and Y. Rubin (2004), Estimating flow parameter distributions using ground-penetrating radar and hydrological measurements during transient flow in the vadose zone, *Adv. Water Resour.*, *27*(6), 583–599, doi:10.1016/j.advwatres.2004.03.003.

Kowalsky, M. B., S. Finsterle, J. Peterson, S. Hubbard, Y. Rubin, E. Majer, A. Ward, and G. Gee (2005), Estimation of field-scale soil hydraulic and dielectric parameters through joint inversion of GPR and hydrological data, *Water Resour. Res.*, *41*(11), W11425, doi:10.1029/2005WR004237.

Kowalsky, M. B., E. Gasperikova, S. Finsterle, D. Watson, G. Baker, and S. S. Hubbard (2011), Coupled modeling of hydrogeochemical and electrical resistivity data for exploring the impact of recharge on subsurface contamination, *Water Resour. Res.*, *47*(2), W02509, doi:10.1029/2009WR008947.

Kvamme, K. L. (2006), Integrating multidimensional archaeological data, *Archaeol. Prospect.*, *13*, 57–72, doi: 10.1002/arp.268.

Lambot, S., E. C. Slob, M. Vanclooster, and H. Vereecken (2006), Closed loop GPR data inversion for soil hydraulic and electric property determination, *Geophys. Res. Lett.*, *33*, L21405, doi:10.1029/2006GL027906.

Linde, N., and A. Revil (2007), A comment on "Electrical tomography of La Soufrière of Guadeloupe volcano: Field experiments, 1D inversion and qualitative interpretation" by Nicollin, F. *et al.* [*Earth Planet. Sci. Lett.* 244 (2006) 709–724], *Earth and Planetary Science Letters*, *258*(3–4), 619–622, doi:10.1016/j.epsl.2007.04.008.

Linde, N., and J. A. Vrugt (2013), Distributed soil moisture from crosshole ground-penetrating radar travel times using stochastic inversion, *Vadose Zone Journal*, *12*(1), 0, doi:10.2136/vzj2012.0101.

Linde, N., A. Binley, A. Tryggvason, L. B. Pedersen, and A. Revil (2006a), Improved hydrogeophysical characterization using joint inversion of cross-hole electrical resistance and ground-penetrating radar traveltime data, *Water Resour. Res.*, *42*(12), W12404, doi:10.1029/2006WR005131.

Linde, N., S. Finsterle, and S. Hubbard (2006b), Inversion of tracer test data using tomographic constraints, *Water Resour. Res.*, *42*(4), W04410, doi:10.1029/2004WR003806.

Linde, N., J. Chen, M B. Kowalsky, and S. Hubbard (2006c), Hydrogeophysical parameter estimation approaches for field scale characterization, in *Applied Hydrogeophysics*, H. Vereecken *et al.*, eds., Springer, New York, pp. 9–44.

Linde, N., A. Tryggvason, J. E. Peterson, and S. S. Hubbard (2008), Joint inversion of crosshole radar and seismic traveltimes acquired at the South Oyster bacterial transport site, *Geophysics*, *73*(4), G29–G37, doi:10.1190/1.2937467.

Linde, N., J. Doetsch, D. Jougnot, O. Genoni, Y. Dürst, B. J. Minsley, T. Vogt, N. Pasquale, and J. Luster (2011), Self-potential investigations of a gravel bar in a restored river corridor, *Hydrol. Earth Syst. Sci.*, *15*(3), 729–742, doi:10.5194/hess-15-729-2011.

Linder, S., H. Paasche, J. Tronicke, E. Niederleithinger, and T. Vienken (2010), Zonal cooperative inversion of crosshole P-wave, S-wave, and georadar traveltime data sets, *J. Appl. Geophys.*, *72*(4), 254–262, doi:10.1016/j.jappgeo.2010.10.003.

Lochbühler, T., J. Doetsch, R. Brauchler, and N. Linde (2013), Structure-coupled joint inversion of geophysical and hydrological data, *Geophysics*, *78*(3), ID1–ID14, doi:10.1190/geo2012-0460.1.

Looms, M. C., A. Binley, K. H. Jensen, L. Nielsen, and T. M. Hansen (2008), Identifying unsaturated hydraulic parameters using an integrated data fusion approach on cross-borehole geophysical data, *Vadose zone J.*, *7*, 238–248.

Mboh, C. M., J. A. Huisman, N. Van Gaelen, J. Rings, and H. Vereecken (2012a), Coupled hydrogeophysical inversion of electrical resistances and inflow measurements for topsoil hydraulic properties under constant head infiltration, *Near Surf. Geophys.*, *10*(5), 413–426, doi:10.3997/1873-0604.2012009.

Mboh, C. M., J. A. Huisman, E. Zimmermann, and H. Vereecken (2012b), Coupled hydrogeophysical inversion of streaming potential signals for unsaturated soil hydraulic properties, *Vadose Zone J.*, *11*(2), 0, doi:10.2136/vzj2011.0115.

Misiek, R., A. Liebig, A. Gyulai, T. Ormos, M. Dobroka, and L. Dresen (1997), A joint inversion algorithm to process geoelectric and surface wave seismic data. Part II: applications, *Geophys. Prospect.*, *45*(1), 65–85, doi:10.1046/j.1365-2478.1997.3190241.x.

Moghadas, D., F. Andre, E. C. Slob, H. Vereecken, and S. Lambot (2010), Joint full-waveform analysis of off-ground zero-offset ground penetrating radar and electromagnetic induction synthetic data for estimating soil electrical properties, *Geophys. J. Int.*, *182*(3), 1267–1278, doi:10.1111/j.1365-246X.2010.04706.x.

Moorkamp, M., B. Heincke, M. Jegen, A. W. Roberts, and R. W. Hobbs (2011), A framework for 3-D joint inversion of MT, gravity and seismic refraction data, *Geophys. J. Int.*, *184*(1), 477–493, doi:10.1111/j.1365-246X.2010.04856.x.

Moysey, S., K. Singha, R. Knight, and Pt (2005), A framework for inferring field-scale rock physics relationships through numerical simulation, *Geophys. Res. Lett.*, *32*(8), 4, doi:10.1029/2004gl022152.

Musil, M., H. R. Maurer, and A. G. Green (2003), Discrete tomography and joint inversion for loosely connected or unconnected physical properties: Application to crosshole seismic and georadar data sets, *Geophys. J. Int.*, *153*(2), 389–402, doi:10.1046/j.1365-246X.2003.01887.x.

Nowak, W., and O. A. Cirpka (2006), Geostatistical inference of hydraulic conductivity and dispersivities from hydraulic heads and tracer data, *Water Resour. Res.*, *42*(8), W08416, doi:10.1029/2005WR004832.

NRC (2001), *Basic Research Opportunities in Earth Science*, National Academies Press, Washington, DC.

NRC (2012), *Challenges and Opportunities in the Hydrologic Sciences*, Washington, DC.

Paasche, H., and J. Tronicke (2007), Cooperative inversion of 2D geophysical data sets: A zonal approach based on fuzzy c-means cluster analysis, *Geophysics*, *72*(3), A35–A39, doi:10.1190/1.2670341.

Parsekian, A. D., K. Singha, B. J. Minsley, W. S. Holbrook, and L. Slater (2015), Multiscale geophysical imaging of the critical zone, *Rev. Geophys.*, doi: 10.1002/2014RG000456.

Pasion, L., S. Billings, and D. Oldenburg (2003), Joint and cooperative inversion of magnetic and time domain electromagnetic data for the characterization of UXO, in *Symposium on the Application of Geophysics to Engineering and Environmental Problems 2003*, Environment and Engineering Geophysical Society, Denver, CO, pp. 1455–1468, doi:0.4133/1.2923153.

Pollock, D., and O. A. Cirpka (2010), Fully coupled hydrogeophysical inversion of synthetic salt tracer experiments, *Water Resour. Res.*, *46*(7), W07501, doi:10.1029/2009WR008575.

Pollock, D., and O. A. Cirpka (2012), Fully coupled hydrogeophysical inversion of a laboratory salt tracer experiment monitored by electrical resistivity tomography, *Water Resour. Res.*, *48*(1), W01505, doi:10.1029/2011WR010779.

Revil, A., M. Karaoulis, S. Srivastava, and S. Byrdina (2013), Thermoelectric self-potential and resistivity data localize the burning front of underground coal fires, *Geophysics*, *78*(5), B258–B272, doi:10.1190/GEO2013-0013.1.

Rockström, J., *et al.* (2009), A safe operating space for humanity, *Nature*, *461*(7263), 472–475, doi:10.1038/461472a.

Rosas Carbajal, M., N. Linde, T. Kalscheuer, and J. A. Vrugt (2014), Two-dimensional probabilistic inversion of plane-wave electromagnetic data: Methodology, model constraints and joint inversion with electrical resistivity data. *Geophys. J. Int.*, *196*, 1508–1524.

Roth, K., R. Schulin, H. Flühler, and W. Attinger (1990), Calibration of time domain reflectometry for water content measurement using a composite dielectric approach, *Water Resour. Res.*, *26*(10), 2267–2273, doi:10.1029/WR026i010p02267.

Rubin, Y., and S. S. Hubbard (2005), *Hydrogeophysics*. Springer, Dordrecht, 523 pp.

Rubin, Y., G. Mavko, and J. Harris (1992), Mapping permeability in heterogeneous aquifers using hydrologic and seismic data, *Water Resour. Res.*, *28*(7), 1809–1816, doi:10.1029/92WR00154.

Santos, F. A. M., S. A. Sultan, P. Represas, and A. L. E. Sorady (2006), Joint inversion of gravity and geoelectrical data for groundwater and structural investigation: Application to the northwestern part of Sinai, Egypt, *Geophys. J. Int.*, *165*(3), 705–718, doi:10.1111/j.1365-246X.2006.02923.x.

Saunders, J. H., J. V. Herwanger, C. C. Pain, M. H. Worthington, and C. R. E. D. Oliveira (2005), Constrained resistivity inversion using seismic data, *Geophys. J. Int.*, *160*(3), 785–796, doi:10.1111/j.1365-246X.2005.02566.x.

Scales, J. A., and R. Sneider (1997), To Bayes or not to Bayes?, *Geophysics*, *62*(4), 1045–1046.

Scholer, M., J. Irving, A. Binley, and K. Holliger (2011), Estimating vadose zone hydraulic properties using ground penetrating radar: The impact of prior information, *Water Resour. Res.*, *47*(10), W10512, doi:10.1029/2011WR010409.

Scholer, M., J. Irving, M. C. Looms, L. Nielsen, and K. Holliger (2012), Bayesian Markov-chain-Monte-Carlo inversion of time-lapse crosshole GPR data to characterize the vadose zone at the Arrenaes Site, Denmark, *Vadose Zone J.*, *11*(4), doi:10.2136/vzj2011.0153.

Sen, M. K., and P. L. Stoffa (2013), *Global Optimization Methods in Geophysical Inversion*, Cambridge University Press, New York.

Singha, K., and S. Moysey (2006), Accounting for spatially variable resolution in electrical resistivity tomography through field-scale rock-physics relations, *Geophysics*, *71*(4), A25–A28, doi:10.1190/1.2209753.

Straface, S., F. Chidichimo, E. Rizzo, M. Riva, W. Barrash, A. Revil, M. Cardiff, and A. Guadagnini (2011), Joint inversion of steady-state hydrologic and self-potential data for 3D hydraulic conductivity distribution at the Boise Hydrogeophysical Research Site, *J. Hydrol.*, *407*(1–4), 115–128, doi:10.1016/j.jhydrol.2011.07.013.

van Genuchten, M. T. (1980), A closed-form equation for predicting the hydraulic conductivity of unsaturated soils, *Soil. Sci. Soc. Am. J.*, *44*(5), 892, doi:10.2136/sssaj1980.03615995004400050002x.

Watters, M. S. (2006), Geovisualization: An example from the Catholme Ceremonial Complex, *Archaeol. Prospect.*, *13*, 282–290, doi:10.1002/arp.290.

Wisen, R., and A. V. Christiansen (2005), Laterally and mutually constrained inversion of surface wave seismic data and resistivity data, *J. Environ. Eng. Geophys.*, *10*(3), 251–262, doi:10.2113/JEEG10.3.251.

8

Integrated Imaging for Mineral Exploration

Peter G. Lelièvre and Colin G. Farquharson

ABSTRACT

The mineral exploration industry has recognized that integrated imaging approaches will be important aids to help keep up with expected demand for raw materials as current reserves dwindle. Our aim for this chapter is to (a) demonstrate the great potential offered by integrated imaging methods for improving the detection, delineation, and discrimination of mineral deposits and (b) encourage further development of the required software. We outline many aspects of mineral exploration that lead to difficulties with geophysical inversion and the need for integrated imaging approaches. We provide recommendations for overcoming these challenges and we review joint and cooperative inversion approaches used in the field to date. The great variety of geological scenarios, geophysical and geological data available, and exploration questions posed make it impossible to suggest any one approach for general problems. There is, however, a growing awareness that a solid understanding of physical property information is critical for directing joint and cooperative inversion strategies. This chapter should help practitioners and researchers find or develop appropriate methods for building fully integrated Earth models for mineral exploration, consistent with all geophysical and geological information available.

8.1. INTRODUCTION

8.1.1. A Summary of Joint Inversion* Challenges in the Industry

The mineral exploration industry has recognized that new geophysical interpretation techniques and tools, including joint and geologically constrained inversion in particular, will be required to help keep up with expected demand for raw materials. *Witherly* [2012] points out that "over the last 60 years many of the readily identifiable mineral deposits (the low-hanging fruit) have been found and finding the next tier of deposits will require resources and ideas which many believe are currently inadequate to meet the challenge." This sentiment has been issued before and likely has existed throughout the history of mineral exploration. For example, by *Schmitt et al.* [2003]: "today, most of the near-surface base-metal deposits have already been found and future developments must seek ever-deeper orebodies." Also by *Debicki* [1996]: "with the known shallow reserves of [metallic ore] declining, however, it has become obvious that new deep exploration tools must be developed if the industry is to remain viable in the future."

In recent decades, the mineral exploration industry has failed to discover new reserves at the same rate as world consumption [*Goodway*, 2012; *Witherly*, 2012; *Wilson et al.*, 2012]. In response to falling rates of shallow discoveries, exploration is beginning to look to greater depths

Department of Earth Sciences, Memorial University of Newfoundland, St. John's, Newfoundland and Labrador, Canada

**To simplify this section, we use the term joint inversion to encompass both joint and cooperative methods.*

Integrated Imaging of the Earth: Theory and Applications, Geophysical Monograph 218, First Edition.
Edited by Max Moorkamp, Peter G. Lelièvre, Niklas Linde, and Amir Khan.

that push the resolution limits of geophysical data collected by standard equipment. As well as discovering new ore bodies, exploration challenges can involve detecting and imaging relatively subtle features that are poorly resolved by geophysical data but can have important economic implications, for example the details of the geometry of a known ore body and possible offshoots. One way forward is to research, develop, and apply joint inversion approaches, which are well placed to play an important part in replacing rapidly diminishing resources as exploration looks deeper, signals get weaker, and the resolving power of each individual data set diminishes [*Oldenburg and Pratt*, 2007].

Regardless of the depth and scale of exploration, even seemingly simple exploration problems are sometimes difficult to answer with standard inversion approaches. Typical "minimum-structure" or "Occam's razor" style inversions generate physical property distributions that exhibit smooth spatial variation. With such approaches it is difficult to recover models that fit geological expectations of sharp and distinct contact boundaries, making it more challenging to interpret the location of such boundaries and assess, for example, how deep one has to drill in order to reach the top of an orebody. Even delineating larger-scale bodies and reliably estimating their physical properties continues to be a challenge for geophysical inversion. Ultimately, these challenges come down to issues of resolution and uncertainty, for which geologically constrained joint inversion methods may provide an important avenue for future improvements. In the remainder of this section we outline the potential advantages of, and challenges involved in, performing joint inversion for mineral exploration. Links to other sections are provided to help readers navigate the chapter.

The physical properties of rocks encountered in mineral exploration can be highly variable: There can be considerable overlap of physical properties between different lithologies, and rock physical properties can change greatly through post-emplacement alteration and weathering processes. Hence, it can be impossible to reliably distinguish a particular lithology based on a single physical property (or analysis of its associated geophysical data). However, it is often possible to distinguish different groups of rocks from one another if multiple physical properties are combined. Integrated imaging approaches therefore become important for solving many exploration problems. It is exceedingly rare that an economic deposit is mined before several types of geophysical data have been collected and analyzed. Sections 8.2.2 and 8.2.3 go into more detail on these issues.

Sometimes, the different physical property distributions associated with a deposit are tied to different genetic phenomena (that created the deposit), making it far more challenging to design appropriate joint coupling strategies based on either structural or property-based considerations (an example is provided in Section 8.2.3). A sophisticated understanding of the different geophysical lithologies (i.e., rocks that can be distinguished by differences in physical properties) is essential. This requires a solid petrophysical understanding of the geology. There is a growing awareness that obtaining a better understanding of physical property information is a critical step for mineral exploration and that this information should be used to direct joint inversion strategies. While physical property information is usually collected during an exploration program, obtaining reliable information from depth is problematic and the data collected are challenging to utilize in inversion studies: Sections 8.3.5 and 8.3.6 go into more detail.

Even if the potential exists to discriminate the different rocks of interest by applying joint inversion techniques, imaging can be hampered by a lack of quality survey data. There can be serious resolution issues inherent to the specific geophysical data types or the survey designs used. The economic realities of mineral exploration commonly mean that one must work with what is available, and the only geophysical data available for a site may be of considerable age and have been collected using a survey design that is no longer optimal for answering the current critical exploration questions. Nevertheless, there are often several types of geophysical data available and this data needs to be used to its full potential, regardless of its quality. Hence, there is plenty of opportunity to apply joint inversion approaches to improve the success of mineral exploration projects, but with incomplete or inadequate data it becomes important to carefully assess the reliability of the models constructed. Sections 8.3.2 and 8.3.3 discuss survey related issues.

The geological scenarios encountered in mineral exploration cover a large variety of possible shapes and scales of ore deposits (covered in Section 8.2.1). Because of the generally complicated 3D geology, and the sometimes difficult exploration questions at play, it is common for mesh-based inverse problems of interest to contain hundreds of thousands of data measurements and a million or more model parameters. Currently, these sizes are challenging enough that inversion of individual geophysical datasets (i.e., one type of data measurement and one type of physical property) can take considerable computing time. Although probabilistic inversion approaches are attractive because they can provide information regarding the reliability of different model features, they are not yet feasible for such large problems. Linear or linearized deterministic inversions using local descent-based optimization methods have been the only viable option. While computing power is continually improving, the established strategy in the industry has been to run single dataset deterministic inversions with sizes that challenge the computing limits existing at the time. Joint inversion typically has much higher computational requirements, so performing joint inversion with problem sizes of interest may be infeasible and could require unattractive concessions. Section 8.3.4 summarizes inversion approaches applicable to mineral exploration and these associated computational challenges.

8.1.2. Chapter Objectives and Outline

Despite much theoretical research in the area of joint inversion, the methods have seen relatively little application to mineral exploration problems in the scientific literature. This is due, in part, to a lack of capacity in existing software, or availability of such software. Our aim for this chapter is to (a) demonstrate the great potential offered by joint inversion methods for improving the detection, delineation, and discrimination of mineral deposits and (b) encourage further development and dissemination of the required software tools and application of those tools.

This chapter is organized as follows. Section 8.2 discusses the structural and petrophysical characteristics of mineral deposits, focusing on the aspects that lead to difficulties with geophysical inversion and the need for integrated imaging approaches. Section 8.3 examines the role and use of geophysics within mineral exploration programs, again pointing out challenges relating to integrated imaging. Sections 8.2 and 8.3 cover pertinent background information that is important to place joint inversion for mineral exploration in the required context before discussing the approaches in detail. Readers familiar with that introductory material may wish to start at Section 8.4, which reviews joint inversion approaches used in the field to date. The review is focused towards, but not limited to, contributions that invert multiple geophysical data types responsive to different physical properties. We conclude in Section 8.5 with recommendations for practitioners and pose challenges for future work in the field.

8.2. CHARACTERISTICS OF MINERAL DEPOSITS

8.2.1. Shapes and Structural Settings of Mineral Deposits

Evans [1993] provides an overview of common orebody shapes and sizes. Orebodies can have dimensions ranging from meters to thousands of meters. Tabular ore bodies can occur in association with veins, for example from the infilling of spaces associated with faults or sedimentary bedding planes. Tubular ore bodies can occur, for example, in association with intrusive pipes from partial dissolution of host rock. Volcanic massive sulfide (VMS) deposits are generally lens-shaped to sheet-like, developing at the interfaces between rock units. Disseminated deposits can be irregularly shaped, with ore minerals scattered throughout the host rock or situated in a stockwork of interlacing veins, the overall shape being anywhere from cylindrical to cap- or pear-shaped. The mineralization in disseminated deposits can fade gradually outwards, leaving no clear boundary to delineate an orebody, making assessment of the economically feasible volume more challenging. Skarn deposits can be even more irregularly shaped, with ore tongues projecting along joints or faults and terminating at structural controls.

This large variety of possible shapes and scales of ore deposits presents a challenge to the use of geophysics for mineral exploration. Although the general bulk shape or trend of a deposit may sometimes be determined through simple analyses of geophysical data or geological observations, important smaller-scale features are usually more difficult to discern without more intensive data collection and analysis. For highly irregularly shaped deposits, the overall shape and location of the mineralization may be well determined but the exploration challenge is to find smaller-scale offshoots from a main body. To resolve smaller features, it is often necessary to constrain geophysical inversion with as much a priori information as possible and to combine multiple sources of geophysical data through coupled (joint or cooperative) inversion.

A further challenge to geophysics is presented by tabular and tubular ore bodies that are very thin compared to their other dimensions. Geophysical numerical modeling methods may have difficulty representing such bodies adequately on discretized meshes without refining to a point that is beyond computational feasibility, thereby precluding the use of one or more types of geophysical data that already present high computational challenges. For inverse modeling, there are limits to the spatial resolution of the geophysical data and refining a mesh beyond a particular limit may not be helpful. However, the underlying forward modeling methods often require particular mesh designs that increase the computational requirements—for example, to move mesh boundaries far away from anomalous material in order to honor approximated boundary conditions, or to refine the mesh around sources and receivers in order to reduce numerical modeling errors (e.g., see *Rücker et al.* [2006] and *Jahandari and Farquharson* [2014]). Hence, for the large problems of interest in mineral exploration, there are presently computational limitations to the geophysical data combinations that can be combined in a joint inversion.

Another possible complicating factor for mineral deposits is that alteration halos can exist around various structural features. Physical properties can change across these alteration zones and affect the geophysical response. We now provide further discussion on the physical properties of mineral deposits.

8.2.2. Physical Properties of Mineral Deposits

Petrophysics provides the link between geological and geophysical investigations of the Earth: Realistic geological interpretations of geophysical inversion results can only be made alongside an understanding of the geological controls on rock physical properties. *Dentith and Mudge* [2014] emphasize this link through their discussion of the general use of geophysics for mineral exploration; we point readers to that reference for more information on the physical properties of mineral deposit rocks and issues related to their measurement. Here we summarize some of the controls on the physical properties of rocks. We emphasize the variability of rock physical properties before tying this to the detectability of mineral deposits using geophysics (Section 8.2.3) and the associated challenges for integrated imaging.

The density of rock can be quite variable, affected by mineral composition but also by texture, porosity, and pore fluid (see *Wohlenberg* [1982]). The magnetic character of igneous rocks and metamorphic rocks is also variable (see *Keller* [1988], *Clark et al.* [1992], and *Hunt et al.* [1995]). The amount of magnetite in a rock depends on the concentrations of iron and oxygen and the pressure and temperatures at which mineralization occurs. The direction and strength of remanent magnetization depends on (a) the geomagnetic field at the time the rock cooled past the Curie temperature, (b) the time duration since that occurred, and (c) thermal, pressure, and chemical changes throughout the history of the rock [*Butler*, 1992]. The direction of remanent magnetization also depends on any geological events that may have physically moved the rock since it was emplaced. The conductivity of rocks is affected by the presence of native metals and pore fluids. Hence, the conductivity of rock is greatly affected by porosity and fissures and is another highly variable physical property (see *Palacky* [1988]). Chargeability is also dependent on mineralogy, porosity, and the type of pore fluid. For metamorphic rocks, the mineralogical composition is the most important factor in determining the elastic properties. However, even a small volume fraction of cracks in a rock can dramatically impact seismic rock properties [*Schmitt et al.*, 2003].

Magnetic, electrical, and seismic rock properties can exhibit anisotropic behavior. Few minerals and rocks have perfectly symmetrical structure or composition, making anisotropy the norm [*Keller*, 1988]. Rocks that have been strongly metamorphosed or tectonized can display significant seismic anisotropy, as can some igneous and sedimentary rocks [*Schmitt et al.*, 2003]. Physical properties may also show nonlinearity: For example, conductivity may be a function of current density or electric field intensity [*Keller*, 1988]. Self-demagnetization effects for materials of high magnetic susceptibility also lead to nonlinear responses and anisotropic-like effects where the direction of the total magnetization depends on the shape of the body (see *Clark and Emerson* [1999]). Most geophysical inversion methods treat physical properties as linear scalar quantities because this simplifies the numerical methods. Use of such methods is inappropriate unless the assumptions of isotropy and linearity are valid or lead to modeling errors acceptable for the particulars of the exploration problem. The appropriateness of such assumptions can be assessed through forward modeling experiments, but this requires some knowledge of the source distribution and access to forward modeling methods that allow for anisotropy or nonlinearity. Such forward modeling software is not common.

8.2.3. Detectability of Mineral Deposits Using Geophysics

To detect a mineral deposit, there must be enough physical property contrast between the host and ore rocks that the orebodies can be sensed by the associated geophysical surveys. The physical properties of metalliferous mineral deposits are often significantly different from those of the surrounding host rock. This is especially true of conductivity: For massive sulfides hosted in igneous or metamorphic rocks, the conductivity contrast in mS/m is typically at least three orders of magnitude, and the only other naturally occurring mineral that substantially overlaps the conductivity range of massive sulfides is graphite (*Palacky*, [1988]; see our Figure 8.1). However, other conductive units such as weathered layers and saline aquifers can complicate geophysical data interpretation. Seismic properties of massive sulfides have also been shown to differ significantly from those of most common host rocks [*Salisbury et al.*, 1996, 2000, 2003; *Salisbury and Snyder*, 2007].

In many exploration scenarios, there can be considerable overlap of physical properties between different rock types (geologically defined lithologies) and it can be impossible to identify a particular lithology from one physical property alone or from the analysis of one related geophysical data type alone (see Figure 8.1; *Dentith and Mudge* [2014] provide similar helpful plots for other physical properties, taken from various sources). However, if enough physical properties are combined, then it becomes possible to use geophysics to distinguish different groups of rocks from one another through the use of integrated imaging approaches (see Figure 8.2). A nice example comes from *Persson et al.* [2011], who analyzed magnetic, electromagnetic (EM), and geoelectrical survey data to study a region of high potential for economic deposits in Orrivaara, Northern Sweden. All geophysical methods provided valuable information, but only in combination was it possible to distinguish between the

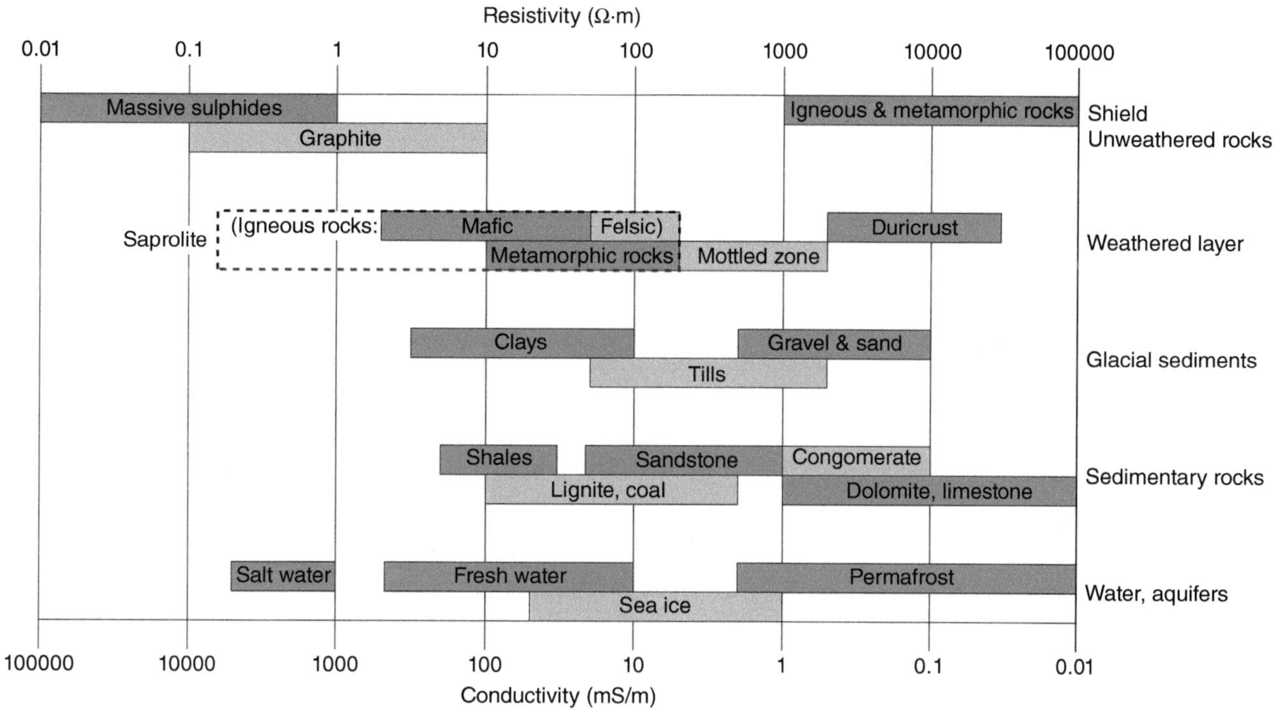

Figure 8.1 Typical ranges of resistivities of Earth materials. Reproduced from *Palacky* [1988].

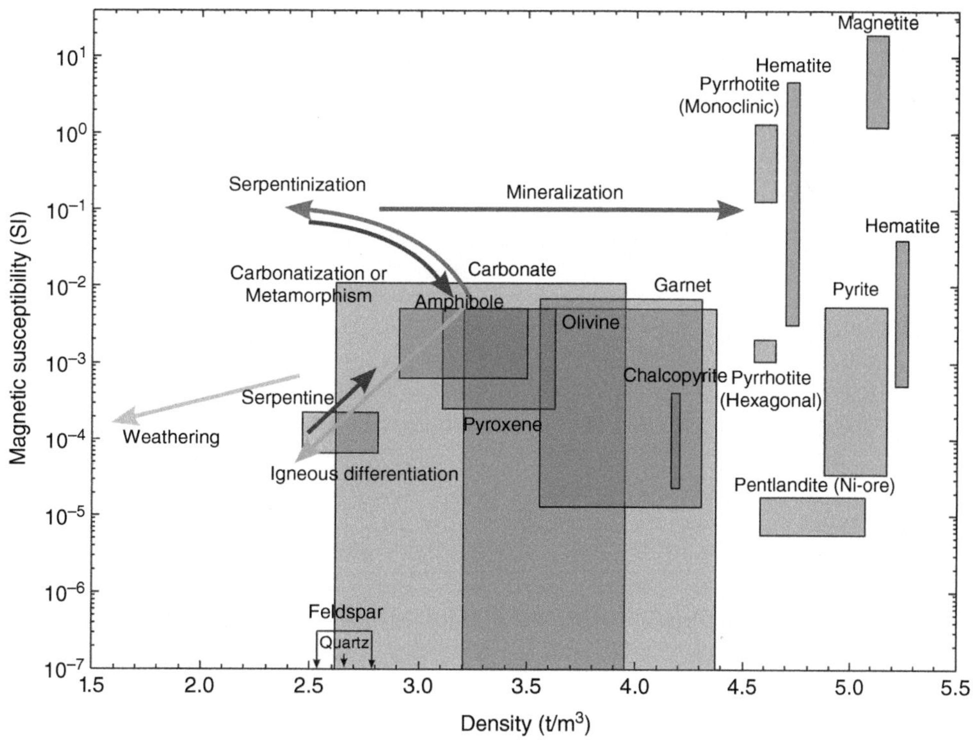

Figure 8.2 Densities and magnetic susceptibilities of common silicate, sulfide, and oxide minerals associated with Archean greenstone-hosted nickel deposits. Reproduced from *Williams* [2008].

different types of rocks: the pyrrhotite-bearing sediments being conductive and strongly magnetic, mafic rocks being resistive and highly magnetic, and brittle deformation zones being conductive and showing low magnetic signatures.

Lithological classifications are typically made though assessment of the mineral proportions in rocks. This may overlook important petrophysical characteristics such as porosity and pore contents. Where geophysics is involved, such geological rock classifications must be replaced with physical classifications such that each "geophysical lithotype" is distinguishable from others by its physical properties. That is, all geophysical lithotypes should present separated clusters on physical property crossplots (see Figure 8.7e). Such classifications have also been termed "petrophysical domains" or "rock property domains" [*Gerrie et al.*, 2014]. *Mahmoodi and Smith* [2014] provide an illustrative example from the Victoria Cu–Ni property in which ten rock types were identified by geologists from drillcore but quantitative analysis of the logged physical properties indicated the data could be best explained using only three petrophysical domains: Some geologically based rock types had essentially homogeneous physical properties while others were heterogeneous.

Sometimes, the different physical property distributions associated with a deposit are tied to different genetic phenomena and are not expected to show structural similarity nor provide petrophysical information that could help design an appropriate joint or cooperative coupling strategy. Significant induced polarization (IP) anomalies can be generated by water-filled shear zones and graphite-bearing sediments that are not of economic importance; EM anomalies can also be caused by noneconomic sources of high conductivity such as graphite-bearing rocks, wet clays, saline water-filled shear zones or aquifers, bodies of water, and cultural objects [*Kearey et al.*, 2002]. A specific example comes from the work of *Phillips* [2001] and *Phillips et al.* [2001], who studied the San Nicolás Cu–Zn VMS deposit in Mexico and performed several independent inversions of various data types. This deposit contained a density high centered on the main sulfide body, the density being simply related to the degree of sulfide mineralization. In contrast, physical property measurements showed that magnetic minerals (a) were not distributed evenly throughout the deposit, (b) were also present in the host rocks, and (c) they might have been deposited by hydrothermal events not associated with the sulphide deposit itself. Inversion recovered a magnetic susceptibility distribution with a high centered closer to the western boundary of the main body and with significant susceptible material extending far beyond the northern boundary of the deposit. High conductivities were associated with the sulfide body but also with a thick overburden layer. A high chargeability anomaly was linked to a dipping fault on the boundary of the deposit, at a different location than the magnetic susceptibility high. For such complicated scenarios, it is difficult to find appropriate methods for coupling a joint or cooperative inversion. Geological lithological classifications that simplify this scenario into ore and host rock units are not helpful. A sophisticated description of the geophysical lithologies is essential—for example, to distinguish between magnetic and nonmagnetic sulfide rocks and between magnetic and nonmagnetic host rocks.

Even if appropriately designated rock units can be discriminated based on physical properties and related geophysical data, discrimination can still be difficult based solely on geophysical data and inversion results. This may be due to unfavorable survey designs, resolution issues inherent to the specific geophysical data type, resolution issues related to the geological scenario, or the nonuniqueness inherent to the inverse problem. These issues are discussed in Section 8.3.

8.3. GEOPHYSICS IN MINERAL EXPLORATION

8.3.1. The Role of Geophysics in an Exploration Program

Before minerals can be processed for use, a mining site must be located and developed. First, a regional exploration campaign is undertaken to discover an orebody or ore system. Further information must then be obtained through local exploration programs to systematically discern the most prospective targets and assess the economic feasibility of extracting the resource. This includes steps to improve the delineation of the orebody, determine the mineral grade, improve the understanding of the associated mineral system and geological controls, and help with the development of mining infrastructure. Geophysics can play an important role in all these tasks. While the only way to obtain direct observations of deep subsurface rocks is via drilling, drilling is expensive and slow compared to geophysical surveying. Furthermore, drilling can only provide sparse information, whereas geophysical methods sense the bulk properties of the rocks in the subsurface.

Phillips [2001] discussed how geophysical inversion can be employed in an integrated exploration program including geological, geophysical, and geochemical components. He outlined the possible role of geophysics in a complex flowchart that illustrated how geological and petrophysical information is essential for the optimal use of geophysics. We summarize his ideas here. Regional geophysical inversions can sometimes identify targets immediately, particularly when an ore body is very large and has high physical property contrast to the surrounding rocks. Otherwise, regional inversion results can be used along with physical property information to aid in regional geological interpretations to identify favorable settings for deposits. Once targets have been defined and a local exploration program initiated, geophysical inversion can help define specific details of the subsurface that relate to the exploration questions: for example the depth and size of a body with favorable physical properties, or the geometry of a geological feature, knowledge of which may be critical in understanding the genesis of the deposit. A drilling program may then commence and provide further data: physical property measurements on core samples, the locations of geological contacts, or downhole geophysical data (although collection and utilization of the latter is still uncommon at the time of writing). The amount of drilling necessary to define an orebody before it is mined is substantial. The knowledge of a deposit must be quite precise before a company will make the huge financial commitment to develop a mine and associated infrastructure. There are also regulations on the claims that can be made about the grades and tonnage of an ore deposit. Hence, there is great need to collect helpful information at this stage. That information can help prepare geologically constrained geophysical inversions that generate improved results and are better able to guide further drilling. The information can also help design future geophysical surveys that are optimal for addressing the exploration questions. These new inversion results or geophysical data feed back into the multidisciplinary workflow and the process continues iteratively until, hopefully, the exploration goals have been achieved. As such, geophysical inversion is not a task that should be undertaken in isolation but is, rather, an important part of an iterative multi-disciplinary process [*Dentith and Mudge*, 2014].

8.3.2. Geophysical Data Collection

Since the 1950s, EM geophysical methods have been of paramount importance for detecting and imaging metallic orebodies [*Palacky*, 1988] because of the large conductivity contrasts involved. Gravity and magnetic measurements are common because they are relatively cheap and simple to take: They require no active source and the receiver instruments are relatively small. Geoelectrical methods are less common because of material costs and logistics associated with laying electrical cables. Seismic surveys are not a traditional method used in mineral exploration, mainly due to their high cost, but they have seen increased interest in recent years (see Section 8.3.3 for further discussion on the challenges of using seismic data for mineral exploration).

Airborne gravity, magnetic, and EM surveys are commonly flown early in exploration for ore deposits because large survey areas can be covered rapidly and economically. Smaller-scale ground-based surveys may follow, including geoelectrical and seismic surveys that require physical contact with the Earth. Ground-based gravity, magnetic, and EM measurements may be collected to provide more densely sampled data—for example, to fill in spaces between flight lines in regional surveys. For ground-based EM measurements, survey layout and geometry (source and receiver size, orientation, and separation) are no longer restricted by the constraints presented by aircraft and can obtain better coupling with the anticipated target. Later in an exploration program, drilling and subsurface mining infrastructure provide opportunities to collect data below the surface, allowing different parts of the volume of interest to be illuminated.

All geophysical survey methods can theoretically be deployed in drillholes and downhole devices exist for collecting all major categories of geophysical data. Downhole gravity and magnetic measurements are similar to their surface counterparts. EM and seismic measurements are restricted in terms of what can fit into a drillhole: A common approach is to position a large source on the surface and place smaller receivers downhole. Geoelectrical methods use small electrodes, reducing such concerns for downhole measurements; an important aspect is that an efficient electrical connection is required between the electrodes and the rocks. Drillhole geophysics is much more costly, relying on expensive and slow drilling campaigns, and carries the possibility of losing expensive survey probes if, for example, holes close during operation [*Smith et al.*, 2012]. Drillhole blockages may also hamper the collection of downhole data by not permitting a tool to travel as far as desired.

While single component surveys are the norm for potential field data (vertical gravity and total field magnetic data), vector and tensor measurements have shown some theoretical advantages (see *Davis and Li* [2011], *Li and Oldenburg* [2000a], *Pilkington* [2012], and *Rim and Li* [2012]). However, as *Li and Oldenburg* [2000a] point out, the successful use of multicomponent data requires that the orientation of the measurement device be accurately known at the measurement locations, and this can be difficult when data are collected in drillholes (see Section 8.3.6 for more).

Geographical land features can interfere with surveying. Rugged, wet, or heavily forested land may be inaccessible to survey equipment. Large or steep topographical features can interfere with the flight of aircraft and therefore with the collection of quality data during airborne surveys (explained further by *Dentith and Mudge* [2014]). Cultural land aspects can also affect surveying. Often, seismic and other ground surveys are only measured along roads that simplify access, providing little power for 3D resolution, yet the roads may be too sinuous to interpret the data with

2D processing even if the geology allows. Cultural objects such as railway lines, roads, fencing, pipelines, buildings, and powerlines can affect magnetic, geoelectrical, and EM measurements, leading to gaps in surveys when such objects are avoided or when unusable data are removed. Land access issues can also lead to gaps in data coverage.

Often the only geophysical data available were originally collected for specific purposes and are therefore not ideal for others. Sometimes this means having to make the most of sparse regional survey data to image smaller-scale targets (e.g., see *Ash et al.* [2006]), before the collection of denser data measurements in more detailed surveys can be justified. Before the advent of high-power computing and the increase in interpretation methods it enabled, many surveys were designed with particular interpretation tools in mind, rather than the more appropriate strategy of designing the surveys based on the exploration questions. For example, geoelectrical and seismic methods can be deployed along profile lines with the intention of interpreting the data in 2D, but such data only illuminate a tiny fraction of the full 3D volume. While this was perhaps more common historically, often the best geophysical data available for a site is of considerable age. A similar problem can occur with crosshole tomography surveys: The positioning of sources and receivers is usually limited to existing drillholes, which may not be distributed densely and uniformly enough to image the entire volume of interest. Even if time and money allow data to be collected between all pairs of drillholes, this essentially provides several vertical 2D sections which may still be too sparse to recover the full 3D volume (e.g., see *Carter-McAuslan* [2014]).

Geophysical data often pass through several processing stages before they arrive in the hands of the inversion practitioner. For example, gridding is a common operation, allowing for easier visual inspection of data maps and the application of interpretation tools that require gridded information. Such processing can introduce artifacts (see *Dentith and Mudge* [2014]). Hence, geophysical inversion is best applied to minimally processed data; however, sometimes such data are not easy to obtain, nor are the details of the processing.

These challenges aside, on advanced-stage mineral exploration projects there is generally a vast amount of useful geophysical data available. On greenfield projects (undeveloped sites) there may only be regional data, usually potential field data, with wider line spacings. Still, this provides plenty of opportunity for performing joint or cooperative inversion and for those coupled inversions to make a significant contribution. However, all geophysical surveys cost money, and the economic realities of mineral exploration mean that one can not expect to have access to high-quality data from a survey perfectly designed to resolve the target of current interest. One must work with what is available, which often involves inadequate data coverage or quality to satisfactorily illuminate smaller-scale features. There can be different data densities for different data types and they may cover different areas, perhaps not even overlapping laterally. Joint or cooperative inversion approaches that combine such incomplete or inadequate data must consider data resolution issues when assessing the reliability of the models constructed (see Section 8.3.7 for more). The flip side of this issue is that, although geophysical data collection does have a significant cost, it is still fundamentally cheaper than drilling, so any data available should be used to their full potential.

8.3.3. Depth of Investigation and Resolution

Reemphasizing our statement ending Section 8.2, even if different rock units can be discriminated based on physical properties, imaging can be difficult based solely on geophysical inversion results because of unfavorable survey designs, resolution issues inherent to the specific geophysical data type, or resolution issues related to the geological scenario. Here we discuss data resolution considerations for different geophysical methods. Emphasis is given to the depths to which geophysical data can provide information about the Earth. Mining at depths greater than two kilometers is currently rare. It can be difficult to justify mine development below 1 km, but that depth is a real target. A thorough understanding of the emplacement of an ore deposit and its surrounding rocks may require even deeper geophysical investigation.

The "depth of investigation" (DOI) is the depth to which a particular set of geophysical data are sensitive enough to distinguish signal from noise and therefore have the potential to resolve the subsurface features. A major consideration here is the depth of penetration (DOP) of the energy associated with a particular geophysical survey.

The DOP for geoelectrical surveys increases with electrode spacing. Because electrodes must make physical contact with the ground and be electrically connected to other parts of the surveying equipment, material costs and logistics typically limit the DOP to relatively small depths compared to other data types. For DC resistivity surveys, the generation of sufficient power to get enough energy to greater depths is also a significant consideration. DOPs are limited to about 1 km for typical equipment but can reach a maximum of about 2 km for specialized systems. However, DOPs can drop considerably where highly resistive near-surface layers exist, below which it is difficult to establish current flow.

The DOP for EM fields decreases with source frequency; while this allows controlled source electromagnetic (CSEM) survey equipment and design to be tuned to desired DOIs, practical DOIs are limited by (a) the amount of power that can safely be generated, (b) geometrical fall-off of the transmitted signal away from the source (e.g., 1 over distance squared for a dipole source), and (c) the frequency range that can be generated and detected. DOPs for typical CSEM surveys are on the order of 1 km but, again, specialized systems can reach greater depths. Use of natural fields, as in magnetotelluric (MT) surveys, helps to improve DOPs to a few kilometers. As with electrical methods, the conductivity of the subsurface rocks is an important consideration for EM DOPs but it is now conductive overburdens that pose a challenge to imaging deeper targets.

Energy generation is not an issue for passive gravity and magnetic surveys, but the concept of DOI doesn't make sense for potential fields in the same way it does for EM/electrical methods. The crux is whether the target signal can be separated from the noise, which depends on the geological scenario and noise levels. A more important consideration for inversion is that gravity and magnetic data display a lack of depth resolution owing to the fundamental nature of potential field methods (see *Blakely* [1996]).

Seismic reflection has been suggested as an ideal tool for deep exploration because it has the theoretical potential for high resolution to the current depth limits of most mining, about 3 km [*Salisbury et al.*, 2000]. Resolution decreases with depth and increases with the dominant frequency. Seismic methods are well suited for investigating layered sedimentary sequences and therefore are by far the most widely used geophysical method for hydrocarbon exploration (see Chapter 9). However, they are problematic for mineral exploration where diverse, strongly deformed and fractured igneous and metamorphic hard-rock terrains dominate, making for an extremely challenging data acquisition and processing environment (see *Eaton et al.* [2003], *Hobbs* [2003], and *Schmitt et al.* [2003]). Unlike geophysical methods traditionally used in mineral exploration, for which the measurements are responsive to bulk properties of the rock, seismic measurements can be significantly affected by small-scale variability in the background rock. Even for a simple geological scenario with an orebody hosted in a lithologically homogeneous background unit, fractures in the background unit may present seismic reflectivity of a similar strength as that at the edges of the orebody. Seismic interpretation is difficult to impossible through traditional means developed by the hydrocarbon exploration industry, and methods must be adapted to suit hard-rock terrains [*L'Heureux et al.*, 2005]. Recent work presented in *Eaton et al.* [2003] has gone some distance to move seismic methods forward for use in mineral exploration, and use of seismic methods for mineral exploration in hard-rock terrains has increased in recent decades [*Salisbury et al.*, 2003]. However, significant knowledge gaps still exist [see *Salisbury and Snyder*, 2007]. Sound interpretation of seismic data requires a fairly detailed understanding of the relevant physical properties of target and host rocks (e.g., see *Pretorius et al.* [2003]), which is often lacking. Seismic methods are significantly more expensive than traditional geophysical survey methods and are often far outside mineral exploration budgets. In general, reliable seismic investigation of the shallow crust for mineral exploration should utilize drillhole geophysical data [*Eaton et al.*, 2003], adding further expense.

An important consideration for improving the DOI for any geophysical survey is increasing the signal-to-noise ratio. Noise associated with vehicle motion can be eliminated in ground-based measurements where sources and receivers can sit solidly on the ground, and reducing such motion can enable the use of more sensitive instruments. For EM measurements, which are affected by a variety of atmospheric and cultural noise sources, more current can be passed through larger ground-based transmitter loops than on airborne platforms to obtain stronger signals. Moving data measurements closer to an orebody is also a simple way to increase the signal. Moving from airborne to ground surveys can help, but geographical and cultural land features can interfere with survey design and implementation. Collecting geophysical data downhole is another option, but that imposes even more constraints on survey design. *Li and Oldenburg* [2000a] simultaneously inverted surface and downhole magnetic data, showing superior synthetic inversion results when downhole data were introduced. *Hattula and Rekola* 2000] presented a sulfide exploration example where downhole EM and mise-a-la-masse (MAM) data were used to complement data collected at the surface; the downhole data provided pivotal information for locating new off-hole orebodies.

DOPs quoted above are merely general estimates for typical survey instruments and configurations. Such simple rules of thumb for estimating DOIs are not generally applicable because DOIs are highly dependent on the relevant physical property distributions, survey geometry, instrument noise, and data collection or processing errors. Practical DOIs for particular survey scenarios can be assessed through various methods; some examples are given by *Oldenburg and Li* [1999], *Christiansen and Auken* [2012], and *Deceuster et al.* [2014]. Ideal survey configurations depend on the estimated characteristics of the target and host, as well as the exploration questions being asked. Survey optimization approaches can help design such ideal configurations; for example, see *Stummer et al.* [2004], *Djikpesse et al.* [2012], *Rawlinson et al.* [2012], *Wilkinson et al.* [2012], and *Yu et al.* [2013].

Often, part of the modeling volume is poorly resolved by one or more geophysical surveys, and multiple datasets must be combined via joint or cooperative inversion to accurately image the entire volume. An example comes from *Roy and Clowes* [2000], who used seismic and potential field methods to image the Guichon Creek batholith, British Columbia, Canada, which contains several economically important copper deposits. Regional-scale seismic reflection data helped define the edges and other features of the batholith, but resolution limitations associated with acquisition geometry meant that seismic data was not able to image the top 1.5 km of the region. However, gravity and magnetics data were able to complement the seismic data and provide information where the latter could not, providing a clear example of where integrated imaging was required to provide an accurate model recovery.

8.3.4. Inversion Approaches for Mineral Exploration

Here, we summarize the more common geophysical inversion methods used for mineral exploration that have been extended to joint inversion, or could easily be so extended. By far the most common inversion approach applied to mineral exploration is a "mesh-based" or "voxel" approach. The subsurface is discretized into a mesh of many small cells, the physical property (or properties) of interest being homogeneous within each cell. The inversion determines one or more piecewise physical property distributions that could have given rise to the data. The allowed physical property values may lie across a continuous range (as by *Li* and *Oldenburg* [1996] and many others); they may take discrete values determined using a priori petrophysical information (e.g., see *Camacho et al.* [2000]; or their distribution may be prescribed in a probabilistic framework (e.g., see *Bosch* [1999]; see also Chapter 2). It is not uncommon for these mesh-based inverse problems of interest in mineral exploration studies to contain tens or hundreds of thousands of data measurements and a million or more model parameters. Problems with hundreds of millions or even a billion model parameters are currently being considered (e.g., see *Čuma et al.* [2012] and *Čuma and Zhdanov* [2014]). Although probabilistic optimization approaches to solving mesh-based inverse problems are attractive—for example, to determine the properties of the subsurface that all possible solutions share—such approaches require extensive objective function evaluations—for example, through Markov Chain Monte Carlo (MCMC) sampling. Hence, they are not feasible for such large inverse problems, which may require tens of minutes, or even hours, for a single objective function evaluation. Linear or linearized deterministic inversions using local descent-based optimization methods have been the only viable option and are expected to be so for the near future. The exceptions involve probabilistic inversions that deal with 2D problems or linear potential field problems that can be efficiently parallelized; for example, see *Bosch and McGaughey* [2001], *Lane et al.* [2008], and Section 8.4.6.

A few other inversion approaches have been used that work with models comprising surfaces representing contacts between rock units. These contact surfaces are parameterized and their locations determined in the inversion algorithm. *Tanner* [1967] developed a 2D inversion approach where the Earth volume of interest was represented as a set of tightly packed vertical rectangular prisms, the top or bottom of these representing a contact between two units. Many others have extended this basic approach to allow for more complicated scenarios. *Fullagar et al.* [2000] (see also *Fullagar et al.* [2004, 2008]) generalized the approach in 3D by allowing for multiple-layered rock units: Vertical rectangular prisms have internal contacts that divide each prism into homogeneous layers. The physical properties of each rock unit in a starting model can remain fixed while the inversion controls the position of the contacts, resulting in what is essentially a "geometry inversion". While it is also possible to allow smooth variations within each unit, allowing for too much flexibility in the model means less ability to reduce the nonuniqueness of the inverse problem. The methods of *Fullagar et al.* [2000] require that stacked layers be appropriate for the geological scenario, although some flexibility is provided because layers can pinch out horizontally, allowing for geological scenarios where smaller lenses can exist between layers.

Instead of working with vertical prisms, *Oliveira and Barbosa* [2013] approximated an isolated geological body by an ensemble of vertically stacked thin tiles with irregular polygonal horizontal outlines. The vertices of the polygonal outlines are parameterized in a polar coordinate system. For each tile, the outline vertices are allowed to move radially away from some central point inside the polygon. This approach requires that the geological scenario involve one or more isolated bodies that can be represented by this parameterization.

Richardson and MacInnes [1989] developed a rudimentary inversion approach with the model consisting of a wireframe surface built of tessellated triangles or other polygonal planar facets. The inversion changes the model via the coordinates of the facet vertices that define the wireframe contact surface. *Lelièvre et al.* [2015] took this approach further, their goal being to work directly with 3D geological models of any complexity, with possibly multiple wireframe surfaces connected together and no limitation to a layered model. A helpful aspect of their methods is that by applying stochastic sampling they are able to obtain information regarding the likelihood of particular features in the model. This inversion approach is best utilized in brownfield scenarios (e.g., detailed exploration of a known deposit) to investigate the viability of a proposed deposit model and present statistical information to help assess that viability. However, the methods could also be safely applied to simpler greenfield problems, for example with isolated sources and few rock units. The computational requirements are significant and the methods require further research before they become computationally feasible for the challenging inverse problems of interest to the exploration community.

Another option that deals with contacts between rock units are level set methods (e.g., see *Zheglova et al.* [2013]). Although an underlying mesh-based discretization of the subsurface is used in the forward modeling, a contact between two units is represented as the zero level of a higher-dimensional function which is perturbed by the inversion algorithm. The physical properties of each cell in the underlying mesh are determined simply by assessing which cells are inside an inclusion (the level set function is positive or zero) and which are in the background (the level set function is negative). For example, a 3D level set function is required to model 2D interfaces, as illustrated in *Lelièvre et al.* [2012a]. While the methods of *Zheglova et al.* [2013], only allow for two units in the model (the background and an inclusion), extension to allow multiple units should be possible. The methods should also be extended to allow for important geological constraints—for example, to tie an interface to a particular pierce point.

8.3.5. Mitigating Uncertainty in Inversion

In mesh-based inversions, there are typically many more model parameters than there are data measurements (commonly two or more orders of magnitude), leading to an underdetermined, nonunique inverse problem for which there are an infinite number of solutions that can fit the data exactly. Instrument noise in data measurements, or errors associated with data positioning and processing, lead to uncertainties in the data measurements. This adds further nonuniqueness to the inverse problem because the data may now be fit inexactly, provided that the fit is within error bars consistent with estimated data uncertainties. To solve nonunique inverse problems deterministically, additional information must be incorporated to restrict the space of admissible solutions to a single and, hopefully, usable solution. That a priori information can lead to different regularization approaches.

When regularization is applied in a deterministic mesh-based inversion, the regularization functional should not be treated as a purely mathematical attempt at dealing with the nonuniqueness of the problem. Instead, regularization functionals should be developed to facilitate the incorporation of geological information and ideas [*Silva et al.*, 2001; *Lelièvre and Farquharson*, 2013]. The ambiguity in the inverse problem should be moderated by any available constraints on the nature and form of the causative physical property distributions, whether those constraints stem from hard factual data observations, intuitive expectations or hypothetical conjectures. This will result in geologically feasible and sensible models with expected characteristics. Several authors have developed different regularization approaches to encourage different model characteristics: models close to a weighted reference model (e.g., see *Li and Oldenburg* [1996]); smooth model features consistent with, for example, an alteration halo extending out from a mineralized zone into country rock (e.g., see *Constable et al.* [1987], *deGroot-Hedlin and Constable* [1990], *Li and Oldenburg* [1996], and *Lelièvre and Farquharson* [2013]); sharp contrasts between more distinct rock units (e.g., see *Farquharson and Oldenburg* [1998], *Farquharson* [2008], and *Portniaguine and Zhdanov* [1999]); dipping elongated bodies (e.g., see *Li and Oldenburg* [2000b] and *Lelièvre and Oldenburg* [2009]); or compact bodies (e.g., see *Last and Kubik* [1983], *Guillen and Menichetti* [1984], *Barbosa and Silva* [1994], and *Boulanger and Chouteau* [2001]).

Adding more geophysical data to the problem through joint or cooperative inversion methods can help reduce the nonuniqueness, provided that the various datasets sense the subsurface in different ways and therefore present complementary information. The hope is that a jointly recovered subsurface model consistent with geophysical data from multiple surveys will be more likely to represent the true subsurface than a model consistent with only a single type of data. There are many possible mathematical methods for coupling a joint inversion (e.g., see Chapters 3 and 4; *Lelièvre et al.* [2012b]) but, as for regularization, the choice of coupling measure used should not be treated as a purely mathematical attempt at dealing with the nonuniqueness of the problem. Instead, the coupling approach should be driven by the geological and petrophysical information available and by the exploration goals [*Lelièvre et al.*, 2012b]. As such, we view coupled inversion as a subset of geologically constrained inversion approaches: A priori information can be incorporated into inversion through the choice of model parameterization, the choice of regularization approach or the setting of associated weights, the setting of numerical bound constraints, or the choice of a joint or cooperative coupling strategy.

8.3.6. Petrophysical Data Collection

There is a growing awareness that obtaining a better understanding of physical property information is a critical step for mineral exploration. *Smith et al.* [2012], *Visser and Lajoie* [2012], *Wilson et al.* [2012], and many others encourage the acquisition of physical property data to constrain geophysical inversions and thereby integrate geological and geophysical knowledge into 3D "common-Earth models" consistent with all available information. Any physical property information, or other a priori information, should be used to direct a joint or cooperative inversion strategy. In practice, petrophysical information is often available, along with other types of helpful geological information such as expected depths, shapes and orientations of targets, and topological rules (e.g., one unit must lie above another). Density, radioactivity, magnetic, and electrical property measurements are common. The exception is measurement of rock velocities, which is critical in hydrocarbon exploration but uncommon in mineral exploration because the benefits have not yet been adequately demonstrated [*Smith et al.*, 2012].

Physical property measurements can be taken on drillcore and outcrop samples. Some measurements can be taken in the field while others must be performed in a lab. Obtaining reliable information from depth is particularly important because geophysical data collected at the surface have decreased resolving power at depth. Diamond drilling to obtain core is more expensive and time-consuming than drilling methods that chip the rock—for example, reverse circulation drilling. Still, many physical rock properties are commonly measured on drillcore samples during a mineral exploration project. Whether or not core is recovered, properties can also be measured in situ with downhole probes, but these measurements can be affected by drillhole conditions—for example, small-scale alteration caused by the drilling.

Physical property measurements taken on hand samples in a lab, and even those taken on outcrop or down a drillhole, are unlikely to represent the true in situ value of those physical properties, for a variety of reasons discussed by *Smith et al.* [2012]. Rock samples can be fractured when extracted from their original location, affecting the accuracy of measurements of density, conductivity, chargeability, and seismic rock properties. The drilling process can also cause changes to the rock surrounding the drillhole through mechanical or drilling fluid invasion, so even physical property measurements taken downhole can be unreliable. Many physical property measurements can be unreliable if in situ conditions are not reproduced during measurement. For example, density, conductivity, and chargeability measurements are affected by the amount and type of fluid in a sample, and seismic velocities measured at low pressures in laboratories are not usually representative of the properties of the same material at depth [*Schmitt et al.*, 2003]. Knowing the original in situ orientation of a sample is important when the physical property information contains directional content such as anisotropic properties or magnetic remanence vectors. While drill-core can be oriented, complete orientations including azimuthal information are usually not measured because of the costs associated with downhole orientation tools [*Pinto and McWilliams*, 1990]. Scales involved in laboratory and downhole physical property measurements are much smaller than those of the bulk properties measured by geophysical methods, and the measurements are then unhelpful unless they can be upscaled. Conductivity measurements are of particular difficulty here because conductivity can be influenced by large-scale structures in the rock, such as fractures, that are absent in smaller rock samples. Magnetic properties are also problematic because they can be distributed heterogeneously at smaller scales, leading to variable local measurements.

Another issue with petrophysical data collection is sampling bias. Measurements of different properties are usually taken in different locations for different purposes: for example, densities in ore zones for resource estimates, magnetic susceptibilities in early exploration holes for mapping purposes, and conductivity and chargeability only by necessity after performing electrical or EM surveys.

Gravity and magnetic methods pose a further challenge because the data are generally processed relative to some arbitrary regional level. While data processing can attempt to remove this level, it is impossible to remove exactly. This means that the associated background density or susceptibility value is unknown, and even accurately determined physical property measurements taken on rock samples cannot be compared to the values in an inversion model in an absolute sense. Only a relative comparison of the values is useful.

As a result of all these issues, many physical property data obtained are problematic to utilize in inversion to help guide regularization and coupling choices. *Austin and Foss* [2014] provide a framework for reconciling rock property measurements with estimates obtained from analysis of geophysical anomalies, and they discuss implications for mineral exploration best practise.

8.3.7. Model Appraisal

Following from Section 8.3.3, even if different rock units can be discriminated based on physical properties, and the geophysical survey data are of sufficient quality to potentially resolve those physical property contrasts, geophysical inversion is inherently nonunique and finding a single acceptable solution via a deterministic inversion approach is not sufficient to provide reliable information regarding subsurface targets. Joint or cooperative inversion studies, especially those that combine incomplete or inadequate data, must carefully assess the reliability of the models constructed. Without such assessments, one cannot say with certainty whether a particular model feature is required by one or more geophysical datasets or if the feature is simply a consequence of the regularization or coupling used in the inversion. Ultimately, the importance of analyzing the uncertainty in a recovered model depends on exploration risk tolerances.

There are several possible approaches to assess models recovered from inversion, many reviewed by *Caterina et al.* [2013]. In Section 8.3.3 we mentioned methods for calculating practical depths of investigation—that is, depths to which geophysical data is particularly sensitive. More detailed information about the relative resolution in different parts of the modeling domain can be obtained by investigating model resolution matrices (e.g., see *Alumbaugh and Newman* [2000] and *Friedel* [2003]): Areas with poor sensitivity or resolution can be considered less reliable.

When a clear exploration question is posed, simple forward modeling or multiple inversion runs can help provide insight into the reliability of different features of the model. An illustrative example comes from *Phillips* [2001], who performed forward modeling tests on a detailed geological model of the San Nicolás deposit to assess the resolvability of a small deep "keel" orebody extension from the actual field survey configurations; he then went further to design survey configurations that could yield improved signals from the keel. Such an approach could utilize any representative model of the subsurface, which could be a geological model (transformed into physical properties) or a physical property model recovered from inversion. Another approach is to run several different inversions with different choices in each—for example, altering regularization measures and associated weights—to investigate model space. This is the essence of the approach taken by *Oldenburg and Li* [1999] to assess DOIs for geoelectrical surveys (they applied a cross-correlation measure to different inversion results). *Lelièvre and Oldenburg* [2009] performed two magnetic inversions that encouraged different orientations to assess whether or not the recovered orientation of a deep fault-related model feature was required by the data.

The most thorough approach to investigating model space is to employ stochastic sampling to provide accurate likelihood information regarding the different model features (see Chapter 2). However, this is generally computationally infeasible without making approximations or interpolations (e.g., see *Tompkins et al.* [2011, 2013]) or working with smaller 2D linear problems (e.g., as by *Bosch and McGaughey* [2001] and *Lane et al.* [2008]; see Section 8.4.6).

8.4. JOINT AND COOPERATIVE INVERSION APPROACHES FOR MINERAL EXPLORATION

8.4.1. Section Overview

There have been few mineral exploration studies published that employ simultaneous joint inversions of multiple geophysical data types responsive to different physical properties. Because of the large computing requirements for the inverse problems typically performed for mineral exploration applications, such simultaneous joint inversion has been infeasible for many. Adequate software tools for joint inversion have not been made widely available to the industry. Also, with mineral exploration being a money-making enterprise, there is important work that remains unpublished for reasons of confidentiality and competitive advantage. Nevertheless, much work has been published in the overall theme of integrated imaging. In this review we focus our discussion on joint and cooperative inversion work that involves multiple geophysical data types responsive to different physical properties. We do not limit the review to that category since much can be learned from other work.

We do not provide a thorough review of joint interpretation studies, where several different types of geophysical data are inverted separately and the results integrated via interpretation to yield an improved understanding of the mineral deposits of interest. Nor do we thoroughly review single-property joint inversion studies, involving multiple types of geophysical data responsive to the same physical property, where no coupling strategy is required between different models. Still, there are lessons regarding resolution and data uncertainties that can be learned from such work and interested readers may benefit from the publications included in Section 8.4.2.

The work reviewed after Section 8.4.2 can be categorized in several ways. The subsurface parameterization may involve an underlying mesh of many cells or may involve structural interfaces between rock units. The inversion framework may be deterministic or probabilistic. There are sequential cooperative inversion strategies where each dataset is inverted independently and the similarity of the physical property models is achieved via the regularization functionals, and there are joint inversions where the different datasets are inverted simultaneously and coupling measures or other strategies are applied to control the similarity of the physical property models. The coupling strategy may be structural or property-based, it may be related to the geostatistical concept of cokriging, or it may simply be dealt with via the choice of subsurface parameterization. Table 8.1 indicates how the material falls into these different categories.

We are not aware of any joint inversions applied to mineral exploration problems where the coupling is based on a mathematical formula derived from the physics of the physical properties involved. Such relationships are rarely applicable in mineral exploration because of the considerations regarding physical properties discussed in Sections 8.2.2, 8.2.3 and 8.3.6. For example, compressional wave velocity is generally related to density for sedimentary rocks but not for metamorphic rocks, where

Table 8.1 Categorization of the Work Reviewed in Sections 8.4.3 through 8.4.8

Section	8.4.3	8.4.4	8.4.5	8.4.6	8.4.7	8.4.8
Mesh-based parameterizations	•	•	•	•		•
Interface-based parameterizations					•	
Deterministic methods	•	•	•	•	•	•
Stochastic methods[a]				•	•	
Sequential cooperative strategies	•					
Simultaneous joint strategies		•	•	•	•	•
Structural coupling	•	•				
Property-based coupling	•		•	•		
Coupling by common interfaces					•	
Coupling by cokriging						•

[a] A fully probabilistic inversion framework or stochastic optimization method is employed.

mineralogical composition is the most important factor controlling velocities. We do, however, discuss some work involving empirically derived relationships: Section 8.4.3 mentions work by *Kamm et al.* [2015] in which an empirical petrophysical relationship was derived from laboratory measurements and used in a cooperative inversion; Section 8.4.5 discusses work by *Lelièvre et al.* [2012b], who applied a linear coupling relationship in a synthetic joint inversion example containing two rock units, where the linear relationship simply connects two known points on a density-versus-slowness plot.

Before we present our review, the readers should be made aware of two important points. First, for the joint and cooperative inversion work we discuss below, the level of data fit obtained is comparable for the different inversions, be they independent or joint. Any differences in the models (e.g., smooth spatial changes versus sharp interfaces) is a result of changes in regularization or constraints associated with particular joint or cooperative inversion approaches. We suggest that readers unfamiliar with the basics of inversion theory, particularly the different forms of regularization and joint coupling, should refer to the theory chapters of this book and other references provided in this chapter. Second, it is important that the readers understand by what metric the different results (e.g., independent and joint) are compared in the works below. For synthetic examples we forgo explanation since the goal is to obtain models closer in character to the true synthetic model by way of spatial features or physical properties. Without ground truth, one cannot make such comparisons and it suffices to mention that the two results (e.g., independent and joint) are different and lead to different interpretations. One can not evaluate the two results without ground truth, but simply obtaining a result that reconciles multiple datasets must be considered an improvement in itself; this is, after all, the underlying philosophy of integrated imaging.

8.4.2. Joint Interpretation and Single-Property Joint Inversion

Oldenburg et al. [1997] jointly interpret models recovered via independent inversion of magnetic, DC resistivity, IP and airborne EM data from the Mt. Milligan Cu–Au porphyry deposit in British Columbia, Canada. Their work clearly illustrates inconsistencies in the different physical property models recovered from independent inversions and, hence, the need for joint or cooperative inversion.

Raiche et al. [1985] combined coincident loop transient EM and Schlumberger sounding to resolve layered conductivity structures. Their work provides a clear description of a complementary data scenario where each geophysical method has weaknesses ameliorated by the strengths of the other.

Sasaki et al. [2014] invert ZTEM and AMT data to recover a mutually consistent conductivity model. They considered a 2D model relating to unconformity-type uranium exploration in the Athabasca Basin, Canada. They showed that simultaneous inversion of both data types was able to identify some structural features that were difficult to resolve from individual data sets alone because of resolution issues.

Li and Oldenburg [2000a] inverted surface total field magnetic data and three-component downhole magnetic data collected from an ironstone deposit in Australia, showing similar improvements over independent inversions. Their major hurdle was designing appropriate weighting functions to counteract the decay of the magnetic kernels, since the standard depth weighting is no longer applicable to the joint problem. Another important issue they discussed related to understanding the sources of error in the two data sets such that appropriate uncertainties could be applied.

Commer and Newman [2009] combined 3D CSEM and MT data to recover conductivity. They suggest approaches for balancing the data weights such that no one dataset dominates the outcomes.

8.4.3. Cooperative Inversion

Iterative cooperative inversion strategies that separate the two or more individual inverse problems have been attractive for mineral exploration problems because they allow the use of existing inversion software without modification. Also, they can avoid computational problems encountered by some joint inversion methods—for example, numerical issues associated with joint coupling measures and having to determine appropriate weights for the different geophysical datasets. Utility software is easily developed to feed the results from one inversion into the input files for the next. There are many design options for cooperative inversion strategies, and structural or property-based coupling methods can be designed. Methodologies can be designed to address specific exploration problems and exploit the benefits of existing inversion codes.

At their simplest, cooperative inversion investigations take the results of one inversion and incorporate information from the recovered model into another. An early example comes from *Lines et al.* [1988], who inverted seismic and gravity data to recover a 1D model comprising stacked layers. Their strategy first used seismic data to determine layer interface depths and layer velocities. A subsequent gravity inversion calculated densities for each layer and updated the depth of the shallowest interface, but the deeper interfaces were fixed at the depths determined from the seismic inversions. In this way, information regarding the interface depths was moved from the seismic inversion to the gravity inversion. This strategy took into account the ability of each dataset to resolve particular parts of the subsurface; that is, the shallowest layer interface depth was not well-defined by the seismic data. Hence, deciding how to carry the information over requires considerations that are similar to deciding the most appropriate coupling strategy in a simultaneous joint inversion; that is, the choice should be made based on the a priori geological and petrophysical information. Any coupling measure designed for simultaneous joint inversion could be used for cooperative inversion. In a sense, this simplified cooperative inversion strategy is akin to simultaneous joint inversion, with mathematical coupling between the physical property models but with only a single model free to change at each iteration.

An inversion example with aspects of such a cooperative inversion strategy is presented in *García Juanatey* [2012] and *García Juanatey et al.* [2013], who studied the Skellefte District, a rich metallogenic mining area in Sweden. The results of 2D interpretation of seismic reflection profiles were used to constrain a subsequent 2D MT inversion. Prominent features in the seismic sections were used to encourage sharp discontinuities by removing the smoothness regularization between mesh cells for the MT inversion. The integrated inversion result was able to reconcile both datasets and lead to different interpretations of the subsurface geology compared to the independent MT result. This is not truly a cooperative inversion, given that no seismic inversion was performed, but it has the same flavor and illustrates one approach for carrying information between inversions: using the structural features in one recovered model to weight the smoothness regularization across the inversion mesh for another model. Coupling the inversions through the smoothness regularization weights was an approach also taken by *Günther and Rücker* [2006] and *Lelièvre* [2009] (discussed below).

Takougang et al. [2015] developed a cooperative inversion approach for seismic and MT data that used a coupling approach similar to that of *García Juanatey et al.* [2013] plus additional measures. They applied their methods to field data from a hard-rock, Carlin-style gold mineralization province in Nevada and provided a synthetic example representative of a subbasalt environment. Their idea was to (a) use MT data to provide information on large-scale (low frequency) features not directly recoverable from the band-limited seismic data and (b) use seismic data to provide high-resolution structural information unobtainable from MT data. They used a variety of means to couple the resistivity and seismic rock property models. To provide structural coupling, they selected horizons on 2D migrated seismic sections and encouraged sharp boundaries at those locations in their MT inversions by removing or reducing the weight of the smoothness regularization between particular mesh cells (as by *García Juanatey et al.* [2013]). In their synthetic example, these constraints improved the definition of a subsurface mineralized zone and reduced artifacts below that zone. To incorporate property-based coupling into their workflow, they converted their recovered resistivity models into acoustic seismic impedance (AI) models using petrophysical relationships (mathematical functions with coefficients derived from borehole data). They used the converted AI models as background models for their subsequent 2D seismic inversions. They compared their final AI models from their cooperative inversion to those from independent seismic inversions. In both their synthetic and real survey examples, the two AI models showed significant differences. For their synthetic example, the cooperative inversion recovered large important features that the independent inversion did not. For their field data example, the cooperative inversion successfully recovered the main geologic units expected. The suitability of their strategy relies on the validity of the petrophysical relationships used. Although they show that it is possible to obtain accurate results with minimal borehole information, it is recommended to use more plentiful borehole information to obtain more reliable results.

A different coupling approach was used by [*Oldenburg et al.* [1997], who inverted several different types of geophysical data from the Mt. Milligan Cu–Au porphyry deposit. They developed a cooperative inversion strategy to follow a hypothesis regarding a general anticorrelation between chargeability and magnetic susceptibility indicated by geochemical analysis. They first inverted total field magnetic data and then built a weighting function valued 1.0 at the smallest recovered susceptibility and increasing to 10.0 at the highest susceptibility. This weighting function was applied in the smallness term of their objective function for a subsequent inversion of IP data, thereby encouraging the IP inversion to place higher chargeabilities away from regions of higher magnetic susceptibility. The weighted inversion result showed significant differences compared to the independent IP inversion result: The high chargeabilities moved towards the surface and one single deeper anomalous feature was split into two shallower parts.

The cooperative inversion procedure carried out by *Oldenburg et al.* [1997] could easily have been performed in the opposite direction, using the result from an independent IP inversion to weight a subsequent magnetic inversion. However, these two opposite cooperative procedures would likely have recovered significantly different models. The sensitivities of the two datasets to their respective physical property distributions are different, meaning that initial independent inversions of the two datasets would likely yield physical property models with different characteristics. Depending on the details of the implementation, their strategy to cooperatively couple the two physical property models via the smallness weighting may strongly force the second model to follow the first. With different starting models, one may then expect different results depending on the order of the two inversions. Hence, a simple two-stage cooperative inversion strategy, such as that by *Oldenburg et al.* [1997], should be designed carefully, with the exploration goals and a priori geological and petrophysical information in mind, and it is advisable to analyze the resolution of each dataset and investigate the strength of the coupling approach to assess the reliability of the results.

Instead of an essentially two-stage cooperative inversion procedure such as those described above, it may be preferable to perform a multistage iterative procedure that should converge such that the two models no longer change in subsequent iterations. It is understandable that *Oldenburg et al.* [1997] did not so do, seeing that computers of that age were such that even 2D inversions were considerably time-consuming. *Lelièvre* [2009] (and see also *Lelièvre* et al. [2009]) developed a ladder-style multistage iterative procedure in which two datasets were inverted separately at each iteration, after which the two recovered physical property models were used to calculate the smoothness weighting on each pair of inversions performed in the next iteration. They inverted gravity and cross-well tomography data for 2D synthetic models, one based on a nickel exploration scenario in the Eastern Goldfields of the Yilgarn Craton in Western Australia. After each pair of inversions, the spatial model gradients for both models were scaled and combined together. High- and low-smoothness weights were set based on thresholding the combined gradients. The threshold level was adjusted at each iteration of the cooperative procedure. A similar procedure worth comparing, despite their work not being specific to mineral exploration, is that of *Günther and Rücker* [2006]. Their weighting strategy was based on the principles of robust modeling, with the smoothness weight for a particular cell determined by comparing contributions to the ℓ_1 and ℓ_2 smoothness model norms. They inverted DC resistivity and refraction tomography data from 2D synthetic models. In their implementation, the resistivity model was used to alter the smoothness weights for the tomography inversion and the velocity model was used to alter the weights for the resistivity inversion. Both *Günther and Rücker* [2006] and *Lelièvre* [2009] achieved significant improvements with their cooperative procedures compared to their original independent inversions, obtaining models that better delineated the contacts between approximately homogeneous rock units. However, such strategies are fairly ad hoc, lacking rigorous mathematical substantiation of the convergence properties, and therefore come with no guarantee of success. If care is not taken, structural information relating to inversion artifacts may propagate between models and such spurious structure could be enhanced in both. Nevertheless, such approaches show promise and are fairly simple to implement given existing inversion software.

For a particular application it may be advantageous to insert the cooperative linking strategy into the inversion algorithms rather than between complete inversion runs. An example comes from *McMillan and Oldenburg* [2014], who evaluated a method for cooperatively inverting multiple EM and geoelectrical datasets to recover a consistent conductivity model. Their workflow performed a partial minimization of their objective function for one dataset and then carried the resulting conductivity model over to become the initial and reference model for a partial minimization of the objective function for a second dataset, continuing as such, circling through each dataset, until all models converged. An additional feature was added to their algorithm to ensure that the different data misfits were reduced to their prescribed target values such that one dataset would not overly influence the result.

They applied their algorithm to field data from the Antonio gold deposit in Peru, producing 3D conductivity models that recovered known subsurface features more accurately than by inverting the individual data sets separately.

Although the work of *McMillan and Oldenburg* [2014] used different geophysical datasets responsive to the same physical property, a similar cooperative approach could be applied to scenarios with multiple physical property models, provided that petrophysical information exists to help link the different properties. *Kamm et al.* [2015] (see also *Kamm* [2014]) did just this, performing a cooperative inversion with similar details to that of *McMillan and Oldenburg* [2014], including adaptive reference models changing throughout the minimization process and steps undertaken to prevent either dataset from dominating the results. They applied their strategy to invert ground-based gravity data and airborne magnetic data to image a gabbro intrusion in Boden, Sweden, relevant for possible Cu, Ni, and Platinum Group Element mineralization. To help couple the two physical property models, they derived a petrophysical relationship between density and susceptibility from laboratory measurements on rock samples (Figure 8.3a). After each inversion iteration, the density model in the gravity inversion was converted to a susceptibility model using the closest value in the laboratory-derived petrophysical relationship; this susceptibility model became the reference model for the next iteration of the magnetic inversion, and a density reference model was obtained from the susceptibility model in the same way. They found that their cooperative inversions improved the distinction between the intrusion and background in their scenario: A sharper contrast allowed for a clear geological interpretation and the recovered physical property values were in agreement with the petrophysical observations (see Figure 8.3b–e). However, they note that even the cooperative inversion was not able to reliably reproduce the lower boundary of the intrusion because of the lack of resolution in the potential field data. An important unanswered practical detail from *McMillan and Oldenburg* [2014] and *Kamm et al.* [2015] is how strongly to weight the reference models throughout the iterative process such that the convergence is smooth and guaranteed.

To summarize, there are many design options for cooperative inversion strategies, and a variety of different approaches have been implemented with varying levels of success. While such approaches show some promise, published methodologies are typically designed to address specific exploration problems and have often been limited by the availability and functionality of existing inversion codes. In some cases, the approaches have been applied to a single exploration example without rigorous testing to assess their robustness or applicability. Hence, practitioners should proceed carefully when considering applying published cooperative inversion methodologies, or versions thereof, to their own exploration scenarios.

8.4.4. Joint Inversion with Structure-Based Coupling

Haber and Oldenburg [1997] applied their curvature-based structural joint inversion approach (see Chapter 4 for details) to radio imaging tomography data acquired between two drillholes. Their goal was to image extensions of mineralization thought to exhibit high attenuation. While this field data example was used solely as a numerical test of their methods, the models recovered from joint inversion showed improved structural similarity and differed substantially from those recovered from individual inversions. The joint coupling approach of *Haber and Oldenburg* [1997] has seen little use since its introduction, with applications showing preference to the cross-gradient coupling approach of *Gallardo and Meju* [2004] and *Fregoso and Gallardo* [2009]. This is likely because the methods of *Haber and Oldenburg* [1997] rely on curvature scaling parameters, the choice of which is crucial and tuning them appropriately requires (1) knowledge of how the physical properties might vary and (2) a decision on the part of the interpreter regarding what model curvatures are considered important. In contrast, the cross-gradient coupling measure is free of scaling parameters, making it a more simple method to apply.

While cross-gradient joint inversion has shown promise in other applications, it has seen less use for mineral exploration applications. *Lelièvre et al.* [2012b] investigated a joint inverse problem involving gravity and seismic tomography data over the "Ovoid" body in the Voisey's Bay Ni–Cu–Co magmatic massive sulphide deposit in Labrador, Canada. Their example was synthetic but based on the true geology of the area as determined from an extensive drilling campaign. Application of the cross-gradient measure in their joint inversion provided little improvement over individual inversion results. This was understood to be due in part to resolution issues associated with the data types used and the geological scenario. First, there is an inherent lack of depth resolution in gravity data. Second, the Ovoid sulfide body is dense and slow compared to the surrounding rocks and, consequently, most first arrival ray paths miss the Ovoid such that the travel times contain no first-order information about the slowness of the Ovoid.

The cross-gradient measure makes very few assumptions regarding the underlying distributions, which makes the method widely applicable but provides less constraining power; that is, it is less able to reduce the model space of acceptable solutions. With poor resolution in one or both datasets—for example, as in the scenario presented by *Lelièvre et al.* [2012b]—the cross-gradient coupling

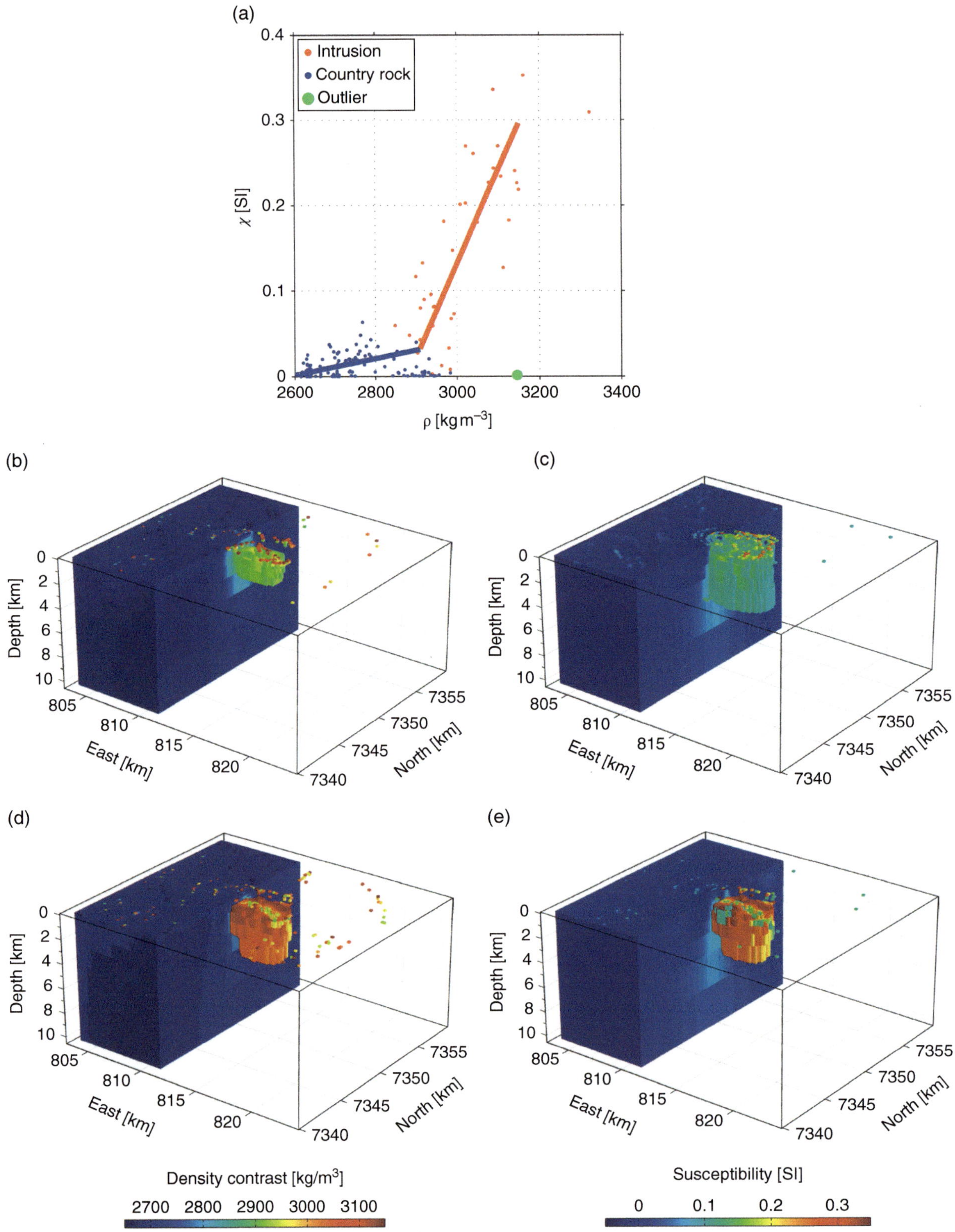

Figure 8.3 **(a)** Petrophysical data and the relationship used in their cooperative inversion. **(b, c)** Independent inversion results. **(d,e)** Cooperative inversion results. In the central and eastern part of the modelling volume, cells with density values above 2900 kg m^{-3} or susceptibility values above 0.1 SI are shown to delimit the intrusion. In the western part, all model cells are displayed to illustrate the background structure. The scattered high property dots at the surface are small near-surface artifacts mostly underneath the gravity measurement locations. From *Kamm et al.* [2015].

constraint may not be able to overcome the resolution issues and reduce the nonuniqueness of the problem enough to provide significant improvements. Another example is provided in Chapter 9 for a hydrocarbon exploration study; *Takougang et al.* [2015] discuss problematic scenarios for the combination of seismic and MT data.

Another consideration with the poor performance of the cross-gradient coupling for the work of *Lelièvre et al.* [2012b] was the optimization approach used. The cross gradient constraint introduces multiple local minima into the optimization problem. A challenge for the cross-gradient constraint is to encourage solutions away from a null solution—for example, a solution where one model changes but the other does not. This requires developing an optimization approach that can find particular local minima of the problem without getting stuck in other, less desirable minima. *Gallardo and Meju* [2004] designed a local descent-based optimization approach to coax their cross-gradient joint inversion towards such desirable solutions. *Lelièvre et al.* [2012b] made no such attempt, taking the conservative approach of heating the cross-gradient coupling measure slowly through several iterations while maintaining the desired level of fit to each dataset. The two different local optimization strategies taken by *Gallardo and Meju* [2004] and *Lelièvre et al.* [2012b] may provide different solutions for the same inverse problem. The optimization approach used must be designed to attempt to find an appropriate local minimum, as determined through a priori knowledge of the subsurface, but without using computationally intensive global optimization methods one cannot guarantee that such a local minimum will be visited.

To conclude, joint inversions based on the cross-gradient coupling approach can suffer from the existence of multiple minima and from issues related to poor resolution in one or more of the datasets. More investigation is required to fully understand these problems before the cross-gradient approach can be reliably applied to mineral exploration problems.

8.4.5. Joint Inversion with Property-Based Coupling

Significant physical property information can be available for mineral exploration problems, suggesting that property-based coupling approaches might be helpful for joint inversion. However, as discussed in Sections 8.2.2, 8.2.3, and 8.3.6, obtaining and utilizing valuable physical property information can be challenging. When working in a deterministic inversion framework requiring derivatives of the objective function, it is difficult to design property-based coupling approaches when the physical property relationships are nonlinear, complex, or uncertain. Furthermore, there might be several physical property relationships existing in one area, but often the knowledge of the spatial applicability of these relationships is lacking. Nevertheless, significant work has been done with this coupling approach.

Continuing their work from where we left of in Section 8.4.4, *Lelièvre et al.* [2012b] applied several property-based coupling strategies to their gravity-tomography joint inverse problem. First, an implicit linear relationship was prescribed via a coupling measure based on cross-correlation from statistics, measuring the degree of the implicit (unknown or unspecified) linear relationship between the two physical properties (as used by *Oldenburg and Li* [1999] for assessing DOI). This measure can be applied whenever only two major units exist in the subsurface: recovered values will lie along a linear continuum, which is inconsistent with a two-unit scenario, but the result is an improvement over uncoupled independent inversions where the recovered values scatter throughout physical property space (on a density-versus-slowness plot for this example). In the work of *Lelièvre et al.* [2012b], constraining the physical properties to display a linear trend lead to significant improvements in the recovery of the Ovoid body, better delineating both the lateral and depth extents.

Next, *Lelièvre et al.* [2012b] applied an explicit linear relationship to couple the models. The parameters of the linear relationship were calculated based on the known physical properties of the two units (i.e., drawing a line through the two points on a density-versus-slowness plot). This led to further improvements in the delineation of the Ovoid body and provided two physical property models with coincident structural features (see Figure 8.4b). This could equally have been a nonlinear functional relationship (as used by *Kamm et al.* [2015]; see Section 8.4.3 and Figure 8.3) but would have introduced additional nonlinearity into the optimization problem, possibly affecting convergence. In contrast to cross-gradient coupling, coupling based on a simple mathematical relationship is by far the most effective and numerically safe coupling method for reducing the non-uniqueness of the joint inverse problem. However, we stress that an empirical mathematical relationship between physical properties is unlikely to exist for a complicated mineral exploration scenario with several rock types (but see *Kamm et al.* [2015]). Even if such a relationship does exist, it might still happen that the physical property data are not of sufficient quality to assign such a relationship.

Finally, *Lelièvre et al.* [2012b] used the fuzzy *c*-means measure (FCM) of *Paasche and Tronicke* [2007] to couple the physical property models in their problem. The cluster centers (in physical property space) were prescribed a priori. In essence, this clustering approach takes the result with explicit linear coupling and asks intermediate values to move towards one point or another on the linear relationship.

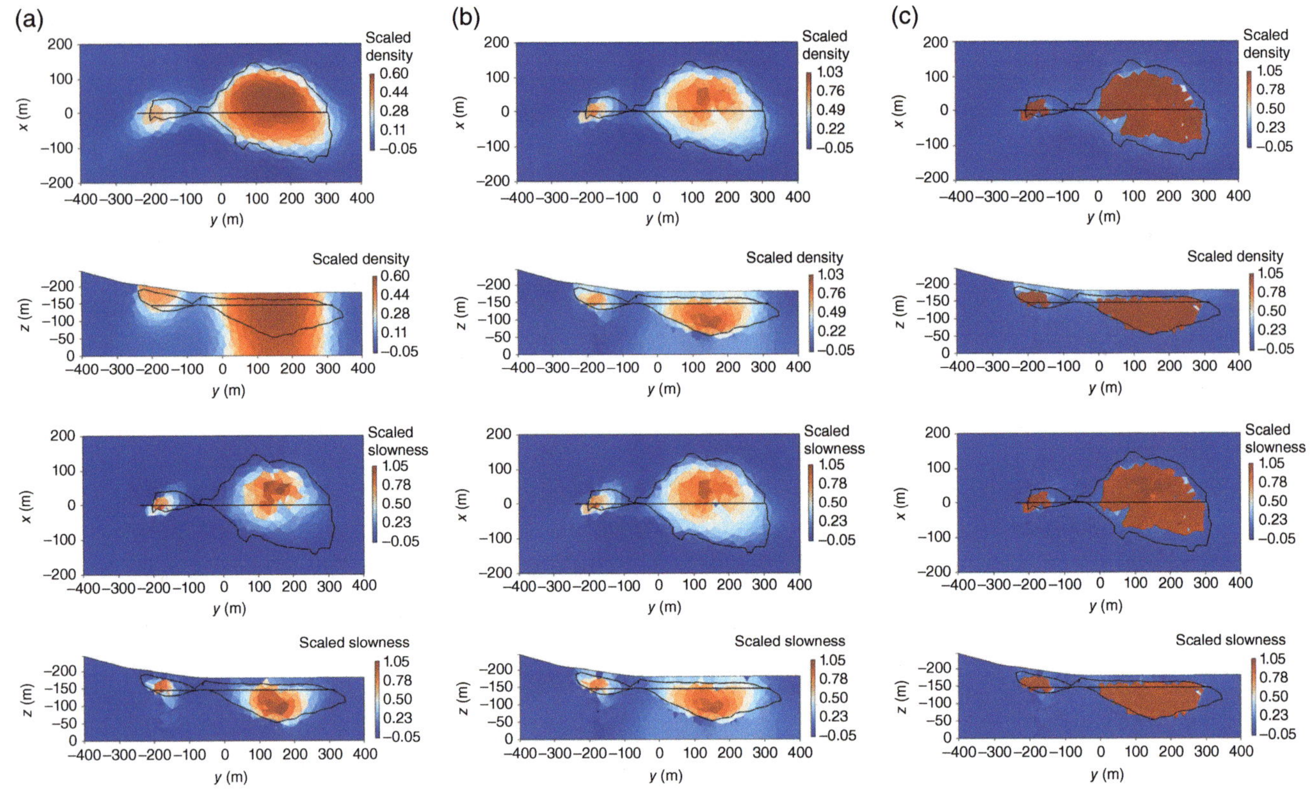

Figure 8.4 Horizontal (z = 45 m) and vertical (x = 0 m) cross sections through the 3D density and slowness models. **(a)** Independent inversion results. **(b)** Joint inversion results with explicit linear relationship. **(c)** Joint inversion results with FCM coupling. Cross-sectional outlines of the true ovoid surface are indicated by black lines. From *Lelièvre et al.* [2012b].

The result was a dramatic improvement on the recovery of the Ovoid body, obtaining consistent models that better represented homogeneous rock units separated by sharp contacts (see Figure 8.4c). However, as for the cross-gradient coupling approach, the FCM approach introduces a numerical concern in the form of multiple minima in the objective function. This means that the inversion results may depend on the initial model used in the iterative descent-based optimization approach.

Carter-McAuslan et al. [2015] followed the methods of *Lelièvre et al.* [2012b] and investigated joint inversion of gravity and cross-well tomography data for a geologically realistic synthetic model based on magmatic massive sulfide deposits. The scenario involved three different rock units that did not present a linear relationship, compared to only two in *Lelièvre et al.* [2012b], making for a more complicated joint inverse problem. They conducted a suite of joint inversion tests and performed some simple zonation and clustering analyses on the results (see Chapter 5 for more sophisticated methods). Their results (some shown in Figure 8.5) clearly demonstrate the potential benefits of joint inversion using FCM coupling. Their work also demonstrates the effects of including inaccurate a priori petrophysical information and they suggest approaches to assess whether such inaccurate information may have been used.

Joint coupling based on FCM was also used by *Sun and Li* [2014] (see also *Sun and Li* [2013]), although they implemented the FCM measure differently than *Lelièvre et al.* [2012b]. *Sun and Li* [2014] also investigated a sulphide deposit exploration scenario but inverted magnetic and IP data. Their scenario included four different geological units and the a priori petrophysical information precluded the use of a simple functional relationship, making theirs an even more difficult problem than those of *Lelièvre et al.* [2012b] and *Carter-McAuslan et al.* [2015]. *Sun and Li* [2014] concluded that only the jointly recovered models could reliably represent the complicated geological structures in their scenario. Figure 8.6 shows some of their results.

The published work discussed above indicates the potential power of property-based coupling approaches to reduce the nonuniqueness of joint inverse problems and provide improved recovered models. There are numerous methods that are suited to a wide range of a priori petrophysical information and physical property relationships. However, obtaining quality physical property information can be challenging and practitioners

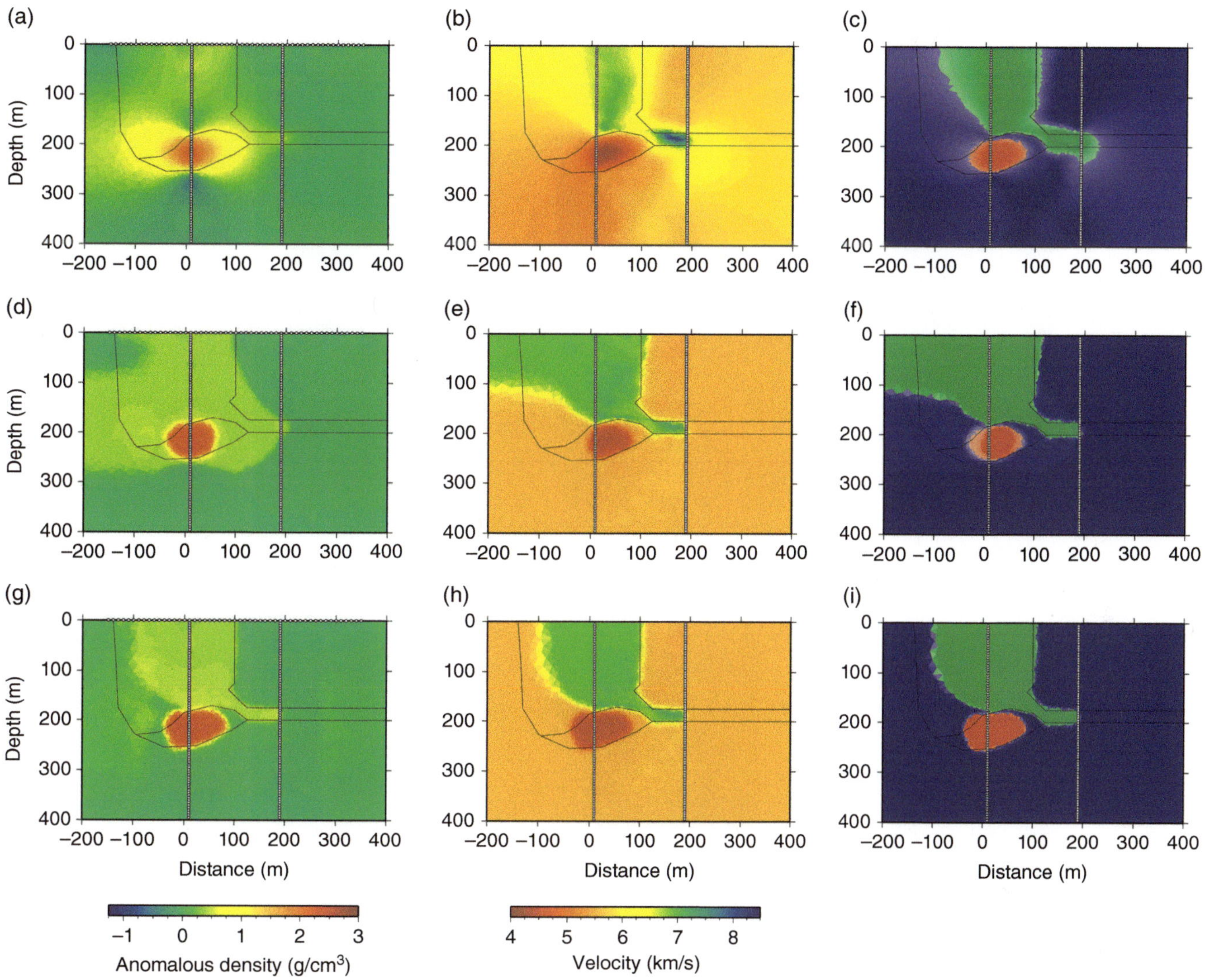

Figure 8.5 Density (*left column*), slowness (*center column*), and zoned (*right column*) models recovered from **(a–c)** independent inversions without FCM clustering, **(d–f)** independent inversions with FCM clustering, and **(g–i)** joint inversion with FCM clustering. The colours in the zoned model are arbitrary but use blue for the background unit, green for the intrusive and red for the lens. In the zoned models, the cells with lower membership values are plotted more lightly (closer to white). White dots indicate the locations of the surface and downhole gravity data or the locations of the downhole transmitters and receivers. The outlines of the rock units in the true model are black. The results in the areas outside of the drillholes at 10 and 190 m, where the seismic tomography data provides no information, should be disregarded. From *Carter-McAuslan et al.* [2015].

should carefully assess the reliability of their a priori information before and after including it in a joint inversion. Also, as for other coupling approaches, there are possible numerical issues and practitioners should understand the potential pitfalls.

8.4.6. Joint Inversion with Common Lithological Units

A different approach to the inverse problem, based on the work of *Bosch* [1999] (see also Chapter 3), deals with lithology as the primary variable, while physical properties are treated as secondary variables linked to lithology through a priori petrophysical information. The approach begins with a discretized lithological model where each cell in the mesh is assigned one of several possible lithologies. The lithological model is altered via stochastic sampling methods while obeying topological rules. The lithologies are linked to physical properties via petrophysical information so joint inversion is vastly simplified, requiring no additional mathematical coupling measure in the objective function. The method is best utilized to investigate the viability of a proposed lithological model

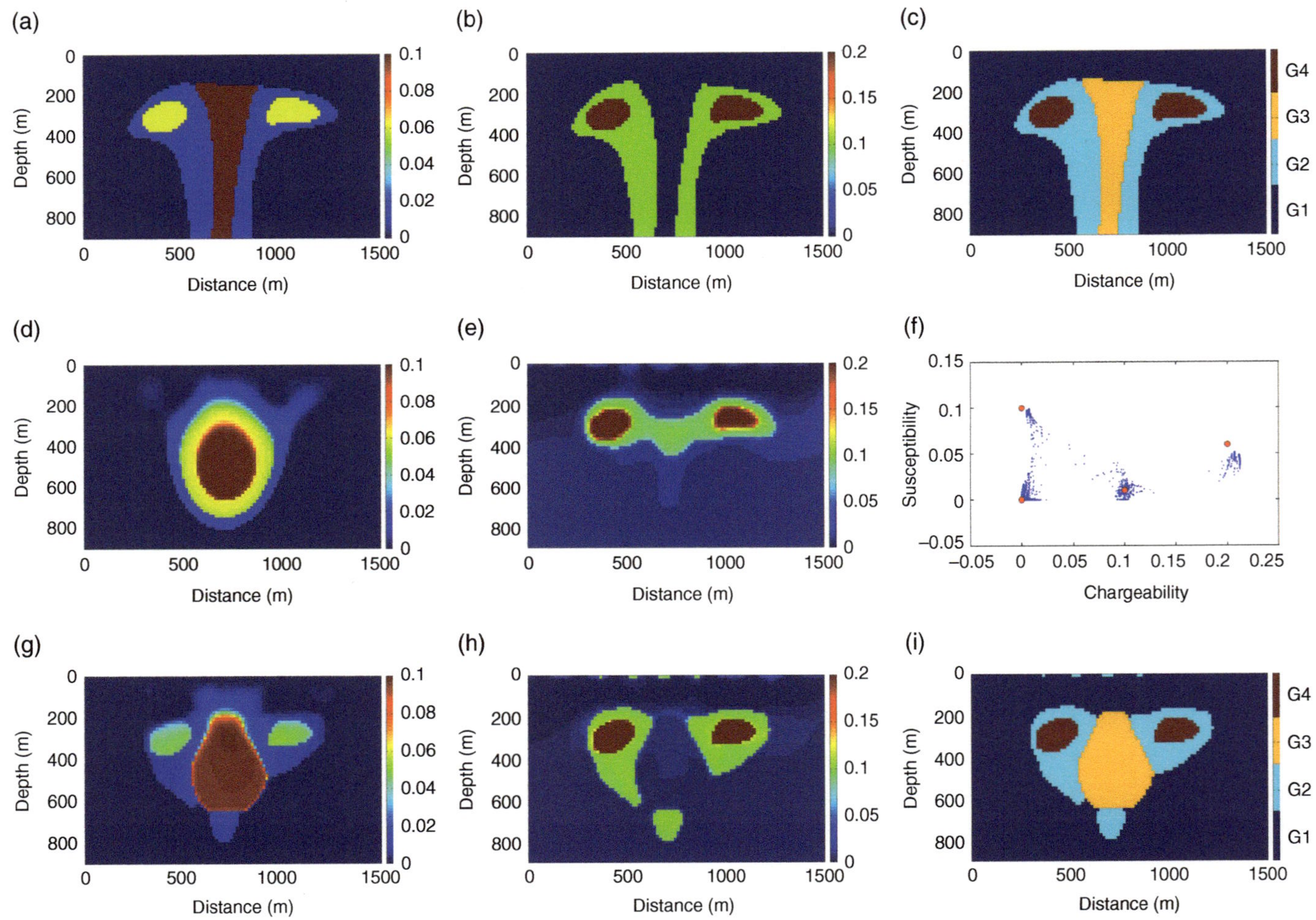

Figure 8.6 **(a)** True susceptibility model. **(b)** True chargeability model. **(c)** True geological model. **(d)** Recovered susceptibility model from clustered independent inversion. **(e)** Recovered chargeability model from clustered independent inversion. **(f)** Susceptibility-versus-chargeability plot for clustered joint inversion results (red dots mark the true physical property values). **(g)** Recovered susceptibility model from clustered joint inversion. **(h)** Recovered chargeability model from clustered joint inversion. **(i)** Geology differentiation from clustered joint inversion results. From *Sun and Li* [2014].

and identify model features that require significant changes to become consistent with the geophysical data. An advantage of the approach is that, because of the stochastic nature of the procedure, probability thresholds can be presented for the location of ore bodies. Such information can be helpful when planning drilling.

Using this approach requires that all lithologies are defined a priori. It must be possible to change the distribution of lithologies in the initial model to that of the true subsurface scenario via the transformations allowed by the inversion algorithm; alternatively, the algorithm should be able to assess when this is not possible. For example, a starting model allowing only two rock units will not be able to recover a true scenario involving three or more units; a starting model with one unit erroneously above another may not be able to recover the true flipped scenario unless the inversion methods are able to flip or swap the units. As such, the starting model should be based on as much a priori information as possible. In later stages of an exploration program, ore deposit models are constructed based on abundant geological data and geophysical inversion results. At such stages, the exploration questions become more specific and focused, and better able to be answered through application of a lithological joint inversion.

The method is computationally intensive and has so far only been applied to inversions with linear forward problems. *Bosch and McGaughey* [2001] applied the method to a joint inversion of gravity and magnetic data over the Kiglapait region in Labrador, Canada: a region of mineral exploration interest in the general region of the Voisey's Bay nickel deposit. Their results (see Figure 8.7) showed significant differences between the initial and recovered models and provide insight into the features of a major intrusive body. *Lane et al.* [2008] applied the method to a joint inversion of gravity and magnetic data over the San Nicolás deposit. Their results were able to refine the source geometry in accordance with the known location of the massive sulfides as determined by drilling.

Another inversion approach that we have yet to discuss is the growing body approach of *Camacho et al.* [2000] and similar work by *Uieda and Barbosa* [2012]. These approaches perturb an initial anomalous body by adding or removing adjacent prismatic mesh elements with prescribed physical properties. At each iteration, only the single adjacent neighboring prism that best reduces the misfit is altered. *Uidea and Barbosa* [2012] inverted airborne gravity gradiometry data flown over the iron ore province of Quadrilátero Ferrífero in southeastern Brazil. Their inversion approach estimated a compact iron ore body in agreement with geologic information and previous interpretations. Their approach has a flavor similar to that of *Bosch and McGaughey* [2001]: The background and several growing bodies can represent different lithological units that are perturbed following strict topological rules. However, the physical properties assigned to these lithologies are single-valued in *Uieda and Barbosa* [2012] whereas *Bosch and McGaughey* [2001] use statistical relationships. Also, the search of the solution space is fundamentally different between the two works: The growing body approach of *Uieda and Barbosa* [2012] does not perform a stochastic search and cannot provide probability information.

8.4.7. Joint Inversion with Common Structural Interfaces

All the inversion approaches discussed above discretize the Earth volume of interest into a mesh of cells. Another approach is to work with models that comprise surfaces representing interfaces (contacts) between rock units. Several such inversion approaches were discussed in Section 8.3.4: Those of *Fullagar et al.* [2000], *Oliveira and Barbosa* [2013], *Lelièvre et al.* [2015], and *Zheglova et al.* [2013] are the most current, advanced, and generally relevant. We are not aware of any joint inversions published using these fairly recent methods. However, where a priori physical property information is able to assign appropriate values for each unit (as in the approach of *Bosch* [1999]), the extension of these methods to joint inversion is relatively simple, requiring no additional mathematical coupling measure in the objective function and only requiring the relative level of fit to each data type to be determined.

Application of such geometry inversion methods requires that the underlying parameterization of the contact surface geometry (e.g., stacked layers) be appropriate based on what is known of the geology. As with the lithological inversion methods discussed above, another requirement is that it be possible to change the geometry of the initial model to that of the true subsurface scenario via the transformations allowed by the inversion algorithm. Again, the starting model should be based on as much a priori information as possible, and geometry joint inversion therefore has a higher potential to be of use when applied in later stages of an exploration program.

8.4.8. Joint Inversion Using Geostatistical Methods

Shamsipour et al. [2012] developed a joint inversion method based on the geostatistical concept of cokriging (see Chapter 6 for more on geostatistical methods). The physical property auto- and cross-covariance are assumed to follow a linear model of coregionalization, the parameters of which are estimated from a v–v plot fitting of experimental covariance. *Shamsipour et al.* [2012] inverted

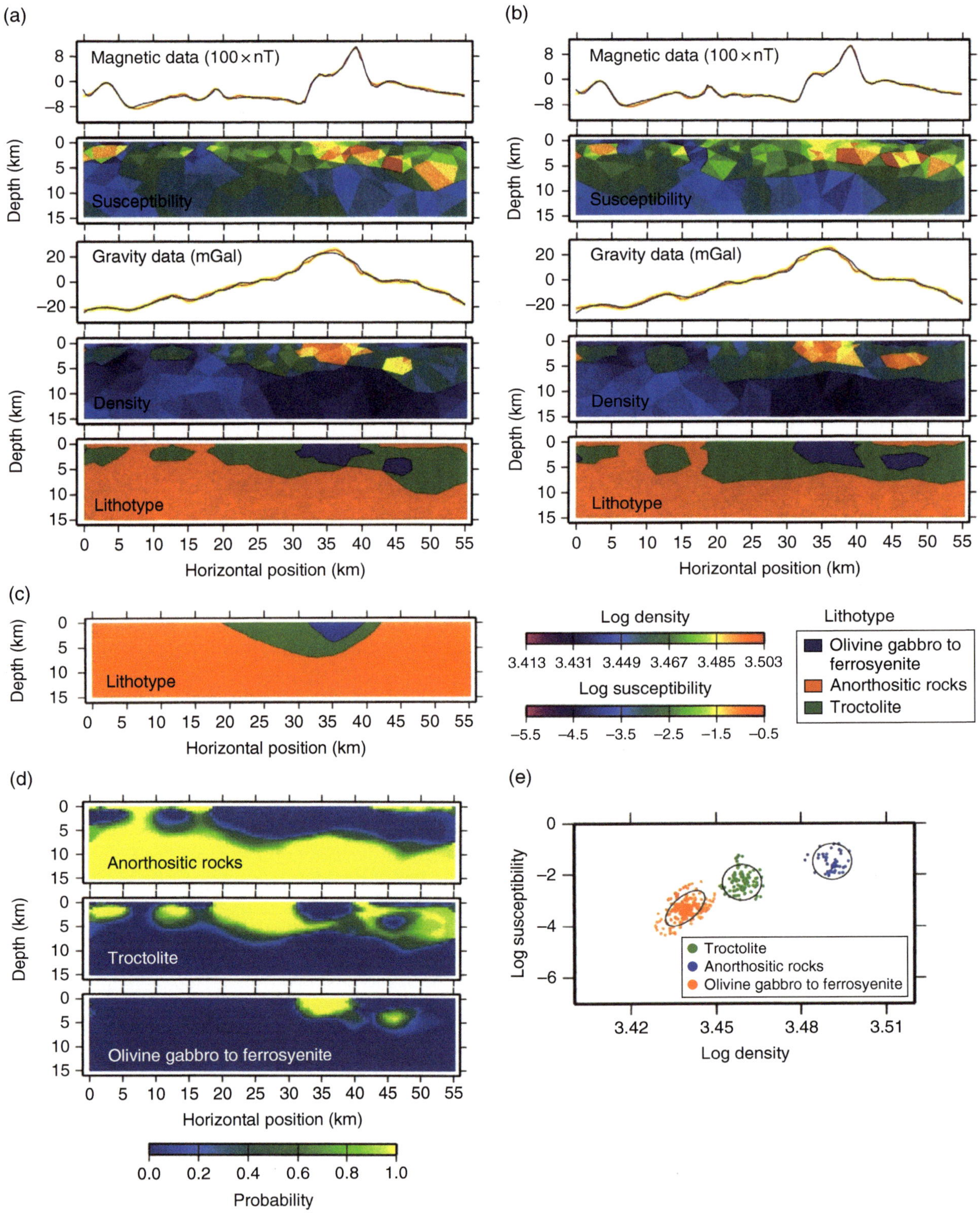

Figure 8.7 **(a,b)** Two stochastic simulations for the real data example of *Bosch and McGaughey* [2001]: In the data profiles, observed data are red, predicted data are blue, and the yellow band indicates one standard deviation of the data uncertainties. **(c)** Initial model. **(d)** Probability maps for each lithotype. **(e)** Petrophysical information used, including two standard deviation ellipsoids. From *Bosch and McGaughey* [2001].

gravity and magnetic data for two synthetic examples and from the Perseverance zinc mine in Quebec, Canada. Their joint inversions better recovered the known models in the synthetic cases and they provided better delineation of the known massive sulphide deposits in the real data example.

Cokriging is a mathematical interpolation and extrapolation tool that uses the spatial correlation between a secondary variable (here, the measured geophysical data) and a primary variable (here, the physical properties in the modeling mesh) to improve estimation of the primary variable at unsampled locations [*Gloaguen et al.*, 2005, 2007]. Therefore, conceptually, the inversion approach of *Shamsipour et al.* [2012] can be compared to data interpolation, but instead of, for example, trying to interpolate between gravity data observations, it is the density in the subsurface that is being interpolated based on how the gravity data are varying. The joint coupling is not rooted in hard geological a priori observations but in a somewhat philosophical argument regarding the spatial positioning of survey data and physical property distributions. One must assume that the spatial statistical properties of the data and model are applicable to the entire problem domain. This approach shows some promise and it can easily be extended to include several data types. However, the validity of the assumptions and estimations made by the method must be assessed for a particular application.

8.5. CONCLUSION

Many joint and cooperative inversion approaches have been applied to mineral exploration problems with varying levels of success but typically performing better than individual independent inversions. The geological scenarios encountered in mineral exploration are varied and complicated. The geophysical data available can also be varied with regard to type, quantity, and quality. So, too, can geological and petrophysical a priori information that might be used to constrain inversions through the design of either regularization or coupling strategies. The exploration questions asked are also varied and change throughout the life of an exploration program or mining operation. This great variety of scenarios makes it impossible to generally declare one joint or cooperative inversion approach superior to another. Each scenario presents its own challenges and questions, so practitioners must take all information into consideration before choosing which integrated imaging strategies might best help meet their exploration goals.

Practitioners should also understand that the joint inversion approaches presented in this chapter suffer from computational and numerical issues. While probabilistic inversion approaches will likely move to the forefront as computing power evolves, they have not yet been able to make significant gains on more traditional deterministic methods. Where deterministic approaches are concerned, most joint inversion algorithms are far more computationally intensive than their single data-type counterparts. Increased nonlinearity from the introduction of a numerically poorly behaved coupling measure can lead to convergence problems. Multiple minima can also be introduced into the joint inverse problem, making it impossible to guarantee that the desired solution will be obtained unless global optimization methods are employed, which generally means an infeasible increase in computational requirements. Numerically problematic coupling measures must be applied within a workflow designed to mitigate such problems. Possible examples include: designing the inversion algorithm to search for particular minima, as done by *Gallardo and Meju* [2004]; careful heating of the joint coupling, as performed by *Lelièvre et al.* [2012b]; and performing several runs with different starting models.

At this point we should remind practitioners that detailed individual dataset inversion studies should be considered essential prerequisites for any joint inversion. This can provide an understanding of the uncertainties in, resolution of, and subsurface features sensed by each dataset. It can also help determine appropriate model coupling, data weighting, and optimization strategies for subsequent joint inversions. Chapter 9 reissues this point and provides some further detail.

The joint coupling approaches applied by *Lelièvre et al.* [2012b], discussed throughout Section 8.4, made increasing assumptions about the physical property relationship. Their results demonstrated that with increased assumptions comes an increased ability to constrain the model space of appropriate solutions and recover improved physical property models. In this sense, the joint coupling should be treated in the same way as regularization measures that specify the desired model features—for example, a smooth model or a model with elongated features in a particular direction. As emphasized by *Lelièvre et al.* [2012b], for any particular mineral exploration scenario, the petrophysical and geological information available should be forefront in the choice of the model regularization and the joint inversion strategies employed—for example, to select a coupling approach. One should apply the methods best able to constrain the inversion, reduce the model space of acceptable solutions, and thereby provide solutions that are most consistent with a priori information, making those solutions more reliable.

The outstanding challenges presented in Chapter 7 are also relevant to mineral exploration applications. We will not duplicate the technical information here, but we will emphasize what we see as the most important practical issue. There is a growing awareness that obtaining a better understanding of physical property information is a critical step for mineral exploration, as is the incorporation of petrophysical and geological information into joint or cooperative geophysical inversions to integrate geological and geophysical data into 3D models. Detailed geological and petrophysical site characterization should be a significant part of any exploration program, with physical property databases developed and expanded throughout.

While academic research software currently exists to solve many of the important joint inverse problems of interest to the mineral exploration community, there has been less development of software packages, commercial or otherwise, available to the industry. This comes from either a lack of effort, time, personnel, or funding and will only be ameliorated by increased cooperation between the parties involved. It is important to note that past developments in geophysical inversion have been enthusiastically adopted in the exploration industry and their use has become widespread [*Paine,* 2007]. As such, we encourage academic, government and industry groups to cooperate in continued research, development, and dissemination of integrated imaging methods for mineral exploration.

While the challenges involved in integrated imaging for mineral exploration should not be ignored or treated lightly, joint and cooperative inversion approaches have proven their worth and are a promising field of research for the future. Joint and cooperative inversion research will offer greater opportunity to integrate different types of data into the interpretation procedure [*Oldenburg and Pratt*, 2007]. The growing integration of geophysical and geological information into common-Earth-type models—for example, as aided by joint and cooperative inversion techniques—has and should continue to change the face of exploration geophysics [*Vallée et al.*, 2011].

ACKNOWLEDGMENTS

We thank two anonymous reviewers for their careful, thorough, positive, and helpful reviews which have improved this chapter. We are also grateful to Jamin Cristall and Nick Williams for their suggestions and for providing their views on the material from an industry perspective.

REFERENCES

Alumbaugh, D. L., and G. A. Newman (2000), Image appraisal for 2-D and 3-D electromagnetic inversion, *Geophysics*, *65*(5), 1455–1467, doi:10.1190/1.1444,834.

Ash, M. R., M. Wheeler, H. Miller, C. G. Farquharson, and A. V. Dyck (2006), Constrained three-dimensional inversion of potential field data from the Voisey's Bay Ni–Cu–Co deposit, Labrador, Canada, in *SEG Technical Program Expanded Abstracts 2006*, pp. 1333–1337, doi:10.1190/1.2369,766.

Austin, J. R., and C. A. Foss (2014), The paradox of scale: Reconciling magnetic anomalies with rock magnetic properties for cost-effective mineral exploration, *J. Appl. Geophys.*, *104*, 121–133, doi:10.1016/j.jappgeo.2014.02.018.

Barbosa, V. C. F., and J. B. C. Silva (1994), Generalized compact gravity inversion, *Geophysics*, *59*(1), 57–68, doi:10.1190/1.1443,534.

Blakely, R. J. (1996), *Potential Theory in Gravity and Magnetic Applications*, Cambridge University Press, New York.

Bosch, M. (1999), Lithologic tomography: from plural geophysical data to lithology estimation, *J. Geophys. Res.: Solid Earth*, *104*(B1), 749–766, doi:10.1029/1998JB900,014.

Bosch, M., and J. McGaughey (2001), Joint inversion of gravity and magnetic data under lithologic constraints, *The Leading Edge*, *20*(8), 877–881, doi:10.1190/1.1487,299.

Boulanger, O., and M. Chouteau (2001), Constraints in 3D gravity inversion, *Geophys. Prospect.*, *49*(2), 265–280, doi:10.1046/j.1365–2478.2001.00,254.x.

Butler, R. F. (1992), *Paleomagnetism: Magnetic Domains to Geologic Terranes*, Blackwell Scientific Publications, Blackwell Scientific Publications, Boston.

Camacho, A. G., F. G. Montesinos, and R. Vieira (2000), Gravity inversion by means of growing bodies, *Geophysics*, *65*(1), 95–101, doi:10.1190/1.1444,729.

Carter-McAuslan, A. (2014), Joint inversion of geologically realistic synthetic Earth models, Master's thesis, Department of Earth Sciences, Memorial University of Newfoundland, St. John's, Newfoundland, Canada.

Carter-McAuslan, A., P. G. Lelièvre, and C. G. Farquharson (2015), A study of fuzzy *c*-means coupling for joint inversion, using seismic tomography and gravity data test scenarios, *Geophysics*, *80*(1), W1–W15, doi:10.1190/geo2014–0056.1.

Caterina, D., J. Beaujean, T. Robert, and F. Nguyen (2013), A comparison study of different image appraisal tools for electrical resistivity tomography, *Near Surface Geophys.*, *11*(6), 639–657, doi:10.3997/1873–0604.2013,022.

Christiansen, A. V., and E. Auken (2012), A global measure for depth of investigation, *Geophysics*, *77*(4), WB171–WB177, doi:10.1190/geo2011–0393.1.

Clark, D. A., and D. W. Emerson (1999), Self-demagnetization, *Preview*, *79*, 22–25.

Clark, D. A., D. H. French, M. A. Lackie, and P. W. Schmidt (1992), Magnetic petrology: Application of integrated rock magnetic and petrological techniques to geological interpretation of magnetic surveys, *Explor. Geophys.*, *23*(2), 65–68, doi:10.1071/EG992,065.

Commer, M., and G. A. Newman (2009), Three-dimensional controlled-source electromagnetic and magnetotelluric joint inversion, *Geophys. J. Int.*, *178*(3), 1305–1316, doi:10.1111/j.1365–246X.2009.04,216.x.

Constable, S. C., R. L. Parker, and C. G. Constable (1987), Occam's inversion: A practical algorithm for generating smooth models from electromagnetic sounding data, *Geophysics*, *52*(3), 289–300, doi:10.1190/1.1442,303.

Čuma, M., and M. S. Zhdanov (2014), Massively parallel regularized 3D inversion of potential fields on CPUs and GPUs, *Comput. Geosci.*, *62*, 80–87, doi:10.1016/j.cageo.2013.10.004.

Čuma, M., G. A. Wilson, and M. S. Zhdanov (2012), Large-scale 3D inversion of potential field data, *Geophys. Prospect.*, *60*(6), 1186–1199, doi:10.1111/j.1365-2478.2011.01,052.x.

Davis, K., and Y. Li (2011), Joint processing of total-field and gradient magnetic data, *Explor. Geophys.*, *42*(3), 199–206, doi:10.1071/EG10,012.

Debicki, E. (1996), MITEC's Exploration Technology Division: helping reverse the trend of declining mineral reserves in Canada, *Can. Inst. Min. Metal. Bull.*, *89*(997), 53–59.

Deceuster, J., A. Etienne, T. Robert, F. Nguyen, and O. Kaufmann (2014), A modified DOI-based method to statistically estimate the depth of investigation of DC resistivity surveys, *J. Appl. Geophys.*, *103*, 172–185, doi:10.1016/j.jappgeo.2014.01.018.

deGroot-Hedlin, C., and S. Constable (1990), Occam's inversion to generate smooth, two-dimensional models from magnetotelluric data, *Geophysics*, *55*(12), 1613–1624, doi:10.1190/1.1442,813.

Dentith, M., and S. T. Mudge (2014), *Geophysics for the Mineral Exploration Geoscientist*, Cambridge University Press, New York.

Djikpesse, H. A., M. R. Khodja, M. D. Prange, S. Duchenne, and H. Menkiti (2012), Bayesian survey design to optimize resolution in waveform inversion, *Geophysics*, *77*(2), R81–R93, doi:10.1190/geo2011-0143.1.

Eaton, D. W., B. Milkereit, and M. H. Salisbury, eds. (2003), *Hardrock Seismic Exploration*, Society of Exploration Geophysicists, Tulsa, OK.

Evans, A. M. (1993), *Ore Geology and Industrial Minerals: An Introduction*, 3rd ed., Wiley-Blackwell, New York.

Farquharson, C. G. (2008), Constructing piecewise-constant models in multidimensional minimum-structure inversions, *Geophysics*, *73*(1), K1–K9, doi:10.1190/1.2816,650.

Farquharson, C. G., and D. W. Oldenburg (1998), Non-linear inversions using general measures of data misfit and model structure, *Geophys. J. Int.*, *134*(1), 213–227, doi:10.1046/j.1365-246x.1998.00,555.x.

Fregoso, E., and L. A. Gallardo (2009), Cross-gradients joint 3D inversion with applications to gravity and magnetic data, *Geophysics*, *74*(4), L31–L42, doi:10.1190/1.3119,263.

Friedel, S. (2003), Resolution, stability and efficiency of resistivity tomography estimated from a generalized inverse approach, *Geophys. J. Int.*, *153*(2), 305–316, doi:10.1046/j.1365-246X.2003.01,890.x.

Fullagar, P. K., N. A. Hughes, and J. Paine (2000), Drilling-constrained 3D gravity inversion, *Explor. Geophys.*, *31*(2), 17–23, doi:10.1071/EG00,017.

Fullagar, P. K., G. Pears, D. Hutton, and A. Thompson (2004), 3D gravity and aeromagnetic inversion for MVT lead-zinc exploration at Pillara, Western Australia, *Explo. Geophys.*, *35*(2), 142–146, doi:10.1071/EG04, 42.

Fullagar, P. K., G. A. Pears, and B. McMonnies (2008), Constrained inversion of geologic surfaces—pushing the boundaries, *The Leading Edge*, *27*(1), 98–105, doi:10.1190/1.2831,686.

Gallardo, L. A., and M. A. Meju (2004), Joint two-dimensional DC resistivity and seismic travel time inversion with cross-gradients constraints, *J. Geophys. Res.: Space Phys.*, *109*(B3), B03,311, doi:10.1029/2003JB002,716.

García Juanatey, M. (2012), Seismics, 2D and 3D inversion of magnetotellurics: Jigsaw pieces in understanding the skellefte ore district, Ph.D. thesis, Department of Earth Sciences, Uppsala University, Uppsala, Sweden.

García Juanatey, M., A. Tryggvason, C. Juhlin, U. Bergström, J. Hübert, and L. B. Pedersen (2013), MT and reflection seismics in northwestern skellefte ore district, sweden, *Geophysics*, *78*(2), B65–B76, doi:10.1190/geo2012-0169.1.

Gerrie, V., C. Drielsma, P. Hooker, R. Leblanc, and P. Patraskovic (2014), Physical rock properties—the quantitative link between geophysics and geology, in *KEGS Geophysical Symposium*, Toronto, Canada.

Gloaguen, E., D. Marcotte, M. Chouteau, and H. Perroud (2005), Borehole radar velocity inversion using cokriging and cosimulation, *J. Appl. Geophys.*, *57*(4), 242–259, doi:10.1016/j.jappgeo.2005.01.001.

Gloaguen, E., D. Marcotte, B. Giroux, C. Dubreuil-Boisclair, M. Chouteau, and M. Aubertin (2007), Stochastic borehole radar velocity and attenuation tomographies using cokriging and cosimulation, *J. Appl. Geophys.*, *62*(2), 141–157, doi:10.1016/j.jappgeo.2006.10.001.

Goodway, B. (2012), Introduction to this special section: Mining geophysics, *The Leading Edge*, *31*(3), 288–290, doi:10.1190/1.3694,894.

Guillen, A., and V. Menichetti (1984), Gravity and magnetic inversion with minimization of a specific functional, *Geophysics*, *49*(8), 1354–1360, doi:10.1190/1.1441,761.

Günther, T., and C. Rücker (2006), A new joint inversion approach applied to the combined tomography of DC resistivity and seismic refraction data, in *Proceedings of SAGEEP*, Seattle, WA.

Haber, E., and D. Oldenburg (1997), Joint inversion: A structural approach, *Inverse Problems*, *13*, 63–77, doi:10.1088/0266-5611/13/1/006.

Hattula, A., and T. Rekola (2000), Exploration geophysics at the Pyhäsalmi mine and grade control work of the Outokumpu Group, *Geophysics*, *65*(6), 1961–1969, doi:10.1190/1.1444,879.

Hobbs, R. W. (2003), 3D Modeling of Seismic-Wave Propagation Using Complex Elastic Screens, with Application to Mineral Exploration, in *Hardrock Seismic Exploration*, Society of Exploration Geophysicists, Tulsa, OK.

Hunt, C. P., B. M. Moskowitz, and S. K. Banerjee (1995), Magnetic properties of rocks and minerals, in *Rock Physics and Phase Relations: A Handbook of Physical Constants*, American Geophysical Union, Washington, DC, pp. 189–204.

Jahandari, H., and C. Farquharson (2014), A finite-volume solution to the geophysical electromagnetic forward problem using unstructured grids, *Geophysics*, *79*(6), E287–E302, doi:10.1190/geo2013-0312.1.

Kamm, J. (2014), Inversion and joint inversion of electromagnetic and potential field data, Ph.D. thesis, Department of Earth Sciences, Uppsala University, Uppsala, Sweden.

Kamm, J., I. Antal Lundin, M. Bastani, M. Sadeghi, and L. B. Pedersen (2015), Joint inversion of gravity, magnetic and petrophysical data—a case study from a gabbro intrusion in boden, sweden, *Geophysics*, *80*(5), B131–B152, doi:10.1190/geo2014-0122.1.

Kearey, P., M. Brooks, and I. Hill (2002), *An Introduction to Geophysical Exploration*, 3rd ed., Wiley-Blackwell, Hoboken, NJ.

Keller, G. V. (1988), Rock and mineral properties, in *Electromagnetic Methods in Applied Geophysics*, Society of Exploration Geophysicists, Tulsa, OK, pp. 52–129.

Lane, R., P. McInerney, R. Seikel, and A. Guillen (2008), Using potential field data and stochastic optimization to refine 3D geological models, in *Prospecters and Developers Association of Canada (PDAC) Geophysics Session*, Toronto, Canada.

Last, B. J., and K. Kubik (1983), Compact gravity inversion, *Geophysics*, *48*(6), 713–721, doi:10.1190/1.1441,501.

Lelièvre, P. G. (2009), Integrating geologic and geophysical data through advanced constrained inversions, Ph.D. thesis, Department of Earth and Ocean Sciences, The University of British Columbia, Vancouver, British Columbia, Canada.

Lelièvre, P. G., and C. G. Farquharson (2013), Gradient and smoothness regularization operators for geophysical inversion on unstructured meshes, *Geophys. J. Int.*, *195*(1), 330–341, doi:10.1093/gji/ggt255.

Lelièvre, P. G., and D. W. Oldenburg (2009), A comprehensive study of including structural orientation information in geophysical inversions, *Geophys. J. Int.*, *178*(2), 623–637, doi:10.1111/j.1365–246X.2009.04,188.x.

Lelièvre, P. G., D. W. Oldenburg, and N. C. Williams (2009), Integrating geological and geophysical data through advanced constrained inversions, *Explor. Geophys.*, *40*, 334–341, doi:10.1071/EG09,012.

Lelièvre, P. G., P. Zheglova, T. Danek, and C. G. Farquharson (2012a), Geophysical inversion for contact surfaces, in *SEG Technical Program Expanded Abstracts 2012*, pp. 1–5, doi:10.1190/segam2012–0716.1.

Lelièvre, P. G., C. G. Farquharson, and C. A. Hurich (2012b), Joint inversion of seismic traveltimes and gravity data on unstructured grids with application to mineral exploration, *Geophysics*, *77*(1), K1–K15, doi:10.1190/geo2011–0154.1.

Lelièvre, P. G., C. G. Farquharson, and R. Bijani (2015), 3D stochastic geophysical inversion for contact surface geometry, in *European Geosciences Union General Assembly*, Vienna, Austria.

L'Heureux, E., B. Milkereit, and E. Adam (2005), 3D seismic exploration for mineral deposits in hardrock environments, *CSEG Recorder*, *30*(9), 36–39.

Li, Y., and D. W. Oldenburg (1996), 3-D inversion of magnetic data, *Geophysics*, *61*(2), 394–408, doi:10.1190/1.1443,968.

Li, Y., and D. W. Oldenburg (2000a), Joint inversion of surface and three-component borehole magnetic data, *Geophysics*, *65*(2), 540–552, doi:10.1190/1.1444,749.

Li, Y., and D. W. Oldenburg (2000b), Incorporating geological dip information into geophysical inversions, *Geophysics*, *65*(1), 148–157, doi:10.1190/1.1444,705.

Lines, L. R., A. K. Schultz, and S. Treitel (1988), Cooperative inversion of geophysical data, *Geophysics*, *53*(1), 8–20, doi:10.1190/1.1442,403.

Mahmoodi, O., and R. Smith (2014), Clustering of down-hole physical properties measurement to characterize rock units at the Victoria Cu-Ni property, in *SEG Technical Program Expanded Abstracts 2014*, pp. 1742–1747, doi:10.1190/segam2014–0190.1.

McMillan, M. S., and D. W. Oldenburg (2014), Cooperative constrained inversion of multiple electromagnetic data sets, *Geophysics*, *79*(4), B173–B185, doi:10.1190/geo2014–0029.1.

Oldenburg, D. W., and Y. Li (1999), Estimating depth of investigation in DC resistivity and IP surveys, *Geophysics*, *64*(2), 403–416, doi:10.1190/1.1444,545.

Oldenburg, D. W., and D. A. Pratt (2007), Geophysical inversion for mineral exploration: A decade of progress in theory and practice, in *Proceedings of Exploration 07: Fifth Decennial International Conference on Mineral Exploration*, B. Milkereit, ed., pp. 61–95.

Oldenburg, D. W., Y. Li, and R. G. Ellis (1997), Inversion of geophysical data over a copper gold porphyry deposit: a case history for Mt. *Milligan, Geophysics*, *62*(5), 1419–1431, doi:10.1190/1.1444,246.

Oliveira, V. C., Jr., and V. C. F. Barbosa (2013), 3-D radial gravity gradient inversion, *Geophys. J. Int.*, *195*(2), 883–902, doi:10.1093/gji/ggt307.

Paasche, H., and J. Tronicke (2007), Cooperative inversion of 2D geophysical data sets: A zonal approach based on fuzzy c-means cluster analysis, *Geophysics*, *72*(3), A35–A39, doi:10.1190/1.2670,341.

Paine, J. (2007), Developments in geophysical inversion in the last decade, in *Proceedings of Exploration 07: Fifth Decennial International Conference on Mineral Exploration*, B. Milkereit, ed., pp. 485–488.

Palacky, G. J. (1988), Resistivity characteristics of geologic targets, in *Electromagnetic Methods in Applied Geophysics*, Society of Exploration Geophysicists, Tulsa, OK, pp. 52–129.

Persson, L., I. A. Lundin, L. B. Pedersen, and D. Claeson (2011), Combined magnetic, electromagnetic and resistivity study over a highly conductive formation in orrivaara, northern sweden, *Geophys. Prospect.*, *59*(6), 1155–1163, doi:10.1111/j.1365–2478.2011.00,998.x.

Phillips, N. D. (2001), Geophysical inversion in an integrated exploration program: examples from the San Nicolas deposit, Master's thesis, Department of Earth and Ocean Sciences, The University of British Columbia, Vancouver, British Columbia, Canada.

Phillips, N., D. Oldenburg, J. Chen, Y. Li, and P. Routh (2001), Cost effectiveness of geophysical inversions in mineral exploration: applications at San Nicolas, *The Leading Edge*, *20*(12), 1351–1360, doi:10.1190/1.1487,264.

Pilkington, M. (2012), Analysis of gravity gradiometer inverse problems using optimal design measures, *Geophysics*, *77*(2), G25–G31, doi:10.1190/geo2011–0317.1.

Pinto, M. J., and M. McWilliams (1990), Drilling-induced isothermal remanent magnetization, *Geophysics*, *55*(1), 111–115, doi:10.1190/1.1442,765.

Portniaguine, O., and M. S. Zhdanov (1999), Focusing geophysical inversion images, *Geophysics*, *64*(3), 874–887, doi:10.1190/1.1444,596.

Pretorius, C. C., M. R. Muller, M. Larroque, and C. Wilkins (2003), A review of 16 years of hardrock seismics on the Kaapvaal Craton, in *Hardrock Seismic Exploration*, Society of Exploration Geophysicists, Tulsa, OK.

Raiche, A. P., D. L. B. Jupp, H. Rutter, and K. Vozoff (1985), The joint use of coincident loop transient electromagnetic and schlumberger sounding to resolve layered structures, *Geophysics*, *50*(10), 1618–1627, doi:10.1190/1.1441,851.

Rawlinson, Z. J., J. Townend, R. Arnold, and S. Bannister (2012), Derivation and implementation of a nonlinear experimental design criterion and its application to seismic network expansion at Kawerau geothermal field, New Zealand, *Geophys. J. Int.*, *191*(2), 686–694, doi:10.1111/j.1365-246X.2012.05,646.x.

Richardson, R. M., and S. C. MacInnes (1989), The inversion of gravity data into three-dimensional polyhedral models, *J. Geophys. Res.: Solid Earth*, *94*(B6), 7555–7562, doi:10.1029/JB094iB06p07,555.

Rim, H., and Y. Li (2012), Single-hole imaging using borehole gravity gradiometry, *Geophysics*, *77*(5), G67–G76, doi:10.1190/geo2012-0003.1.

Roy, B., and R. M. Clowes (2000), Seismic and potential-field imaging of the Guichon Creek batholith, British Columbia, Canada, to delineate structures hosting porphyry copper deposits, *Geophysics*, *65*(5), 1418–1434, doi:10.1190/1.1444,831.

Rücker, C., T. Günther, and K. Spitzer (2006), Three-dimensional modelling and inversion of DC resistivity data incorporating topography—I. Modelling, *Geophys. J. Int.*, *166*(2), 495–505, doi:10.1111/j.1365-246X.2006.03,010.x.

Salisbury, M., and D. Snyder (2007), *Mineral Deposits of Canada: A Synthesis of Major Deposit-Types, District Metallogeny, the Evolution of Geological Provinces, and Exploration Methods*, vol. Special Publication No. 5, Application of Seismic Methods to Mineral Exploration, Geological Association of Canada, Mineral Deposits Division, pp. 971–982.

Salisbury, M. H., B. Milkereit, and W. Bleeker (1996), Seismic imaging of massive sulfide deposits: Part 1. rock properties, *Economic Geol.*, *91*(5), 821–828, doi:10.2113/gsecongeo.91.5.821.

Salisbury, M. H., B. Milkereit, G. Ascough, R. Adair, L. Matthews, D. R. Schmitt, J. Mwenifumbo, D. W. Eaton, and J. Wu (2000), Physical properties and seismic imaging of massive sulfides, *Geophysics*, *65*(6), 1882–1889, doi:10.1190/1.1444,872.

Salisbury, M. H., C. W. Harvey, and L. Matthews (2003), The acoustic properties of ores and host rocks in hardrock terranes, in *Hardrock Seismic Exploration*, Society of Exploration Geophysicists, Tulsa, OK.

Sasaki, Y., M. Yi, and J. Choi (2014), 2D and 3D separate and joint inversion of airborne ZTEM and ground AMT data: Synthetic model studies, *J. Appl. Geophys.*, *104*, 149–155, doi:10.1016/j.jappgeo.2014.02.017.

Schmitt, D. R., C. J. Mwenifumbo, K. A. Pflug, and I. L. Meglis (2003), Geophysical logging for elastic properties in hard rock: A tutorial, in *Hardrock Seismic Exploration*, Society of Exploration Geophysicists, Tulsa, OK.

Shamsipour, P., D. Marcotte, and M. Chouteau (2012), 3D stochastic joint inversion of gravity and magnetic data, *J. Appl. Geophys.*, *79*, 27–37, doi:10.1016/j.jappgeo.2014.02.017.

Silva, J. B. C., W. E. Medeiros, and V. C. F. Barbosa (2001), Potential-field inversion: Choosing the appropriate technique to solve a geologic problem, *Geophysics*, *66*(2), 511–520, doi:10.1190/1.1444,941.

Smith, R., M. Shore, and D. Rainsford (2012), How to make better use of physical properties in mineral exploration: The exploration site measurement, *The Leading Edge*, *31*(3), 330–337, doi:10.1190/1.3694,901.

Stummer, P., H. Maurer, and A. G. Green (2004), Experimental design: Electrical resistivity data sets that provide optimum subsurface information, *Geophysics*, *69*(1), 120–139, doi:10.1190/1.1649,381.

Sun, J., and Y. Li (2013), A general framework for joint inversion with petrophysical information as constraints, in *SEG Technical Program Expanded Abstracts 2013*, pp. 3093–3097, doi:10.1190/segam2013-1185.1.

Sun, J., and Y. Li (2014), Exploration of a sulfide deposit using joint inversion of magnetic and induced polarization data, in *SEG Technical Program Expanded Abstracts 2014*, pp. 1780–1784, doi:10.1190/segam2014-1511.1.

Takougang, E. M. T., B. Harris, A. Kepic, and C. V. A. Le (2015), Cooperative joint inversion of 3D seismic and magnetotelluric data: With application in a mineral province, *Geophysics*, *80*(4), R175–R187, doi:10.1190/GEO2014-0252.1.

Tanner, J. G. (1967), An automated method of gravity interpretation, *Geophys. J. R. Astron. Soc.*, *13*(1–3), 339–347, doi:10.1111/j.1365-246X.1967.tb02,164.x.

Tompkins, M. J., J. L. Fernández Martínez, D. L. Alumbaugh, and T. Mukerji (2011), Scalable uncertainty estimation for nonlinear inverse problems using parameter reduction, constraint mapping, and geometric sampling: marine controlled-source electromagnetic examples, *Geophysics*, *76*(4), F263–F281, doi:10.1190/1.3581,355.

Tompkins, M. J., J. L. Fernández Martínez, and Z. Fernández Muñiz (2013), Comparison of sparse-grid geometric and random sampling methods in nonlinear inverse solution uncertainty estimation, *Geophys. Prospect.*, *61*(1), 28–41, doi:10.1111/j.1365-2478.2012.01,057.x.

Uieda, L., and V. C. F. Barbosa (2012), Robust 3D gravity gradient inversion by planting anomalous densities, *Geophysics*, *77*(4), G55–G66, doi:10.1190/geo2011-0388.1.

Vallée, M. A., R. S. Smith, and P. Keating (2011), Metalliferous mining geophysics—State of the art after a decade in the new millennium, *Geophysics*, *76*(4), W31–W50, doi:10.1190/1.3587,224.

Visser, S. J., and J. J. Lajoie (2012), The discover of the Maria deposit, Mexico, *The Leading Edge*, *31*(3), 296–301, doi:10.1190/1.3694,896.

Wilkinson, P. B., M. H. Loke, P. I. Meldrum, J. E. Chambers, O. Kuras, D. A. Gunn, and R. D. Ogilvy (2012), Practical aspects of applied optimized survey design for electrical resistivity tomography, *Geophys. J. Int.*, *189*(1), 428–440, doi:10.1111/j.1365-246X.2012.05,372.x.

Williams, N. C. (2008), Geologically-constrained UBC-GIF gravity and magnetic inversions with examples from the Agnew–Wiluna greenstone belt, Western Australia, Ph.D. thesis, Department of Earth and Ocean Sciences, The University of British Columbia, Vancouver, British Columbia, Canada.

Wilson, G. A., L. H. Cox, M. Čuma, and M. S. Zhdanov (2012), Inverting airborne geophysical data for mega-cell and giga-cell 3D Earth models, *The Leading Edge*, *31*(3), 316–321, doi:10.1190/1.3694,899.

Witherly, K. (2012), The evolution of minerals exploration over 60 years and the imperative to explore undercover, *The Leading Edge*, *31*(3), 292–295, doi:10.1190/1.3694,895.

Wohlenberg, J. (1982), Density, in *Physical Properties of rocks, Landolt-Börnstein: Numerical Data and Functional Relationships in Science and Technology, Group V: Geophysics and Space Research*, Vol. *1a*, Springer, Berlin, pp. 66–183.

Yu, C., Z. Fu, H. Zhang, H. M. Tai, and X. Zhu (2013), Transient process and optimal design of receiver coil for small-loop transient electromagnetics, *Geophys. Prospect.*, *62*(2), 377–384, doi:10.1111/1365-2478.12,093.

Zheglova, P., C. G. Farquharson, and C. A. Hurich (2013), 2-D reconstruction of boundaries with level set inversion of traveltimes, *Geophys. J. Int.*, *192*(2), 688–698, doi:10.1093/gji/ggs035.

9

Joint Inversion in Hydrocarbon Exploration

Max Moorkamp,[1] Björn Heincke,[2,3] Marion Jegen,[2] Richard W. Hobbs,[4] and Alan W. Roberts[4,5]

ABSTRACT

In this chapter we review and investigate the particular challenges for joint inversion in the context of hydrocarbon exploration. Subsurface imaging for oil and gas exploration is traditionally dominated by seismic reflection as it offers a resolution unparalleled by other methods in layered sedimentary environments, the classic areas for hydrocarbon reservoirs. As exploration is moving into more complex environments—for example, sub-basalt and sub-salt sediments—seismic methods reach their limits and alternative approaches are needed. This has driven the industry to investigate joint inversion and joint interpretation approaches.

In the first part of this chapter we review the current literature and introduce interesting approaches that have been developed in the context of hydrocarbon exploration and could potentially be used in other application areas as well. Commercial providers of joint inversion in particular try to maximise the level of detail in the joint inversion models by combining full-waveform inversion with controlled source electromagnetics in order to meet the expectations of exploration managers used to modern three-dimensional reflection surveys. Joint inversion methods incorporating petrophysical models of the reservoir can directly yield quantities of interest such as porosity and permeability, but have to be carefully tuned to the area under investigation. Probabilistic inversion approaches that permit detailed quantitative specification of prior assumptions and yield uncertainty information on the inversion results are gaining popularity in hydrocarbon exploration even though they are still at a largely conceptual stage. Emulators provide a way to approximate lengthy forward calculations for such statistical inversion approaches and have mathematically well defined properties. They can therefore open the door for full stochastic inversions with high computing requirements that are currently still out of reach.

In the second part we present two detailed joint inversion case studies, one for sub-salt imaging and one for sub-basalt imaging. We investigate the different coupling methods that are required for successful joint inversion and the impact on different parameters on the final results. Although some of the specifics might be related to our particular implementations, they hopefully provide practitioners in any application area with useful recipes and experiences.

9.1. INTRODUCTION

Exploration for hydrocarbons requires accurate imaging of the subsurface at all stages from large-scale reconnaissance to detailed estimation of reservoir properties for well locationing and finally production monitoring. For this reason the hydrocarbon industry has been a strong driver for developing new Earth imaging

[1]*Department of Geology, University of Leicester, Leicester, United Kingdom*
[2]*Geomar, Helmholtz Centre for Ocean Sciences, Kiel, Germany*
[3]*Present address: Geological Survey of Denmark and Greenland, Copenhagen, Denmark*
[4]*Department of Earth Sciences, Durham University, Science Labs, Durham, United Kingdom*
[5]*Present address: Geospatial Research Limited, Durham, United Kingdom*

Integrated Imaging of the Earth: Theory and Applications, Geophysical Monograph 218, First Edition.
Edited by Max Moorkamp, Peter G. Lelièvre, Niklas Linde, and Amir Khan.

approaches. Due to the potentially high economic benefits associated with new and improved techniques, this leading role is not always reflected in the amount of publications. In fact, a substantial amount of the work described below has only been published as extended abstracts and often with limited information on the methods, field area, or both.

Although electromagnetic and potential field data are nowadays often considered in the exploration workflow (e.g., see *Strack* [2014] and *Hackney et al.* [2015]), to date the most important method for hydrocarbon exploration remains seismic reflection imaging (e.g., see *Yilmaz* [2001]). For common geological scenarios (e.g., layered sedimentary sequences), migrated seismic reflection data provide information on geological structure at a resolution unmatched by other geophysical methods because impedance contrasts often coincide with variations in the lithology (e.g., see *Latimer et al.* [2000]). Therefore it is relatively straightforward to pass the geophysical results on to geologists who interpret them in terms of basin stratigraphy and other factors relevant for resource estimation and well location (e.g., see *Singh* [2014]). In contrast, other geophysical methods and most of the joint inversion approaches developed to date provide smooth models that do not always resemble typical geological structures such as faults or sedimentary layering. Thus these models appear to be of limited use to an interpreter not intricately familiar with the geophysical method. For a long time, this has been a considerable barrier to making non-seismic and joint inversion approaches an integral part of the exploration workflow. Interestingly, in crystalline environments typically encountered in mineral exploration (see Chapter 8), seismic methods are of limited use and electromagnetic and potential field methods prevail.

In recent years, oil and gas exploration has moved into areas that contain geological units with similar characteristics as encountered in mineral exploration (e.g., basalt flows). Imaging the extent and structure of sediments beneath basalt layers and below and at the flanks of salt domes is particularly problematic (e.g., see *Jones and Davison* [2014]) and it is for these areas that hydrocarbon exploration companies are looking for alternative imaging methods. In both environments the problems are similar, the high-velocity contrast between the basalt or salt and the sediments above makes it easy to image the top of the structures due to the high reflection coefficient. This high reflection coefficient though has the effect that little energy penetrates below and any topography on this interface will distort the seismic wave. Furthermore, both basalt and salt are often internally heterogeneous causing scattering of the remaining signal that masks the weak reflections from the bottom of the structure and any sediments beneath (e.g., see *Martini et al.* [2005]). Together these problems make reliable seismic imaging difficult, even with sophisticated processing techniques [*Hokstad et al.*, 2011].

As a result, exploration activity in such areas now regularly includes controlled-source electromagnetic (CSEM) (e.g., see *Key et al.* [2006] and *MacGregor and Tomlinson* [2014]) as well as magnetotelluric (MT) and potential field measurements (e.g., see *Adriano et al.* [2014] and *Colombo et al.* [2014]). CSEM and MT are receiving particular attention as their sensitivities complement seismic data and each other well (e.g., see *Hoversten et al.* [1998] and *Constable* [2010]). Both basalt and salt are more resistive than the surrounding sediments, typically by an order of magnitude or more. Magnetotelluric soundings penetrate the resistor and mainly sense the presence of sub-salt/sub-basalt sediments, while CSEM data are sensitive to the resistive structures as well as to the sediments as long as they are not too deep. Depending on the subsurface resistivity, the source strength, and the acquisition geometry, modern CSEM surveys can reach depths of 3500 m [*Mittet and Morten*, 2012].

In addition, both methods have inherent depth resolution circumventing many problems with inverting potential field data which are largely insensitive to the depth of anomalies [*Hinze et al.*, 2013]. One of the main selling points of CSEM in hydrocarbon exploration has been its ability to detect the signature of relatively resistive hydrocarbons within conductive sediments (e.g., see *Constable and Weiss* [2006] and *Avdeeva et al.* [2007]), and it occasionally has been cited as a Direct Hydrocarbon Indicator (DHI) (e.g., see *Johansen et al.* [2005]).

This chapter will start with a review of some of the methodological approaches that have been specifically designed for joint inversion in a hydrocarbon exploration context. We will not discuss the specifics of widely used methods such as the cross-gradient, but refer the reader to the theory section of this book. Instead we will focus on methods that take into account specific properties of hydrocarbon reservoirs (e.g., through petrophysical models) and the data typically acquired in exploration (e.g., seismic reflection data). We will also cover emulation of the forward calculation, a concept that can be applied generally in joint inversion, but so far has only been applied in hydrocarbon exploration and is not covered in any of the theory chapters. In our discussion, we will draw parallels with the more fundamental discussions in the theory chapters and make some comparison to other application areas.

After this we will discuss two case studies that are based on our own work in some more detail. These serve as examples for joint inversion in two different exploration settings and we will compare and contrast the challenges and potential solutions for each setting. We will provide recipes for joint inversion strategies that work in our experience. Even though some of the specifics of these recipes are consequences of the choices we made for our joint inversion software, we feel that the conclusions are sufficiently general to aid practitioners in performing joint inversion not only in an exploration context. Furthermore, many of these aspects (e.g., how to choose the weight for a cross-gradient constraint), are not well covered in the current literature.

9.2. METHODOLOGICAL DEVELOPMENTS

9.2.1. High-Resolution Methods

Most exploration managers, the main evaluators of any new imaging technology in the oil industry, benchmark the results from that technology against 3D seismic reflection images. Therefore considerable effort in improving joint inversion has been focused on obtaining high-resolution images with realistic geometries. Commercial providers of joint inversion in particular have focused on maximizing small-scale detail (e.g., see *Abubakar et al.* [2012b]) by tightly integrating the joint inversion results with the seismic imaging workflow (e.g., see *De Stefano et al.* [2011] and *Tartaras et al.* [2011]). In fact, often the main result of joint inversion in a commercial setting is an improved velocity model which is consequently used to migrate seismic reflection data (e.g., see *Colombo et al.* [2008] and *Mantovani et al.* [2013]) as the broad structures that come out of typical regularized inversions are not deemed to contain enough detail for exploration purposes. Some studies address the issue of resolution by using full-waveform seismic data, which are starting to become more commonplace, and combining them with controlled-source electromagnetic (CSEM) data (e.g., see *Hu et al.* [2009], *Abubakar et al.* [2012b], *Gao et al.* [2012], and *Giraud et al.* [2013]). Due to the high computational cost of CSEM and full-waveform calculations, such joint inversion schemes have to be computationally efficient and highly parallel.

In the context of combining high-resolution methods such as seismic full waveform inversion with lower resolution data (e.g., MT or CSEM), some authors argue that the resolution characteristics of the different methods have to match in terms of their spatial wavelengths. For example, *Um et al.* [2014] use a Laplace transform to reduce the resolution of seismic full waveform data to match the resolution of the CSEM. From our perspective, this is not a necessary requirement under all circumstances. For example, in both case studies presented below, the seismic tomography resolves a step change in velocity that is impossible to resolve with MT; that is, the resolution of the seismic data is locally significantly higher than the MT. In both cases the success of the joint inversion is related to the fact that the information on this step change is transferred to the MT model via the joint coupling. Another example is the joint inversion of seismic receiver functions and surface waves (e.g., *Julia et al.* [2000] and *Moorkamp et al.* [2010]; see also Chapter 10). Receiver functions are strongly sensitive to the position of velocity changes but have very little sensitivity to absolute velocity, while surface waves only recover smooth variations in velocity, but constrain absolute velocities. In combination we can reliably recover models that show step changes in velocity where necessary and reproduce the broad velocity variations in the subsurface [*Moorkamp et al.*, 2010].

In these cases there are clear reasons why imposing a step change on the lower resolution method is reasonable. Receiver functions and surface waves both are sensitive to the same physical parameters and sample similar regions of the Earth. For the sub-basalt and sub-salt case studies, the velocity change is associated with the transition from sediments to basalt/salt, and this transition should be associated with a change from low resistivity to high resistivity. The joint inversion simply helps to locate an interface that is not identified by the other method due to the lack of resolving power. Similar arguments are employed by *Averill et al.* [2006] for the integration of seismic reflection and first arrival tomography data.

In contrast, if we combined seismic reflection data and MT in a joint inversion and associated each reflection with a resistivity change, we could create highly heterogeneous models with fine details and strong variations in resistivity. However, all these variations would be in the nullspace of the inversion and it is highly questionable that all of these resistivity changes are associated with property changes within the Earth. From this perspective, *Um et al.* [2014] make the very important argument, that joint inversion has to be used very carefully to avoid artificially imposing structures that have no physical basis. We are not aware of any studies that systematically investigate how methods with different resolution interact and how one can identify artificially imposed features. Such a study would provide valuable guidance for future joint inversion approaches.

9.2.2. Petrophysical Coupling Approaches

From a joint inversion perspective the so-called petrophysical coupling approaches which have been theoretically discussed in Chapter 3 offer interesting possibilities. In hydrocarbon exploration there is often a wealth of a-priori information available from well logs in the area. Furthermore, for the classic hydrocarbon reservoir, a porous sandstone filled with varying amounts of water, oil, and gas, different theoretical formulations have been developed to link petrophysical quantities such as porosity, permeability, oil/gas/water saturation and so on, to geophysical quantities such as conductivity and seismic velocity (e.g., see *Carcione et al.* [2007]). Thus, when trying to use joint inversion to obtain more detailed information on a known or suspected reservoir, these theoretical relationships provide a natural coupling that can be usefully exploited.

Hoversten et al. [2006] and *Chen et al.* [2007] present 1D joint inversions of seismic amplitude versus angle (AVA) and CSEM data that use a rock physics model to estimate reservoir properties. Outside the reservoir, they invert for standard geophysical parameters (i.e., seismic velocities and electrical conductivities), while for the target zone within the reservoir they parameterize the inversion in terms of fluid saturations for oil S_o, gas S_g and water S_w and porosity ϕ. To translate from these parameters back to velocity and conductivity, they use the well-known Archie's law for conductivity and a combination of Gassmann's equation with a rock physics model developed for sandstone velocity and density (see *Hoversten et al.* [2003]). To a certain degree, such a hybrid parameterization, with geophysical parameters outside the target zone and petrophysical parameters inside the target zone, is similar to the petrological parametrization shown in Chapter 10 for lithospheric studies. There the crust is parameterized differently from the mantle where composition and temperature are sought and used to couple the different geophysical methods. Consequently, these approaches share a lot of the same advantages and disadvantages. Some of these are discussed in Chapter 3 where the theoretical basis of petrophysical coupling approaches is explained.

An obvious prerequisite for using rock property based coupling approaches is that the assumed parameter relationship has to be valid in the region where it is utilized in the inversion. For a simple velocity–conductivity relationship, *Moorkamp et al.* [2011] show how discrepancies in the relationship map into inversion artifacts. This conclusion also applies for more complicated relationships or parametrizations in terms of other quantities. *Hoversten et al.* [2006] conclude that for their study deviations from the rock property model produce comparable relative errors in the estimation of reservoir parameters; that is, an error of 10% in the cross-property relationship will translate to porosity estimates that are systematically 10% too high or too low—thus the definition of the region where the relationship is applied and the appropriate specification of the relationship are of critical importance for the success of such a joint inversion.

Another important issue for petrological coupling approaches is the sensitivity of the geophysical parameters to the different petrological parameters (see also *Swidinsky et al.* [2012]). If the petrophysical parametrization contains a quantity that has negligible influence on the geophysical parameters, then that petrophysical parameter will be poorly determined. However, it is equally important that at least a subset of the petrophysical parameters has significant sensitivity to different geophysical parameters simultaneously. Otherwise if the sensitivity of each petrophysical parameter is largely determined by a single geophysical parameter, we lose the coupling between the different methods as we effectively have just re-scaled the individual parameter axes.

Gao et al. [2012] show sensitivities of water saturation and porosity on seismic velocity, electrical conductivity and density (their Figures 1 and 2). Both velocity and density show high sensitivity to porosity and show little sensitivity to water saturation. Therefore performing a joint inversion of velocity and density using these parameters for coupling would not result in a significantly better retrieval of water saturation. However, the inversion would be strongly coupled as the geophysical parameters are effectively reduced to a single petrophysical parameter (in this case porosity).

Electrical conductivity shows a different pattern of sensitivity to the petrophysical parameters, and thus in combination with seismic data it is possible to constrain porosity and water saturation. The success of the approaches shown in *Hoversten et al.* [2006] for linearized inversion and in *Chen et al.* [2007] for Bayesian inversion is based on this complementary nature of the parameter relationship. This is also the reason why these studies constrain themselves to two phases, water and gas, and fix the oil saturation S_o to some assumed value. For this two-phase system the water saturation S_w and gas saturation S_g are related through the simple relationship $S_w = c - S_g$, here $c = 1 - S_o$. In a similar way a two-phase system of water and oil could be set up, where no gas fill is assumed. As saline water has a very different conductivity to oil and gas, a strong effect on conductivity is observed as soon as the water content increases. In contrast, oil and gas have relatively similar conductivities. Therefore in a three-phase system it is only possible to reliably determine the total amount of gas and oil, but the relative fraction between the two cannot be resolved. *Chen et al.* [2007] show examples of the impact of an incorrectly assumed oil concentration. As we have seen above, neither seismic or gravity data have the necessary sensitivity to distinguish between oil and gas saturation; thus with the most common geophysical techniques, this ambiguity cannot be resolved.

For synthetic data both the linearized [*Hoversten et al.*, 2006], and the Bayesian joint inversion approaches [*Chen et al.*, 2007], show significant improvements in the estimates of porosity and gas concentration in comparison with separately performed inversions. For real datasets the results are less clear, although in both cases they report closer agreement of the joint inversion result with logged values than for the individual inversions. There are a number of sources of uncertainty which the authors discuss. The approximation of a layered Earth is probably not appropriate for the whole region sampled by the geophysical data. The borehole logs that are used to estimate the parameters in the petrophysical relationships are noisy, and the relationships themselves might not be appropriate everywhere.

Gao et al. [2012] and *Abubakar et al.* [2012a] circumvent problems with trying to explain the data with a layered Earth by using a two-dimensional joint inversion approach for seismic full-waveform and CSEM data. Their petrophysical formulation is similar to *Hoversten et al.* [2006], but they avoid the hybrid parameterization as they are considering cross-well measurements and assume that the methods are only sensitive to regions where the petrophysical parametrization can be considered appropriate. They carefully test the impact of error on different petrophysical parameters for a marine exploration case and for a time-lapse reservoir monitoring scenario. In both cases they find that the success of the joint inversion does not depend strongly on the parameters used in Archie's law. In contrast, they emphasize that high-quality information on the seismic rock-physics parameters is needed, particularly the density of the rock matrix to obtain reliable results.

An alternative approach to petrophysical coupling has been proposed by *Bosch* [2004] for petrophysical inversion of seismic data. He uses a geostatistical approach to constructing the relationship between seismic velocity and petrophysical quantities of interest such as porosity. The theoretical basis for this approach is explained in Chapter 3, and a good review of the approach and similar studies is given in *Bosch et al.* [2010]. *Chen and Hoversten* [2012] apply a similar idea to joint inversion of seismic AVA and CSEM data. As in *Bosch* [2004] they use borehole logs from the area of interest to construct a statistical rock physics model. In the subsequent joint inversion this statistical model is then used to both couple the data and directly invert for petrophysical parameters. Compared to the Archie/Gassman coupling approach there are different advantages and trade-offs. For example, the original Archie's law has been developed for clean sand, and a number of modifications exist to deal with clay and changes to the rock matrix (e.g., see *Carcione et al.* [2007]). Depending on which variant is used in the inversion, different assumptions are made about the constituents of the rock and its pore space. The statistical approach in contrast is purely data-driven, and no explicit assumptions about clay content, and so on, are made. Instead the borehole data are considered representative for the whole area under investigation. Depending on lateral heterogeneity in the geology, this assumption might or might not be appropriate. A similarly data-driven approach based on fuzzy *c*-means clustering has recently been published by *Sun et al.* [2013] and shows promising results on a range of synthetic test models. However, to apply any data-driven approach, different boreholes should ideally be available to capture spatial variations in rock properties. In either case, errors in the assumptions on the cross-property relationships are directly mapped into the inversion results.

These results make it clear that such petrophysical coupling approaches are more suited for use in well-explored areas than for front-line exploration in unknown territory. As a minimum, logging data from a borehole or reasonable knowledge of the geology and its spatial variations are required. Still, the underlying idea, inverting for parameters of direct economical interest and using them to couple the different geophysical techniques, is highly appealing and a promising area of future research.

9.2.3. Emulation

Another interesting methodological concept that has been developed in the context of hydrocarbon exploration, but is not limited to such scenarios, is so-called "emulation" of the forward calculations. For models with more than a few tens of model parameters, stochastic inversion approaches typically become prohibitive due to the excessive amount of time needed to sample the model space adequately. This is also known as the "curse of dimensionality." Even in cases where the time for calculating the forward response for a given model is small (e.g., on the order of a second), the necessity to calculate the responses for millions of models means that we quickly reach practical limits.

From a statistical perspective, an emulator is a stochastic model of the function which connects the model parameters with the output data. Emulators may take a variety of forms depending on the system concerned (e.g., *Craig et al.* [1997]); however, the core idea is that a complex forward modeling code is represented by a fast-to-evaluate parametric component with an uncertainty. The power of an emulator-based method rests on the principle that for large areas of a potential model parameter space we do not need the precision of the physics-based forward calculations (the so-called simulator) to determine that a given candidate model is implausible. This is because the resulting output data residual is far greater than the uncertainty of the emulator estimation of the simulator code output. Thus, for the purpose of discerning the plausible model space for an observed dataset, we can replace the simulator with a proxy (the emulator) that can be evaluated several orders of magnitude faster than the full forward operator. A detailed description of the properties of such an emulator and its application to joint inversion for sub-salt imaging is given

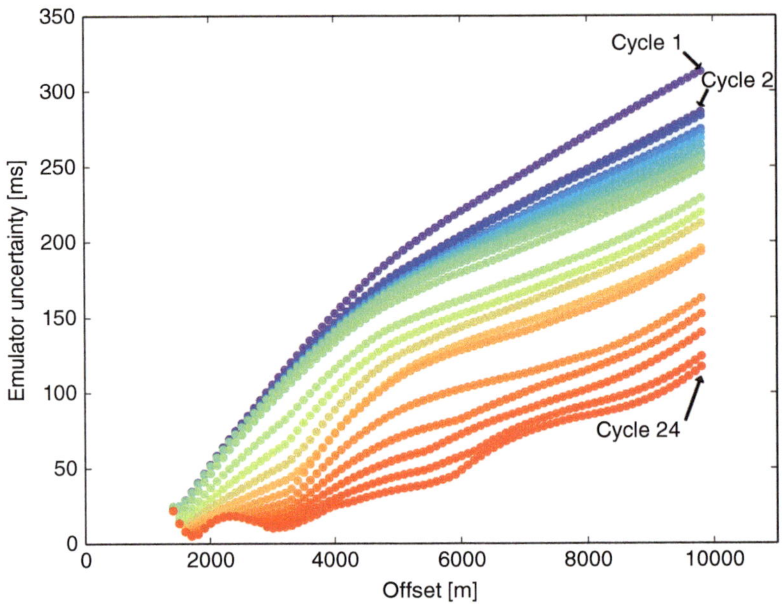

Figure 9.1 Uncertainty for an emulator constructed for seismic tomography data over the Uranus salt-dome (see text for a more detailed description of the dataset). At each cycle we identify models that are incompatible with the data given data error and emulator uncertainty and reduce model space accordingly. This allows us to construct an improved emulator and repeat the process until it converges after 24 cycles. The final emulator uncertainty is on the same order as the assumed data error.

in *Roberts et al.* [2012]. We therefore only sketch the basic ideas of the approach here.

When building an emulator, a number of calibration runs with the full forward algorithm have to be performed. The number of models used for such calibration is moderate (typically 100–1000) and should sample the entire model space under investigation. In order to ensure adequate sampling, designated strategies (e.g., the Latin hypercube [*McKay et al.*, 1979]) exist. From this sparse sampling of model space polynomial coefficients are estimated that relate the model parameters to the resulting data. This process is somewhat analogous to constructing a neural network for geophysical modelling (e.g., see *Meier et al.* [2007]). It is clear that when using a polynomial representation to encapsulate potentially complicated physical processes such as the relationship between a resistivity model and the magnetotelluric impedance data, this representation cannot yield a perfect representation of the full simulator code output. In particular in the early stages of the screening process when the model space is large, the emulated geophysical data will only be a crude representation of the true physical response. However, for the purpose of screening model space, this is not so important. It is more critical to have a reliable estimate of the uncertainty of the emulated output with respect to the full simulator code, even if this uncertainty is quite large, since it is this uncertainty range which will be used to test the proximity of the emulated output data to the observed dataset.

Having built an emulator, including a reliable estimate of the uncertainty of its output, the model space can now be screened, and models that are deemed implausible given the combined observational and emulator uncertainty can be rejected. For a combination of seismic, magnetotelluric, and gravity data, *Roberts et al.* [2012] report an acceptance of 0.05%—that is, 53 out of 100,000 sampled models for a test case. Thus even a relatively crude approximation of the forward response can discard a significant amount of model space as implausible. Having reduced the size of the plausible model space, we can repeat the cycle by building a second emulator over this smaller model space. As the size of this new model space will be smaller, we can expect that an emulator constructed from samples in this reduced space will have a lower uncertainty on the output. As a result, this second emulator can be used to further constrain the region of plausible model space.

Figure 9.1 shows an example of the uncertainty of a seismic tomography emulator. We can see that the uncertainty of the initial emulator (marked cycle 1) is large compared to typical picking errors at all offsets. Still, we can use this approximation to reduce model space and construct a new emulator (marked cycle 2). This new emulator can then be used for another screening round, and the process of emulator building and model space screening can be repeated until the emulator uncertainty converges and new screening rounds do not result in further reduction of model space. In this case we achieve convergence after 24 cycles. At each cycle the emulator uncertainty has been reduced and the final uncertainty is less than 50 ms at offsets smaller than 6000 m which is comparable to typical picking errors. As the samples in the screening process are generated completely independently—in fact we use a simple Monte Carlo process [*Press*, 1968]—the resulting models can be analyzed using standard statistical tools for stochastic inversion to estimate probability density functions and extract models of interest.

At the moment, the use of emulators in geophysical imaging is still in its infancy. The methodology presented in *Roberts et al.* [2012] and further developed in *Roberts et al.* [submitted] is based on one-dimensional models and thus at the moment is more a proof of concept than a readily available tool. Still, we see great promise in emulator-based approaches. They have the potential to alleviate the "curse of dimensionality" at least from a modeling perspective. For joint inversion the reduction in computational time is particularly useful when considering many different datasets simultaneously.

At the moment, little experience exists on suitable strategies for different types of geophysical data and how the parameterization of the data and the model can be extended to higher dimensions. In such cases it might be attractive to combine emulation with alternative approaches to accelerate the forward computations such as model order reduction (e.g., see *Presho et al.* [2014]) or transdimensional Bayesian methods that automatically determine the number of model parameters within the inversion (e.g., see *Ray and Key* [2012] and *Ray et al.* [2014]). This is the subject of current research.

9.3. CASE STUDIES

In the following we will present more detailed case studies together with reviews of some of the main approaches and results generated by other studies for the same areas. We will focus on joint inversion for exploration in marine sub-salt and sub-basalt settings as these appear to attract the main interest judging from the number of recent publications. Other application areas for joint inversion include carbonate platforms [*Colombo et al.*, 2008], arid land [*Colombo et al.*, 2010], deep water basins [*Tartaras et al.*, 2011], and production monitoring [*Liang et al.*, 2012; *Lien*, 2013].

9.3.1. Joint Inversion for Sub-salt Imaging

Salt diapirs piercing through sedimentary units have been a prime target for exploration activity for a long time (e.g., see *Schowalter* [1979], *Ladzekpo et al.* [1988], and *Hoversten et al.* [1998]). The impermeable salt acts as a structural trap for hydrocarbons; and depending on the geometry of the salt dome, the target is economically prospective or not (Figure 9.2). Overhanging salt, in particular, forms an ideal trap for hydrocarbons, and imaging the presence of these overhangs is often the target of geophysical exploration activity (e.g., *Hoversten et al.* [2000], *Key et al.* [2006], and *Avdeeva et al.* [2012]). However, with seismic reflection data the geometry of the lower parts of the salt dome is difficult to image [*Ji et al.*, 2011], as the high-velocity contrast between the surrounding sediments and the rough top of the salt scatters the seismic energy. In the presence of intra-salt inclusions, so-called dirty salt, the scattering is further amplified [*Haugen et al.*, 2009], and in some areas even modern seismic migration algorithms cannot image the base of the salt [*Hokstad et al.*, 2011].

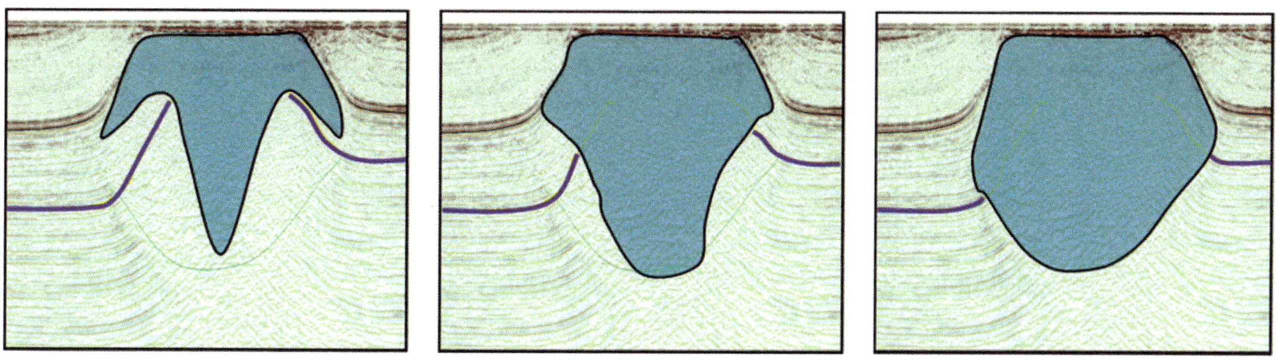

Figure 9.2 Three different possible interpretations of the shape of the Uranus salt dome based on seismic data alone. While the top of the salt dome is well constrained, scattering and low amplitudes make it difficult to define the base of the salt well, leading to very different possible assessments of the prospectivity. Reproduced from *Hokstad et al.* [2011].

The Uranus salt dome in the Barent Sea offshore northern Norway and the Gulf of Mexico (e.g., see *De Stefano et al.* [2011]) are particularly well investigated and have been used as examples for new joint inversion approaches. In the Nordkapp Basin in the Barent Sea, the failure of a well drilled on a conceptual model generated from seismic data highlighted the limitations of current seismic reflection imaging and sparked initiatives using nonseismic data [*Hoksatd et al.*, 2011; *Wiik et al.*, 2013]. The resulting dataset of seismic, magnetotelluric, CSEM, gravity, and gravity gradiometry data from this region has been made available to several groups and has resulted in inversions of different combinations of that data. A good overview of the area and a brief description of the geological setting is given by *Stadtler et al.* [2014], and *Fanavoll et al.* [2012, 2014] give an overview of exploration activity in the area. *Stadtler et al.* [2014] also present a joint interpretation workflow where they invert gravity, gravity gradient, and magnetic data based on constraints obtained from seismic, electromagnetic, well data, and known geology. These constraints are necessary because potential field data, even when used in combination, have limited depth resolution [*Gallardo-Delgado et al.*, 2003]. They demonstrate that through incorporating these constraints, they can test the different scenarios for the base of salts suggested by other geophysical methods. However, they also show how uncertainty on the Moho topography interferes with the recovery of some of the large-scale features and recommend tighter integration with other techniques.

For the same region, *Fanavoll et al.* [2014] show the main use of electromagnetic data in current hydrocarbon exploration workflows. They consider it a supplementary tool to obtain additional information on the properties of the subsurface when co-rendering the resulting conductivity model with migrated seismic reflection data. The smooth models retrieved from inversions of EM data do not contain sufficient structural information to obtain detailed stratigraphic information. Instead the resistive anomalies are visually correlated with the seismic reflection sections and the relationship between the anomalies and structural features is interpreted in terms of potential hydrocarbon trapping mechanisms.

In 2010 we received seismic, magnetotelluric, and gravity data acquired over the Uranus salt dome as a test for a newly developed 3D joint inversion framework [*Moorkamp et al.*, 2011]. One subtle, but important, issue when combining different datasets acquired by different operators at different times are the geodetic parameters of the coordinate system used to locate the measurements. For the Uranus datasets, seismic data were given in UTM coordinates with a ED50 geodetic datum, while the magnetotelluric data were given as latitude/longitude with a WGS84 geodetic datum. The shift between these two coordinate systems is about 150 m with an additional slight rotation and is significant enough to impact on the inversion results when combining the two datasets without correcting for the different coordinate systems.

Compared to the sub-basalt case study described below, the overlap of the different datasets is very good. Even though we only have seismic travel time data along a single transect and are therefore essentially dealing with a two-dimensional situation, the magnetotelluric data are acquired along the same line. The Uranus salt dome, which is the target of interest, is also covered by gravity data. However, a second salt dome in the west which appears in the seismic travel time inversions is covered neither by MT nor gravity data. Still, considering that the data acquisition was not designed for joint inversion, the data coverage is adequate.

In *Moorkamp et al.* [2013] we present results from a structural joint inversion for seismic, gravity, and magnetotelluric data and use seismic reflection and borehole data as benchmarks for the joint inversion results. From our perspective the most important, and somewhat surprising, result is how well the cross-gradient coupled inversion results reflect the velocity–resistivity relationship from the borehole logs and we will discuss this issue in more detail below. We refrained from using these logs which were available to us together with the other geophysical data for two reasons: (i) There are distinctly different relationships for the salt and the surrounding sediments. Thus a simple functional relationship, as we used for our synthetic test studies [*Moorkamp et al.*, 2011] and for the sub-basalt case study below, is not appropriate. (ii) Geophysical inversion measures rock properties over scales of tens to hundreds of meters while borehole logs measure over scale of centimeters. It is therefore not clear if these two techniques measure the same thing. Our results for the Uranus salt dome and for the sub-basalt scenario presented below, indicate that scale is not a major issue and somewhat justify the often cursory treatment given to the different scale lengths in other studies (e.g., see *Panzner et al.* [2014]).

In the following we present updated joint inversion results for the Uranus dataset. Compared to our previously published results, a number of changes have been made. We halved the horizontal cell sizes from 460 m to 230 m and the vertical cell size from 230 m at the top to 115 m. This refined grid should allow a better approximation of the geometry of the salt. We also rotate all datasets so that the seismic and magnetotelluric profiles align with the east–west direction, that is, the x-axis of the model grid. The initial inversions were performed in the original acquisition coordinate system which simplifies data treatment. A disadvantage of this setup is that the profile of highest resolution is at an oblique angle to the inversion mesh and requires interpolation to be displayed.

For this reason, we do not show the inversion mesh in *Moorkamp et al.* [2013].

To perform the joint inversion for seismic, gravity, and MT data using a cross-gradient coupling approach, we have to specify nine weights: one weight to control the influence of each dataset, one weight for each regularization term of the different physical properties, seismic velocity, density, and conductivity, and in total three weights for the cross-gradient terms (see also *Moorkamp et al.* [2011]). In theory, weighting factors between the different geophysical data are superfluous if we have the variance for each datum or even covariance between the data. Multiplying the objective function associated with a certain type of data with a constant factor is equivalent to dividing the variance by the square of that factor, that is, artificially decreasing the assumed measurement error for each datum. Theoretically, introducing such fudge factors is highly undesirable, as the data errors should be simply an estimate of our confidence in the data and determined statistically or through assessing the accuracy of our travel-time picks, for example. Mathematically, data errors derived in this way provide the only correct weighting between the different data. However, all joint inversions of real data that we have performed so far require some form of re-weighting.

In our experience, two factors make some form of re-weighting between the different datasets necessary: (i) Even for methods such as MT where we have data errors derived from the data in a statistically well-defined manner (e.g., see *Chave and Thomson* [2004] and *Chave and Jones* [2012]), the calculated error bars do not fully reflect the uncertainty in the data. Furthermore, different processing algorithms can result in significantly different error estimates [*Jones et al.*, 1989]: (ii) The misfit of the initial models can be quite different for the different datasets. In such a case the joint inversion will focus solely on the dataset with the highest misfit and ignore the other data. As a consequence, the initial iterations will be very similar to a single of inversion of that data. In our experience the best joint inversion results are produced when all data participate in all stages of the inversion, and re-weighting the data or separating the objective functions ensures this.

A procedure to determine the weights that has worked well in practice is to first perform individual inversions for each dataset with varying regularization parameters to determine a good misfit versus smoothness trade-off. We consider individual inversions an essential prerequisite for any joint inversion, as we can develop an understanding of the structures sensed by each dataset and also identify problematic data and other potential issues. We also use these individual inversions to identify regularization parameters that provide the optimum trade-off between smoothness and data misfit when plotted as an ℓ-curve [*Hansen*, 1992]. We then run the first joint inversion with a small cross-gradient coupling constraint. Based on the misfits, we weight the data terms for each method so that the re-weighted objective function is roughly equal for all types of data. After the first few iterations, we can see how the misfit for each method evolves. In our experience, for a successful joint inversion the misfit for all methods decreases in the first few iterations. Within 3–5 of these inversion runs, we can usually find a set of data weights that fulfill this criterion and we multiply the regularization parameter for the corresponding model parameters by the same factor. This ensures that the relative regularization for each method remains comparable to the individual inversions. Once we have found appropriate data weights, we vary the cross-gradient weights. Currently, we keep the weight for all three cross-gradient terms (one for each pair of models) identical to simplify the procedure and we observe that in most cases this choice is a good trade-off between practicality and obtaining the best possible results. It should be possible to optimize the joint inversion further by specifying individual weights for each coupling term. The inversion run with highest possible cross-gradient weight and data misfits comparable to the individual inversions is the final result of the joint inversion.

Figure 9.3 shows the result we obtain with the refined grid using this strategy. Overall, the models are very similar to the results presented in *Moorkamp et al.* [2013]. As expected, the finer grid results in a better match with the boundary of the salt as outlined by the seismic reflection data compared to our previous results. Furthermore, the velocities and densities retrieved are comparable to those results. We observe somewhat lower resistivities in our new models though. There is still a high resistivity contrast between the surrounding sediments and the salt, but we do not obtain some of the very high resistivities (>1000 Ω.m) that we observed previously. The reason for this is associated with the rotation of the magnetotelluric data. In contrast to seismic and gravity data, the magnetotelluric impedance is a tensor quantity. A correct consideration of errors after rotation requires that we include a covariance matrix with nonzero off-diagonal elements in the inversion [*Booker*, 2014; *Moorkamp et al.*, 2014] and several practitioners have observed changes in the inversion results depending on the coordinate system the inversion is performed in *Tietze and Ritter* [2013] and *Kiyan et al.* [2013]. In our case we observe a reduction in initial misfit for identical models after rotating the coordinate system. This indicates an artificial increase of the assumed errors associated with not considering the off-diagonal elements of the covariance matrix. The increase in error is equivalent to a decrease in information content in the data.

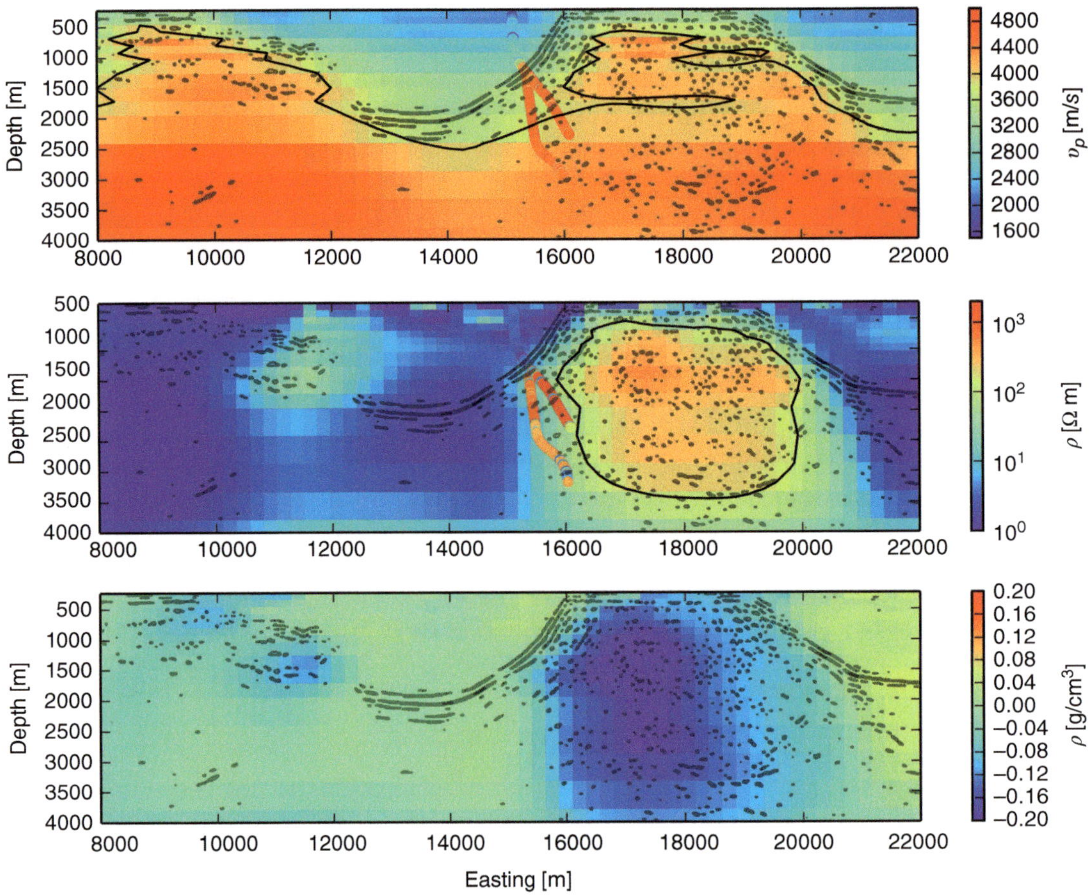

Figure 9.3 Joint inversion of travel-time tomography, magnetotelluric and gravity data for the Uranus salt dome structure using cross-gradient coupling constraints. We show the resulting velocity **(top)**, resistivity **(middle)**, and density **(bottom)** models for a transect across the salt-dome. In addition, we mark the major reflections from a coincident migrated reflection profile (*thin black lines*) and the outline of the salt based on typical values ($v_p > 4500$ m/s and $\rho > 200\Omega \cdot$m) assumed in the literature (*thick black line*). We also show the transects of two boreholes drilled into the flank of the salt-dome colored by the averaged velocities and resistivities from the borehole log.

Comparing the joint inversion results with the individual inversion results (Figure 9.4), we observe that a significant amount of the information about the shape of the salt-dome comes from the magnetotelluric data. For the electrical resistivity model, the differences between the individual inversion and the joint inversion results are relatively small. In contrast, the recovered velocities and densities show substantial differences. As in our previous study, we only recover small, isolated high-velocity anomalies and the geometry of the low-density anomaly is only vaguely associated with the shape of the salt dome for these single inversions.

In *Moorkamp et al.* [2011] we showed the relative insensitivity of the joint inversion results to the value of the weight for the cross-gradient constraint, and a similar experience has been reported by *Linde et al.* [2008]. Still, to our knowledge there have not been any systematic investigations of the impact of the cross-gradient weight for real data. In Figure 9.5 we show the evolution of the different elements of the joint objective function as a function of iteration and for a range of weights κ for the cross-gradient constraints. For simplicity the weight for all three cross-gradient constraints is identical. The development of the RMS is virtually unaffected by the choice of κ; but for the largest value, $\kappa = 10^7$, we start to see slightly higher RMS for gravity and MT data. This suggests that increasing κ further would start to have detrimental effects on convergence as we observed in our synthetic tests [*Moorkamp et al.*, 2011]. As expected, increasing κ is associated with a reduction in the values of the cross-gradient objective function. Increasing κ by a factor of 1000 (i.e., from $\kappa = 10^4$ to $\kappa = 10^6$) results in a decrease of the velocity–density and velocity–resistivity cross-gradient terms by a factor of 4 and reduction of

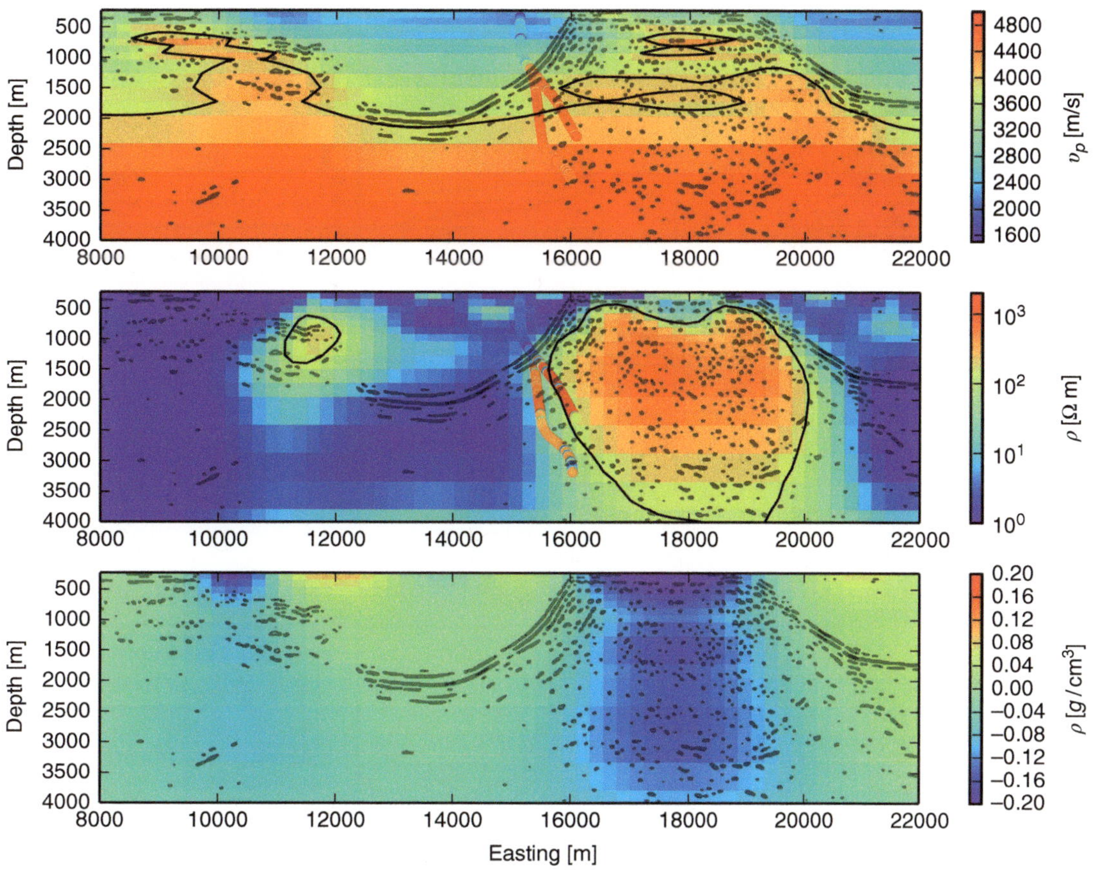

Figure 9.4 Individual inversion results of travel-time tomography, magnetotelluric, and gravity data for the Uranus salt dome structure across the same transect as the joint inversion results in Figure 9.3. We show the resulting velocity **(top)**, resistivity **(middle)**, and density **(bottom)** models. In addition, we mark the major reflections from a coincident migrated reflection profile (*thin black lines*) and the outline of the salt based on typical values ($v_p > 4500$ m/s and $\rho > 200\Omega\cdot$m) assumed in the literature (*thick black line*). We also show the transects of two boreholes drilled into the flank of the salt-dome colored by the averaged velocities and resistivities from the borehole log.

the density–conductivity constraint by a factor of 10 (see Figure 9.5). Finally, model curvature, the measure we chose to regularize the inversion, is unaffected by the value of κ for the velocity model and has little effect on the density model for all but the largest value of κ. For the conductivity model we see a systematic decrease in regularization with increasing κ, again for all but the largest κ. Overall, this effect is relatively small though, a factor of 1000 increase in κ results in a 30% decrease in regularization for conductivity. Still, the differences in model evolution for the density model and the change in cross-gradient constraints demonstrate that there is some complex interaction between the different components of the joint objective. From Figure 9.5 it appears that the effect is most pronounced for the density and conductivity models, while the velocity model is only marginally affected.

Figure 9.6 shows the velocity, density, and resistivity models for cross-gradient weights of $\kappa = 10^4$ and $\kappa = 10^6$, respectively. At first glance there does not seem to be a significant difference between the models generated with different values for κ. Closer inspection reveals, however, that there are small but significant changes. For $\kappa = 10^4$ the low-density anomaly extends straight to the surface, with the strongest anomaly located at the top. This appears to be a remnant of the typical phenomenon when inverting gravity data without further constraints to place the strongest anomalies near the surface [*Li and Oldenburg*, 1998] and similar to the density model from the individual inversions in Figure 9.4. It suggests that this value for the cross-gradient coupling weight does not provide enough constraint on the density model to have a significant impact. For $\kappa = 10^6$, the maximum of the anomaly is located lower within the salt and co-located with the highest velocities and highest resistivities. This is a clear indication of the cross-gradient at work to reconcile the different physical parameters. For this salt-imaging scenario we

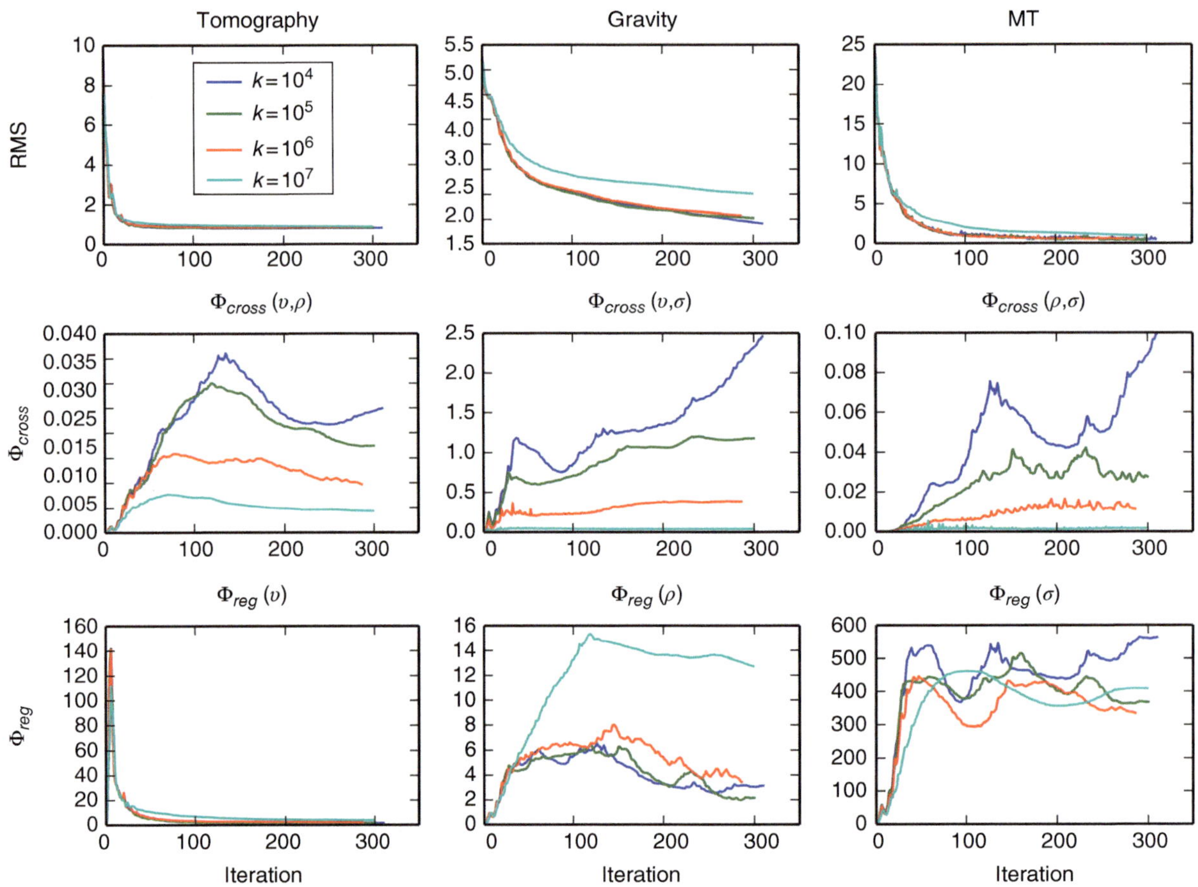

Figure 9.5 Convergence curve for joint inversions with different weights κ for the cross-gradient constraints. We plot the RMS **(top row)**, cross-gradient constraint values for the different parameter combinations **(middle row)**, and regularization **(bottom row)** as a function of frequency.

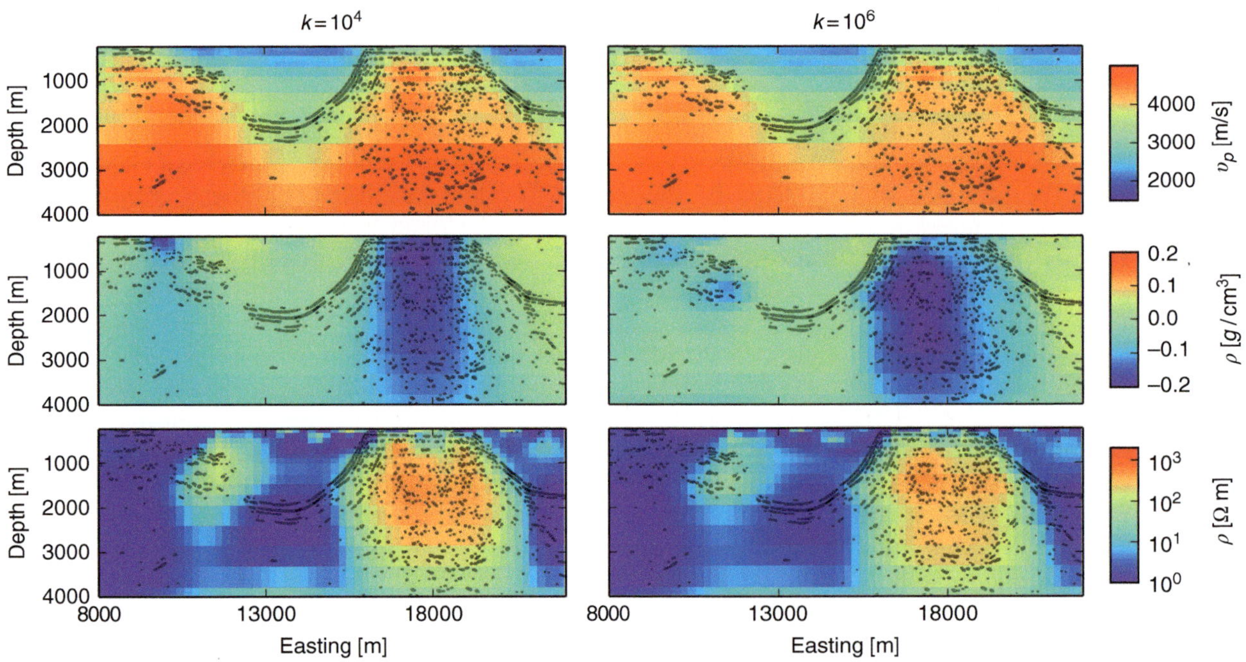

Figure 9.6 Comparison of the velocity models **(top row)**, density models **(middle row)**, and resistivity models **(bottom row)** for two different weights κ for the cross-gradient constraint.

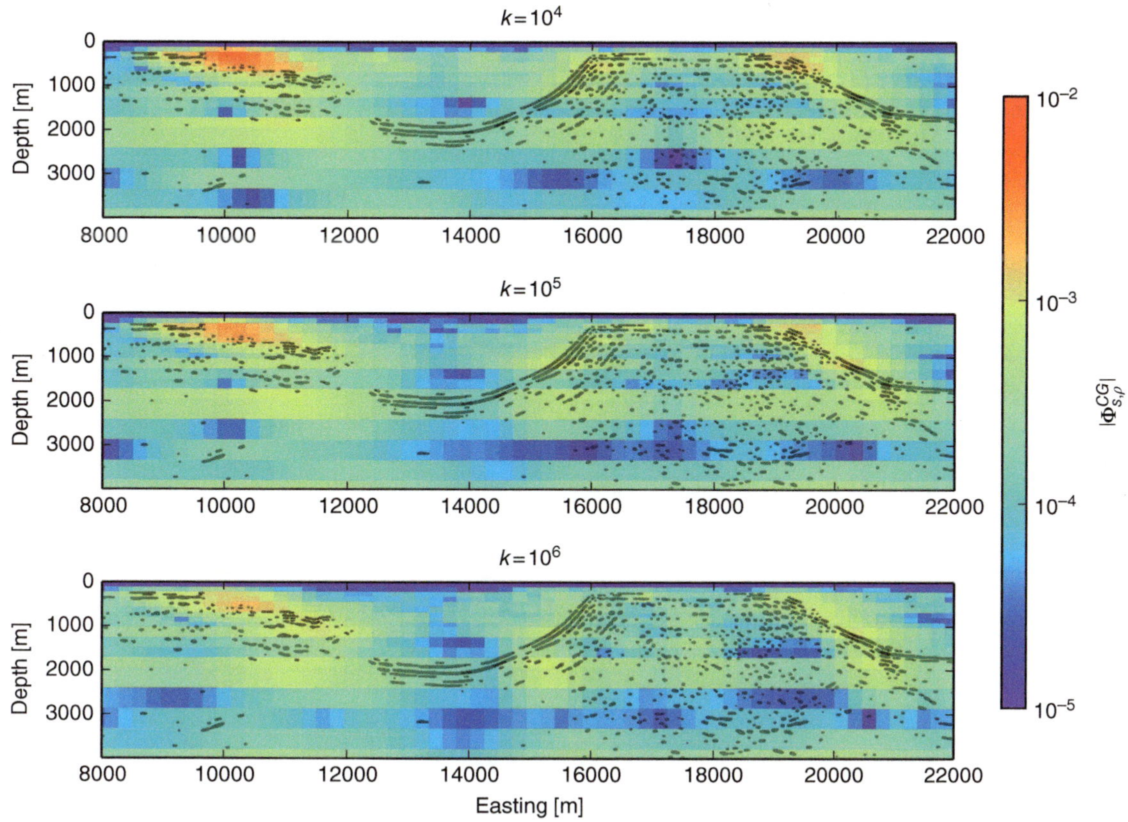

Figure 9.7 Comparison of the magnitude of the cross-gradient constraint $\Phi^{CC}_{s,\rho}$ between slowness s and density ρ for joint inversions with different weights κ for the cross-gradient. We show $\kappa = 10^4$ **(top)**, $\kappa = 10^5$ **(middle)**, and $\kappa = 10^6$ **(bottom)**.

regard the deeper location of the maximum density anomaly as more realistic and thus the model with the higher cross-gradient constraint as better.

Figure 9.7 shows a comparison of the magnitude of the cross-gradient vector between slowness and density for different weights for the constraint. There are some visible differences in magnitude between the lowest weight (top panel in Figure 9.7) and the highest weight (bottom panel in Figure 9.7). The most pronounced difference can be seen near the top of the second salt dome around an easting of 10,000 m. There the high values are gradually reduced with increasing weight for the constraint. However, the overall pattern of values is very similar, particularly around the main salt dome where all datasets have maximum resolution. These observations further confirm that the value for the weight for the cross-gradient constraints can vary by up to an order of magnitude without affecting the interpretation of the results.

9.3.2. Imaging Sub-basalt Sediments

From a seismic perspective the problems associated with imaging sub-basalt sediments are somewhat similar to imaging salt dome structures. The strong velocity contrast between basalt and the overlying sediments makes it easy to image the top of the basalt. However, this strong impedance contrast together with fine scale structure and heterogeneities within the basalt results in strong scattering of seismic waves and considerable internal multiples. For purely seismic based imaging, using very long source receiver offsets can alleviate some of the difficulties in imaging sub-basalt sediments (e.g., see *White et al.* [1999], *Fruehn et al.* [2001], and *White and Gordon* [2003]); however, the logistical effort for such measurements is high.

Similar to salt, basalt is relatively resistive compared to typical sediments and therefore magnetotelluric signals penetrate through the basaltic layer. Thus the properties of the different datasets and the aim of the joint inversion are very similar for salt structures and sub-basalt sediment imaging. A major difference between the two scenarios are the differences in geometry between salt domes and typical basalt structures. For a salt dome, the shape of its near-vertical flanks is one of the main targets of the imaging and thus salt domes are inherently two-dimensional or in many cases even three-dimensional structures. In most geological settings basalt sequences are emplaced as near horizontal layers with relatively

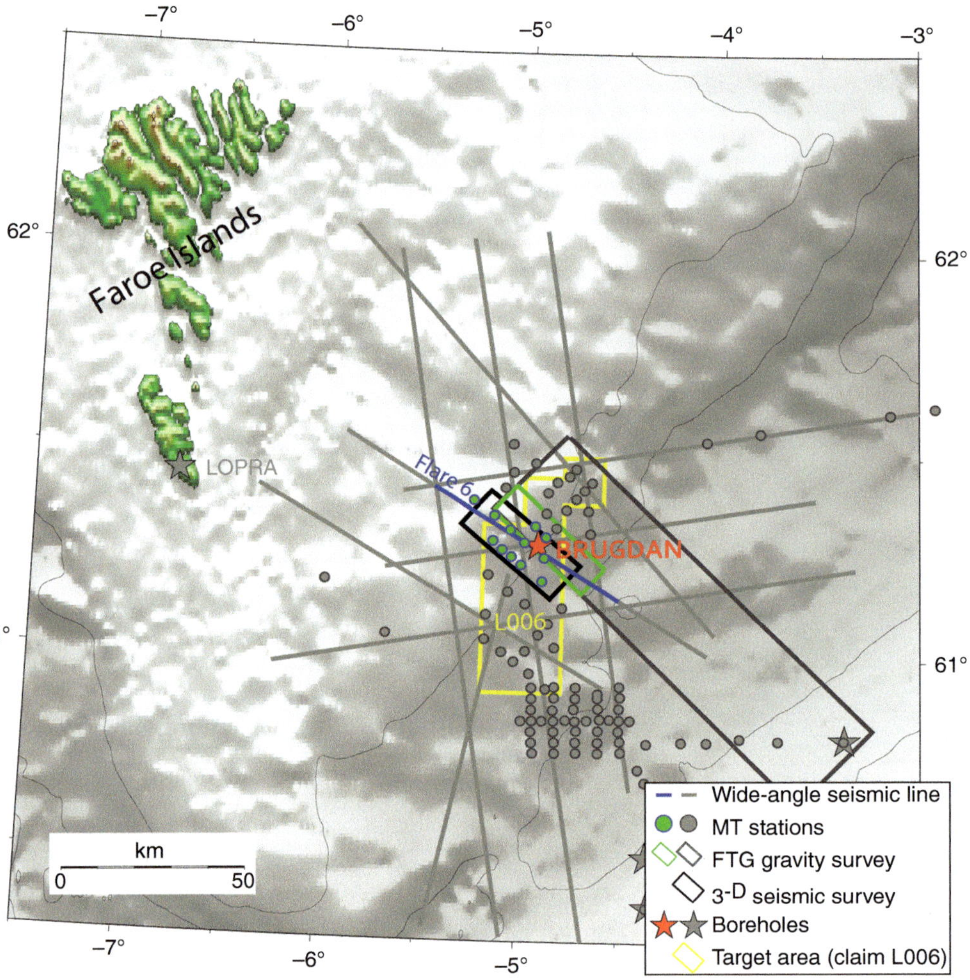

Figure 9.8 Map of the investigation area for the sub-basalt case study. All data that are available for this project are shown. (Data from MT stations and borehole locations with gray symbols are not considered in the inversions.)

smooth lateral variations in thickness and structure. For this reason, earlier work on joint inversion in sub-basalt settings has focused on one-dimensional horizontally layered approaches (e.g., see *Manglik and Verma* [1998]), in some cases stitched together to provide a pseudo two-dimensional view [*Manglik et al.*, 2009; *Jegen et al.*, 2009].

Jegen et al. [2009] apply their joint inversion approach to magnetotelluric and gravity data from the Faroe–Shetland shelf and incorporate constraints from seismic data. Similar to the Barents Sea for salt imaging, the Faroe–Shetland area has become a natural test bed for different sub-basalt imaging approaches due to the large volume of data available. *Christie and White* [2008] summarize some of the approaches to image sub-basalt sediments in the area. While earlier integrated and joint inversion studies focused on the area directly southeast of the Faroe islands along the FLARE10 line [*White et al.*, 1999; *Heincke et al.* 2006], recent activity has moved further south to the so-called License 6 area (e.g., see *Panzner et al.* [2014]; see also Figure 9.8). The interest in this area was sparked by ongoing exploration activity that culminated in the drilling of the BRUGDAN borehole in 2012. Drilling activity was temporarily suspended in 2013, but continued in 2014 with the well reaching a total depth of approximately 4200 m and penetrating into the sub-basalt sediments. However, the well was dry and it was then plugged and abandoned. As seismic velocities, densities, and electrical conductivities were logged in the borehole, it provides valuable constraints for joint inversion in this area [*Schuler et al.*, 2012].

As part of the SINDRI-II project, we received legacy seismic, gravity, and marine magnetotelluric data from the License 6 area in order to assess the potential of joint inversion for sub-basalt imaging. Figure 9.8 shows an overview of the area, in total there are 10 long-offset seismic lines and 43 MT stations in the study area with a relatively good spatial coverage. However, as typical for legacy data that has not been designed for joint inversion, there are some differences in spatial coverage that pose a challenge for joint inversion. For example, the area of highest density MT measurements does not overlap with the gravity measurements and is only crossed by two of the ten seismic lines. Furthermore, some of the older magnetotelluric measurements only span a relatively narrow frequency range of less than two decades resulting in limited depth resolution. With an experimental setup designed for joint inversion, we therefore expect higher resolution results than presented below.

One unique feature of our group of collaborators is that we have two independently developed joint inversion algorithms with different underlying philosophies. The joint inversion framework jif3D, which is used for the previously presented sub-salt study, is geared towards large-scale 3D joint inversion and designed for parallel execution and minimum memory usage even for large models. It therefore uses a memory efficient limited-memory quasi-Newton optimization approach, L-BFGS [*Nocedal and Wright*, 1999]. One consequence of this choice of optimization approach is that the weighting for the different datasets, the regularization, and coupling constraints has to be constant throughout one inversion run in order to guarantee local convergence. This means that the weights have to be determined by trial-and-error through experimentation, and the effectiveness of this experimentation is largely determined by the experience of the user.

In contrast, the second joint inversion algorithm, jinv2D, is limited to two-dimensional magnetotelluric modeling. Thus the joint inversion is always performed along profiles even though seismic and gravity modeling can be performed in three dimensions. However, the relative weighting between each dataset, the regularization, and the coupling constraints are adjusted at each inversion step, reducing the number of user-defined parameters to a minimum. A detailed description of the approach and its implementation are beyond the scope of this contribution, but the general ideas are described in *Heincke et al.* [2010] and *Heincke et al.* [2014].

Two core ideas are used to reduce the number of user-defined weights and ensure smooth convergence. As a first step, instead of specifying a single objective function for all dataset, coupling, and regularization terms as we describe in *Moorkamp et al.* [2011], the problem is split into separate objective functions associated with each geophysical dataset. Using seismic tomography and magnetotellurics as an example, we would have two objective functions

$$\Phi_{\sigma}^{seis}(\mathbf{v}) = \Phi^{data}(\mathbf{v}) + \lambda_{seis}\Phi^{reg}(\mathbf{v}) + \nu_{seis}\Phi^{coupling}(\mathbf{v},\sigma), \quad (9.1)$$

$$\Phi_{\mathbf{v}}^{MT}(\sigma) = \Phi^{data}(\sigma) + \lambda_{MT}\Phi^{reg}(\sigma) + \nu_{MT}\Phi^{coupling}(\sigma,\mathbf{v}). \quad (9.2)$$

Here Φ^{data}, Φ^{reg} and $\Phi^{coupling}$ are the data misfit, regularization, and coupling terms of the objective function, respectively, and $\mathbf{v}$ and σ are the model vectors for velocity and conductivity, respectively. Each objective function is minimized separately, but at each step the conductivity and velocity information is exchanged between the two optimization problems by means of the coupling constraints that depend on both physical parameters at the same time. Such an approach has also been suggested by *Haber and Holtzman Gazit* [2013] to improve the convergence of the inversion and has been used by *Hu et al.* [2009] for their joint inversion of seismic and EM data. As the model information is exchanged at each nonlinear optimization step, this approach is a variant of joint inversion and different from cooperative inversion approaches where information is only exchanged after performing a full inversion on one dataset (e.g., *Um et al.* [2014]).

Apart from the improved convergence properties hypothesized by *Haber and Holtzman Gazit* [2013], splitting the objective function also removes the necessity to weight the different datasets within the joint inversion. Even with the separated objective functions, there are two weights left to be determined for each objective function: the regularization weight λ and the coupling weight ν. For the regularization, weights are typically determined using an ℓ-curve approach [*Hansen*, 1992], generalized cross-validation (e.g., see *Farquharson and Oldenburg* [2004], or within the inversion using an OCCAM style algorithm [*Constable et al.*, 1987]. For the coupling weight, little experience exists, but our experiments with synthetic data [*Moorkamp et al.*, 2011] and the sub-salt imaging example presented above suggest that at least for the cross-gradient a change of a factor of 10 does not create a significant difference.

Our approach used in the 2D inversions shown below compares the reduction in misfit for an individual inversion iteration step with the reduction in misfit including the coupling constraint. The algorithm aims at keeping the improvement of the misfit with the joint inversion within a certain fraction, typically 0.5–0.9, of the individual inversion. Adjusting the coupling weights ν so that this criterion is satisfied and using the history of coupling weights from previous iterations to estimate a weight for the current estimation results in a stable algorithm that works well in practice [*Heincke et al.*, 2010, 2014]. For the regularization we use the algorithm proposed by *Lelièvre et al.* [2012] to adjust the weights λ.

One experience we had with this dataset regardless of the joint inversion approach used is that structural coupling alone does not provide a strong enough connection between the different datasets to make a significant impact on the inversion results. Compared to individual inversions, the joint inversion results using a cross-gradient approach do not differ significantly. This is in stark contrast to our results for the salt-dome data where the cross-gradient works very well. One major difference between the two scenarios is that for the salt dome the seismic tomography data constrains the shape of the salt dome at the top as well as on the side; that is, the structure is, at least partially, laterally constrained. This puts lateral constraints on the resistivity and helps to resolve the shape of the resistor. For the sub-basalt case, the seismic tomography data only constrains the depth to the top of the basalt which varies relatively little within the study area. All structures below have to be resolved by MT and gravity with little lateral constraint because the seismic tomography has no ray coverage here. As the basaltic layer is relatively thin considering the resolution of MT at this depth and the resistivities of the sediments above and below the basalt are comparable, MT is only moderately sensitivity to the location of the lower boundary of the basalt. With no structural constraint from seismic tomography below the top of the basalt, the results from structural joint inversion consequently do not differ much from the individual inversions. A similar issue is shown in Chapter 8 for a mineral exploration study.

In addition, the resistivity of a resistive layer below a conductor is not well constrained by MT, and we can only confidently recover the minimum permissible resistivity [*Parker*, 1983]. The cross-gradient constraint only facilitates the appearance of a coincident boundary in the velocity and conductivity models but is explicitly designed not to prescribe a magnitude of change across the boundary. The travel-time data have sensitivity to the velocity at the top of the basaltic layer. Therefore using a parameter relationship between velocity and resistivity will calibrate the resistivity within the basalt and, in turn, assists in resolving the lower boundary of the basalt with MT.

The resolution to the resistive basalt could further be improved by incorporating CSEM data [*Panzner et al.*, 2014]. For this case study we did not have access to such data and we currently do not have the capability to include CSEM data in the joint inversion. This is planned for the near future though (see *Sommer et al.* [2013]), but will significantly increase the computational demand (e.g., see *Um et al.* [2014]) and require considering electrical anisotropy in the inversion [*Newman et al.*, 2010].

In order to achieve adequate coupling, we use logging data from the BRUGDAN borehole to extract functions that describe the cross-property relationships (Figure 9.9). These are

$$\log_{10}\rho_r = 7.876\cdot 10^{-8} v_p^2 - 0.1512, \tag{9.3}$$

$$\rho_d = 1.737\cdot 10^{-4} v_p + 1.868. \tag{9.4}$$

Here ρ_r is electrical resistivity, v_p is p-wave velocity, and ρ_d is density. As can be seen in Figure 9.9, these relationships capture the general variations of the cross-property plots. Still, there is a considerable amount of scatter and at high velocities the electrical resistivity in the borehole data is larger than estimated from the functional relationship. *Panzner et al.* [2014] use the same resistivity and velocity log data, but derive a different relationship

$$v_p(\rho) = \begin{cases} 2118\log_{10}\rho_r + 1869.5 & : \quad v_p < 5000 \ m/s \\ \dfrac{-1239}{\log_{10}\rho_r - 0.701} + 6604 & : \quad v_p \geq 5000 \ m/s \end{cases}, \tag{9.5}$$

that captures the steep rise in resistivity. These differences in parametrization highlight the ad hoc nature of the currently used coupling approaches. In the absence of a comprehensive theory of cross-property relationships for a wide range of Earth materials, defining suitable parameter relationships is more of an art with a relatively high degree of arbitrariness. As the differences in cross-property relationships will impact on the joint inversion results, more systematic and rigorous approaches to construct such relationships are needed. A possible solution could be the statistical approaches by *Bosch* [2004] and *Chen and Hoversten* [2012] discussed above. In any case, systematic studies are required that quantify the effect of different cross-property relationships on the joint inversion results for real data and complex geometries.

Figure 9.10 shows the joint inversion results along the Flare 6 line using our original parameter relationship and the 2D code jinv2D with automatic weighting. We show electrical resistivity in the area around the BRUGDAN borehole together with a cross section extracted from 3D seismic reflection data. As the seismic reflection data was only provided to us with time, as opposed to depth, as the vertical axis, we convert the depth of each cell in the resistivity model to two-way travel time using the velocity model retrieved from the joint inversion. It is important to note that no information from the seismic reflection data was used in the joint inversion, and thus this comparison provides a benchmark for our results. We see excellent agreement between the reflection associated with the top of the basalt and the change from conductive sediments to resistive basalt in the joint inversion model. This is because the first arrival travel times used in the inversion are also highly sensitive to the position of the basalt and its velocity. Through the velocity–resistivity relationship, this information is transferred to the

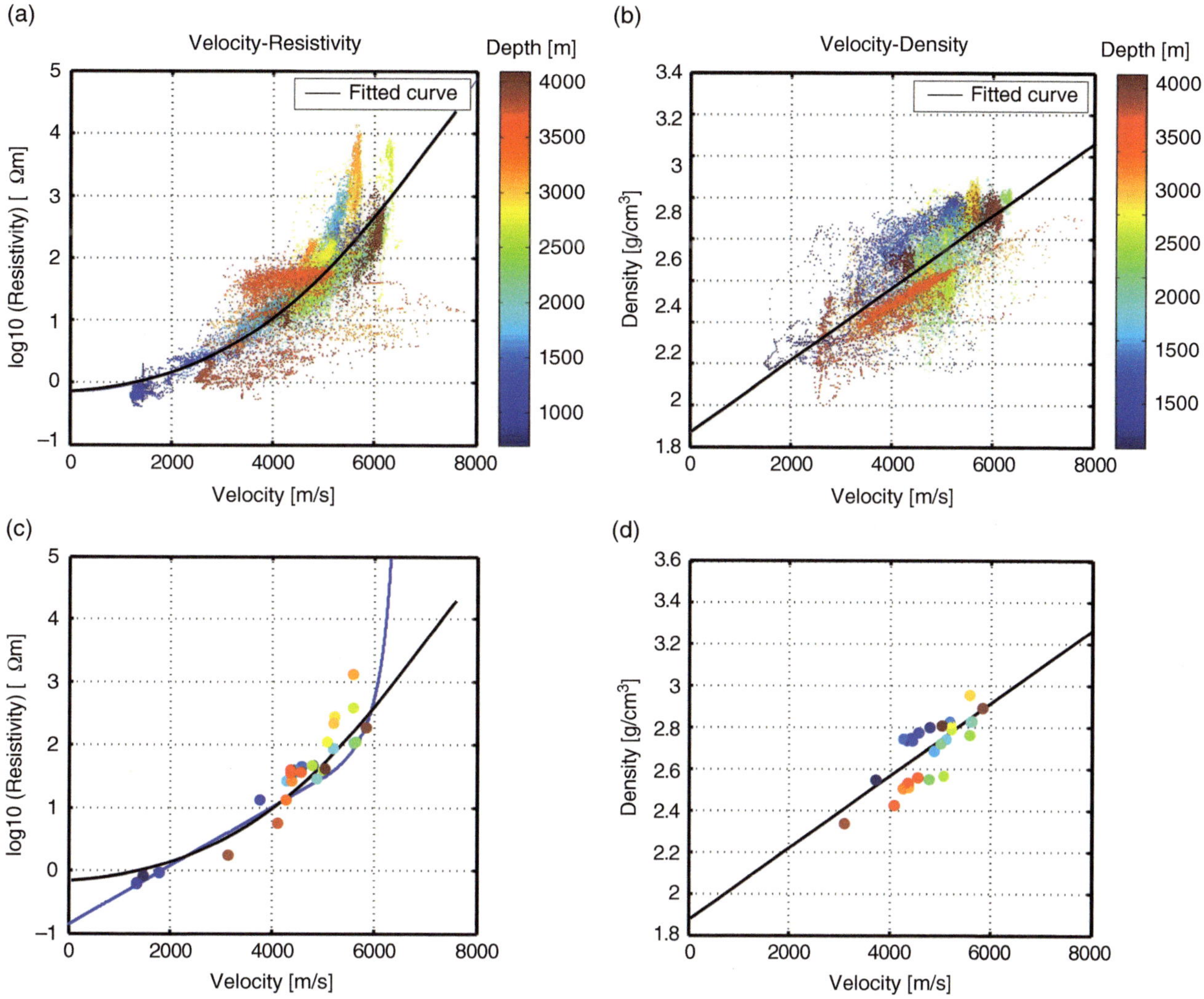

Figure 9.9 Velocity–resistivity **(left)** and velocity–density **(right)** cross plots for logging data from the BRUGDAN borehole. The color of each point corresponds to depth below sea floor. We show the raw logging data **(top)** and the data averaged in intervals of 100 m **(bottom)** to make it comparable to the joint inversion results. We also show our estimated cross-property relationships (*black lines*) and the velocity–resistivity relationship used by *Panzner et al.* [2014] (*blue line*).

resistivity model. The fact that the strong reflection and the top of the resistor coincide after conversion to two-way travel time suggests that the retrieved velocities are representative of the sediments and the top of the basalt.

Below the very top of the basalt layer the travel-time data have no sensitivity and all information about deeper structures is provided by MT and gravity data. Compared to individual inversions of such data, the resistivities and densities have been calibrated with respect to the velocities at the top of the basalt through the parameter relationships. In this region below the top basalt, there is a quasi-horizontal highly resistive, high-velocity, and high-density anomaly with a thickness of about 2 km. The high values of physical parameters of this anomaly are in good agreement with the physical properties from the geophysical logs from the BRUGDAN borehole. We observe that, after average filtering of the logging data (the length of the window for averaging is 100 m and corresponds to the cell sizes of the inversion in z-direction), the values from the logs and from the joint inversion are generally similar (Figure 9.11). We used the borehole log to construct the conductivity–velocity relationship used in the joint inversion, so we can expect some degree of similarity. However, we did not utilize the depth dependency of the borehole log as a constraint, thus the comparison provides a benchmark for the inversion results. Only short-wavelength undulations with $\lambda < 200$–300 m in the logging data cannot be identified in the joint inversion results. The reasons for this are both the limited resolution of the joint inversion and deviations of the

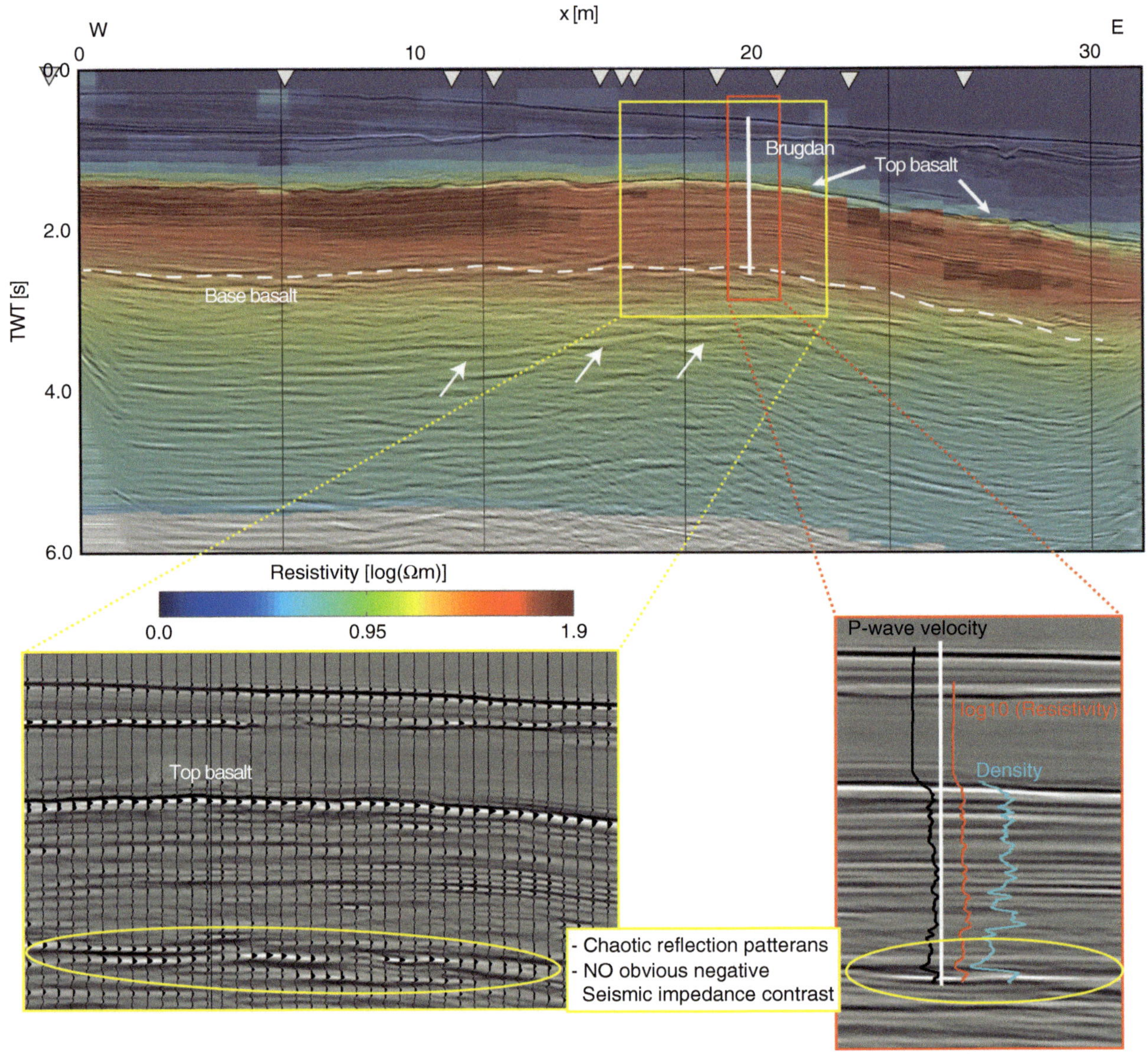

Figure 9.10 Comparison of 3-D reflection seismic data and the joint inversion results along the profile Flare 6. **Top:** A cross section in WNW-SES direction is extracted from the 3-D reflection seismic dataset (see Figure 9.8) and the resistivity distributions from the joint inversion is superimposed. Because of the lack of proper velocity information the *z*-axis of the seismic section is expressed in time. The resistivity model is converted from depth to time using the velocity model from the joint inversion. The *gray triangles* indicate the locations of the used MT stations. Based on the combined interpretation of the borehole data, the reflection seismic section and joint inversion results the base basalt is approximated (see the *white dashed line*). *White arrows* mark a second pattern of discontinuous reflections. **Bottom:** To the left, a part of the seismic section is shown in more detail and to the right logging results from the BRUGDAN borehole are plotted. Note that the base basalt does not have a clear negative seismic impedance contrast as one would expect for a boundary between basaltic and sedimentary rocks.

real rock property relationships from the used parameter relationships for some of the depth intervals (see Figure 9.9).

The lower boundary of the horizontal high resistive anomaly is significantly smoother than the upper boundary (see Figure 9.9), indicating that the resolution of the joint inversion is lower here. Nonetheless, the location of the lower boundary fits well with the depth of the base basalt in the BRUGDAN borehole. In addition, a pattern of discontinuous reflections is present in the 3D seismic dataset at TWTs of about 2.2 to 3.0 s (see Figure 9.10) that coincides well with the lower boundary of the highly

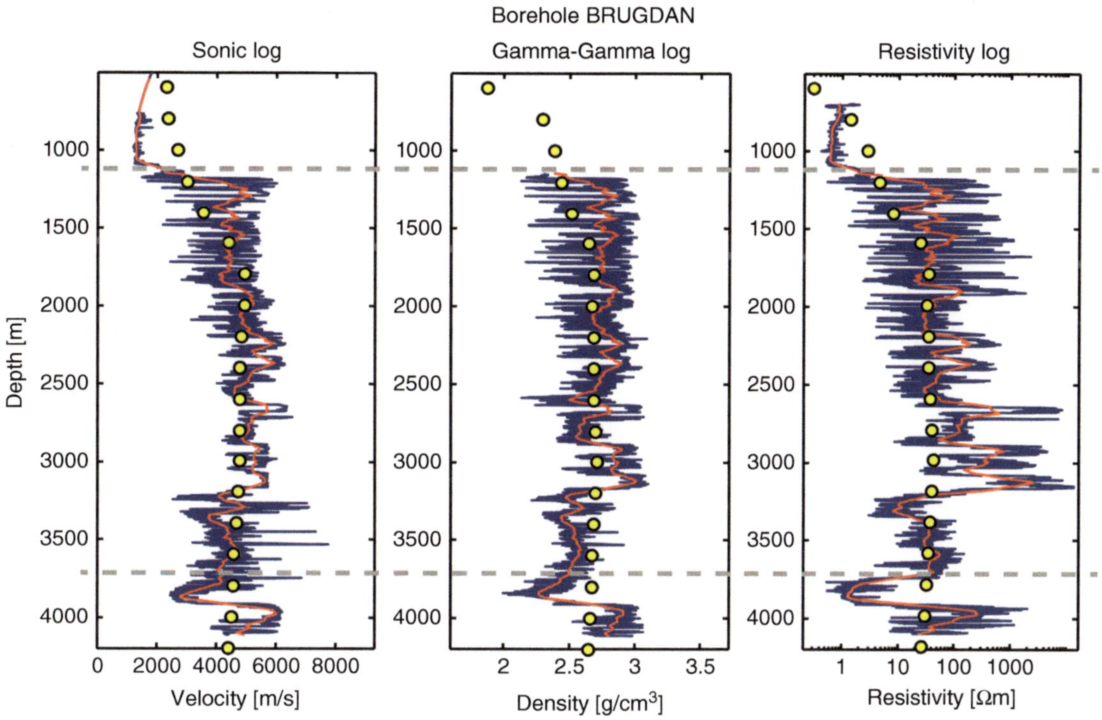

Figure 9.11 Comparison of the raw logging data from the BRUGDAN borehole (*blue line*), the logging data averaged with a 100-m window (*red line*), and the joint inversion results along the borehole trace (*yellow dots*). Considering the differences in scale, we generally observe a good agreement between the joint inversion results and the averaged borehole log (see also discussion in text).

resistive anomalies. *Planke et al.* [2000] mention that it is common that the base basalt is characterized by such discontinuous features in reflection seismic data. Finally, other wide-angle seismic studies [*Fliedner and White*, 2003; *Spitzer et al.*, 2003] and the CSEM-MT joint inversion study of *Panzner et al.* [2014] obtain similar shapes and thickness for the basalt in this region. To evaluate the reliability of the location of the lower boundary, our joint inversion was tested with different starting models. Results from these tests show that the uncertainty of the lower boundary is in the range of about 300–500 m, provided that the parameter relationships are a good representation of the true rock property relationships. Using this combination of data, it is not possible to reliably resolve the thickness of the underlying sediment sequence and the depths of the pre-rifted basement. No highly resistive structure could be identified at larger depths that could be associated with the pre-rifted basement. Based on these results, we conclude that it is required to add more information to the joint inversion—for example, considering wide angle reflection onsets from the pre-rifted basement to increase the resolution in the depth range > 4000 m or having larger periods for the magnetotelluric data.

9.4. CONCLUSIONS AND OUTLOOK

As the previous discussion demonstrates, there is considerable activity in developing new joint and cooperative inversion approaches for hydrocarbon exploration fueled by the need of extracting resources from complex areas. Commercially, maximizing resolution and extracting velocity information to migrate seismic reflection data appears to be most profitable. At the same time the use of cross-property relationships, petrophysical models, and the development of new modeling approaches such as emulation has an impact on joint inversion in other areas of application. The potentially high economic value of new joint inversion approaches has the advantage that significant investment goes into developing them. As a consequence, highly advanced algorithms combining full waveform seismic data with CSEM for time-lapse inversion, for example, have emerged. A disadvantage of this high economic value is that the development activity is not always reflected in the number of publications and systematic studies with real datasets are surprisingly sparse. Still, it is probably fair to say that joint inversion techniques have established themselves in hydrocarbon exploration although they are still far from being mainstream. A major obstacle to more widespread adoption of joint inversion approaches is the considerable effort required in obtaining high-quality results. For a successful joint inversion, it is necessary to invert each dataset, construct different coupling approaches, and evaluate the impact on the final results. This is often incompatible with the timescales for making a drilling decision. Thus one of the main challenges for joint inversion in the near future is to develop practical workflows that yield reliable results for a wide range of environments. Such a development needs to be based on experience with different datasets from different areas, and thus we expect to see the focus of activity shift from new technique developments to case studies that address some of the issues discussed above. In the longer term and with the experience gained through such studies, it should be possible to integrate all acquired data: borehole logs, potential field, and electromagnetic and seismic data. With appropriate workflows in place, the increase in information gained through such an integrated analysis has the potential to transform the way hydrocarbon exploration is performed in the future.

ACKNOWLEDGMENTS

We would like to thank the sponsors of the JIBA project Chevron, ExxonMobil, Nexen, RWE Dea, Shell, Statoil, and Wintershall for supporting the work shown in the sub-salt case study. The sub-basalt study was supported by the SINDRI consortium and data provided by Statoil. The joint inversions were run using the ALICE High-Performance Computing Facility at the University of Leicester.

REFERENCES

Abubakar, A., G. Gao, T. M. Habashy, and J. Liu (2012a), Joint inversion approaches for geophysical electromagnetic and elastic full-waveform data, *Inverse Problems*, *28*(5), 055,016.

Abubakar, A., M. Li, Y. Lin, and T. Habashy (2012b), Compressed implicit jacobian scheme for elastic full-waveform inversion, *Geophys. J. Int.*, *189*(3), 1626–1634.

Adriano, L., P. Menezes, and A. Cunha (2014), Tectonic framework of the barra de Sao Joa graben, campos basin, Brazil: Insights from gravity data interpretation, *Interpretation*, *2*(4), SJ65–SJ74, doi:10.1190/INT-2014-0011.1.

Avdeeva, A., M. Commer, and G. A. Newman (2017), Hydrocarbon reservoir detectability study for marine CSEM methods: Time domain versus frequency domain, *SEG Technical Program Expanded Abstracts*, *26*(1), 628–632, doi:10.1190/1.2792497.

Avdeeva, A. D., D. B. Avdeev, and M. Jegen (2012), Detecting a salt dome overhang with magnetotellurics: 3D inversion methodology and synthetic model studies, *Geophysics*, *77*(4), E251–E263.

Averill, M., G. Keller, K. Miller, P. Sroda, T. Bond, and A. Velasco (2006), Data fusion in geophysics: Seismic tomography and crustal structure in poland as an example, *Special Paper of the Geological Society of America*, *397*, 153–168, doi:10.1130/2006.2397(11).

Booker, J. R. (2014), The magnetotelluric phase tensor: A critical review, *Surv. Geophys.*, *35*(1), 7–40.

Bosch, M. (2014), The optimization approach to lithological tomography: Combining seismic data and petrophysics for porosity prediction, *Geophysics*, *69*, 1272–1282, doi:10.1190/1.1801944.

Bosch, M., T. Mukerji, and E. Gonzalez (2010), Seismic inversion for reservoir properties combining statistical rock physics and geostatistics: A review, *Geophysics*, *75*(5), 75A165–75A176, doi:10.1190/1.3478209.

Carcione, J. M., B. Ursin, and J. I. Nordskag (2007), Cross-property relations between electrical conductivity and the seismic velocity of rocks, *Geophysics*, *72*(5), E193–E204, doi:10.1190/1.2762224.

Chave, A., and A. Jones (2012), *The Magnetotelluric Method*, Cambridge University press, Cambridge, United Kingdom.

Chave, A. D., and D. J. Thomson (2004), Bounded influence magnetotelluric response function estimation, *Geophys. J. Int.*, *157*, 988–1006.

Chen, J., Hoversten, G. M., Vasco, D., Rubin, Y., & Hou, Z. (2007). A Bayesian model for gas saturation estimation using marine seismic AVA and CSEM data. *Geophysics*, *72*(2), WA85–WA95.

Chen, J., and G. M. Hoversten (2012), Joint inversion of marine seismic ava and csem data using statistical rock-physics models and markov random fields, *Geophysics*, *77*(1), R65–R80.

Christie, P. A., and R. S. White (2008), Imaging through Atlantic margin basalts: An introduction to the sub-basalt mini-set, *Geophys. Prospect.*, *56*(1), 1–4, doi:10.1111/j.1365-2478.2007.00676.x.

Colombo, D., M. Cogan, S. Hallinan, M. Mantovani, M. Virgilio, and W. Soyer (2008), Near-surface p-velocity modelling by integrated seismic, em, and gravity data: Examples from the middle east, *First Break*, *26*, 91–102.

Colombo, D., T. Keho, et al. (2010), The non-seismic data and joint inversion strategy for the near surface solution in Saudi Arabia, in 2010 SEG Annual Meeting, Society of Exploration Geophysicists, Tulsa, OK.

Colombo, D., G. McNeice, N. Raterman, M. Zinger, D. Rovetta, and E. Sandoval Curiel (2014), Exploration beyond seismic: The role of electromagnetics and gravity gradiometry in deep water subsalt plays of the Red Sea, *Interpretation*, *2*(3), SH33–SH53.

Constable, S. (2010), Ten years of marine csem for hydrocarbon exploration, *Geophysics*, *75*(5), 75A67–75A81, doi:10.1190/1.3483451.

Constable, S., and C. J. Weiss (2006), Mapping thin resistors and hydrocarbons with marine EM methods: Insights from 1D modeling, *Geophysics*, *71*, G43–G51.

Constable, S. C., R. L. Parker, and C. G. Constable (1987), Occam's inversion: A practical algorithm for generating smooth models from electromagnetic sounding data, *Geophysics*, *52*(3), 289–300.

Craig, P., M. Goldstein, A. Seheult, and J. Smith (1997), Pressure matching for hydrocarbon reservoirs: A case in the use of Bayes linear strategies for large computer experiments (and discussion), in *Case Studies in Bayesian Statistics*, Springer-Verlag, New York, pp. 37–93.

De Stefano, M., F. Golfr Andreasi, S. Re, M. Virgilio, and F. Snyder (2011), Multiple-domain, simultaneous joint inversion of geophysical data with application to subsalt imaging, *Geophysics*, *76*(3), R69–R80, doi:10.1190/1.3554652.

Fanavoll, S., S. Ellingsrud, P. T. Gabrielsen, R. Tharimela, and D. Ridyard (2012), Exploration with the use of em data in the barents sea: The potential and the challenges, *First Break*, *30*(4), 89–96.

Fanavoll, S., P. T. Gabrielsen, and S. Ellingsrud (2014), CSEM as a tool for better exploration decisions: Case studies from the barents sea, Norwegian continental shelf, *Interpretation*, *2*(3), SH55–SH66.

Farquharson, C. G., and D. W. Oldenburg (2004), A comparison of automatic techniques for estimating the regularization parameter in non-linear inverse problems, *Geophys. J. Int.*, *156*(3), 411–425.

Fliedner, M. M., and R. S. White (2003), Depth imaging of basalt flows in the Faeroe–Shetland Basin, *Geophys. J. Int.*, *152*(2), 353–371.

Fruehn, J. R., M. M. Fliedner, and R. S. White (2001), Integrated wide-angle and near-vertical subbasalt study using large-aperture seismic data from the Faroe–Shetland region, *Geophysics*, *66*, 1340–1348.

Gallardo-Delgado, L. A., M. A. Pérez-Flores, and E. Gómez-Treviño (2003), A versatile algorithm for joint 3d inversion of gravity and magnetic data, *Geophysics*, *68*(3), 949–959.

Gao, G., A. Abubakar, and T. Habashy (2012), Joint petrophysical inversion of electromagnetic and full-waveform seismic data, *Geophysics*, *77*(3), WA3–WA18.

Giraud, J., M. De Stefano, and F. Miotti (2013), Simultaneous joint inversion of electromagnetic and seismic full-waveform data—A sensitivity analysis to biot parameter, in *75th EAGE Conference & Exhibition Incorporating SPE EUROPEC 2013*.

Haber, E., and M. Holtzman Gazit (2013), Model fusion and joint inversion, *Surv. Geophys.*, *34*(5), 675–695, doi:10.1007/s10712-013-9232-4, 2013.

Hackney, R., J. Goodwin, L. Hall, K. Higgins, N. Holzrichter, S. Johnston, M. Morse, G. Nayak, and P. Petkovic (2015), Potential-field data in integrated frontier basin geophysics: Successes and challenges on Australia's continental margin, *Marine Petrol. Geol.*, *59*(0), 611 – 637.

Hansen, P. C. (1992), Analysis of discrete ill-posed problems by means of the ℓ-curve, *SIAM Rev.*, *34*(4), 561–580.

Haugen, J. A., B. Arntsen, and J. Mispel (2009), Modeling of dirty salt, *SEG Technical Program Expanded Abstracts 2008*, *426*, 2127–2131, doi:10.1190/1.3059308.

Heincke, B., M. Jegen, and R. Hobbs (2006), Joint inversion of MT, gravity and seismic data applied to sub-basalt imaging, *SEG Technical Program Expanded Abstracts*, *25*(1), 784–789, doi:10.1190/1.2370374.

Heincke, B., M. Jegen, Moorkamp, J. M., Chen, and R. Hobbs (2010), Adaptive coupling strategy for simultaneous joint inversions that use petrophysical information as constraints, *SEG Technical Program Expanded Abstracts*, *29*(1), 2805–2809.

Heincke, B., M. Jegen, M. Moorkamp, and R. W. Hobbs (2014), Joint-inversion of magnetotelluric, gravity and seismic data to image sub-basalt sediments offshore the Faroe-Islands, *SEG Technical Program Expanded Abstracts 2014*, 147, 770–775, doi:10.1190/segam2014-1401.1.

Hinze, W. J., R. R. B. von Frese, and A. H. Saad (2013), *Gravity and Magnetic Exploration*, Cambridge University Press, New York.

Hokstad, K., et al. (2011), Joint imaging of geophysical data: Case history from the Nordkapp Basin, Barents Sea, *SEG Technical Program Expanded Abstracts*, *30*(1), 1098–1102.

Hoversten, G., R. Gritto, J. Washbourne, and T. Daley (2003), Pressure and fluid saturation prediction in a multicomponent reservoir using combined seismic and electromagnetic imaging, *Geophysics*, *68*(5), 1580–1591, doi:10.1190/1.1620632.

Hoversten, G. M., H. F. Morrison, and S. C. Constable (1998), Marine magnetotellurics for petroleum exploration, Part II: Numerical analysis of subsalt resolution, *Geophysics*, *63*, 826–840.

Hoversten, G. M., S. C. Constable, and H. F. Morrison (2000), Marine magnetotellurics for base-of-salt mapping: Gulf of Mexico field test at the Gemini structure, *Geophysics*, *65*, 1476–1488.

Hoversten, G. M., F. Cassassuce, E. Gasperikova, G. A. Newman, J. Chen, Y. Rubin, Z. Hou, and D. Vasco (2006), Direct reservoir parameter estimation using joint inversion of marine seismic AVA and CSEM data, *Geophysics*, *71*, C1+, doi:10.1190/1.2194510.

Hu, W., A. Abubakar, and T. M. Habashy (2009), Joint electromagnetic and seismic inversion using structural constraints, *Geophysics*, *74*, R99–R109, doi:10.1190/1.3246586.

Jegen, M. D., R. W. Hobbs, P. Tarits, and A. Chave (2009), Joint inversion of marine magnetotelluric and gravity data incorporating seismic constraints. Preliminary results of sub-basalt imaging off the Faroe Shelf, *Earth and Planetary Sci. Lett.*, *282*, 47–55, doi:10.1016/j.epsl.2009.02.018.

Ji, S., T. Huang, K. Fu, and Z. Li (2011), Dirty salt velocity inversion: The road to a clearer subsalt image, *Geophysics*, *76*(5), WB169–WB174.

Johansen, S. E., H. E. F. Amundsen, T. Rosten, S. Ellingsrud, T. Eidesmo, and A. H. Bhuyian (2005), Subsurface hydrocarbons detected by electromagnetic sounding, *First Break*, *23*, 31–36.

Jones, A. G., A. D. Chave, G. Egbert, D. Auld, and K. Bahr (1989), A comparison of techniques for magnetotelluric response function estimation, *J. Geophys. Res.*, *94*, 14,201–14,213.

Jones, I., and I. Davison (2014), Seismic imaging in and around salt bodies, *Interpretation*, *2*(4), SL1–SL20, doi:10.1190/INT-2014-0033.1.

Julia, J., C. J. Ammon, R. B. Herrmann, and A. M. Correig (2000), Joint inversion of receiver function and surface wave dispersion observations, *Geophys. J. Int.*, *143*(1), 99–112.

Key, K., S. C. Constable, and C. Weiss (2006), Mapping 3D salt using the 2D marine magnetotelluric method: Case study from Gemini Prospect, Gulf of Mexico, *Geophysics*, *71*, B17–B27.

Kiyan, D., A. G. Jones, and J. Vozar (2013), The inability of magnetotelluric off-diagonal impedance tensor elements to sense oblique conductors in three-dimensional inversion, *Geophys. J. Int.*, p. ggt470.

Ladzekpo, D. H., K. Sekharan, G. H. Gardner, et al. (1988), Physical modeling for hydrocarbon exploration, in *1988 SEG Annual Meeting*, Society of Exploration Geophysicists, Tulsa, OK.

Latimer, R. B., R. Davidson, and P. Van Riel (2000), An interpreter's guide to understanding and working with seismic-derived acoustic impedance data, *The Leading Edge*, *19*(3), 242–256.

Lelièvre, P. G., C. G. Farquharson, and C. A. Hurich (2012), Joint inversion of seismic travel-times and gravity data on unstructured grids with application to mineral exploration, *Geophysics*, *77*, K1, doi:10.1190/geo2011-0154.1.

Li, Y., and D. W. Oldenburg (1998), 3-D inversion of gravity data, *Geophysics*, *63*, 109–119, doi:10.1190/1.1444302.

Liang, L., A. Abubakar, T. Habashy, et al. (2012), Joint inversion of time-lapse crosswell electromagnetic, seismic, and production data for reservoir monitoring and characterization, in *2012 SEG Annual Meeting*, Society of Exploration Geophysicists.

Lien, M. (2013), Simultaneous joint inversion of amplitude-versus-offset and controlled-source electromagnetic data by implicit representation of common parameter structure, *Geophysics*, *78*(4), ID15–ID27.

Linde, N., A. Tryggvason, J. E. Peterson, and S. S. Hubbard (2008), Joint inversion of crosshole radar and seismic traveltimes acquired at the South Oyster Bacterial Transport Site, *Geophysics*, *73*, G29–G37, doi:10.1190/1.2937467.

MacGregor, L., and J. Tomlinson (2014), Marine controlled-source electromagnetic methods in the hydrocarbon industry: A tutorial on method and practice, *Interpretation*, *2*(3), SH13–SH32.

Manglik, A., and S. K. Verma (1998), Delineation of sediments below flood basalts by joint inversion of seismic and magnetotelluric data, *Geophys. Res. Lett.*, *25*(21), 4015–4018, doi:10.1029/1998GL900063, 1998.

Manglik, A., S. Verma, and K. Singh (2009), Detection of sub-basaltic sediments by a multi-parametric joint inversion approach, *J. Earth Syst. Sci.*, *118*(5), 551–562, doi:10.1007/s12040-009-0043-4.

Mantovani, M., M. Clementi, and F. Ceci (2013), Use of simultaneous joint inversion as a maximum concordance solver for statics, in *75th EAGE Conference & Exhibition incorporating SPE EUROPEC 2013*.

Martini, F., R. Hobbs, C. Bean, and R. Single (2005), A complex 3d volume for sub-basalt imaging, *First Break*, *23*(7).

McKay, M. D., R. J. Beckman, and W. J. Conover (1979), Comparison of three methods for selecting values of input variables in the analysis of output from a computer code, *Technometrics*, *21*(2), 239–245.

Meier, U., A. Curtis, and J. Trampert, Global crustal thickness from neural network inversion of surface wave data, *Geophys. J. Int.*, *169*(2), 706–722, 2007.

Mittet, R., and J. P. Morten (2012), Detection and imaging sensitivity of the marine csem method, *Geophysics*, *77*(6), E411–E425.

Moorkamp, M., A. G. Jones, and S. Fishwick (2010), Joint inversion of receiver functions, surface wave dispersion and magnetotelluric data, *J. Geophys. Res.*, *115*, B04,318.

Moorkamp, M., B. Heincke, M. Jegen, A. W. Roberts, and R. W. Hobbs (2011), A framework for 3-D joint inversion of MT, gravity and seismic refraction data, *Geophys. J. Int.*, *184*, 477–493, doi:10.1111/j.1365-246X.2010.04856.x.

Moorkamp, M., A. W. Roberts, M. Jegen, B. Heincke, and R. W. Hobbs (2013), Verification of velocity-resistivity relationships derived from structural joint inversion with borehole data, *Geophys. Res. Lett.*, *40*(14), 3596–3601, doi:10.1002/grl.50696.

Moorkamp, M., A. Avdeeva, and K. Tietze (2014), Some thoughts on measures of misfit in magnetotelluric inversion, in *22nd EM Induction Workshop Weimar, Germany*.

Newman, G. A., M. Commer, and J. J. Carazzone (2010), Imaging CSEM data in the presence of electrical anisotropy, *Geophysics*, *75*, 51–61, doi:10.1190/1.3295883.

Nocedal, J., and S. J. Wright (1999), *Numerical optimization*, Springer-Verlag, New York.

Panzner, M., W. W. Weibull, and J. P. Morten (2014), Sub-basalt imaging in the Faroe-Shetland Basin using CSEM and MT data to constrain the velocity model, *SEG Technical Program Expanded Abstracts 2014*, *727*, 3806–3810, doi:10.1190/segam2014-0715.1.

Parker, R. L. (1983), The magnetotelluric inverse problem, *Geophys. Surv.*, *6*, 5–25, doi:10.1007/BF01453993.

Planke, S., P. A. Symonds, E. Alvestad, and J. Skogseid (2000), Seismic volcanostratigraphy of large-volume basaltic extrusive complexes on rifted margins, *J. Geophys. Res.: Solid Earth (1978–2012)*, *105*(B8), 19,335–19,351, 2000.

Presho, M., A. Protasov, and E. Gildin (2014), Local–global model reduction of parameter-dependent, single-phase flow models via balanced truncation, *J. Comput. Appl. Mathe.*, *271*, 163–179, doi:10.1016/j.cam.2014.03.022.

Press, F. (1968), Earth models obtained by Monte Carlo inversion, *J. Geophys. Res.* *73*(16), 5223–5234, doi:10.1029/JB073i016p05223.

Ray, A., and K. Key, Bayesian inversion of marine CSEM data with a trans-dimensional self parametrizing algorithm, *Geophys. J. Int.*, *191*(3), 1135–1151, doi:10.1111/j.1365-246X.2012.05677.x, 2012.

Ray, A., K. Key, T. Bodin, D. Myer, and S. Constable (2014), Bayesian inversion of marine csem data from the Scarborough gas field using a transdimensional 2-d parametrization, *Geophys. J. Int.*, *199*(3), 1847–1860, doi:10.1093/gji/ggu370.

Roberts, A., R. Hobbs, M. Goldstein, M. Moorkamp, M. Jegen, and B. Heincke (2012), Crustal constraint through complete model space screening for diverse geophysical datasets facilitated by emulation, *Tectonophysics*, *572*, 47–63.

Roberts, A., R. Hobbs, M. Goldstein, M. Moorkamp, M. Jegen, and B. Heincke (2015), Joint stochastic constraint of a large dataset from a salt-dome, *Geophysics*, accepted.

Schowalter, T. T. (1979), Mechanics of secondary hydrocarbon migration and entrapment, *AAPG Bull.*, *63*(5), 723–760.

Schuler, J., P. A. F. Christie, and R. S. White (2012), Seismic attenuation of flood basalts in the Brugdan and William Wells and stratigraphic correlation on the Faroe Shelf, *EAGE 74th Conference & Exhibition, Copenhagen, Denmark*.

Singh, R. (2014), Exploration application of seismic amplitude analysis in the Krishna–Godavari Basin, east coast of India, *Interpretation*, *2*(4), SP5–SP20, doi:10.1190/INT-2013-0197.1.

Sommer, M., S. Hlz, M. Moorkamp, A. Swidinsky, B. Heincke, C. Scholl, and M. Jegen (2013), GPU parallelization of a three dimensional marine CSEM code, *Comput. Geosci.*, *58*(0), 91–99.

Spitzer, R., R. S. White, and P. A. Christie (2003), Enhancing subbasalt reflections using parabolic τ–p transformation, *The Leading Edge*, *22*(12), 1184–1201.

Stadtler, C., C. Fichler, K. Hokstad, E. A. Myrlund, S. Wienecke, and B. Fotland (2014), Improved salt imaging in a basin context by high resolution potential field data: Nordkapp Basin, Barents Sea, *Geophys. Prospect.*, *62*(3), 615–630, doi:10.1111/1365-2478.12101.

Strack, K. (2014), Future directions of electromagnetic methods for hydrocarbon applications, *Surv. Geophys.*, *35*(1), 157–177.

Sun, J., Y. Li, et al. (2013), A general framework for joint inversion with petrophysical information as constraints, in *2013 SEG Annual Meeting*, Society of Exploration Geophysicists, Tulsa, OK.

Swidinsky, A., C. Scholl, and M. Jegen (2012), Determining reservoir and rock physics parameters with a petrophysically coupled joint inversion of CSEM and ava data, in *74th EAGE Conference & Exhibition*.

Tartaras, E., et al. (2011), Multi-property earth model building through data integration for improved subsurface imaging, *First Break*, *29*(4), 83–88.

Tietze, K., and O. Ritter (2013), Three-dimensional magnetotelluric inversion in practicethe electrical conductivity structure of the San Andreas fault in central California, *Geophys. J. Int.*, doi:10.1093/gji/ggt234.

Um, E., M. Commer, and G. Newman (2014), A strategy for coupled 3d imaging of large-scale seismic and electromagnetic data sets: Application to subsalt imaging, *Geophysics*, *79*(3), ID1–ID13, doi:10.1190/geo2013-0053.1.

White, M., and R. Gordon (2003), Deep imaging: New technology lowers cost of discovery, *Canadian Mining J.*, *124*(3), 27–28.

White, R. S., J. Fruehn, K. R. Richardson, E. Cullen, W. Kirk, J. R. Smallwood, and C. Latkiewicz (1999), Faroes large aperture research experiment (flare): Imaging through basalts., in *Geology of the Northwest European Continental Margin*, Geological Society of London.

Wiik, T., K. Hokstad, B. Ursin, and L. Mütschard (2013), Joint contrast source inversion of marine magnetotelluric and controlled-source electromagnetic data, *Geophysics*, *78*(6), E315–E327.

Yilmaz, O. (2001), *Seismic Data Analysis*, Society of Exploration Geophysicists, Tulsa, OK.

10

Imaging the Lithosphere and Upper Mantle: Where We Are At and Where We Are Going

Juan Carlos Afonso,[1] Max Moorkamp,[2] and Javier Fullea[3,4]

"…It will become clear that the simplicity of the inner Earth is only apparent; with the progress of [experimental and observational] techniques, we may perhaps expect that someday "physics of the interior of the Earth" will make as little sense as "physics of the crust"…"

J.-P. Poirier (1991, *Introduction to the Physics of the Earth's Interior*)

ABSTRACT

Hypotheses and conclusions concerning the physical state of the interior of the Earth are under constant debate. At least part of the controversy lies in the fact that traditionally studies of different nature (i.e., seismic, geochemical, electromagnetic, etc.), with very different spatial and temporal resolutions and sensitivities to the thermochemical structure of the Earth's interior, have been used in isolation to explain the same phenomena (e.g., temperature or velocity anomalies, magmatism, plate motion, strain partitioning, etc.). There is no a priori reason, however, why the results from these diverse studies should be strictly comparable, consistent, or compatible, despite sampling the same physical structure. In recent years, however, advances on computational power, inversion methods, and laboratory experimental techniques, as well as the dramatic increase on both quality and quantity of multiple geophysical and geochemical datasets, have created great interest on integrated (joint) multidisciplinary analyses capable of exploiting the complementary benefits of different datasets/methods. This chapter endeavors to provide a comprehensive review of the current state of the art in such integrated studies of the lithosphere and sublithospheric upper mantle, as well as of their benefits and limitations. Although important stand-alone (single-data) methods are briefly discussed, the emphasis is on forward, inverse, and probabilistic techniques that integrate two or more datasets into a formal joint analysis. The role of emerging trends for imaging the Earth's interior and their potential for elucidating the physical state of the planet are also discussed.

[1]*CCFS—Department of Earth and Planetary Sciences, Macquarie University, Sydney, New South Wales, Australia*

[2]*Department of Geology, University of Leicester, Leicester, United Kingdom*

[3]*Institute of Geosciences (CSIC, UCM), Madrid, Spain*

[4]*Dublin Institute for Advanced Studies, Dublin, Ireland*

10.1. INTRODUCTION

Most information on the physical state of the Earth's interior comes from the application of imperfect geophysical and geochemical theories to sparse observations made at the Earth's surface. Such theories must be

Integrated Imaging of the Earth: Theory and Applications, Geophysical Monograph 218, First Edition.
Edited by Max Moorkamp, Peter G. Lelièvre, Niklas Linde, and Amir Khan.

constantly refined to accommodate an ever-increasing amount of data arriving in the form of field observations and laboratory experiments. At least in part, the substantial increase in the quality and quantity of available datasets and processing power over the past few decades has been responsible for major changes in how we perceive and conceptualize the nature of the Earth's interior. For instance, the advent and development of global tomography has clearly demonstrated that the Earth's upper mantle, once thought to be relatively homogeneous, is highly heterogeneous at various length scales. Comprehensive studies involving large compilations of mantle samples (e.g., xenoliths, ophiolites, abyssal peridotites, etc.) and mantle-derived volcanic rocks provide further support for a complex thermochemical structure of the upper mantle and lithosphere. As a result, it is now widely accepted that the lithospheric and sublithospheric upper mantle are complex physicochemical systems that interact via mass and energy transfer processes over various length and time scales and that these interactions largely control the evolution of important tectonic and geological phenomena. The detection of these interactions and the imaging of the thermochemical structure of the Earth's interior, however, is far from straightforward and currently represent two of the most important and challenging goals of modern geophysics.

In this chapter, we will discuss some of the most relevant methods that are currently used to image the physical state of the lithosphere and upper mantle. While some single-data approaches will be mentioned due to their significant contribution to our understanding of the Earth's interior, we will focus on joint approaches that integrate complementary datasets. Due to length limitations, we cannot hope to include all relevant references in one chapter. We apologize to those authors whose work could not be directly cited.

10.2. GENERAL CONSIDERATIONS ABOUT THE LITHOSPHERE

The Earth's lithosphere consists of the entire crust (oceanic and continental) and a portion of the uppermost upper mantle. It is critical to humans, as most tectonic and biological activities on which modern society depends take place either within or at the boundaries of lithospheric plates. Examples are volcanic and seismic activity, mineralization events, and water and CO_2 recycling, among others. Although many definitions of lithosphere have been proposed in the literature (see Table 10.1), all of them can ultimately be related to the thermochemical state (i.e., temperature, stress field and composition) of the actual rocks making up the crust and uppermost mantle. This stems from the simple fact that important rock properties such as elastic moduli, electrical conductivity, strength, and viscosity are largely controlled by temperature, pressure, lithology (i.e., composition), and H_2O content (cf. *Ranalli* [1995] and *Karato* [2008]). Although rock fabric can in principle affect the strength of the lithosphere, the relative effect of fabric is of second-order compared to that of temperature and/or H_2O content. Therefore, most of the confusion around the use and meaning of different definitions of lithosphere is, in most cases, only apparent. Several review monographs have recently addressed the question of how to best define the lithosphere and the so-called lithosphere–asthenosphere boundary (LAB), as well as the relation between different definitions (e.g., see *Eaton et al.* [2009], *Artemieva* [2009, 2011], and *Kind et al.* [2012]). We emphasize,

Table 10.1 Commonly Used Definitions of "Lithosphere"[a].

Definition	Main Distinctive Feature
Mechanical	Outer part of the Earth where there are no significant vertical gradients in horizontal strain rate (i.e., no internal deformation) and effectively isolated from the underlying convective mantle over geological time scales (cf. *Burov* [2011] and *Turcotte and Schubert* [2014])
Seismological (LID)	High-velocity material that overlies the upper mantle Low Velocity Zone (LVZ) (cf. *Anderson* [1989] and *Fisher et al.* [2010])
Thermal	Material above a critical isotherm where heat is transferred primarily by conduction (cf. *Turcotte and Schubert* [2014] and *Artemieva* [2011])
Elastic	Strong outer shell of the Earth that can support applied loads elastically and without permanent deformation (cf. *Watts* [2001])
Electrical	Outer, generally resistive, layer of the Earth overlying more conductive material (cf. *Jones* [1999] and *Jones and Craven* [2004])
Geochemical	Material that preserve distinct geochemical and isotopic signatures for longer periods than the underlying convecting mantle (cf. *Griffin et al.* [1999] and *O'Reilly and Griffin* [2010])
Petrological	Uppermost portion of the upper mantle where amphibole (magnesian pargasite) is stable (cf. *Green and Fallon* [1998] and *Green et al.* [2010])

[a] We note that different authors assign slightly different characteristics to some of these definitions.

however, that within the plate tectonics paradigm, there is only one strict definition, namely the mechanical or rheological definition (e.g., see *Isacks et al.* [1968] and *Le Pichon et al.* [1973]). Thus, the lithosphere represents the Earth's rigid/strong/viscous outermost shell, which can sustain and transmit relatively large stresses over geologic time scales. It forms relatively rigid plates that move over a hotter and rheologically weaker layer (the asthenosphere) that is characterized by pervasive ductile deformation (solid-state creep) and multiscale convection. Although its formal definition is therefore clear, the impossibility of directly probing the lithospheric mantle gives rise to a number of indirect "technique-based" definitions (Table 10.1).

Given the strong dependence of rock strength on temperature, the so-called "thermal definition" of the lithosphere has some practical and conceptual advantages over other proposed definitions. Being a thermal boundary layer itself, the lithosphere is a nonconvecting region of relatively high temperature gradient (controlled by conduction of heat) between its lower boundary and the Earth's surface. This represents a well-established and important tenant of lithospheric modeling that allows practical estimations of "thermal" lithospheric structures provided appropriate boundary conditions and physical parameters are chosen (cf. *Jaupart and Mareschal* [2011], *Hasterok and Chapman* [2011], *Artemieva* [2011], and *Furlong and Chapman* [2013]). Importantly, through the use of temperature-dependent rheological laws appropriate for lithospheric rocks, a formal relation between the thermal and mechanical definitions can be defined (cf. *Ranalli* [1995], *Kohlstedt et al.* [1995], *Burov* [2011], and *Turcotte and Schubert* [2014]). On the downside, the viscosity of rocks is dependent not only on temperature, but also on pressure, composition, melt content, fluid content, strain rate, and so on (e.g., *Ranalli* [1995], *Kohlstedt and Zimmerman* [1996], *Mei and Kohlstedt* [2000a,b], *Karato* [2008], and *Tasaka et al.* [2013]), and some important parameters such as the activation volumes and energies of the aggregate are still subject to large uncertainties. Thus, a single temperature would not correspond to a single viscosity value everywhere. Whereas the pressure effect is negligible when considering actual uncertainties in temperature estimations from geophysical and/or thermobarometric methods, volatile content, on the other hand, can significantly complicate the relation between temperature and viscosity (cf. *Hirth and Kohlstedt* [1996], *Karato and Jung* [1998], *Mei and Kohlstedt* [2000a,b], *Karato* [2008], and *Burov* [2011]). Also, episodic magmatism and or fluid circulation can locally and temporarily alter conductive geotherms [*Furlong and Chapman*, 2013]. However, significant departures (of the order of 150–250°C at scales of 50–100 km) from a conductive profile are only important in regions that experienced extended tectonothermal events (rifting, orogenesis, subduction) less than ~80 Ma ago [*Furlong and Chapman*, 2013]. With these caveats in mind, the thermal definition of the lithosphere has the advantages of (a) being a practical and reliable working definition, (b) having a formal relation to the mechanical definition, (c) having a direct connection with other popular definitions (e.g., electrical, seismic, etc.) through laboratory-based equations of state for mineral properties (e.g., elastic moduli, electrical conductivity, etc.), and (d) eliminating ambiguities associated with other methods which struggle to locate the LAB beneath some tectonic settings, such as Archean cratons (i.e., the thermal definition always outputs a specific isotherm that can be related to the LAB). All other definitions of lithosphere (Table 10.1) respond to *specific features* related to (but not strictly defining) the nature and evolution of the lithosphere and typically predict significantly different lithospheric thicknesses (e.g., see *Eaton et al.* [2009] and *Jones et al.* [2010]). For instance, the so-called *seismic lithosphere* or lid, is characterized by a layer of relatively high velocities and/or frozen anisotropy (cf. *Anderson* [1989] and *Plomerova et al.* [2002]). These specific features are a consequence, rather than the cause, of the cold and viscous nature of lithospheric rocks compared to the underlying mantle. Likewise, the distinct geochemical signatures used to define a *chemical lithosphere* (e.g., see *Griffin et al.* [1999]) can only exist due to the highly viscous, nonconvecting nature (i.e., thus, no homogenization) of lithospheric rocks.

One of main advantages of adopting the seismic definition of the lithosphere is that one can use global seismic models to estimate the structure of the lithosphere (e.g., *Priestley and McKenzie* [2006], *Conrad and Lithgow-Bertelloni* [2006], *Thybo* [2006], *Lebedev and van der Hilst* [2008], *Fischer et al.* [2013], and *Artemieva* [2011]). *Pasyanos et al.* [2014] have recently applied this principle to construct a 1° tessellated global model of the lithosphere (LITHO1.0) by inverting Love and Rayleigh (group and phase) dispersion data over a wide period range (up to 200 sec). Similarly, *Priestley and McKenzie* [2006] and *Priestley and Tilmann* [2009] also inverted dispersion data to obtain the seismic velocity structure of the upper mantle, which is subsequently used to estimate geotherms and the base of the lithosphere. Despite some minor differences in the data and approach used, these models, as many others before, reveal a great variability in lithospheric thicknesses (either seismological or thermal) worldwide, with a marked long-wavelength correlation between lithospheric thickness and surface tectonics. This correlation has been highlighted before in numerous studies (e.g., *Jordan* [1988], *Anderson* [1989], *Grand* [1994], *Nolet et al.* [1994], and *Zhang and Tanimoto* [1991]) and it represents a robust seismological feature of the upper

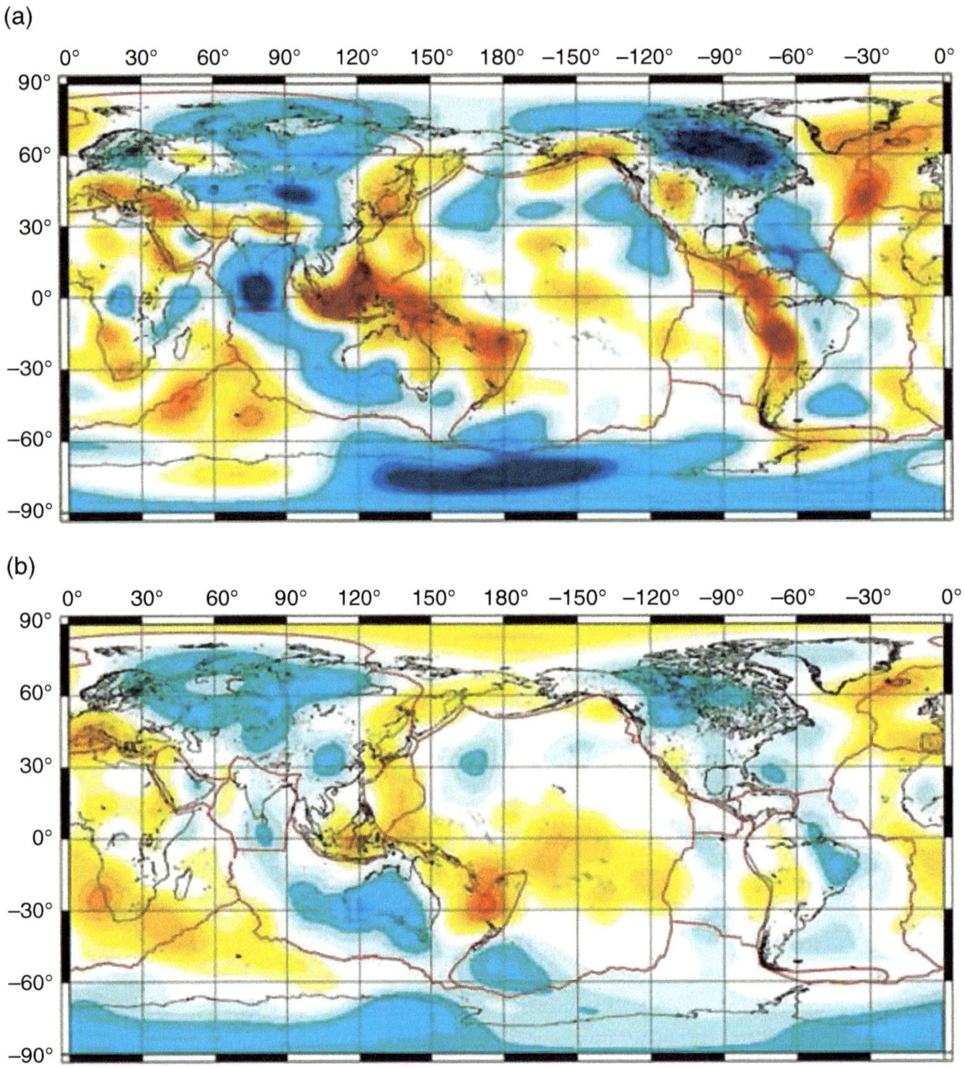

Figure 10.1 **(a)** Observed EGM96 global free-air gravity anomalies. **(b)** The "mean" free-air gravity anomalies derived from global seismic models (mean of five different seismic models). Both observed and predicted anomalies shown are synthesized from spherical harmonics up to degree and order 20. Modified from *Forte* [2007].

mantle. A difficulty with such studies, however, is that when velocity anomalies are converted into density anomalies, the gravity anomalies, induced mantle flow, topography, and plate velocities predicted by the model tend to be poor representations of the real observations (Figure 10.1; *Forte* [2007, 2000] and *Simmons et al.* [2010]). This is not surprising, since the magnitude of the temperature anomalies necessary to significantly affect geodynamic observables is of the same order as the uncertainties associated with temperatures derived from global seismic models. Approaches that include the simultaneous inversion of potential fields and/or geodynamic observables offer an attractive solution (e.g., see *Simmons et al.* [2010]).

Although the base of the lithosphere is commonly thought of as a first-order boundary or structural discontinuity, there are no strong a priori physical arguments why it should be a global sharp boundary. Factors that can contribute to produce a relatively sharp boundary at the base of the lithosphere include the presence of melt lenses, sharp change in fluid content (mainly H_2O), and strong shearing due to the horizontal motion of the plate, among others. Some or all of these factors are likely to be more common in oceanic environments (e.g., *Afonso et al.* [2008b], *Karato* [2012], *Kawakatsu et al.* [2009], *Rychert and Shearer* [2011], *Olugboji et al.* [2013], and *Naif et al.* [2013]), where evidence for sharp seismic discontinuities (which in *some* cases *may* coincide with the base of the lithosphere) is clear (e.g., *Kawakatsu et al.* [2009] and *Olugboji et al.* [2013]). However, geochemical, geophysical, and numerical evidence suggest that, at least beneath some continents, this boundary is more like a gradual "transitional zone" and subject to complex multi-scale physicochemical interactions between the overlying lithospheric plate and the sublithospheric upper mantle. Some of these interactions include the episodic transfer of mass and energy from the asthenosphere to the base of the lithosphere (e.g., percolation of low-degree melts, refertilization, underplating) and from the lithosphere to the asthenosphere (e.g., lithospheric downwellings or drips, small-scale convection), which can significantly modify the properties of this complex region both in space and time. The term lithosphere–asthenosphere boundary, or LAB, therefore seems to be a bit of a misnomer, especially when used to define lithospheric structure. Perhaps it would more appropriate to simply use lithospheric thickness when referring to lithospheric structure and lithosphere–asthenosphere transition zone (LATZ) or lithosphere–asthenosphere system (LAS) when referring to the complex transitional region between the lithosphere and sublithospheric upper mantle.

10.3. COMPOSITION OF THE UPPER MANTLE

It is commonly agreed that the bulk composition of the upper mantle can be represented as that of a peridotite *sensu lato*. Although this model has been regularly challenged (cf. *Anderson* [1989]), abundant evidence collected in the past 30 years from experimental phase equilibria, mineral physics studies, seismological observations, composition of mantle-derived magmas, and studies of exhumed mantle support a bulk peridotitic upper mantle (cf. *McDonough and Sun* [1995], *Pearson et al.* [2003], and *Bodinier and Godard* [2003]). The four main mineral phases are olivine, clinopyroxene, orthopyroxene, and an aluminum-rich phase. The latter can be either garnet, spinel, or plagioclase, depending on the equilibration pressure (some of these phases can coexist at certain PT conditions). The dominant Al-rich phase present in the rock typically defines the "facies" from which the samples have been recovered (e.g., garnet versus spinel facies). According to the IUGS classification nomenclature, the term "peridotite" is restricted to ultramafic rocks with more than 40% modal olivine (Figure 10.2). Peridotites are further subdivided into lherzolites (abundant amounts of clino- and orthopyroxene), harzburgites (mostly orthopyroxene), dunites (>90% olivine), and wherlites (mostly clinopyroxene), with the first three making up over 90% of all recovered mantle samples (see below). Secondary phases, such as apatite (phosphate), rutile (TiO2), zircon (ZrSiO4), monazite (phosphate), phlogopite (Mg-rich mica), and amphiboles (hydrous silicate), may also be present, especially around localized veins where metasomatizing fluids/melts percolated through mantle rocks [*Pearson et al.*, 2003; *O'Reilly and Griffin*, 2013]. Moreover, solidified melts within the lithospheric mantle coupled with fluid–rock interaction processes can result in local, but significant, lithological contrasts (e.g., eclogite and/or pyroxenite bodies; SiO_2-enriched harzburgitic domains, etc. [*Kelemen et al.*, 1998; *Jacob*, 2004; *Pearson et al.*, 2003; *Bodinier et al.*, 2008]). The actual spatial distribution and abundance of ultramafic rocks other than peridotites within the upper mantle is debated and

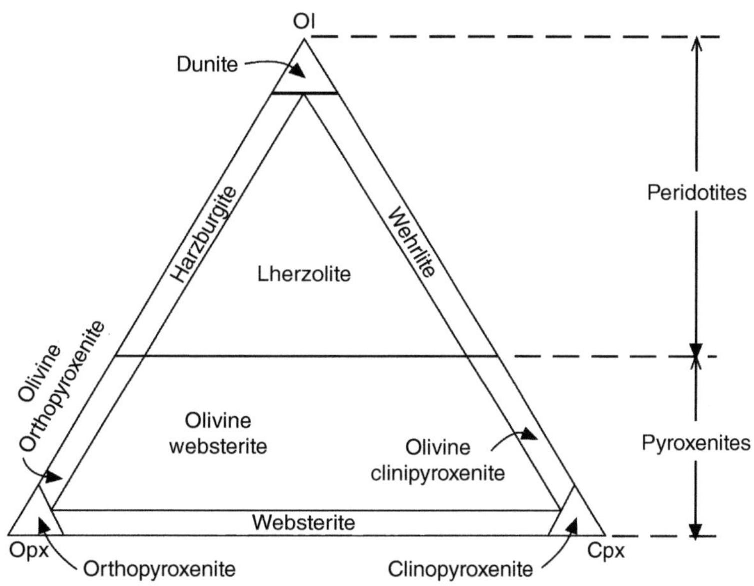

Figure 10.2 IUGS modal classification of major rock types in the lithospheric mantle. After *Streckeisen* [1979].

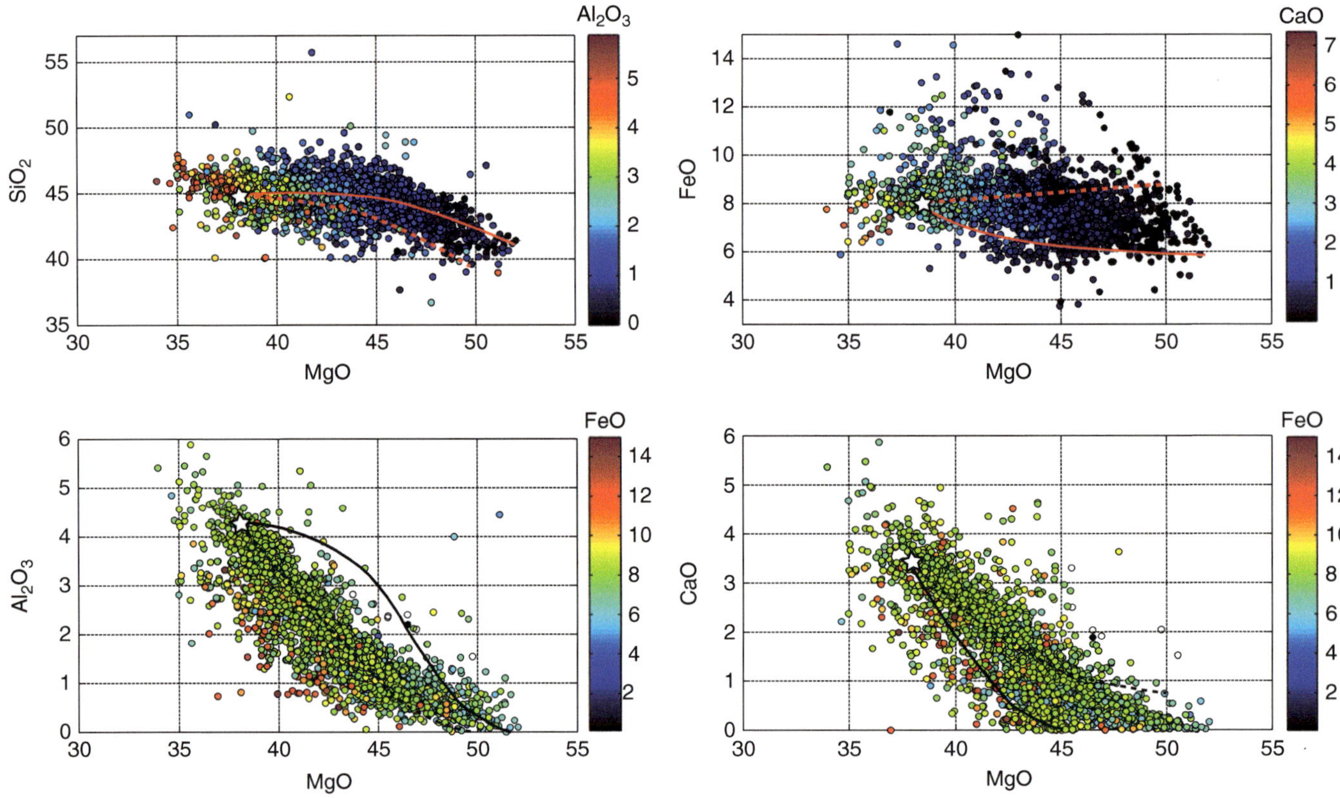

Figure 10.3 Covariation plots for a large database of mantle samples (e.g., xenoliths, abyssal peridotites, orogenic peridotites, etc.) showing the wide compositional range observed in natural samples. The color scales represent the content of a third oxide. Two polybaric perfect fractional melting paths [*Herzberg*, 2004] with different initial pressures of melting are included for comparison (*dashed line* = 2 GPa, *solid line* = 7 GPa). The initial fertile composition for these melting paths is indicated by the white star. After *Afonso et al.* [2013a].

difficult to constrain, but there is some agreement that their *average* volumetric proportions should not be higher than a few percent (e.g., see *Schulze* [1989], *Griffin and O'Reilly* [2007], and *Downes* [2007]). Even in such small amounts, eclogites can have a significant effect on some geophysical observables and mechanical properties of the lithosphere if they are concentrated at specific depths [*Griffin and O'Reilly*, 2007]. Unfortunately, their detection with geophysical techniques is still subject to large uncertainties.

Besides the presence of other ultramafic rocks, there is now abundant evidence of large compositional variability at different scales (vertical and lateral) within the peridotitic component of the lithospheric mantle [Figure 10.3; cf. *Carlson et al.* [2005], *Griffin et al.* [2009], *Bodinier and Godard* [2003], *Pearson et al.* [2003], and *Walter* [2003]). Since peridotites are thought to be the most important component by volume in the upper mantle, and their composition can inform on the origin, evolution, and bulk physical properties of the lithosphere, compositional variability within this group is of particular relevance. The terms "depleted" and "fertile" have become standard when describing the degree to which the composition of a mantle peridotite has been modified, relative to some assumed starting composition, by melt extraction and/or metasomatism. Highly depleted rocks can be significantly less dense and generally seismically faster than their fertile counterparts (cf. *Jordan* [1988], *Lee* [2003], *Anderson and Isack* [1995], *Schutt and Lesher* [2006], *Matsukage et al.* [2005], *Afonso et al.* [2010], and *Afonso and Schutt* [2012]). This is due to the transfer of specific chemical elements from the solid aggregate to the melt phase (i.e., incompatible elements), which is subsequently removed from the system. For instance, SiO_2, CaO, and Al_2O_3 are preferentially removed from the solid aggregate when melting occurs, while MgO tends to remain in the solid residue, and therefore its relative abundance increases in the solid. The behavior of FeO is more intricate, particularly at low pressures ($P < 3$ GPa; see *Afonso and Schutt* [2012]), but it generally decreases or increases at a smaller rate than MgO [*Herzberg*, 2004; *Kinzler and Grove*, 1992] during partial melting. Despite second-order discrepancies between different melting models/experiments, the general "depletion pattern" of the five main oxides SiO_2–Al_2O_3–MgO–FeO–CaO is well understood, and their atomic ratios are typically used to quantify the degree of depletion in peridotites (cf. *Pearson et al.* [2003] and *Walter* [2003]). Amongst these, the ratio between MgO and FeO or "magnesium number" (Mg# = MgO/[MgO + FeO]), is of particular interest because (i) together with SiO_2, FeO and MgO typically account for more than 95% by weight of peridotites and thus exert a major control on modal compositions, (ii) these elements have a distinctive behavior during melting episodes that informs about the extent of partial melting (see below), (iii) MgO and FeO's strong influence on the relative abundances of mineral end-members of volumetrically dominant mineral phases (e.g., olivine and pyroxenes), and (iv) MgO and FeO's affect significantly the physical properties of the aggregate. During melt extraction, the Mg# of the residue increases almost linearly with degree of melting, regardless of whether melting is a batch or a fractional process and/or whether it occurs under wet or dry conditions [*Hirose and Kawamoto*, 1995; *Herzberg and O'Hara*, 2002; *Herzberg*, 2004]. Also, mainly due to (i) and (iii) above, there is a strong correlation between the residue's Mg# and its bulk density, shear-wave velocity, electrical conductivity, and compressional shear velocity (e.g., see *Speziale et al.* [2005], *Matsukage et al.* [2005], *Jones et al.* [2009], and *Afonso et al.* [2010]). The reader is referred to *Afonso and Schutt* [2012] and *Afonso et al.* [2013a] for thorough discussions of these topics in the context of geophysical studies.

The interpretation of the Mg# is less clear when pervasive metasomatism by infiltration of mafic melts take place [*Griffin et al.*, 2009]. This is expected to occur in the deeper parts of the lithospheric mantle, which can be affected and modified by interaction with small amounts of melts produced in the asthenosphere (e.g., see *Tang et al.* [2006] *Piccardo* [2008], and *O'Reilly and Griffin* [2010, 2013]). When this occurs, FeO can be re-introduced into the solid assemblage during metasomatism by infiltration of mafic melts [*Griffin et al.*, 2009], resulting in an overall reduction of the residue's Mg#. CaO and Al_2O_3 are also commonly added to the system by such processes. The metasomatic reintroduction of incompatible elements back into the depleted residue is referred to as "refertilization". Indeed, there is geochemical and geophysical evidence suggesting that large volumes of lithospheric mantle have been refertilized through percolation of melts (cf. *Chen et al.* [2009], *Pinto et al.* [2010], *Le Roux et al.* [2007], *Zheng et al.* [2007], *Griffin et al.* [2009], *Tang et al.* [2013], and *O'Reilly and Griffin* [2013]). Therefore, the Mg# of peridotites should not be seen as a measure of melt depletion only, but as an overall indicator of depletion and refertilization processes over time. Moreover, cautions should be exercised when making assumptions about the average Mg# of the subcontinental mantle based on the age of the overlying crust, as it is now well-known that this is not a global pattern [*Griffin et al.*, 2009; *O'Reilly and Griffin*, 2013].

If the lithosphere is indeed isolated from the homogenizing process of high-temperature convection, it should accumulate and preserve distinct geochemical and isotopic signatures for longer periods than the underlying convecting mantle. Indeed, trace-element and isotopic studies in mantle peridotites have provided crucial information on the nature and time scale of melting and metasomatic events (e.g., see *Hofmann* [1997] and *Stracke and Bourdon* [2009]). However, the effects of trace elements on geophysically relevant properties of mantle rocks is negligible. A significant exception is water, which exerts a major influence on some important properties of mantle rocks, even when present in trace (ppm) amounts (e.g., see *Karato and Jung* [1998], *Karato* [2008], and *Yoshino and Katsura* [2013]). Water is two to three orders of magnitude more soluble in melts than in mantle minerals (cf. *Hirschmann* [2006]). Consequently, considering typical water contents in mantle rocks (<200–300 ppm wt%), melt extractions of only <10% can effectively "dry out" the residue, significantly increasing its electrical resistivity and viscosity relative to its hydrated counterpart (e.g., see *Karato* [1986], *Hirth and Kohsltedt* [1996], *Karato* [2008], and *Jones et al.* [2012]). Furthermore, it has been recently suggested that water can also significantly affect seismic attenuation and velocities by increasing the effect of the dissipation peak due to grain-boundary sliding [*Karato*, 2012]. This hypothesis is consistent with a limited number of experimental results (e.g., *Aizawa et al.* [2008]) and helps reconciling some contradictory seismological observations that are difficult to explain in terms of our current state of knowledge regarding the temperature and water effects on seismic observables.

In summary, there is now abundant evidence that the lithospheric mantle is highly heterogeneous, both in composition and structure. Although single-data or stand-alone approaches have been successful in providing the long-wavelength pattern of lithospheric structure (see above), imaging the fine-scale thermochemical structure of the lithosphere with geophysical techniques remains a challenging, yet crucial, problem. Some of the difficulties are technical/methodological in nature (e.g., insufficient computer power, inefficient algorithms, biased prior information, etc.) or related to incomplete/poor data (e.g., lack of dense high-resolution datasets, incomplete mineral physics data, etc.) and are therefore amendable in principle. However, a significant part of the problem is related to its actual physical nature and thus intrinsic to the problem of retrieving the physical state of the Earth's interior from indirect measurements (Section 10.5).

10.4. SINGLE-DATA AND MULTI-OBSERVABLE APPROACHES

Geophysical methods used to image the lithospheric and upper mantle structure have improved dramatically in the past 30 years due to rapid advances in both computing power and data acquisition/processing techniques (cf. *Romanowicz* [2003], *Rawlinson et al.* [2010], *Liu and Gu* [2012], *Kuvshinov and Semenov* [2012]). Quite naturally, these methods evolved largely in isolation from each other, each focusing on the treatment of a specific dataset. Some of the most common single-data approaches used to study the lithosphere and upper mantle include teleseismic tomography (e.g., see *Evans and Achauer* [1993], *Granet et al.* [1995], and *Rawlinson et al.* [2006], surface-wave tomography (e.g., see *Pasyanos and Nyblade* [2007], *Yang et al.* [2008], *Fishwick et al.* [2008], and *Agius and Lebedev* [2013]), gravity modeling (e.g., see *Zeyen and Fernàndez* [1994], *Torne et al.* [2000], *Ebbing et al.* [2006], *Chapell and Kusznir* [2008], *Tašárová et al.* [2009]), electromagnetic methods (e.g., see *Heinson* [1999], *Jones* [1999], *Jones et al.* [2009a], *Evans et al.* [2005], *Evans et al.* [2011], and *Meqbel et al.* [2014]), local earthquake tomography (e.g., *Aki and Lee* [1976], *Eberhart-Phillips* [1990], and *Kissling et al.* [1994]), and receiver function studies (e.g., *Yuan et al.* [2006], *Kawakatsu et al.* [2009], *Rychert and Shearer* [2011], and *Kind et al.* [2012]). The first four are of particular relevance to us, as they are most sensitive to the shallow mantle structure, they offer complementary information and resolutions, their datasets are amenable to well-known inversion/modeling methods, and they are not restricted to seismogenic regions (as is local earthquake tomography). In addition, a number of studies have successfully combined them (in pairs) in joint inversions for crustal, lithospheric, and upper mantle structure in general. The most common example, and perhaps the most mature in terms of methodology, is that of joint inversion of body-wave and surface-wave data (e.g., see *West et al.* [2004], *Rawlinson and Fishwick* [2011], and *Obrebski et al.* [2011]). However, attempts to combine magnetotelluric (MT) data and surface waves [*Moorkamp et al.*, 2010; *Roux et al.*, 2011; *Vozar et al.*, 2014], receiver functions and MT data [*Moorkamp et al.*, 2007], surface waves and receiver functions [*Julia et al.*, 2000; 2003; *Tkalčić et al.*, 2006], and gravity and body waves (e.g., see *Zeyen and Achauer* [1997] and *Tiberi et al.* [2003]) have also yielded promising results (see Section 10.4.1) and will undoubtedly become the subject of further research.

The need for combining different datasets when studying the structure of the lithosphere and upper mantle is straightforward to understand and justify. *Afonso et al.* [2013a] recently provided an extensive discussion of the benefits (and difficulties) of working with multiple observables for this purpose. The main objective of inverting/modeling multiple geophysical observables is twofold: Firstly, we aim at minimizing the range of acceptable models by adding additional constraining data (i.e., reduce the space of valid solutions). Secondly, we try to gain additional information on the problem at hand (i.e., obtain "more comprehensive" models) by adding observations with different sensitivities to different aspects of the problem. For instance, while short-period surface-wave data offers good resolution at shallow depths (where teleseismic ray paths do not cross) and provide information on the absolute velocity structure beneath the array, body-wave data can add critical information on deeper features (provided the array has a large aperture) and improve lateral resolution [*Rawlinson and Fishwick*, 2011; *Obrebski et al.*, 2011]. It is therefore not only desirable but also necessary that each new observable added to the inversion/modeling have a different sensitivity to the model parameters of interest. This seemingly obvious requirement is crucial when the uncertainties in observations are formally considered. Every new observable carries uncertainties that are propagated to the final result (cf. *Tarantola* [2005]). Therefore, if a new observable is added to the inversion, but its sensitivity is similar to or less than that of the existent data, the net effect could be a degradation of the results (uncertainty is added, but not new constraints).

10.4.1. Topography and Gravity Field Modeling

To the first order, and disregarding dynamic contributions, the elevation (i.e., topography and bathymetry) of the Earth's surface is a measure of the buoyancy (average density) of the lithosphere [*Lachenbruch and Morgan*, 1990]. Local isostasy is one of the oldest principles in geophysics and has been proven to be a suitable approximation for explaining topographic loads with wavelengths $\gtrsim$100 km [*Turcotte and Schubert*, 2014; *Lachenbruch and Morgan*, 1990; *Watts*, 2011; *Hasterok and Chapman*, 2007]. According to the principle of isostasy in its most popular form, all regions of the Earth with identical elevation must have the same buoyancy (mass per unit area) when referenced to a common compensation level. Although there is no perfect compensation level in the Earth's mantle (temperature and pressure disturbances associated with convective flow prevent its existence), it can be shown that in a mantle-like fluid, surface topography is primarily controlled by temperature (density) variations in the vicinity of the upper thermal boundary layer (i.e., lithosphere) and is relatively insensitive to thermal (density) anomalies below a certain critical isotherm [*Parsons and Daly*, 1983]. Therefore, absolute elevation can be used as a constraint to possible density structures within the lithosphere (e.g., see *Sclater and Francheteau*, [1970], *Turcotte and Schubert* [2014],

Lachenbruch and Morgan [1990], and *Hasterok and Chapman* [2007, 2011]). This principle is commonly referred to as "lithospheric" or "thermal isostasy".

Gravity-related observables are sensitive to subsurface mass anomalies, but their inversion alone for the density distribution within the Earth is an ill-posed and non-unique problem. Further constrains can be added to the modeling/inversion of gravity considering the temperature dependence of mantle density through thermal expansion. This normally involves joint modeling/inversion of gravity data together with an extra observable directly related to thermal structure and independent from density variations: the surface heat flow (e.g., see *Zeyen and Fernàndez* [1994], *Fullea et al.* [2007], and *Jiménez-Munt et al.* [2010]). Unfortunately, surface heat flow tends to be dominated by crustal effects (e.g., radiogenic heat production, groudwater flow, etc.), which need to be properly taken into account when modeling lithospheric geotherms [*Jaupart and Mareschal*, 2011; *Furlong and Chapman*, 2013]. Also, since both topography and gravity field observables (i.e., gravity and geoid anomalies, gravity gradients) depend upon the density distribution within the Earth, in principle they can be jointly modeled (e.g., see *Fernàndez et al.* [2004] and *Zeyen et al.* [2005]) or inverted (e.g., see *Fullea et al.* [2007] and *Kumar et al.* [2014]) to obtain estimates of the lithospheric structure. However, care must be taken in regions where large sublithospheric dynamic effects may be present (e.g., see *Hager and Richards* [1989], *Forte* [2007], *Afonso et al.* [2008b], *Khan et al.* [2013], and *Shan et al.* [2014], as they can introduce systematic errors in the modeled lithospheric structure.

Gravity anomalies (Bouguer, free air) are typically used to constrain crustal densities and geometries (e.g., see *Ebbing et al.* [2006], *Chapell and Kusznir* [2008], *Torne et al.* [2000], and *Fullea et al.* [2008]). Geoid anomalies are affected by mass/density anomalies over a wider range of depths than gravity anomalies (e.g., see Bowin [2000]). Therefore, appropriate wavelength filtering of the geoid anomaly allows studying the Earth at different scales, from the crust (e.g., see *Doin et al.* [1996] and *Sandwell and MacKenzie* [1989]) to the deep mantle (e.g., see *Hager* [1984], *Ricard et al.* [1984], *Forte and Mitrovica* [2001], and *Deschamps et al.* [2001]). Satellite gravity data are a unique source of information on the density structure of the Earth due to its global and relatively uniform coverage. The recent GOCE (Gravity field and steady-state Ocean Circulation Explorer) satellite mission (launched in 2009) included, for the first time, an on-board three-axis gradiometer able to measure the Earth's gravity gradients at satellite height (e.g., see *Pail et al.* [2011]). This new type of gradiometric satellite measurement is more sensitive to spatial structure and directional properties of the Earth's internal density distribution than classic observations on gravitational intensity (gravity and geoid anomalies). In contrast to terrestrial gravity gradients (characterized by a high-frequency content related to near-surface structures), the gradiometric tensor measured in the local satellite framework (255 km on average for GOCE) is naturally filtered, retaining the lithospheric and underlying mantle density signal. Pioneering recent studies have showed very promising results in imaging the Earth's crust [*Reguzzoni et al.*, 2013] and mantle density structure using this new type of data [*Panet et al.*, 2014; *Fullea et al.*, 2015], although its full potential is yet to be fully exploited.

10.4.2. Joint Inversions of Seismic Data

Teleseismic travel time tomography is one of the most widely used techniques to image the 3D velocity structure of the lithosphere and uppermost upper mantle (cf. *Evans and Achauer* [1993] and *Rawlinson et al.* [2010]). Its popularity resides in the relatively low cost of passive seismic deployments, the high horizontal resolution (on the scale of tens of kilometers) that can be achieved, its applicability to regions with no local seismicity, and the high-quality data that can be extracted from the array using cross-correlation techniques (e.g., see *VanDecar and Crosson* [1990], *Evans and Achauer* [1993], and *Rawlinson and Kennett* [2008]). However, despite the many benefits of teleseismic travel-time tomography, there are some important limitations that need to be considered when interpreting tomographic images obtained with this technique. Two of the most important limitations when studying the structure of the lithosphere and upper mantle are the lack of information on absolute velocities and the lack of resolution of the shallow structure beneath the array (cf. see *Evans and Achauer* [1993] and *Foulger et al.* [2013]). The first of these limitations is due to the fact that, in order to mitigate biases associated with uncertainties in source parameters (i.e., hypocenter location and origin time) and large-scale heterogeneities along the path between the source and the receiver, only *relative* travel time residuals are used in the inversion. As a consequence, only *relative* wave speeds can be recovered and the actual amplitudes of the variations are typically underestimated. The second limitation is a consequence of using teleseismic events, which results in ray paths that are sub-vertical when they arrive at the array, resulting in poor ray crossing at shallow depths beneath the array.

One way to overcome (or at least minimize) these limitations is to include information from surface wave data. Several variations of this general recipe have been used in the literature to obtain high-resolution models of velocity structure above the mantle transition zone (e.g., see *West et al.* [2004], *Rawlinson and Fishwick* [2011], and

Obrebski et al. [2011]). Phase-velocity dispersion maps obtained from the combination of ambient noise tomography with teleseismic surface wave tomography constitute one of the best sources of information for retrieving absolute Vs values in the lithosphere and sublithospheric upper mantle (e.g., see *Yang et al.* [2008]). The main advantage of this method is that while ambient noise tomography (ANT) provides detailed information at short periods (shallow levels), teleseismic earthquake methods such as multiple-plane-wave tomography provide the complementary information at longer periods (deep levels). The ability to recover "good" values of Vs, however, depends on the uncertainties associated with the phase-velocity dispersion maps generated by each method. The combination of surface and body wave data in joint inversions will likely be boosted by (1) the recent advances in ambient noise tomography to obtain short and long period phase-velocity information, (2) the popularization/optimization of Monte Carlo approaches to invert for velocity structure, and (3) the rapidly growing seismic infrastructure and deployment of dense arrays around the globe (e.g., USArray, IberArray, ChinArray, etc.). This is an area with great (proven) potential to study the fine-scale velocity structure of the lithosphere and upper mantle. The inclusion of surface wave overtones into such joint inversions provides not only better resolution with depth but also additional (albeit weak) sensitivity to the Vp structure [*van Heijst and Woodhouse*, 1999]. This (information on both Vs and Vp) is particularly useful when making inferences about the thermochemical structure of the Earth [cf. *Khan et al.*, 2011; *Afonso et al.*, 2010, 2013a].

Joint inversions of surface wave and receiver function (RF) data are an attractive alternative to constrain the velocity structure of the first ~300 km of the Earth. While surface wave data provide essential constrains on the absolute 3D velocity structure, it cannot constrain simultaneously the strength and location of shear velocity discontinuities (e.g., Moho–mantle transition). This problem results in velocity models with high uncertainties around regions of sharp discontinuities. RFs, on the other hand, are highly sensitive to velocity discontinuities beneath the station, particularly to the Moho discontinuity, but not to the absolute velocity structure due to a depth–velocity trade-off [*Ammon et al.*, 1990]. Following the pioneering work of *Julia et al.* [2000] using P-wave RFs and dispersion data, many authors have successfully used this approach to study crustal and shallow mantle structures (e.g., see *Lawrence and Wiens* [2004], *Tkalčić et al.* [2006], and *Shen et al.* [2013]). The extension of this technique to constrain the whole lithospheric structure, however, still is in its infancy, and further studies are needed to show its general applicability and reliability. To this respect, the joint inversion of S-wave RFs and long period dispersion data represents a promising, yet understudied, approach. The advantage of using S-wave RFs is that deeper signals associated with LAB structure are less susceptible to contamination by crustal and deeper reverberations [*Yuan et al.*, 2006], which constitute a major difficulty when working with P-wave RFs (cf. *Kind et al.* [2012] and *Chen et al.* [2009]). Despite this advantage, S-waves usually have a larger noise–signal ratio than P-waves and are characterized by lower frequencies (due to stronger attenuation at higher frequencies). Therefore, S-wave RFs are not well suited to resolve details about the crustal structure beneath the station. Therefore, while the use of S or P-wave RFs in joint inversions is dependent on the goals of the study, the joint inversion of S and P-wave RFs with surface wave data is an attractive possibility that holds some potential to obtain detailed seismic images of the lithosphere.

10.4.3. Joint Inversions of Electromagnetic (EM) and Seismic Data

Traditionally, deep EM studies have used time series of the magnetic field variations recorded on the global network of geomagnetic observatories to infer the Earth's electrical conductivity. 1D geomagnetic deep soundings and horizontal spatial gradient studies [e.g., *Fuji and Schultz*, 2002; *Olsen*, 1999] and 3D inversion of geomagnetic observatory data [e.g., *Kelbert et al.*, 2009; *Tarits and Mandea*, 2010; *Semenov and Kuvshinov*, 2012] constrain the conductivity distribution in the mantle in the depth range 400–1600 km (i.e., periods of few days to few months). Long-period magnetotelluric (MT) responses (magnetic plus horizontal electric field component) cover complementary periods up to few days, sampling upper mantle depths (<400 km). However, high-quality long-period MT data (periods <10 days) are difficult to obtain due to technical and intrinsic limitations of the method [*Shimizu et al.*, 2010]. On top of that, the biggest limitation of observatory data is their poor coverage of the Earth's surface (sparse distribution over continents and almost nothing over oceans) that severely hampers the robustness and accuracy of global EM models. Satellite measurements represent an attractive alternative to the sparse terrestrial observatory data, providing high-precision and high-resolution magnetic field measurements with uniform global coverage (e.g., see *Sabaka et al.* [2004] and *Olsen et al.* [2013]). Recent advances in global 3D EM (spherical) forward modeling, the increasing availability of computational resources, and satellite data have made it possible to obtain a new generation of global electrical conductivity models of the mantle. Electromagnetic perturbations of ionospheric source (Sq) derived from satellite data are sensitive to the lithosphere–uppermost mantle conductivity structure (depth: 100–400 km; cf. *Koch and Kuvshinov* [2013]). The conductivity structure in the lower mantle can be derived from the inversion of time series of internal (induced) and external (inducing) spherical harmonic expansion coefficients related to disturbed storm-time variations of magnetospheric ring current origin (Dst) (time and frequency approaches; cf. *Velímský* [2013] and *Püthe and Kuvshinov* [2014]). A common caveat associated with most EM inversions is that while they predict perfectly valid resistivity models, these models are not necessarily directly related to the actual thermochemical structure inside the Earth due to resolution and uncertainty issues. For example, although for a 1D Earth there exists a uniqueness theorem for perfect data at all frequencies [*Bailey*, 1970], data insufficiency and inaccuracy lead to highly nonlinear resolution of model parameters. In particular, the true resistivity of a resistive layer or region beneath a more conducting one is very difficult to resolve due to the screening effect of the upper layer. Thus, only a lower bound on the resistivity of the mantle directly below the crust can usually be set [*Jones*, 1999].

More than 15 years ago, *Jones* [1998] showed a variety of examples where joint interpretation of collocated seismic and magnetotelluric (MT) measurements led to more robust inferences about structures and processes within the lithosphere, partially mitigating the inherent uncertainties associated with EM inversion. Since then there has been a steady, but relatively small, number of studies that formally make use of both datasets (e.g., see *Snyder et al.* [2004], *Jones et al.* [2009b], *Jones* [1998] and *McGary et al.* [2014]). New interdisciplinary initiatives such as Earthscope [*Meltzer*, 2003] provide ample opportunity for more formal combinations of seismic and electromagnetic measurements.

So far, there have been successful examples of joint modeling (see below), and the widespread use of structural constraints (see Chapter 4) utilized in near-surface (Chapter 7), mining (Chapter 8), and hydrocarbon applications (Chapter 9) demonstrates that it is possible to generate useful joint inversion results even without precise knowledge of the physical properties of the different minerals. In the following we will give an overview of the first attempts that have been made to combine MT and seismic data to study the lithosphere–asthenosphere system. We will focus on quantitative approaches that move beyond simple comparisons of images. These include both simultaneous joint inversion and modeling as well as structural classification based on independent inversions.

To the best of our knowledge, the only joint inversion study of MT and seismic data with real data that focuses on the structure of the crust, and that has not been covered in any of the other chapters, has recently been published by *Bennington et al.* [2015]. These authors combine double-difference tomography [*Zhang and Thurber*, 2003] with a 2D MT modeling algorithm based on OCCAM2D [*Constable et al.*, 1987]. To couple velocity and conductivity they use a modified version of the cross-gradient constraint they call the "normalized cross-gradient" where they normalize the components of the gradient by the total length of the gradient vector before calculating the cross product. This normalization enhances the influence of regions in the model where one or both of the gradients in velocity and conductivity are small. Their synthetic tests with realistic acquisition geometries and noise comparable with field data demonstrate that the joint inversion produces significantly better results even though it might not be visually obvious at first glance. Similarly, the overall appearance of the individual inversion results and the joint inversion results is quite similar.

Bedrosian et al. [2007] outline a possible strategy for identifying lithological variations across a fault zone based on a post-inversion structural classification of resistivity and velocity models. Some of the technical details of their approach are discussed in Chapter 5. Even though the two models have been obtained independently and thus similarity between the models has not been enforced in the inversion, they demonstrate that their method produces spatially coherent clusters that correspond to known and assumed geological formations. Compared to a full joint inversion of the seismic and MT data, the advantage of the post-inversion approach is that it can be applied to models generated by different researchers at different times. Also, different classification approaches could be used without having to modify the existing inversion algorithms. A potential disadvantage compared to joint inversion is that there is no exchange of information on structural boundaries. In some cases this could result in overly blurry boundaries between different lithological classes. For the results presented by *Bedrosian et al.* [2007], however, this does not seem to be a major issue.

Most studies combining seismic and MT data in a joint inversion and looking at the whole lithosphere and the asthenosphere are based on the methodology described in *Moorkamp et al.* [2007, 2010] and are limited to simple one-dimensional geometries. *Moorkamp et al.* [2010] invert magnetotellurics, receiver functions, and surface wave dispersion data to determine conductivities, shear wave velocities, and layer thicknesses for a layered Earth (Figure 10.4). In Figure 10.4 we can see how the layer interfaces for the electric and seismic model coincide for each site as coincident layer interfaces couple the electric and seismic parameters. For this type of approach, this is an essential step to ensure adequate coupling. If there are too few layers, it is not possible to explain both datasets

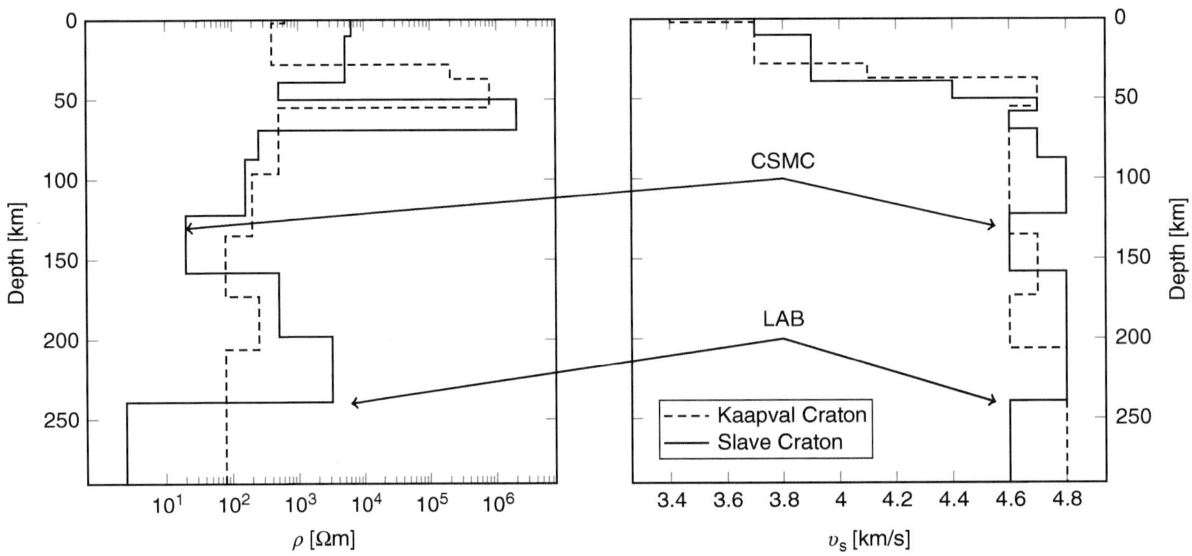

Figure 10.4 Comparison of the joint inversion results for the Kaapvaal Craton (*dashed line*) and the Slave Craton (*solid line*). We show resistivity on the left and velocity on the right-hand side. The results for the Slave Craton show a clear lithosphere–asthenosphere boundary (LAB) and a mantle conductor (CSMC) associated with a low-velocity zone. The Kaapvaal Craton shows some similar features; however, their expression is less pronounced, possibly because of lower data quality.

simultaneously. If there are too many layer, there is no coupling between the seismic and electromagnetic methods.

Roux et al. [2011] show an extension of the approach discussed above to 1D anisotropic structures. The motivation for developing this approach was the observation of coincident directions of anisotropy at similar depths in independently generated surface wave [*Lebedev et al.*, 2007] and MT [*Gatzemeier and Moorkamp*, 2005] models. Thus one of the aims of this study was to quantitatively investigate the hypothesis that the directions and depth ranges of the electric and seismic anisotropy can be modelled as coincident within the resolution of the data. This is one of the predictions made by the hydrogen diffusion argument [*Gatzemeier and Tommasi*, 2006] sometimes employed to explain electrical anisotropy in the asthenosphere (e.g., see *Poe et al.* [2010] and *Dai and Karato* [2014]). The joint inversion results agree with this hypothesis and retrieve two main regions of anisotropy. The resistivities in the asthenosphere differ by a factor of ten between the most conductive and most resistive direction. This is one order of magnitude lower than the previously inferred factor of 100 [*Gatzemeier and Moorkamp*, 2005] and agrees with the values expected from hydrogen diffusion [*Simpson and Tommasi*, 2005]. A subsequent study by *Mandolesi and Jones* [2014] using MT inversion constrained by a seismic model yields slightly different results, but confirms the essential conclusions of an anisotropic asthenosphere.

The above discussion has illustrated some of the strengths, but also some severe limitations of joint inversion of EM and seismic data for the crust and mantle structure in its current form. Structural constraints can be employed, as a priori information, to retrieve coincident structures (i.e., electrical conductivity and seismic velocity) under general assumptions as a first-order estimate. The more powerful, but also more computationally restrictive, geophysical–petrological approach is based on the experimental and thermodynamic relationship between physical properties of rocks (seismic, electric) and the subsurface thermochemical structure (e.g., see *Fullea et al.* [2011], *Khan and Shankland* [2012], and *Vozar et al.* [2014]). The electrical conductivity of the major upper mantle mineral phases (e.g., olivine, pyroxenes, garnet) can be described as an activated process in a semiconductor and, therefore, characterized by an Arrhenius-type power law. The conduction mechanism changes at around 1300–1500°C from small polaron at $T < 1300$°C to ionic conduction. The bulk or whole rock conductivity can be determined according to the individual contribution of its mineral phase constituents (including the presence of interconnected amounts of water and melt) using an appropriate mixing theory (e.g., see *Xu et al.* [2000], *Ledo and Jones* [2005], *Fullea et al.* [2011],and *Khan and Shankland* [2012]). In order to model consistent amounts of water in the mantle, it is important to constrain, based on laboratory studies, (i) the water storage capacity or solubility for the major mineral phases and (ii) the partition coefficient or solubility ratio of water for the different minerals. The solubility reflects the maximum possible water content that a mineral, or an assemblage of minerals, can accommodate within their structures without saturating or producing a water-rich fluid or hydrous melt at a given pressure, temperature, and composition (e.g., see *Mierdel et al.* [2007] and *Ferot and Bolfan-Casanova* [2012]). The partition coefficients characterize the relative proportion of water among the different phases in the assemblage (e.g., see *Dai and Karato* [2009], *Ferot and Bolfan-Casanova* [2012], and *Novella et al.* [2014]). Values for partition coefficients estimated from measurements in natural samples (e.g., see *Kovács et al.* [2012] and *Grant et al.* [2007]) and laboratory studies (e.g., see *Dai and Karato* [2009], *Aubaud et al.* [2004, 2008], and *Kovács et al.* [2012]) show considerable variability. Experimental studies also show, with considerable scatter in the values, a significant variation (up to two orders of magnitude) of partition coefficients with variations in temperature (e.g., see *Dai and Karato* [2009]), pressure (e.g., see *Novella et al.* [2014]) and water content (e.g., see *Dai and Karato* [2009]).

In contrast to seismic velocities, which are mostly sensitive to bulk physical properties, electrical conductivity is strongly affected by interconnected minor constituents (e.g., water, volatiles, or melt) with a large influence on the rheological–dynamic behavior of the Earth. The effect of water, in particular, is a matter of heated debate (e.g., see *Karato* [2011] and *Dai and Karato* [2014]), with estimates of its distribution within nominally anhydrous magnesium-silicate mantle minerals (NAMs) also uncertain (e.g. see *Jones et al.* [2012]). In spite of their characterization as anhydrous, NAMs (e.g., olivine, wadsleyite, ringwoodite, pyroxenes) can host significant amounts of (structurally bounded) water in their crystal structure based on experimental solubility studies (e.g., see *Bolfan-Casanova et al.* [2000]).

Vozar et al.'s [2014] 1D joint modeling of long-period surface-wave phase velocities, MT responses and surface topography in central Tibet (Qiangtang and Lhasa terranes) is based on a self-consistent petrological–geophysical thermodynamic framework where mantle properties are calculated as a function of temperature, pressure, and composition. Long-period surface-wave phase velocities are most sensitive to seismic wave speeds and, through them, to temperature within the lithosphere and sublithospheric mantle. In contrast, surface waves are only weakly sensitive to small amounts of water, which has a large impact in MT data along with temperature. Surface topography senses the average lithospheric density distribution, the latter being affected by the geotherm and bulk composition. In the northern Qiangtang terrane, these authors find a dry 80-120-km-thick lithosphere associated with a relatively fertile composition (garnet lherzolite), as commonly present in Asian mantle xenolith data (Figure 10.5). In contrast, in the southern Lhasa terrane, the LAB depth is about 180 km, with an additional requirement of a moderate amount of water in the lithospheric mantle (<0.02 ppm wt%). The mantle composition modeled in Lhasa is compatible with phlogopite-bearing Fe-rich spinel harzburgite rocks from xenoliths erupted in West Lhasa, suggesting metasomatism (e.g., Fe-enrichment of a previously depleted mantle) and rehydration as evidenced by the hydrous mineral phases [*Vozar et al.*, 2014]. The complementary sensitivities of the different data types modeled by *Vozar et al.* [2014] allows putting tight constraints on the thermochemical structure and hydrous state of the lithosphere, overcoming the limitations, restrictions, and inconsistencies of separate modeling of the different datasets.

The current 1D joint approaches integrating EM and seismic datasets can only capture a crude picture of the true structure of the Earth's subsurface. Although valid under some conditions, the assumption of a layered Earth is less appropriate in areas where large lateral heterogeneity plays an important role. This dimensional limitation will be overcome once 3D joint inversion approaches on this scale become available. Significant efforts have been devoted to connect EM and other geophysical datasets with the subsurface thermochemical structure (e.g., see *Fullea et al.* [2011], *Jones et al.* [2012], *Khan and Shankland* [2012], *Koyama et al.* [2014], *Vozar et al.* [2014], *Khan et al.* [2015], and *Koyama et al.* [2014]), yet a full 3D joint inversion and/or modeling of global EM data—in particular, new satellite magnetic data with global and uniform coverage—for the thermochemical structure in the lithosphere is still to come. Combining a thermochemical/experimental EM approach with other geophysical techniques in a relatively mature state (e.g., seismic, gravity field modeling/inversion), exploiting complementary sensitivities, is an important strand of research that will likely receive attention in the near future.

10.4.4. Integrated Modeling

Most of the preceding methods can be called "classic" or "traditional" in the sense that their main objective is to construct models of the distribution and magnitude of physical *parameters* inside the Earth, such as wave velocity, electrical conductivity, wave attenuation and bulk density, which are then used to make a posteriori (indirect) inferences about the physical state of the Earth's interior (i.e., temperature, pressure, composition). We refer to these methods as "*parametric methods*". Unfortunately, the interpretation of these parameters in terms of the physical state of the Earth's interior is often contentious and ambiguous. Indeed, this remains a major stumbling block in understanding exactly what is being imaged with tomography methods and what is the real geological significance of the imaged boundaries and 3D architecture

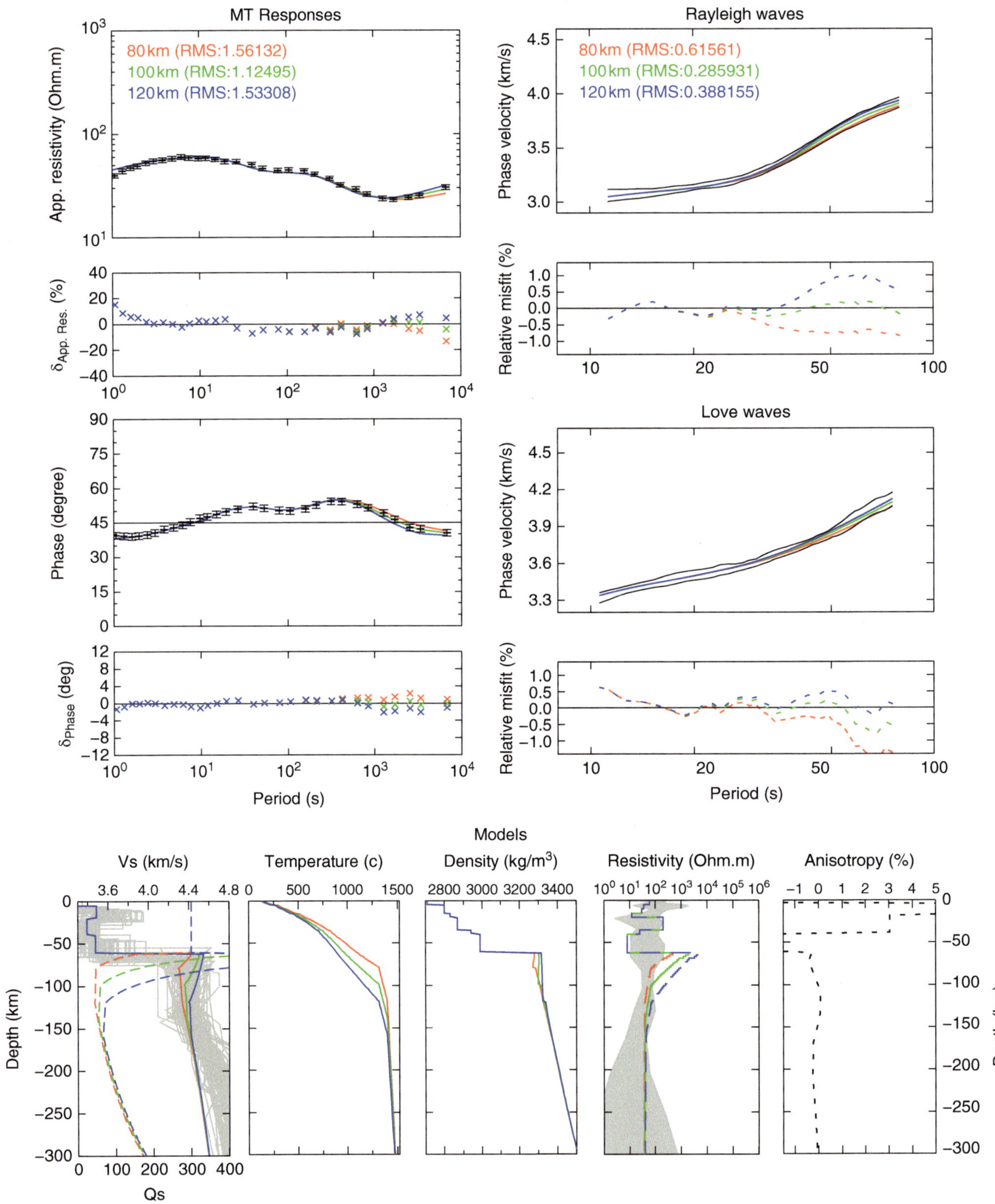

Figure 10.5 Integrated geophysical–petrological modeling of surface waves and MT dat. **(Top)** Fits of MT and seismic data for three (red 80 km, green 100 km, and blue 120 km) different LAB depths. **(Bottom)** Corresponding calculated models of Vs velocities (*dashed lines* represent shear-wave quality factors Qs, *gray lines* represent all possible models based only on seismic modeling), temperature, density, resistivity (*gray area* represents the range of all resistivity models based only on MT data), and anisotropy. After *Vozar et al.* [2014]).

(e.g., see *Afonso et al.* [2010, 2013a,b], *Foulger et al.* [2013], *Becker* [2012], and *Khan et al.* [2015]). In a strict sense, however, there is no real pressing need for knowing the distribution of, for example, wave velocity or electrical conductivity inside the Earth. What we ultimately require is the thermochemical structure of the planet, as this controls important near-surface processes (e.g., seismicity, magma emplacement, mineralization events, etc.), the expression of most geophysical observables (e.g., gravity, travel times, etc.), and the behavior and evolution of tectonic plates (e.g., plate velocity, strain partitioning, etc.).

An integrated approach specifically designed for forward thermochemical modeling of the lithosphere and uppermost mantle, including most available geophysical–petrological constraints, is represented by the LitMod suite [*Afonso et al.*, 2008a; *Fullea et al.*, 2009]. The LitMod suite is a set of 1D, 2D, and 3D self-consistent and interactive codes where all relevant mantle properties (e.g., density, electrical conductivity and seismic velocities) are functions of temperature, pressure, and bulk composition and obtained by Gibbs free energy minimization (e.g., *Connolly* [2009]) using well-known thermodynamic formalisms and internally consistent databases (e.g., see *Stixrude and Lithgow-Bertelloni* [2005, 2011] and *Holland and Powell* [2011]. These codes solve all the required forward problems (e.g., heat transfer, geopotential, electromagnetic, etc.) allowing to simultaneously fit multiple geophysical observations (e.g., elevation, heat flow, gravity potential fields, seismic and magnetotelluric data). When available, petrological data (e.g., amount of partial melting, compositional data from xenoliths, etc.) can be input into the model as well. Besides allowing for a better control of the lateral and vertical variations of the bulk properties, this method reduces the uncertainties associated with fitting each observable alone or in pairs, as commonly done in the literature, and explicitly considers modal (mineral volume fractions) and mineral compositional effects, complex solid–solid mineral phase transitions (and their associated entropy and latent heat changes), hydration effects/state, and, in some cases, melting. The reader is referred to *Afonso et al.* [2008a] and *Fullea et al.* [2009] for further details. This integrated forward methodology has been successfully applied to produce models of the lithosphere at a regional scale in different tectonic scenarios, from young Phanerozoic active zones (e.g., see *Fullea et al.* [2010], *Gradmann et al.*, [2013], *Carballo et al.* [2015], *Vozar et al.* [2014], and *Pedreira et al.* [2015]) to old cratonic areas (e.g., see *Fernàndez et al.* [2010] and *Fullea et al.* [2011]) (Figure 10.6). As an example of a recent application of this approach, we refer to the study of *Gradmann et al.* [2013]. These authors studied the southwestern Fennoscandian thermochemical lithospheric structure using observed seismic velocities, gravity anomalies and topography data. Their results

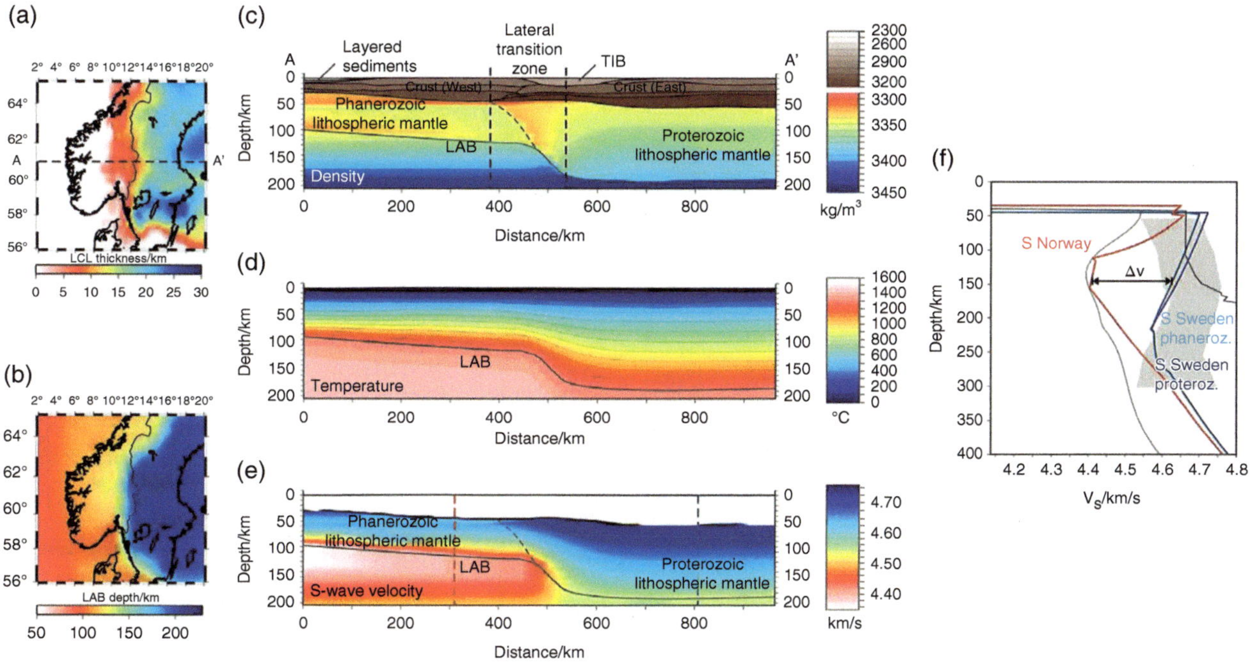

Figure 10.6 **(a)** Thickness of lower crustal layer (LCL). **(b)** Depth to thermal lithosphere–asthenosphere boundary (LAB). **(c)** Cross section at 61 °N showing the subsurface density distribution and general model structure (TIB, Transscandinavian Igneous Belt). **(d)** Temperature structure along cross section (c). **(e)** S-wave velocity along cross section (c) showing an overall velocity increase to the east. **(f)** Comparison of measured (*black lines*) and modeled (*red, blue*) velocities of southern Norway and southern Sweden. The *gray band* represents velocities from other Archean and Proterozoic cratons. After *Gradmann et al.* [2013].

confirmed that the topography in the Scandinavian orogen is isostatically compensated, in absence of a significant crustal root, by lateral changes in the lower crust and lithospheric mantle structure. They also showed that (i) a step-like increase in lithosphere thickness exists from southern Norway towards Sweden and (ii) a chemical change from fertile Phanerozoic mantle in southern Norway to a more depleted Proterozoic composition exists beneath southern Sweden (Figure 10.6). This lateral transition zone roughly separates Meso- to Neo-Proterozoic and Palaeo-Proterozoic tectonothermal domains, delineating the long-lived edge of the Baltic craton and the Fennoscandia domain. A somewhat similar methodology has been more recently introduced by *Kaban et al.* [2014] and *Tesauro et al.* [2014] and used to study the North American lithosphere.

10.5. THERMOCHEMICAL STRUCTURE FROM MULTI-OBSERVABLE PROBABILISTIC METHODS

The modeling approaches described in the previous section can provide useful first-order information on the thermochemical structure of the lithosphere provided enough information is available to the modeler. However, such methods are not well prepared to deal with one or more of the following problems:

1. Strong nonlinearity of the system. Traditional linearized inversions do not generally provide reliable estimates of the thermochemical structure of the Earth.
2. The temperature effect on geophysical observables is in most cases greater than the compositional effect; therefore the latter is significantly more difficult to isolate.
3. Nonuniqueness of the compositional field. Different compositions can fit equally well seismic and potential field observations (Figure 10.7).
4. Strong correlations between physical parameters and geophysical observables complicate the inversion procedure and their effects are poorly understood.
5. Trade-off between temperature and composition in wave speeds.
6. Uncertainties affecting the results are difficult to estimate.

Taking into account all these considerations, a robust characterisation of the 3D thermochemical structure of the upper mantle would require six basic ingredients:

• Representative and general a priori information on compositional space. Given that the problem of retrieving a single composition from geophysical information has

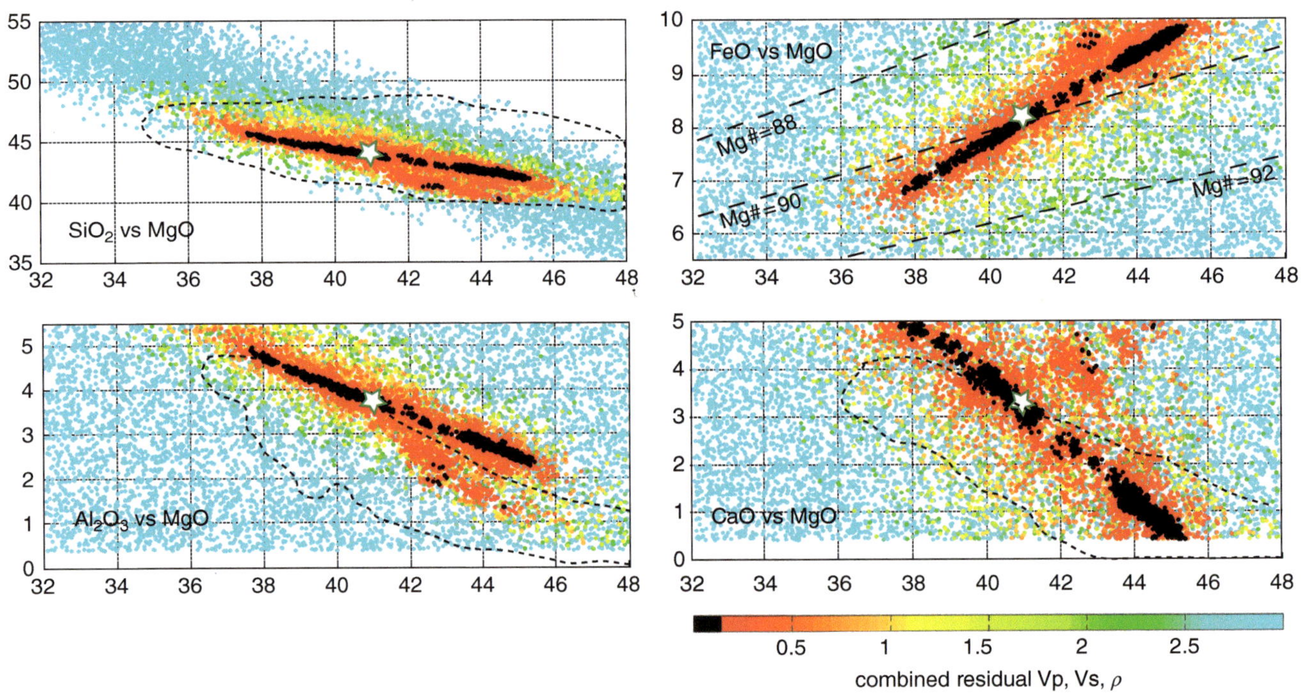

Figure 10.7 Monte Carlo search within the "peridotitic" compositional space of samples that fit a specific set of seismic velocities and bulk density (indicated by the *white star*). All *black samples* are identical, for all practical purposes, in terms of their physical properties and therefore a unique best solution is difficult (impossible?) to justify based on physical properties only. The *dashed envelopes* encircle > 95% of the natural samples shown in Figure 10.2. Provided that this database is representative of the natural variability in the lithospheric mantle, it can be used as prior information to minimize the number of possible solutions (e.g., *Afonso et al.* [2013a,b]).

no unique solution [*Afonso et al.*, 2013a], the role of a priori information is critical. This prior information should be general enough to cover the entire spectrum of observed/expected bulk compositions in mantle peridotites, from fertile (or refertilized) lithospheric mantle to highly depleted harzburgites and dunites.

• Thermodynamic consistency between thermal and compositional parameters. This means that parameters such as bulk density, bulk modulus, or shear-wave velocity are not free to vary independently, but are linked together by the fundamental laws of thermodynamics; the only independent variables in the system are temperature, pressure, and composition. This guarantees that no violation of thermodynamic principles will occur during the modeling or inversion (something that is impossible when using assumed ratios such as ρ/Vs; e.g., see *Forte* [2007] and *Simmons et al.* [2010]).

• Internal consistency between observables and the thermochemical state of the mantle. This means that all calculated observables (e.g., dispersion curves, travel times, etc.) are only and ultimately dependent on temperature, pressure, and composition (the governing independent variables) while being linked together by robust and sound (typically nonlinear) physical theories. This guarantees that a local change in properties such as density or Vs, which may be required to improve the fitting of a particular observable, will be also reflected in all other observables in a consistent manner. It also implies that no linearity between observables needs to be assumed; each observable responds according to its own governing physical theory (e.g., sound propagation).

• The use of multiple observables, each sensitive to different aspects of the problem. This minimizes the range of acceptable models by increasing the number of constraining datasets. This requires that each observable has a different sensitivity to the main parameters (T, P, composition). This seemingly obvious requirement is crucial when the uncertainties in observations are considered. Every new observable carries uncertainties that are propagated to the final result. Therefore, if a new observable is added to the inversion, but its sensitivity is equal to or less than other previous observables, the net effect could be a degradation of the results.

• Representative estimation of uncertainties, preferably as full posterior probability density functions. This is critical when working with different datasets subject to different uncertainties (both in nature and magnitude). In this context, probabilistic methods provide a natural way to deal with both theoretical/methodological and observational uncertainties into the inversion problem. The final posterior distribution (i.e., result of the inversion) is in itself a measure of the uncertainties affecting the results.

• Generality and objectivity. This requirement ensures the inverse framework is applicable to a large number of scenarios (e.g., different tectonic settings, different terrestrial bodies) and to a multitude of data with the minimum (necessary) input of subjective information.

A number of recent efforts have focused on developing methods honoring these ingredients and capable of directly inverting geophysical, geochemical, and/or geodetic data sets for the fundamental thermodynamic variables inside the Earth, namely temperature, pressure, and composition [*Khan et al.*, 2009, 2011; *Kuskov et al.*, 2011, 2014; *Afonso et al.*, 2013a,b, 2014]. One approach that holds great promise for obtaining reliable images of the temperature and compositional structure of the lithosphere and upper mantle is based on the joint inversion of multiple datasets within a probabilistic (Bayesian) framework. In the past 20 years, Bayesian methods applied to inverse problems have become standard and powerful tools in geophysics, and many review papers and textbooks are now available on the subject (cf. *Mosegaard* [1998], *Bosch* [1999], *Mosegaard and Tarantola* [1995], *Mosegaard and Sambridge* [2002], *Tarantola* [2005], *Idier* [2008], *Biegler et al.* [2011]). *Khan et al.* [2009, 2011] and *Afonso et al.* [2013a,b] provide detailed discussions on the advantages of probabilistic over traditional (nonprobabilistic) inversion methods when studying the thermochemical structure of the mantle. In addition, probabilistic methods can easily deal with multiscale parameterizations (multiple resolutions) needed to simultaneously invert datasets of different nature, and they allow to work with realistic (complex) Earth models. When combined with robust thermodynamic formalisms (e.g., see *Stixrude and Lithgow-Bertelloni* [2005] and *Ganguly et al.* [2009]) to solve the forward problems (i.e., prediction of observables), probabilistic methods offer the most general, and arguably the best, solution to invert for possible compositional structures within the upper mantle. On the downside, probabilistic inversion methods rely on long Markov Chain Monte Carlo (MCMC) simulations, in which the forward problems are solved typically on the order of 10^4 to 10^8 times, making them much more time-consuming and computationally expensive than traditional matrix-based methods.

In order to relate seismic structure to the actual/absolute physical state of the lithosphere and upper mantle using thermodynamic formalisms (i.e., absolute temperature, bulk composition, etc.), we need information on absolute rather than relative wave speeds. Thus, surface wave tomography seems to be more adequate to this end than teleseismic tomography. In this context, phase–velocity dispersion maps obtained from the combination of ambient noise tomography with teleseismic surface wave tomography constitute one of the best sources of information for retrieving absolute Vs values in the lithosphere and sublithospheric upper mantle (see Section 10.4.2). The recent extension of ambient noise techniques to obtain long-period (up to 150 sec) dispersion maps [*Yang*, 2014] bypasses the need to complement short-period ambient noise dispersion data with teleseismic earthquake data and represents a potentially excellent source of absolute Vs values at lithospheric depths. However, surface wave data from fundamental mode observations are not sensitive to structure at depths below ~300 km and their lateral resolution is poorer than body wave data (in dense arrays). Although the inclusion of overtones can in principle increase the sensitivity of surface waves to deeper structure (e.g., see *van Heijst and Woodhouse* [1999], *Visser et al.* [2008], *Khan et al.* [2011], and *Schaeffer and Levedev* [2013]), the limited lateral resolution remains an issue in regional studies.

A number of authors have combined the individual strengths of surface wave and body wave teleseismic data to map the absolute velocity structure of the upper mantle (e.g., see *West et al.* [2004], *Rawlinson and Fishwick* [2011], and *Obrebski et al.* [2011]. Similarly, teleseismic tomography and surface wave tomography can be combined under a thermodynamically constrained probabilistic inversion scheme to obtain direct estimates of the temperature and compositional structure of the lithosphere and upper mantle. In this approach, Vp, Vs, and bulk density are linked through sound thermodynamic relationships, and therefore any Vs structure used to fit, for example, dispersion data has an equivalent and thermodynamically consistent Vp structure that can be used to solve the P-wave teleseismic problem (accounting for frequency-dependent attenuation). Given the time involved in computing teleseismic travel times in a 3D volume (a nonlinear problem), the implementation of grid-based eikonal solvers within probabilistic inversions is challenging. *Afonso et al.* [2014] have recently tackled this problem with a strategy that, albeit simple, provides reliable results. The authors do not solve the nonlinear problem at every step of the MCMC random walk, but only every n predefined steps instead. In between these n nonlinear solutions, travel-time residuals are computed by a simple integration of slowness along each fixed ray path. This locally linear approximation therefore assumes that the ray-path geometry is independent of the actual velocity structure. This is a reasonable approximation for teleseismic paths as long as the model does not change significantly between each nonlinear step. Moreover, by updating the paths every n steps as the inversion proceeds, the nonlinear behavior of the system is ultimately explicitly considered. The choice of n depends on the region under study, and it can be chosen based on preliminary tests [*Afonso et al.*, 2014]. One difficulty associated with this approach is the assumption of a spherically symmetric Earth outside the target volume. Several studies have shown that heterogeneities in the lower mantle and transition zone can affect the amplitude of the velocity anomalies in standard regional tomography methods (e.g., see *Masson and Trampert* [1997], *Bijwaard et al.* [1998], and *Zhao et al.* [2013]). However, the general pattern of anomalies is not significantly affected [*Zhao et al.*, 2013]. Since the method in *Afonso et al.* [2014] does not rely on recovered amplitudes, it is unclear whether the improvement that would be gained by considering a global velocity model outside the target volume justifies the significant increase in computational time. Further tests are needed to evaluate this possibility.

An important difficulty that needs to be emphasized when estimating temperature and composition in the mantle is the fact that no unique composition can be generally found; an entire population of compositions can generally fit the data equally well. This is neither related to the primitive uncertainties associated with the datasets nor related to the theoretical uncertainties affecting forward solvers. It is an intrinsic feature displayed by mantle rocks/minerals [*Afonso et al.*, 2013a]. This is illustrated in Figure 10.7, which shows the results of an extensive parameter space search using the Neighbourhood Algorithm of *Sambridge* [1999]. The search is guided by a simple (L1) misfit function of the form

$$\text{misfit} = \left|\rho^p - \rho^o\right| + \left|Vp^p - Vp^o\right| + \left|Vs^p - Vs^o\right|, \qquad (10.1)$$

where the superscripts ^{p} and ^{o} refer to "predicted/theoretical" and "observed/true," respectively, and $|\cdots|$ brackets indicate absolute values. Predicted values of bulk density and sound velocities are computed at $T = 900°C$, $P = 1.2$ GPa by energy minimization [*Connolly*, 2009] following the thermodynamic formalism of *Stixrude and Lithgow-Bertelloni* [2005] and the thermodynamic database of *Xu et al.* [2008]. For all practical purposes, all samples in black are characterized by identical bulk ρ, Vp, and Vs, and therefore they explain the original data equally well. Note that this example represents the best possible scenario, where we have true values of the actual physical parameters. In a real case scenario, however, we would have to invert real data (e.g., gravity anomalies, travel times, etc.) subject to observational uncertainties, which would further expand the space of valid solutions.

The example above highlights the critical role of a priori information on the compositional space. Two common approaches to reducing the population of valid solutions are either assigning tighter bounds to the compositional space or working with a linear combination of basaltic and harzburgitic/dunitic end-member components to represent the fertile-depleted trend of peridotites. The first option usually uses local xenolith information (e.g., see *James et al.* [2004], *Roy et al.* [2005], and *Fullea et al.* [2010]) or empirical correlations between the age of the overlying crust and the composition of the lithospheric mantle (e.g., see *Carballo et al.* [2015]). While practical and convenient, this approach also represents a highly subjective option that cannot be used to explore the compositional heterogeneity of the lithospheric mantle. The second option constitutes a more general approach, especially in large-scale studies (e.g., see *Khan et al.* [2011] and *Ritsema et al.* [2009]), but it still leaves a large portion of the naturally occurring compositional variability unrepresented (Figure 10.3). One way of including a priori compositional information into multi-observable inversions without compromising generality and representativeness has been presented in *Afonso et al.* [2013a]. The approach is based on a large database of natural and well-studied mantle samples (from xenoliths, abyssal peridotites, ophiolites, etc.) and on robust correlations between major oxides. This database is used to derive multidimensional probability density functions in the "peridotitic" compositional space that describe the natural variability observed in samples from many different tectonic settings (Figure 10.3). However, other minor mantle lithologies such as eclogites and pyroxenites were not included explicitly as part of the a priori PDFs used in *Afonso et al.* [2013a], and therefore its application is strictly valid if the assumption of an upper mantle dominated by peridotite holds true. If large amounts of basaltic or eclogitic material are present as mechanically and chemically independent lithologies (e.g., see *Xu et al.* [2008], *Khan et al.* [2011], and *Ritsema et al.* [2009]), the above approach should be extended to include these lithologies. However, more work towards subjecting the mixture hypothesis to rigorous tests (using multiple geophysical observables), as well as to geochemical/geodynamic arguments, is still needed.

10.6. THE PROMISING FUTURE

The last 20 years have witnessed an outstanding increase in both the amount and quality of geophysical data pertinent to the lithosphere and upper mantle. The data and images obtained with dense broadband arrays such as USArray, ChinaArray, and IberArray have clearly demonstrated their value and effectiveness in characterizing the fine-scale structure of the lithosphere and upper mantle. Although difficult to predict, the deployment of such arrays in other regions of the world surely will become more common in the near future. Likewise, satellite missions are providing cost-effective global databases of the topography, magnetic, and gravity fields with ever-increasing resolution and accuracy. The gradiometric and magnetic measurements provided by the ESA GOCE and Swarm missions represent a recent example of the utility and potential of global satellite data in studying the lithosphere [e.g. *Bouman et al.*, 2013; *Fullea et al.*, 2015]. Dense magnetotelluric arrays, often coincident with seismic broadband arrays (e.g., USArray, IberArray) are also becoming more popular (e.g., see *Meqbel et al.* [2014], *Dong et al.* [2011], and Stolz [2013]), and this is a trend likely to continue.

Taking full advantage of all these datasets (and others not mentioned, e.g., GPS, A-DInSAR surface motion) requires methods capable of inverting/modeling multiple observables, with different resolutions and sensitivities to different features of the Earth's interior. We have argued in this chapter that probabilistic methods offer a viable, and perhaps the best, approach to inverting multiple datasets for the physical state of the lithosphere and upper mantle. However, multi-observable probabilistic inversions still is a relatively understudied field, and a number of conceptual and technical difficulties remain to be addressed and circumvented. For instance, a realistic and formal treatment of both observational and theoretical uncertainties is challenging, yet critical (e.g., *Afonso et al.* [2013a,b], *Fournier et al.* [2013], and *Rawlinson et al.* [2014]). Also, more efficient and robust parallel MCMC algorithms to jointly invert massive multiple datasets will be needed. In this context, working with significantly different forward problems (different resolutions and sensitivities) that cannot be related to each other using a unique physical theory can create conflict between predictions that may be difficult to reconcile. In other words, while one observable guides the solution towards one set of model parameters, another observable points towards a different set. Although this can create an internal conflict that poses a serious challenge to the modeler as well as to common MCMC optimization algorithms, the fact that the inversion gives conflicting solutions does not necessarily mean that the observables have an intrinsic conflicting nature. At least in some cases, this conflicting behavior is rooted in the parameterization/design of our forward models. Therefore, the detection of such inconsistencies can inform the user about aspects of the model that are not well understood or dealt with, which in itself represents a gain of information. Regardless of whether the inconsistency lies in the design of the forward problem (or model) or in the nature of the observables, the problem is typically exacerbated when single-objective MCMC algorithms are used. Strictly, multiobservable inversion is a multiobjective or vector optimization problem (cf. *Marler and Arora* [2004]) and standard scalarization (or weighted-sum) single-objective approaches, typically used in geophysics, are not guaranteed to offer a good sampling of the complete pareto-optimal set (e.g., see *Das and Dennis* [1998] and *Marler and Arora* [2004]). The latter is an important ingredient of multiobjective optimization that provides critical and systematic information on the trade-offs between model parameters and the multiobjective functions used in the inversion. Moreover, the selection of weights to form the single-objective (or misfit) function is typically problem-dependent and involves a great deal of calibration and subjective input by the modeler. This can make the inversion process inefficient. However, an exploration of full multiobjective methods in the context of multiobservable inversions for the physical state of the Earth is not only nonexistent, but also beyond the scope of the present paper. We leave this issue for future work.

Working under a scheme that allows taking full advantage of complex forward problems is also important to move towards a more holistic approach to imaging the Earth's interior. Clearly, the generation of thermophysical models that are simultaneously constrained by multiple geophysical and geodynamic observables would bring a qualitative leap in our understanding of the nature and evolution of the lithosphere, its interaction with the sublithospheric mantle, and the driving forces responsible for plate motion. Several approaches have been proposed for solving the joint geodynamic-gravity-seismic inversion/modeling problem; all of them are intended for low-resolution, large-scale models (e.g., see *Forte* [2007], *Forte* [2000] *Simmons et al.* [2006, 2010], and *Cammarano et al.* [2011]). One of the most popular and advanced approaches is that of *Simmons et al.* [2010], in which shear and compressional wave velocities are linked to bulk density through empirical mineral-physics parameters. In contrast to other similar approaches, *Simmons et al.* [2010] perform a real inversion of seismic data. However, this method still relies on two somewhat restrictive assumptions: (i) All responses of the system are assumed to be linear and (ii) temperature and compositional signatures are decoupled during the inversion. Future studies coupling consistently the dynamic behavior of the mantle with multiple surface observations, honoring the nonlinear and thermodynamic consistency of the system, will undoubtedly shed critical information about the internal workings of the Earth, particularly on the complex interaction between plates and the underlying convecting mantle. Some preliminary steps have been recently taken [*Afonso et al.*, 2014], demonstrating the feasibility and potential of such inversions.

Similarly, the migration from approaches based on ray theory to finite frequency and full waveform approaches seems to be an obvious next step in multiobservable inversions. Both tomography methods hold great potential to improve on the quality of thermochemical and seismic images of the upper mantle, particularly at regional and continental scales (e.g., see *Chevrot and Zhao* [2007], Nolet [2008], *Liu and Gu* [2012], *Fitchner et al.* [2013], and *Yuan et al.* [2014]). Likewise, inclusion of global information into the computation of teleseismic travel times (to explicitly account for the effect of deep anomalies) can be of benefit [*Zhao et al.*, 2013]. However, since the use of more traditional ray-based approaches in multiobservable probabilistic inversion methods still is in its initial stages of development [*Afonso et al.*, 2014], it is unclear at the moment to what extent results will be improved by implementing the above approaches or whether the additional benefits will outweigh the extra computational cost. Based on current trends, these issues are expected to be addressed in the next decade.

10.7. FINAL REMARKS

Integrated modeling and joint inversions for the structure of the lithosphere and upper mantle are becoming increasingly popular, partly because of the advances reviewed in this chapter. The recognition that working with complementary datasets provides a substantially more instructive view of the Earth is not new, but surprisingly practical implementations of such joint or multiobservable approaches are still far less common than single-data approaches. Working with multiple datasets demands a multidisciplinary framework and a diverse set of skills that in most cases go beyond geophysics, well into geochemistry, thermodynamics, and petrology. The more we learn about our planet, the more we recognize the importance of understanding it as a complex, dynamic physical entity rather than a static object described by stiff physical parameters such as mean density, seismic velocity, coefficients of thermal expansion, and so on. And when we pay attention to these complexities, to the great variety of links between physical and chemical processes within the Earth, we realize the importance of the connections between traditional disciplines. Unfortunately, traditional geophysics university programs tend to underestimate these connections, something that we believe will change in the next decade or so. This book is testament to the increasing interest in the solid Earth community regarding such integrated studies.

ACKNOWLEDGMENTS

Much of the material presented here is the result of many discussions with many colleagues over the years. We are particularly indebted to N. Rawlinson, Y. Yang, A. G. Jones, J. A. D. Connolly, W. Griffin, S. O'Reilly, B. Kennett, H. Yuan, S. Karato, U. Faul, S. Zlotnik, M. Fernandez, D. Schutt, S. Lebedev, H. Zeyen, M. Sambridge, A. Khan, S. Thiel, G. Heinson, and M. Qashqai. We thank S. I. Karato and F. Deschamps for their detailed and most valuable comments on an earlier version of this chapter. The work of JCA has been supported by an Australian Research Council Discovery Grant (DP120102372). JF was supported by the JAE-DOC programme (CSIC-Spain) co-founded by ESF, as well as by Spanish Ministry of Economy and Competitiveness grant CGL2012-37222, and funded by the People Programme (Marie Curie Actions) of the European Union's H2020-MSCA-IF-2014 programme under REA grant agreement n° 657357. This is contribution 619 from the ARC Centre of Excellence for Core to Crust Fluid Systems (http://www.ccfs.mq.

edu.au) and 1006 in the GEMOC Key Centre (http://www.gemoc.mq.edu.au).

REFERENCES

Afonso, J. C., and D. Schutt (2012), The effects of polybaric partial melting on the density and seismic velocities of mantle restites, *Lithos*, *134–135*, 289–303.

Afonso, J. C., M. Fernàndez, G. Ranalli, W. L. Griffin, and J. A. D. Connolly (2008a), Integrated geophysical–petrological modelling of the lithospheric-sublithospheric upper mantle: methodology and applications. *Geochem. Geophys. Geosyst. 9*, Q05008, doi:10.1029/2007GC001834.

Afonso, J. C., S. Zlotnik., and M. Fernandez (2008b), The effects of compositional and rheological stratifications on small-scale convection under the oceans: implications for the thickness of oceanic lithosphere and seafloor flattening, *Geophys. Res. Lett.*, *35*, doi:10.1029/2008GL035419.

Afonso, J. C., G. Ranalli, M. Fernandez, W. L. Griffin, S. Y. O'Reilly, and U. Faul (2010), On the Vp/Vs–Mg# correlation in mantle peridotites: Implications for the identification of thermal and compositional anomalies in the upper mantle. *Earth Planet. Sci. Lett.*, *289*, 606–618.

Afonso, J. C., J. Fullea, W. L. Griffin, Y. Yang, A. G. Jones, J. A. D. Connolly, and S. Y. O'Reilly (2013a), 3-D multi-observable probabilistic inversion for the compositional and thermal structure of the lithosphere and upper mantle. I: A priori petrological information and geophysical observables, *J. Geophys. Res. Solid Earth*, *118*, 25862617, doi:10.1002/jgrb.50124.

Afonso, J. C., J. Fullea, Y. Yang, J. A. D. Connolly, and A. G. Jones (2013b), 3-D multi-observable probabilistic inversion for the compositional and thermal structure of the lithosphere and upper mantle. II: General methodology and resolution analysis, *J. Geophys. Res. Solid Earth*, *118*, 16501676, doi:10.1002/jgrb.50123.

Afonso, J.C., Rawlinson, N., Yang, Y., Schutt, D., Jones, A., Fullea, J. (2014), Multi-observable probabilistic tomography reveals the thermochemical structure of Central-Western US. AGU Fall Meeting, Abstract S51B-4467.

Agius, M. R., and Lebedev, S. (2013), Tibetan and Indian lithospheres in the upper mantle beneath Tibet: Evidence from broadband surface-wave dispersion. *Geochem. Geophys. Geosyst.*, *14*, 42604281, doi:10.1002/ggge.20274.

Aizawa, Y., Barnhoorn, A., Faul, U.H., Fitz Gerald, J.D., Jackson, I., Kovcs, I. (2008) Seismic properties of Anita Bay Dunite: An exploratory study of the influence of water. *J. Petrol.*, *49*, 841–855.

Aki, K., and Lee, W. H. K. (1976), Determination of three-dimensional velocity anomalies under a seismic array using first P-arrival times from local earthquakes, 1, homogeneous initial model. *J. Geophys. Res.*, *81*, 4381–4399.

Alasoniati-Tasarova, S., J. C. Afonso, M. Bielik, H-J. Goetze, and J. Hok. (2009), The lithospheric structure of the Western Carpathian–Pannonian Basin region based on the CELEBRATION 2000 seismic experiment and gravity modelling. *Tectonophysics*, *475*, 454–469.

Ammon, C. J., G. E. Randall, and G. Zandt (1990), On the non-uniqueness of receiver function inversions, *J. Geophys. Res.*, *95*(B10), 1530315318, doi:10.1029/JB095iB10p15303.

Anderson, D. (1989), *Theory of the Earth*, Blackwell Scientific Publications, Oxford, UK.

Anderson, O. L., and D.G. Isaak (1995), Elastic constants of mantle minerals at high temperature, in *Mineral Physics and Crystallography: A Handbook of Physical Constants*, J. Ahrens, ed., AGU reference shelf 2, Washington, D.C.

Artemieva, I. M. (2009), The continental lithosphere: Reconciling thermal, seismic, and petrologic data, *Lithos 109*, 23–46.

Artemieva, I.M. (2011), *The Lithosphere: An Interdisciplinary Approach*, Cambridge University Press, Cambridge, UK.

Aubaud, C., E. H. Hauri, and M. M. Hirschmann (2004), Hydrogen partition coefficients between nominally anhydrous minerals and basaltic melts, *Geophys. Res. Lett.*, *31*, L20611, doi:10.1029/2004GL021341.

Aubaud, C., M. M. Hirschmann, A. C. Withers, and R. L. Hervig (2008), Hydrogen partitioning between melt, clinopyroxene, and garnet at 3 GPa in a hydrous MORB with 6 wt.% H_2O, *Contrib. Min. Petrol.*, *156*, 607–625.

Bailey, R. C. (1970), Inversion of the geomagnetic induction problem, *Proc. R. Soc. London*, *315*, 185–194.

Becker, T. W. (2012), On recent seismic tomography for the western United States, *Geochem. Geophys. Geosyst.*, *13*, Q01W10, doi:10.1029/2011GC003977.

Bedrosian, P. A. (2007), MT+ integrating magnetotellurics to determine earth structure, physical state, and processes, *Surv. Geophys.*, *28*, 2–3, 121–167.

Bedrosian, P. A., N. Maercklin, U. Weckmann, Y. Bartov, T. Ryberg, and O. Ritter (2007), Lithology-derived structure classification from the joint interpretation of magnetotelluric and seismic models, *Geophys. J. Int.*, *170*, 737–748.

Bennington, N. L., H. Zhang, C. H. Thurber, and P. A. Bedrosian (2015), Joint inversion of seismic and magnetotelluric data in the Parkfield region of California using the normalized cross-gradient constraint, *Pure Appl. Geophys.*, *172*, 1033–1052.

Biegler, L., et al. (eds.) (2011), *Large-Scale Inverse Problems and Quantification of Uncertainty*, John Wiley & Sons, UK.

Bijwaard H., W. Spakman, and E. R. Engdahl (1998), Closing the gap between regional and global traveltime tomography, *J. Geophy. Res.*, *103*, 3005530078.

Bodinier, J.-L., C. J. Garrido, I. Chanefo, O. Bruguier, and Gervilla, F. (2008), Origin of pyroxeniteperidotiteveined mantle by refertilization reactions: Evidence from the ronda peridotite (Southern Spain), *J. Petrol.*, *49*, 999–1025.

Bodinier, J.-L., and M. Godard (2003), Orogenic, ophiolitic, and abyssal peridotites, in *The Mantle and Core*, R. W. Carlson, ed., Vol. *2*, *Treatise on Geochemistry*, H. D. Holland, and K. K Turkeian, eds., Elsevier-Pergamon, Oxford.

Bolfan-Casanova, N., H. Keppler, and D.C. Rubie (2000), Water partitioning between nominally anhydrous minerals in the $MgO–SiO_2–H_2O$ system up to 24 GPa: Implications for the distribution of water in the Earth's mantle, *Earth Planet. Sci. Lett.*, *182*, 209–221.

Bosch, M. (1999), Lithologic tomography: From plural geophysical data to lithology estimation, *J. Geophys. Res.*, *104*(B1), 749766, doi:10.1029/1998JB900014.

Bowin, C. (2000), Mass anomaly structure of the Earth, *Rev. Geophys.*, *38*(3), 355–387.

Burov, E. B. (2011), Rheology and strength of the lithosphere, *Marine Petrol. Geol.*, *28*, 1402–1443.

Cammarano F., P. Tackley, and L. Boschi (2011), Seismic, petrological and geodynamical constraints on thermal and compositional structure of the upper mantle: Global thermochemical models, *Geophys. J. Int.*, *187*, 1301–1318.

Carballo, A., M. Fernandez, M. Torne, I. Jimnez-Munt, and A. Villaseor (2015), Thermal and petrophysical characterization of the lithospheric mantle along the northeastern Iberia geo-transect, *Gondwana Res.*, http://dx.doi.org/10.1016/j.gr.2013.12.012.

Carlson, R. W., D. G. Pearson, and D. E. James (2005), Physical, chemical, and chronological characteristics of continental mantle. *Rev. Geophys.*, *43*, RG1001, 1–24.

Chappell, A. R., and N. J. Kusznir (2008), Three-dimensional gravity inversion for Moho depth at rifted continental margins incorporating a lithosphere thermal gravity anomaly correction, *Geophys. J. Int.*, *174*(1) (2008), 113.

Chen, C.-W., S. Rondenay, R. L. Evans, and D. B. Snyder (2009), Geophysical detection of relict metasomatism from an Archean (3.5 Ga) subduction zone, *Science*, *20*, 326, 1089–1109.

Chevrot, S., and L. Zhao (2007), Multiscale finite-frequency Rayleigh wave tomography of the Kaapvaal craton. *Geophys. J. Int.*, *169*, 201–215.

Connolly, J. A. D. (2009), The geodynamic equation of state: What and how, *Geochem. Geophys. Geosyst.*, *10*, Q10014, doi:10.1029/2009GC002540.

Conrad, C. P., and C. Lithgow-Bertelloni (2006), Influence of continental roots and asthenosphere on platemantle coupling, *Geophys. Res. Lett.*, *33*, doi:10.1029/2005GL025621.

Constable, S. C., and R. L. Parker, and C. G. Constable (1987), Occam's inversion: A practical algorithm for generating smooth models from electromagnetic sounding data, *Geophysics*, *52*, 289–300.

Dai, L., and S. I. Karato (2009), Electrical conductivity of orthopyroxene: Implications for the water content of the asthenosphere, *Proc. Japan Acad., Ser. B, Phys. Biol. Sci.*, *85*, 466.

Dai, L., and S.-I. Karato (2014), High and highly anisotropic electrical conductivity of the asthenosphere due to hydrogen diffusion in olivine, *Earth Planet. Sci. Lett.*, *408*, 79–86.

Das, I., and J. E. Dennis (1998), Normal-boundary intersection: A new method for generating the Pareto surface in nonlinear multicriteria optimization problems, *SIAM J. Optim*, *8*, 631657.

Deschamps, F., R. Snieder, and J. Trampert (2001), The relative density-to-shear velocity scaling in the uppermost mantle. *Phys. Earth Planet. Inter.*, *124*(3), 193–212.

Doin, M. P., L. Fleitout, and D. P. McKenzie (1996), Geoid anomalies and the structure of continental and oceanic lithospheres, *J. Geophys. Res.*, *101*, 16119–16135.

Dong, S., T. Li, R. Gao, H. Hou, Y. Li, S. Zhang, and M. Liu (2011), A multidisciplinary earth science research program in China, Eos, *Transactions American Geophysical Union*, *92*(38), 313–314.

Downes, H. (2007), Origin and significance of spinel and garnet pyroxenites in the shallow lithospheric mantle: Ultramafic massifs in orogenic belts in Western Europe and NW Africa. *Lithos*, *99*, 1–24.

Eaton, D. W., F. Darbyshire, R. L. Evans, H. Grutter, A. G. Jones, and X. Yuan (2009), The elusive lithosphere–asthenosphere boundary (LAB) beneath cratons, *Lithos*, *109*, 1–22.

Ebbing, J., C. Braitenberg, and H.-J. Gtze (2006), The lithospheric density structure of the Eastern Alps, *Tectonophysics*, *414*, 145–155. doi:10.1016/j.tecto.2005.10.015.

Eberhart-Phillips, D. (1990), Three-dimensional P and S velocity structure in the Coalinga Region, California. *J. Geophys. Res.*, *95*, 15343–15363.

Evans, J. R., and Achauer, U. (1993), Teleseismic velocity tomography using the ACH method: Theory and application to continental-scale studies, in *Seismic Tomography*, H. M. Iyer and K. Hirahara, eds., Chapman & Hall, London, pp. 319–360.

Evans, R. L., et al. (2011), Electrical lithosphere beneath the Kaapvaal craton, southern Africa, *J. Geophys. Res.*, *116*, B04105, doi:10.1029/2010JB007883.

Evans, R. L., and G. Hirth, K. Baba, D. Forsyth, A. Chave, and R. Mackie (2005), Geophysical evidence from the MELT area for compositional controls on oceanic plates, *Nature*, *437*, 249–252.

Fernandez, M., M. Torne, D. Garcia-Castellanos, J. Verges, W. Wheeler, and R. Karpuz (2004), Deep structure of the Vring Margin: The transition from a continental shield to a young oceanic lithosphere, *Earth Planet. Sci. Lett.*, *221*, 131–144.

Fernàndez, M., J. C. Afonso, and G. Ranalli (2010), The deep lithospheric structure of the Namibian volcanic margin, *Tectonophysics*, *481*, 68–81.

Ferot, A. and Bolfan-Casanova, N. (2012), Water storage capacity in olivine and pyroxene to 14GPa: Implications for the water content of the Earth's upper mantle and nature of seismic discontinuities, *Earth Planet. Sci. Lett.*, *349*, 218–230.

Fichtner A., J. Trampert, P. Cupillard, E. Saygin, T. Taymaz, Y. Capdeville, and A. Villasenor (2013), Multiscale full waveform inversion. *Geophys. J. Int.*, *194*, 534–556.

Fischer, K. M., H. A. Ford, D. L. Abt, and C. A. Rychert (2010), The lithosphere-asthenosphere boundary, *Annu. Rev. Earth Planet. Sci.*, *38*, 551–575.

Fischer, K. M., H. A. Ford, D. L. Abt, and C. A. Rychert (2013), The lithosphere–asthenosphere boundary, *Annu. Rev. Earth Planet. Sci.*, *38*, 551–575.

Fishwick, S., M. Heintz, B. L. N. Kennett, A. Reading, and Y. Yoshizawa (2008), Steps in lithospheric thickness within eastern Australia, evidence from surface wave tomography. *Tectonics*, *27*, doi:10.1029/2007TC002116.

Forte, A. M. (2000), Seismic–geodynamic constraints on mantle flow: Implications for layered convection, mantle viscosity, and seismic anisotropy in the deep mantle, in *Earth's Deep Interior: Mineral Physics From the Atomic to the Global Scale, Geophysics Monograph Series 117*, by S. I. Karato et al., eds., AGU, Washington, DC, p. 336.

Forte, A. M. (2007), Constraints on seismic models from other diciplines—Implication for mantle dynamics and composition, in *Treatise of Geophysics*, Vol. *1*, B. Romanowicz and A. M. Dziewonski, eds., Elsevier, Oxford, pp. 805–854.

Forte, A. M., and J. X. Mitrovica (2001), Deep-mantle high-viscosity flow and thermochemical structure inferred from seismic and geodynamic data, *Nature*, *410* (6832), 1049–1056.

Foulger, G., et al. (2013), Caveats on tomographic images, *Terra Nova*, *25*, 259–281.

Fournier, A., et al. (2013), Assessing uncertainty in geophysical problems—Introduction, *Adv. Geophys.*, *78*, 3, WB1–WB2.

Fullea, J., J. Fernandez, H. Zeyen, and J. Vergs (2007), A rapid method to map the crustal and lithospheric thickness using elevation, geoid anomaly and thermal analysis. Application to the Gibraltar Arc System, Atlas Mountains and adjacent zones, *Tectonophysics*, *430*, 97–117.

Fullea, J., M. Fernandez, and H. Zeyen (2008), FA2BOUG-a FORTRAN 90 code to compute Bouguer gravity anomalies from gridded free air anomalies: Application to the Atlantic–Mediterranean transition zone, *Comput. Geosci.*, *34*, 1665–1681, doi:10.1016/j.cageo.2008.02.018.

Fullea, J., J. C. Afonso, J. A. D. Connolly, M. Fernàndez, D. Garcia-Castellanos, and H. Zeyen (2009), LitMod3D: An interactive 3-D software to model the thermal, compositional, density, seismological, and rheological structure of the lithosphere and sublithosheric upper mantle. *Geochem. Geophys. Geosyst.*, *10*, Q08019, doi:10.1029/2009GC002391.

Fullea, J., M. Fernàndez, J. C. Afonso, J. Vergés, and H. Zeyen (2010), The structure and evolution of the lithosphere–asthenosphere boundary beneath the Atlantic–Mediterranean transition region. *Lithos*, *120*, 74–95.

Fullea, J., M. R. Muller, and A. G. Jones (2011), Electrical conductivity of continental lithospheric mantle from integrated geophysical and petrological modeling: Application to the Kaapvaal Craton and Rehoboth Terrane, southern Africa, *J. Geophys. Res.*, *116*, B10202, doi:10.1029/2011JB008544.

Fullea, J., J. Rodríguez-González, M. Charco, Z. Martinec, A. Negredo, and A. Villaseñor (2015), Perturbing effects of sub-lithospheric mass anomalies in GOCE gravity gradient and other gravity data modelling: Application to the Atlantic–Mediterranean transition zone, *Int. J. App. Earth Obs. Geoinf.*, *35*, 54–69.

Furlong, K. P., andD. S. Chapman (2013), Heat flow, heat generation, and the thermal state of the lithosphere, *Ann. Rev. Earth Planet. Sci.*, *41*, 385–410.

Fujii, I., and A. Schultz (2002), The 3D electromagnetic response of the Earth to ring current and auroral oval excitation, *Geophys. J. Int.*, *151*, 689–709.

Gallardo, L. A., and M. A. Meju (2007), Joint two-dimensional cross-gradient imaging of magnetotelluric and seismic traveltime data for structural and lithological classification, *Geophys. J. Int.*, *169*, 1261–1272.

Ganguly, J., A. M. Freed, and S. K. Saxena (2009), Density profiles of oceanic slabs and surrounding mantle: Integrated thermodynamic and thermal modeling, and implications for the fate of slabs at the 660 km discontinuity, *Phys. Earth Planet. Int.*, *172*, 257–267.

Gatzemeier, A., and M. Moorkamp (2005), 3D modelling of electrical anisotropy from electromagnetic array data: Hypothesis testing for different upper mantle conduction mechanisms, *Phys. Earth Planet. Int.*, *149*, 225–242.

Gatzemeier, A., and A. Tommasi (2006), Flow and electrical anisotropy in the upper mantle: Finite-element models constraints on the effects of olivine crystal preferred orientation and microstructure, *Phys. Earth Planet. Int.*, *158*, 92–106.

Gradmann, S., J. Ebbing, and J. Fullea (2013), Integrated geophysical modelling of a lateral transition zone in the lithospheric mantle under Norway and Sweden, *Geophys. J. Int.*, *194*, 1359–1374.

Grand, S. P. (1994), Mantle shear structure beneath the Americas and surroundings oceans, *J. Geophys. Res.*, *99*, 11591–11621.

Grant, K., J. Ingrin, J. P. Lorand, and P. Dumas (2007), Water partitioning between mantle minerals from peridotite xenoliths, *Contributions to Mineralogy and Petrology*, *154* (1), 15–34.

Granet, M., M. Wilson, and U. Achauer (1995), Imaging a mantle plume beneath the French Massif Central, *Earth Planet. Sci. Lett.*, *136*, 281–296.

Green, D. H., and T. J. Fallon (1998), Pyrolite: A Ringwood concept and its current expression, In *The Earth's Mantle edited by* I. Jackson, Cambridge University Press, pp. 311–378.

Green, D. H., W. O. Hibberson, I. Kovács, and A. Rosenthal (2010), Water and its influence on the lithosphere-asthenosphere boundary, *Nature*, *467* (7314), 448–451.

Griffin, W. L., S. Y. O'Reilly, and C. G. Ryan (1999), The composition and origin of sub-continental lithospheric mantle. In: Fei, Y. et al., (eds.), Mantle petrology: Field observations and high-pressure experimentation: A tribute to Francis R. (Joe) Boyd, Vol. 6, The Geochemical Society Houston, pp. 13–45.

Griffin, W. L., and S. Y. O'Reilly (2007), Cratonic lithospheric mantle: Is anything subducted? *Episodes*, *30*(1), 43–53.

Griffin, W. L., S. O'Reilly, J. C. Afonso, and G. Begg (2009), The composition and evolution of lithospheric mantle: A Re-evaluation and its tectonic implications, *J. Petrol.*, *50*(7), 1185–1204.

Hager, B. H. (1984), Subducted slabs and the geoid: Constraints on mantle rheology and flow, *J. Geophys. Res.*, *89*(B7), 60036015, doi:10.1029/JB089iB07p06003.

Hager, B. H., and Richards, M. A. (1989), Long-wavelength variations in Earth's geoid: Physical models and dynamical implications, *Phil. Trans. R. Soc. London, Math. Phys. Eng. Sci.*, *328*, 309–327.

Hasterok, D., and D. S. Chapman (2007), Continental thermal isostasy: 1. Methods and sensitivity, *J. Geophys. Res.*, *112*, B06414, doi:10.1029/2006JB004663.

Hasterok, D., and D. S. Chapman (2011), Heat production and geotherms for the continental lithosphere, *Earth Planet. Sci. Lett.*, *307*, 59–70, doi:10.1016/j.epsl.2011.04.034.

Heinson, G. (1999), Electromagnetic studies of the lithosphere and asthenosphere, *Surv. Geophys.*, *20*, 229–255.

Herzberg, C. (2004), Geodynamic information in peridotite petrology, *J. Petrol.*, *45*, 2507–2530.

Herzberg, C., and M. J. O'Hara (2002) Plume-associated ultramafic magmas of Phanerozoic age, *J. Petrol.*, *43*, 18571883.

Hirose, K., and T. Kawamoto (1995), Hydrous partial melting of lherzolite at 1 GPa: The effect of H_2O on the genesis of basaltic magmas, *Earth Planet. Sci. Lett.*, *133*, 463473.

Hirschmann, M. M. (2006), Water, melting, and the deep Earth's H_2O cycle, *Annu. Rev. Earth Planet. Sci.*, *34*, 629653.

Hirth, G., and D. L. Kohlstedt (1996), Water in the oceanic upper mantle: Implications for rheology, melt extraction and the evolution of the lithosphere, *Earth Planet. Sci. Lett.*, *144*, 93–108.

Hofmann, A. W. (1997), Mantle geochemistry: The message from oceanic volcanism, *Nature*, *385*, 219–229, doi:10.1038/385219a0.

Holland, T. J. B., and R. Powell (2011), An improved and extended internally consistent thermodynamic dataset for phases of petrological interest, involving a new equation of state for solids, *J. Metamorphic Geolo.*, *29*: 333383. doi: 10.1111/j.1525-1314.2010.00923.x

Idier, J. (2008), *Bayesian Approach to Inverse Problems*, John Wiley & Sons, UK.

Isacks, B. L., J. Oliver, and L. R. Sykes (1968), Seismology and the new global tectonics, *J. Geophys. Res.*, *73*, 5855–5899.

Jacob, D. E (2004), Nature and origin of eclogite xenoliths, *Lithos*, *77*, 295–316.

James, D. E., F. R. Boyd, D. Schutt, D. R. Bell, and R. W. Carlson (2004), Xenolith constraints on seismic velocities in the upper mantle beneath southern Africa, *Geochem. Geophysics. Geosys.*, *5*(1), Q01002.

Jaupart, C., and J-C. Mareschal (2011), *Heat Generation and Transport in the Earth*, Cambridge University Press, Cambridge, UK.

Jiménez-Munt, I., M. Fernández, J. Vergés, J. C. Afonso, D. Garcia-Castellanos, and J. Fullea (2010), Lithospheric structure of the Gorringe Bank: Insights into its origin and tectonic evolution, *Tectonics*, *29*, TC5019, doi:10.1029/2009TC002458.

Jones, A. G. (1999), Imaging the continental upper mantle using electromagnetic methods, *Lithos*, *48*, 57–80.

Jones, A. G. (1998), Waves of future: Superior inferences from collocated seismic and electromagnetic experiments, *Tectonophysics*, *286*, 273–298.

Jones, A. G., and J. A. Craven (2004), Area selection for diamond exploration using deep-probing electromagnetic surveying, *Lithos*, *77* (1), 765–782.

Jones, A. G., Evans, R. L., and Eaton, D. W. (2009a), Velocity–conductivity relationships for mantle mineral asssemblages in Archean cratonic lithosphere based on a review of laboratory data and Hashin–Shtrikman extremal bounds, *Lithos*, *109*, 131–143.

Jones, A. G., et al. (2009b), Area selection for diamonds using magnetotellurics: Examples from southern Africa, *Lithos*, *112*, 83–92.

Jones, A. G., Plomerova, J., Korka, T., Sodoudi, F., and Spakman, W. (2010), Europe from the bottom up: A statistical examination of the central and northern European lithosphere–asthenosphere boundary from comparing seismological and electromagnetic observations, *Lithos*, *120*, 14–29.

Jones, A. G., Fullea, J., Evans, R. L., and Muller, M. R. (2012), Calibrating laboratory-determined models of electrical conductivity of mantle minerals using geophysical and petrological observations, *Geochem. Geophys. Geosyst. 13*, Q06010, doi:10.1029/2012GC004055.

Jordan, T. H. (1988), Structure and Formation of the Continental Tectosphere. *J. Petrol.*, *1*, 11–37.

Julia, J., C. J. Ammon, R. B. Herrmann, and A. M. Correig (2000), Joint inversion of receiver functions and surface-wave dispersion observations, *Geophys. J. Int.*, *143*, 99–112.

Julia, J., C. J. Ammon, and R. B. Herrmann (2003), Lithospheric structure of the Arabian Shield from the joint inversion of receiver functions and surface wave group velocities. *Tectonophysics*, *371*, 1–21.

Kaban, M. K., M. Tesauro, W. D. Mooney, and S. A. P. L. Cloetingh (2014), Density, temperature, and composition of the North American lithosphereNew insights from a joint analysis of seismic, gravity, and mineral physics data: 1. Density structure of the crust and upper mantle, *Geochem. Geophys. Geosyst.*, *15*, 4781–4807, doi:10.1002/2014GC005483.

Karato, S.-I. (2011), Water distribution across the mantle transition zone and its implications for global material circulation, *Earth Planet. Sci. Lett.*, *301*, 413–423.

Karato, S.-I., and Wang, D. (2013), Electrical conductivity of minerals and rocks, in S.-I. Karato, ed., *Physics and Chemistry of the Deep Earth*, John Wiley & Sons, Hoboken, NJ, 145–182.

Karato, S-I. (1986), Does partial melting reduce the creep strengh of the upper mantle? *Nature*, *319*, 309–310.

Karato, S.-I. (2008), *Deformation of Earth materials: An Introduction to the Rheology of the Solid Earth*, Cambridge University Press, Cambridge, UK.

Karato, S. I., and H. Jung (1998), water, partial melting and the origin of the seismic low velocity and high attenuation zone in the upper mantle, *Earth Planet. Sci. Lett.*, *157*, 193–207.

Karato, S-I. (2012) On the origin of the asthenosphere, *Earth Planet. Sci. Lett.*, *321*, 95–103.

Kawakatsu, H., P. Kumar, Y. Takei, M. Shinohara, T. Kanazawa, E. Araki, and K. Suyehiro (2009), Seismic evidence for sharp lithosphere-asthenosphere boundaries of oceanic plates. *Science*, *324*, 499–502.

Kelbert, A., A. Schultz, and G. Egbert (2009), Global electromagnetic induction constraints on transition-zone water content variations, *Nature*, *460*, 1003–1006.

Kelemen, P.B, Hart, S.T., and S. Bernstein (1998), Silica enrichment in the continental upper mantle via melt/rock reaction, *Earth Planet. Sci. Lett.*, *164*, 387–406.

Khan, A., L. Boschi, and J. A. D. Connolly (2009), On mantle chemical and thermal heterogeneities and anisotropy as mapped by inversion of global surface wave data, *J. Geophys. Res.*, *114*, B09305, doi:10.1029/2009JB006399.

Khan, A., L. Boschi, and J. A. D. Connolly (2011), Mapping the Earth's thermochemical and anisotropic structure using global surface wave data, *J. Geophys. Res.*, *116*, B01301, doi:10.1029/2010JB007828.

Khan, A., and Shankland, T. J. (2012), A geophysical perspective on mantle water content and melting: inverting electromagnetic sounding data using laboratory-based electrical conductivity profiles, *Earth Planet. Sci. Lett.*, *317*, 27–43.

Khan, A., A. Zunino, and F. Deschamps (2013), Mantle thermochemical and anisotropic variations imaged beneath Australia from Bayesian inversion of surface-wave phase velocities, *J. Geophys. Res. Solid Earth*, *118*, 5285–5306, doi:10.1002/jgrb.50304.

Khan, A., S. Koch, T. J. Shankland, A. Zunino, and J. A. D. Connolly (2015), Relationships between seismic wave-speed, density, and electrical conductivity beneath Australia from seismology, mineralogy, and laboratory-based conductivity profiles, in *The Earths Heterogeneous Mantle—A Geophysical, Geodynamical, and Geochemical Perspective*, A. Khan and F. Deschamps, eds., Springer, New York, pp. 145–171.

Kind, R., X. Yuan, and P. Kumar (2012), Seismic receiver functions and the lithosphere–asthenosphere boundary, *Tectonophysics*, *536–537*, 25–43.

Kinzler, R. J., and T. L. Grove (1992) Primary Magmas of Mid-Ocean Ridge Basalts 1. Experiments and Methods. *J. Geophys. Res.*, *97*(B5), 68856906, doi:10.1029/91JB02840.

Kissling, E., W. L. Ellsworth, D. Eberhart-Phillips, and U. Kradolfer (1994), Initial reference models in local earthquake tomography, *J. Geophys. Res.*, *99*, 19635–19646.

Koch, S., and A. Kuvshinov (2013), Global 3-D EM inversion of Sq variations based on simultaneous source and conductivity determination: Concept validation and resolution studies, *Geophys. J. Int.*, ggt227.

Kohlstedt, D. L., B. Evans, and S. J. Mackwell (1995), Strength of the lithosphere: Constraints imposed by laboratory experiments, *J. Geophys. Res.*, *100*, 17587–17602.

Kohlstedt, D. L., and M. E. Zimmerman (1996), Rheology of partially molten mantle rocks, *Annu. Rev. Earth Planet. Sci. 24*, 41–62.

Koyama, T., Khan, A., and Kuvshinov, A. (2014), Three-dimensional electrical conductivity structure beneath Australia from inversion of geomagnetic observatory data: Evidence for lateral variations in transition-zone temperature, water content and melt, *Geophys. J. Int.*, *196*, 1330–1350.

Kovács, I., D. H. Green, A. Rosenthal, J. Hermann, H. St C. O'Neill, W. Hibberson, and O. Udvardi (2012), An experimental study of water in nominally anhydrous minerals in the upper mantle near the water-saturated solidus, *J. Petrol.*, egs044.

Kumar, N., H. Zeyen, and A. P. Singh (2014), 3D lithosphere density structure of southern Indian shield from joint inversion of gravity, geoid, and topography data, *J. Asian Earth Sci.*, *89*, 98–107.

Kuskov, O. L., V. A. Kronrod, and A. A. Prokofev (2011), Thermal structure and thickness of the lithospheric mantle underlying the Siberian Craton from the Kraton and Kimberlite superlong seismic profiles, *Izvestiya, Physics of the Solid Earth*, *47*, 155–175.

Kuskov, O. L., V. A. Kronrod, and N. I. Pavlenkova. (2014), Thermodynamic–geophysical modeling of the interior structure of the Siberian cratonic mantle along cross-cutting profiles kimberlite and meteorite, *J. Geol. Geosci.*, *3*(155), 2.

Kuvshinov, A., and Semenov, A. (2012), Global 3-D imaging of mantle electrical conductivity based on inversion of observatory C-responses—I. An approach and its verification, *Geophys. J. Int.*, *189*, 13351352, doi:10.1111/j.1365-246X.2011.05349.x

Lachenbruch, A. H., and P. Morgan (1990), Continental extension, magmatism, and elevation: Formal relations and rules of thumb, *Tectonophysics*, *174*, 3962.

Lawrence, J. F., and D. A. Wiens (2004), Combine receiver-function and surface wave phase-velocity inversion using a niching genetic algorithm: Application to Patagonia, *Bull. Seismol. Soc Am.*, *94*(3), 977–987.

Ledo, J. and Jones, A.G. (2005), Upper mantle temperature determined from combining mineral composition, electrical conductivity laboratory studies and magnetotelluric field observations: Application to the intermontane belt, Northern Canadian Cordillera, *Earth Planet. Sci. Lett.*, *236*, 258–268.

Le Pichon, X., J., Francheteau, and J. Bonnin (1973), *Plate Tectonics, Developments in Geotectonics*, No. 6, Elsevier, Amsterdam, p. 300.

Le Roux, V., J.-L. Bodinier, A. Tommasi, O. Alard, J.-M. Dautria, A. Vauchez, and A. J. V. Riches (2007) The Lherz spinel lherzolite: Refertilized rather than pristine mantle, *Earth and Planet. Sci. Lett.*, *259*, 599–612, doi:10.1016/j.epsl.2007.05.026.

Lebedev, S., and R. D. van der Hilst (2008), Global upper-mantle tomography with the automated multimode inversion of surface and S-wave forms. *Geophys. J. Int.*, *173*, 505–518.

Lebedev, S., B. Endrun, M. Bischoff, and T. Meier (2007), Layering of seismic anisotropy and the past and present deformation on the lithosphere and asthenosphere beneath Germany, XXIV General Assembly of the IUGG, Perugia, Italy.

Lee, C.-T. A. (2003), Compositional variation of density and seismic velocities in natural peridotites at stp conditions: Implications for seismic imaging of compositional heterogeneites in the upper mantle, *J. Geophys. Res.*, *108*, doi:10.1029/2003JB002,413.

Liu, Q., and Gu, Y. J. (2012), Seismic Imaging: From classical to adjoint tomography, *Tectonophysics*, 566–567, 31–66.

Mandolesi, E., and A. G. Jones (2014), Magnetotelluric inversion based on mutual information, *Geophys. J. Int.*, *199*, 242–252.

Marler, R. T., and J. S. Arora (2004) Survey of multi-objective optimization methods for engineering, *Struct. Multidisc. Optim.*, *26*, 369–395.

Masson, F., and Trampert, J. (1997), On ACH, or how reliable is regional teleseismic delay time tomography? *Phys. Earth. Planet. Int.*, *102*, 21–32.

Matsukage, K. M., Y. Nishihara, and S. I. Karato (2005), Seismological signature of chemical differentiation of Earth's upper mantle, *J. Geophys. Res.*, *110*, B12305, doi:10.1029/2004JB003.

McDonough, W. F., and Sun, S.-S. (1995), Composition of the Earth, *Chemical Geology*, *120*:223–253. doi: 10.1016/0009-2541(94)00140-4

Meltzer, A. (2003), EarthScope: Opportunities and challenges for earth-science research and education, *The Leading Edge*, *22*, 268–271.

McGary, R. S., and R. L. Evans, P. E. Wannamaker, J. Elsenbeck, and S. Rondenay (2014) Pathway from subducting slab to surface for melt and fluids beneath Mount Rainier, *Nature*, *511*(7509), 338–340.

Mei, S., and D. L. Kohlstedt (2000a), Influence of water on plastic deformation of olivine aggregates. 1. Diffusion creep regime, *J. Geophys. Res.*, *105*, B9, 21457–21469.

Mei, S., and D. L. Kohlstedt (2000b), Influence of water on plastic deformation of olivine aggregates. 2. dislocation creep regime, *J. Geophys. Res.*, *105*, B9, 21471–21481.

Meqbel, N. M., Egbert, G. D., Wannamaker, P. E., A. Kelbert, and A. Schultz (2014). Deep electrical resistivity structure of the northwestern US derived from 3-D inversion of USArray magnetotelluric data. *Earth Planet. Sci. Lett.*, *402*, 290–304.

Mierdel, K., H. Keppler, J. R. Smyth, and F. Langenhorst (2007), Water solubility in aluminous orthopyroxene and the origin of Earth's asthenosphere, *Science*, *315*, 364–368.

Mosegaard, K. (1998), Resolution analysis of general inverse problems through inverse Monte Carlo sampling, *Inverse Problems*, *14*, 405–426.

Mosegaard, K., and A. Tarantola (1995), Monte Carlo sampling of solutions to inverse problems, *J. Geophys. Res.*, *100*(B7), 12431–12447.

Mosegaard, K., and M. Sambridge (2002), Monte Carlo analysis of inverse problems, *Inverse Problems 18*, R29–R54.

Moorkamp, M., A. G. Jones, and D. W. Eaton (2007), Joint inversion of teleseismic receiver functions and magnetotelluric data using a genetic algorithm: Are seismic velocities and electrical conductivities compatible?, *Geophys. Res. Lett.*, *34*, L16311.

Moorkamp, M., and A. G. Jones, and S. Fishwick (2010), Joint inversion of receiver functions, surface wave dispersion and magnetotelluric data, *J. Geophys. Res.*, *115*, B04318.

Naif, S., K. Key, S. Constable, and R. L. Evans (2013), Melt-rich channel observed at the lithosphere–asthenosphere boundary, *Nature*, *495*, 356–359.

Nolet, G. (2008), *A Breviary of Seismic Tomography: Imaging the interior of the Earth and the Sun*, Cambridge University Press, Cambridge, UK.

Nolet, G., S. P. Grand, and B. L. N. Kennett (1994), Seismic heterogeneity in the upper mantle, *J. Geophys. Res.*, *99*, 23753–23766.

Novella, D., D. J. Frost, E. H. Hauri, H. Bureau, C. Raepsaet, and M. Roberge (2014), The distribution of H_2O between silicate melt and nominally anhydrous peridotite and the onset of hydrous melting in the deep upper mantle, Earth, *Planet. Sci. Lett.*, *400*, 1–13.

Obrebski, M., R. M. Allen, F. Pollitz, and S. H. Hung, S. H. (2011), Lithosphere–asthenosphere interaction beneath the western United States from the joint inversion of body-wave traveltimes and surface-wave phase velocities, *Geophys. J. Int.*, *185* (2), 1003–1021.

Olugboji T. M., S. Karato, and J. Park (2013), Structures of the oceanic lithosphere-asthenosphere boundary: Mineral-physics modeling and seismological signatures, *Geochem. Geophys. Geosyst.*, *14*, 880901, doi:10.1002/ggge.20086.

O'Reilly, S. Y. and W. L. Griffin (2010), The continental lithosphere–asthenosphere boundary: Can we sample it? *Lithos*, *120*, pp. 1–13.

O'Reilly, S. Y., and W. L. Griffin (2013), Mantle metasomatism, in *Metasomatism and the Chemical Transformation of Rock*, Springer, New York, pp. 471–533.

Olsen, N. (1999), Induction studies with satellite data, *Surv. Geophys.*, *20*, 309–340.

Olsen, N., et al. (2013), The Swarm satellite constellation application and research facility (SCARF) and Swarm data products, *Earth, Planets and Space*, *64*, 1189–1200.

Pail R., S. Bruinsma, F. Migliaccio, C. Frste, H. Goiginger, W.-D. Schuh, E. Hck, M. Reguzzoni, J. M. Brockmann, O. Abrikosov, M. Veicherts, T. Fecher, R. Mayrhofer, I. Krasbutter, F. Sans, and C. C. Tscherning (2011), First GOCE gravity field models derived by three different approaches; *J. Geodesy*, *85*(11), 819–843.

Panet, I., G. Pajot-Métivier, M. Greff-Lefftz, L. Métivier, M. Diament, and M. Mandea (2014), Mapping the mass distribution of Earth's mantle using satellite-derived gravity gradients, *Nature Geosci.*, *7*, 131–135.

Parsons, B., and S. Daly (1983), The relationship between surface topography, gravity anomalies, and temperature structure of convection, *J. Geophys. Res.*, *88*(B2), 11291144, doi:10.1029/JB088iB02p01129.

Pasyanos, M. E., and Nyblade, A. A. (2007), A top to bottom lithospheric study of Africa and Arabia, *Tectonophysics*, *444*, 27–44.

Pasyanos, M. E., T. G. Masters, G. Laske, Z. Ma (2014), LITHO1.0: An updated crust and lithospheric model of the Earth, *J. Geophys. Res.*, *119*, 2153–2173.

Pearson, D. G., D. Canil, and S. B. Shirey (2003), Mantle samples included in volcanic rocks: Xenoliths and diamonds, in *The Mantle and Core*, R. W. Carlson, ed., Vol. *2*, *Treatise on Geochemistry*, H. D. Holland, and K. K Turkeian, eds., Elsevier-Pergamon, Oxford, UK.

Pedreira, D., J. C. Afonso, J. A. Pulgar, J. Gallastegui, A. Carballo, M. Fernàndez, and O. García-Moreno (2015), Geophysical-petrological modeling of the lithosphere beneath the Cantabrian Mountains and the North-Iberian margin: Geodynamic implications, *Lithos*, *230*, 46–68.

Piccardo, G. B. 2008, The Jurassic Ligurian Tethys, a fossil ultraslow-spreading ocean: The mantle perspective, in *Metasomatism in Oceanic and Continental Lithospheric Mantle*, M. Cottorti and M. Gregoire, eds., Geological Society, London, Special Publications, 293, p. 1133.

Pinto, L. G. R., M. B. de Padua, N. Ussami, I. Vitorello, A. L. Padilha, and C. Braitenberg (2010) Magnetotelluric deep soundings, gravity and geoid in the south Sao Francisco craton: Geophysical indicators of cratonic lithosphere rejuvenation and crustal underplating, *Earth Planet. Sci. Lett.*, *297* 423434.

Plomerova, J., D. Kouba, and V. Babuska (2002), Mapping the lithosphere–asthenosphere boundary through changes in surface-wave anisotropy, *Tectonophysics*, *358*, 175–185.

Püthe, C., and Kuvshinov, A. (2014), Mapping 3-D mantle electrical conductivity from space: A new 3-D inversion scheme based on analysis of matrix Q-responses, *Geophys. J. Int.*, *197*, 768–784.

Poe, B. T., C. Romano, F. Nestola, and J. R. Smyth (2010), Electrical conductivity anisotropy of dry and hydrous olivine at 8GPa, *Phys. Earth Planet. Int.*, *181*, 103–111.

Priestley, K. F., and D. P. McKenzie (2006), The thermal structure of the lithosphere from shear wave velocities, *Earth Planet. Sci. Lett.*, *244*, 285–301.

Priestley, K., and F. Tilmann (2009), Relationship between the upper mantle high velocity seismic lid and the continental lithosphere, *Lithos*, *109*, 112124, doi:10.1016/j.lithos.2008.10.021.

Ranalli, G., *Rheology of the Earth*, 2nd ed., Chapman & Hall, London, 1995.

Rawlinson, N., S. Pozgay, and S. Fishwick (2010), Seismic tomography: A window into deep Earth, *Phys. Earth Planet. Int.*, *178*, 101–135.

Rawlinson, N., A. M. Reading, and B. L. N. Kennett (2006), Lithospheric structure of Tasmania from a novel form of teleseismic tomography, *J. Geophys. Res.*, *111*, B02301, doi:10.1029/2005JB003803.

Rawlinson, N., and S. Fishwick (2011), Seismic structure of the southeast Australian lithosphere from surface and body wave tomography, *Tectonophysics*, *572–573*, 111–122.

Rawlinson, N., and B. L. N. Kennett (2008), Teleseismic tomography of the upper mantle beneath the southern Lachlan Orogen, *Phys. Earth Planet. Int. 167*, 84–97

Rawlinson, N., A. Fichtner, M. Sambridge, and M. Young (2014), Seismic tomography and the assessment of uncertainty. *Adv. Geophys.*, *55*, 1–76, 10.1016/bs.agph.2014.08.001

Reguzzoni, M., Sampietro, D., Sansò, F. (2013), Global Moho from the combination of the CRUST2.0 model and GOCE data, *Geophys. J. Int.*, ggt247.

Ricard, Y., L., Fleitout, and C. Froidevaux (1984), Geoid heights and lithospheric stresses for a dynamic Earth, *Ann. Geophys.*, *2*, 267–286.

Richards, M. A., and B. H. Hager (1989), Effects of lateral viscosity variations on long-wavelength geoid anomalies and topography. *Journal of Geophysical Research 94*: doi: 10.1029/89JB00718. issn: 0148–0227.

Ritsema, J., W. Xu, L. Stixrude, and Lithgow-Bertelloni, C. (2009), Estimates of the transition zone temperature in a mechanically mixed upper mantle, *Earth Planet. Sci. Lett.*, *277*, 244–252, doi:10.1016/j.epsl.2008.10.024

Romanowicz, B. (2003), Global mantle tomography: progress status in the past 10 years, *Annu. Rev. Earth Planet. Sci.*, *31*, 303–328.

Roux, E., M. Moorkamp, A. G. Jones, and M. Bischoff, B. Endrun, S. Lebedev, and T. Meier (2011), Joint inversion of long-period magnetotelluric data and surface-wave dispersion curves for anisotropic structure: Application to data from Central Germany, *Geophys. Res. Lett.*, *38*(5), L05304.

Roy, M., J. K. MacCarthy, and J. Selverstone (2005), Upper mantle structure beneath the eastern Colorado Plateau and Rio Grande rift revealed by Bouguer gravity, seismic velocities, and xenolith data, Geochem. *Geophys., Geosys.*, *6*(10), doi:10.1029/2005GC001008.

Rychert, C. A., and P. M. Shearer (2011), Imaging the lithosphere–asthenosphere boundary beneath the Pacific using SS waveform modelling, *J. Geophys. Res.*, *116*, doi: 10.1029/2010JB008070

Sabaka, T. J., Olsen, N., and Purucker, M. E. (2004), Extending comprehensive models of the Earth's magnetic field with Ørsted and CHAMP data, *Geophys. J. Int.*, *159*, 521–547.

Sambridge, M. (1999), Geophysical inversion with a neighbourhood algorithm I. Searching a parameter space, *Geophys. J. Int.*, *138*, 479494.

Sandwell, D. T., and MacKenzie, K. R. (1989), Geoid height versus topography for oceanic plateaus and swells, *J. Geophys. Res. 94*: doi:10.1029/88JB03764. issn: 0148–0227.

Schaeffer, A. J., and S. Lebedev (2013), Global shear speed structure of the upper mantle and transition zone, *Geophys. J. Int.*, ggt095.

Schultz, A., R. D. Kurtz, A. D. Chave, and Jones, A. G. (1993), Conductivity discontinuities in the upper mantle beneath a stable craton, *Geophys. Res. Lett.*, *20*(24), 2941–2944.

Schulze, D. J. (1989), Constraints on the abundance of eclogite in the upper mantle, *J. Geophys. Res.*, *94*, 4205–4212.

Schutt, D. L., and Lesher, C.E. (2006) Effects of melt depletion on the density and seismic velocity of garnet and spinel lherzolite, *J. Geophys. Res.*, *111*, B05401, doi:10.1029/2003JB002.

Sclater, J. G., and J. Francheteau (1970), Implications of terrestrial heat flow observations on current tectonic and geochemical models of crust and upper mantle of earth, *Geophys. J. R. Astron. Soc. 20*, 509–542, 10.1111/j.1365-246X.1970.tb06089.x

Semenov, A., and A. Kuvshinov (2012), Global 3-D imaging of mantle conductivity based on inversion of observatory C-responses—II. Data analysis and results, *Geophys. J. Int.*, *191*(3), 965–992.

Shan, B., J. C. Afonso, Y. Yang, C. J. Grose, Y. Zheng, X. Xiong, and L. Zhou (2014), The thermochemical structure of the lithosphere and upper mantle beneath south China: Results from multiobservable probabilistic inversion, *J. Geophys. Res. Solid Earth*, *119*, 84178441, doi:10.1002/2014JB011412.

Shen, W., M. H. Ritzwoller, and V. Schulte-Pelkum (2013), A 3-D model of the crust and uppermost mantle beneath the central and western US by joint inversion of receiver functions and surface wave dispersion, *J. Geophys. Res.*, doi:10.1029/2012JB009602.

Simpson, F., and A. Tommasi (2005), Hydrogen diffusivity and electrical anisotropy of a peridotite mantle, *Geophys. J. Int.*, *160*, 1092–1102.

Simmons, N. A., Forte, A. M., & Grand, S. P. (2006), Constraining mantle flow with seismic and geodynamic data: a joint approach. *Earth and Planetary Science Letters*, *246*(1), 109–124.

Simmons, N. A., A. M. Forte, L. Boschi, and S.P. Grand (2010), GyPSuM: A joint tomographic model of mantle density and seismic wave speeds, *J. Geophys. Res.*, *115*(12), B12310.

Shimizu, H., Koyama, T., Baba, K., and Utada, H. (2010), Revised 1-D mantle electrical conductivity structure beneath the north Pacific, *Geophys. J. Int.*, *180*, 1030–1048.

Speziale, S., F. Jiang, and T. S. Duffy (2005), Compositional dependence of the elastic wave velocities of mantle minerals: Implications for seismic properties of mantle rocks. Earth's deep mantle: structure, composition, and evolution, in *Geophys. Monograph*, Vol. *160*, ed., AGU, pp. 301–320.

Snyder, D. B., S. Rondenay, M. G. Bostock, and G. D. Lockhart (2004), Mapping the mantle lithosphere for diamond potential using teleseismic methods, *Lithos*, *77*, 859–872.

Stixrude, L., and C. Lithgow-Bertelloni (2011), Thermodynamics of mantle minerals II. Phase equilibria, *Geophys. J. Int.*, *184*, 11801213, doi:10.1111/j.1365-246X.2010.04890.x.

Stixrude, L., and C. Lithgow-Bertelloni (2005), Thermodynamics of mantle minerals: 1. Physical properties, *Geophys. J. Int.*, *162*, 610–632, 2005, doi: 10.1111/j.1365-246X.2005.02642.x

Stracke, A, and B. Bourdon (2009), The importance of melt extraction for tracing mantle heterogeneity, *Geochim. Cosmochim. Acta*, *73*, 218238.

Streckeisen, A. (1979), Classification and nomenclature of volcanic rocks, lamprophyres, carbonatites, and melilitic rocks: Recommendations and suggestions of the IUGS Subcommission on the Systematics of Igneous Rocks, *Geology*, *7* (7), 331–335.

Stolz, N. (2013), Bigger, better, smarter: high performance computers applied to Government geophysics, *ASEG Extended Abstracts 2013*(1), 1–4.

Tang, Y. J., H. F. Zhang, and J. F. Ying (2006), Asthenosphere lithospheric mantle interaction in an extensional regime: Implication from the geochemistry of Cenozoic basalts from Taihang Mountains, North China Craton, *Chem. Geol. 233*, 309327.

Tang, Y.-J., H.-F. Zhang, J.-F. Ying, and B.-X. Su (2007), Widespread refertilization of cratonic and circum-cratonic lithospheric mantle, *Earth Sci. Rev.*, *118*, 45–68.

Tarantola, A. (2005), *Inverse Problem Theory and Model Parameter Estimation*, Society for Industrial and Applied Mathematics, Philadelphia.

Tasaka, M., T. Hiraga, and M. E. Zimmerman (2013), Influence of mineral fraction on the rheological properties of forsterite+enstatite during grain-size-sensitive creep: 2. Deformation experiments, *J. Geophys. Res. Solid Earth*, *118*, 39914012, doi:10.1002/jgrb.50284.

Tašárová, A., J. C. Afonso, M. Bielik, H. J. Götze, and J. Hók (2009), The lithospheric structure of the Western Carpathian–Pannonian Basin region based on the CELEBRATION 2000 seismic experiment and gravity modelling, *Tectonophysics*, *475*(3), 454–469.

Tesauro, M., M. K. Kaban, W. D. Mooney, and S. A. P. L. Cloetingh (2014), Density, temperature, and composition of the North American lithosphere. New insights from a joint analysis of seismic, gravity, and mineral physics data: 2. Thermal and compositional model of the upper mantle, *Geochem. Geophys. Geosyst.*, *15*, 48084830, doi:10.1002/2014GC005484.

Thybo, H. (2006), The heterogeneous upper mantle low velocity zone, *Tectonophysics*, *416*, 53–79.

Tiberi, C., M. Diament, J. Dverchre, C. Petit-Mariani, V. Mikhailov, S. Tikhotsky, and U. Achauer (2003), Deep structure of the Baikal rift zone revealed by joint inversion of gravity and seismology, *J. Geophys. Res.*, *108*, 2133, doi:10.1029/2002JB001880, B3.

Tarits, P., and Mandéa, M. (2010), The heterogeneous electrical conductivity structure of the lower mantle, *Phys. Earth Planet. Int.*, *183*, 115–125.

Tkalcic, H., M. Pasyanos, A. Rodgers, R. Gok, W. Walter, A. Al-Amri (2006), Multistep method in joint modelling of receiver functions and surface waves: Implication for lithospheric structure of the Arabian Peninsula, *J. Geophys. Res.*, *B11311*, doi:10.1029/2005JB004130.

Torne, M., M. Fernandez, M. C. Comas, J. I. Soto (2000), Lithospheric structure beneath the Alboran Basin: Results from 3D gravity modelling and tectonic relevance, *J. Geophys. Res.*, *105*, 3209–3228.

Turcotte, D. L., and G. Schubert (2014), *Geodynamics*, Cambridge University Press, Cambridge, UK.

Unsworth, M., and P. A. Bedrosian (2004), Electrical resistivity structure at the SAFOD site from magnetotelluric exploration, *Geophys. Res. Lett.*, *31*, L12S05.

Van Decar, J. C., and R. S. Crosson (1990), Determination of teleseismic relative phase arrival times using multichannel cross-correlation and least-squares, *Bull. Seism. Soc. Am.*, *80*, 150–169.

van Heijst, H. J., and J. Woodhouse (1999), Global high-resolution phase velocity distributions of overtone and fundamental-mode surface waves determined by mode branch stripping. *Geophys. J. Int.*, *137*(3), 601–620.

Velmskỳ, J. (2013), Determination of three-dimensional distribution of electrical conductivity in the Earth's mantle from Swarm satellite data: Time domain approach, *Earth Planets Space*, *65*, 1239–1246.

Visser, K., J. Trampert, S. Lebedev, and B. L. N Kennett (2008), Probability of radial anisotropy in the deep mantle, *Earth and Planetary Science Letters*, *270*(3), 241–250.

Vozar, J., A. G. Jones, J. Fullea, M. R. Agius, S. Lebedev, F. Le Pape, and W. Wei (2014), Integrated geophysical–petrological modeling of lithosphere-asthenosphere boundary in central Tibet using electromagnetic and seismic data, *Geochem. Geophys. Geosyst.*, *15*, 39653988, doi:10.1002/2014GC005365.

Walter, M. J. (2003), Melt extraction and compositional variability in the mantle lithosphere, in *Treatise on Geochemistry*, Vol *2*, *The Mantle and Core*, R. W. Carlson, H. D. Holland, and K. K. Turekian, eds., Elsevier, Amsterdam, pp. 363–394.

Watts, A. B. (2011), Isostasy, In Encyclopedia of solid earth geophysics, Springer Netherlands, pp. 647–662.

Watts, A. B. (2001), Isostasy and Flexure of the Lithosphere, Cambridge University Press.

Wei, W., et al. (2001), Detection of widespread fluids in the Tibetan crust by magnetotelluric studies, *Science*, *292*, 716–718.

West, M., W. Gao, and S. Grand (2004) A simple approach to the joint inversion of seismic body and surface waves applied to the southwest US. *Geophys. Res. Lett.*, *31*, L15615, doi:10.1029/2004GL020373.

Xu, W., C. Lithgow-Bertelloni, L. Stixrude, and J. Ritsema (2008), The effect of bulk composition and temperature on mantle seismic structure, *Earth Planet. Sci. Lett.*, *275*, 70–79, doi:10.1016/j.epsl.2008.08.012.

Xu, Y., T. J. Shankland, and B. T. Poe (2000), Laboratory-based electrical conductivity in the Earth's mantle, *J. Geophys. Res.*, *105*, 27865–27875.

Yang, Y., and D. W. Forsyth (2006), Regional tomographic inversion of amplitude and phase of Rayleigh waves with 2-D sensitivity kernels, *Geophys. J. Int.*, *166*, 1148–1160.

Yang, Y., M. H. Ritzwoller, F.-C. Lin, M. P. Moschetti, and N. M. Shapiro (2008), Structure of the crust and uppermost mantle beneath the western United States revealed by ambient noise and earthquake tomography, *J. Geophys. Res.*, *113*, B12310, doi:10.1029/2008JB005833.

Yoshino, T., and T. Katsura (2013), Electrical conductivity of mantle minerals: Role of water in conductivity anomalies. *Annu. Rev. Earth Planet. Sci.*, *41*, 601–628.

Yuan, H., R. Kind, X. Li, R. Wang (2006), The S receiver functios: Synthetics and data example. *Geophys. J. Int.*, *165*, 555–564.

Yuan, H., S. French, P. Cupillard, and B. Romanowicz (2014), Lithospheric expression of geological units in central and eastern North America from full waveform tomography, *Earth Planet. Sci. Lett.*, *402*, 176–186.

Zeyen, H., and M. Fernàndez (1994), Integrated lithospheric modeling combining thermal, gravity, and local isostasy analysis: Application to the NE Spanish Geotransect, *J. Geophys. Res.*, *99*(B9), 18,08918,102, doi:10.1029/94JB00898.

Zeyen, H., and U. Achauer (1997), Joint inversion of teleseismic delay times and gravity anomaly data for regional structures: Theory and synthetic examples, in *Upper Mantle Heterogeneities from Active and Passive Seismology,* NATO Workshop, K. Fuchs, ed., Kluwer Academic Publishers, Norwell, MA, pp. 155–168.

Zeyen, H., P. Ayarza, M. Fernàndez, and A. Rimi (2005), Lithospheric structure under the western African-European plate boundary: A transect across the Atlas Mountains and the Gulf of Cadiz, *Tectonics*, *24*, TC2001, doi:10.1029/2004TC001639.

Zhang, H., and C. H. Thurber (2003), Double-difference tomography: The method and its application to the Hayward Fault, California, *Bull. Seis. Soc. Am.*, *93*(5), 1875–1889, doi = 10.1785/0120020190.

Zhang, Y-S., and T. Tanimoto (1991), Global Love wave phase velocity variation and its significance to plate tectonics. *Phys. Earth Planet. Int.*, *66*, 160–202.

Zhao, D., Yamamoto, Y., Yanada, T. (2013), Global mantle heterogeneity and its influence on teleseismic regional tomography, *Gondwana Res.*, *23*, 595–616.

Zheng, Y., Lay, T., Flanagan, M. P., Williams, Q. (2007), Pervasive seismic reflectivity and metasomatism of the Tonga mantle wedge, *Science*, *316*, 855–859.

11

Constitution and Structure of Earth's Mantle: Insights from Mineral Physics and Seismology

Andrea Zunino,[1] Amir Khan,[2] Paul Cupillard,[3] and Klaus Mosegaard[1]

ABSTRACT

Geophysical, geochemical, and petrological methods have played prominent roles in determining constitution and structure of the Earth. Most of Earth's interior remains geochemically unsampled, however, and direct measurements or numerical simulations of physical properties (e.g., elasticity and density) of natural samples and analogs at the physical conditions of the deep Earth allow us to compare with properties obtained from geophysical field models. These sources constitute a large complement of information that helps constrain possible models of Earth's mantle (and core) composition, temperature, and mineralogy. Traditionally, therefore, inference has proceeded by way of estimating "laboratory-based" profiles of physical properties with the intent of comparing these with field-derived estimates. A more definitive approach, however, resides in integrating data and results from experimental measurements, theoretical considerations, and inverse calculations to directly transform geophysical data into the main parameters of interest. In this study, we describe a quantitative approach that integrates data and results from mineral physics, petrological analyses, and geophysical inverse calculations to map geophysical data directly for mantle composition and thermal state. Computation of physical properties using thermodynamic models is described and discussed, and an application of the joint inverse methodology is illustrated in a case study where mantle composition and thermal state beneath continental Australia is determined directly from seismic data.

11.1. INTRODUCTION

Seismic tomography has proved an important tool to image the inaccessible parts of the Earth and has done much to enhance our understanding of deep-Earth processes and their relation to surface geology. For example, upper mantle seismic wave-speed variations at the global and regional scales are observed to correlate strongly with tectonic features. Tectonically old and stable lithosphere shows up in the form of fast seismic wave speeds, whereas regions that have undergone relatively recent tectonic activity are distinguished by slow wave speeds (e.g., see *Fichtner et al.* [2010], *Khan et al.* [2011], *Lekic and Romanowicz* [2011], and *Schaeffer and Lebedev* [2013]). This suggests that lateral variations in the upper mantle are temperature-dominated [e.g., see *Goes et al.* [2000]). But what is the contribution of compositional variations to Earth's present-day structure? Since entities such as temperature and composition ultimately reflect internal processes related to Earth's current state and past dynamics and, hence, its origin and evolution, unraveling Earth's chemical make-up and thermal state is of fundamental

[1] *Solid Earth Physics, Niels Bohr Institute, University of Copenhagen, Copenhagen, Denmark*
[2] *Institute of Geophysics, Swiss Federal Institute of Technology, Zürich, Switzerland*
[3] *Laboratoire GeoRessources, Université de Lorraine, CNRS, Vandœuvre-lès-Nancy, France*

Integrated Imaging of the Earth: Theory and Applications, Geophysical Monograph 218, First Edition.
Edited by Max Moorkamp, Peter G. Lelièvre, Niklas Linde, and Amir Khan.

importance to Earth science (e.g., see *van der Hilst et al.* [2005], *Karato* [2013], and *Khan and Deschamps* [2015]).

In the absence of samples and in situ measurements, geophysical methods (based on, for example, seismology) are insufficient because these can only provide clues about the main parameters of interest (temperature and composition). So, how can we determine these quantities from measurements made at Earth's surface? Since we lack analytical expressions describing the temperature-dependent constitutive relations between stress, strain, and strain rate that are needed to compute the physical properties as functions of pressure and temperature, we rely on the use of thermodynamical theory. This, supplemented with mineral physics data, presents the principal means of establishing a link between geophysical observables and the fundamental thermochemical parameters of interest. As reviewed elsewhere (e.g., see *Stixrude and Jeanloz* [2007]), the main aim of mineral physics is the use of high-pressure devices and computer simulations based on first-principles theory to measure or determine properties of Earth materials at conditions of the deep Earth to help constrain the aforementioned constitutive relations. By comparing the elastic properties measured in the laboratory and computed numerically with those determined seismologically, we are able to place observational constraints on the interior structure. This approach is illustrated with the method of Adams and Williamson, who used seismically determined bulk sound velocity and compressibility measurements to show that Earth is well-approximated by a set of adiabatic and homogeneous layers (e.g., see *Anderson* [1989]).

The remarkable growth—both in quantity and quality—of high-pressure and high-temperature (high P–T) measurements of thermodynamic properties of crust and mantle materials (e.g., see *Holland and Powell* [1998] and *Stixrude and Lithgow-Bertelloni* [2005a, 2011]) has enabled quantitative inferences on the distribution of temperature and composition to be made. Methods employed for this purpose are generally classifiable in terms of forward and inverse approaches. An additional subclassification describes the particular observation used to calculate the "misfit" measure—that is, whether geophysical models or proper geophysical data are employed as a means of achieving model improvement. As an example of the simplified forward modeling approach, where a small number of models are tested for consistency with real data, we would compute, based on a parameterized phase-diagram approach (i.e., fixed composition and geotherm) seismic wave speeds, for example, and compare these to radial profiles of Earth's structure (e.g., see *Jackson* [1998], *Vacher et al.* [1998], *da Silva et al.* [2000], *Goes et al.* [2000], *Deschamps et al.* [2001], *Marton and Cohen* [2002], *Cammarano et al.* [2003], *Stixrude and Lithgow-Bertelloni* [2005a], *Xu et al.* [2008], and *Kuskov et al.* [2011, 2014]) or globally averaged geophysical observables such as travel times, normal modes, and electromagnetic sounding data (e.g., see *Dobson and Brodholt* [2000], *Xu et al.* [2000], *Cammarano et al.* [2005, 2009], and *Fullea et al.* [2011]). Inverse approaches, on the other hand, tend to be more sophisticated in that geophysical datasets are directly inverted for one-dimensional radial profiles or three-dimensional grids of bulk composition and temperature. Given the complexity and nonuniqueness of the problem, the latter approach clearly presents a more appropriate means of analysis and includes studies that focus both on radial variations (e.g., see *Mattern et al.* [2005], *Khan et al.* [2006], *Matas et al.* [2007], *Cobden et al.* [2008], and *Verhoeven et al.* [2009]) and analysis of lateral distributions of temperature and composition on both regional and global scales (e.g., see *Shapiro and Ritzwoller* [2004], *Trampert et al.* [2004], and *Khan et al.* [2009], *Ritsema et al.* [2009], *Simmons et al.* [2010], *Cammarano et al.* [2011], *Khan et al.* [2011], *Fullea et al.* [2012], *Mosca et al.* [2012], *Afonso et al.* [2013a,b], *Drilleau et al.* [2013], *Khan et al.* [2013], and *Kaban et al.* [2014]).

These studies provide considerable evidence for the presence of thermochemical variations in the mantle. Although thermal variations tend to govern the physical structure of the lithosphere and upper mantle as shown in many studies (e.g., see *Shapiro and Ritzwoller* [2004], *van Gerven et al.* [2004], *Goes et al.* [2005], *Schuberth et al.* [2012], and *Kaban et al.* [2014]), compositional effects cannot be excluded even at long wavelengths (e.g., see *Khan et al.* [2013]). Moreover, transition zone structure appears to be complex and decoupled from that of the upper mantle, raising the question of what physical processes or mechanisms are responsible for producing the imaged lateral wave-speed variations (e.g., see *Lee et al.* [2014], *Takeuchi et al.* [2014], *Schmandt et al.* [2014], *Emry et al.* [2015], *Koyama et al.* [2014], and *Thompson et al.* [2015]) and their relation, if any, to mantle dynamics. In this context, the internal dynamics of the Earth are driven by lateral density variations (e.g., see *Forte and Perry* [2000] and *Deschamps et al.* [2001]). Yet, while detailed maps of compressional and shear-wave velocity models exist, there are few constraints on mantle density structure (e.g., see *Kennett et al.* [1998] and *Ishii and Tromp* [2004]), because of the relative insensitivity of seismic data to density (e.g., see *Resovsky and Ritzwoller* [1999], *Romanowicz* [2001], and *Kuo and Romanowicz* [2002]).

Thus, it is our contention that in order to improve our understanding of the observed complexity, it is desirable to invert directly for thermal state and composition of Earth's interior using a multiparameter inversion scheme as laid out in, for example, *Khan et al.* [2009]. There are several advantages to this approach: (1) We address the fundamental parameters of interest; (2) we enable joint inversion of different geophysical datasets (e.g., seismic, electromagnetic, and gravity) and are able to simultaneously constrain P-wave speed, S-wave speed, and density; (3) we can integrate information and/or data from other disciplines (e.g., petrology); and (4) we implicitly take into account structure associated with thermochemically induced mineralogical phase transitions as observed in the laboratory.

The multiparameter joint inversion scheme makes use of a self-consistent thermodynamic methodology in the form of Gibbs free-energy minimization. Gibbs energy minimization is a technique by which rock mineralogy, and ultimately its elastic properties, can be predicted as a function of pressure, temperature, and bulk composition from thermodynamic data (e.g., *Connolly* [2005, 2009]). Although Gibbs energy minimization has long been advocated for geophysical problems (e.g., see *Saxena and Eriksson* [1983], *Wood and Holloway* [1984], *Kuskov and Panferov* [1991], *Sobolev and Babeyko* [1994], *Bina* [1998], and *Fabrichnaya* [1999]), it is only in the last decade that the availability of comprehensive thermodynamic databases (e.g., see *Ricard et al.* [2005], *Stixrude and Lithgow-Bertelloni* [2005b], *Khan et al.* [2006], *Piazzoni et al.* [2007], and *Stixrude and Lithgow-Bertelloni* [2011]) has made the application of such calculations feasible for geophysical models of Earth's entire mantle.

In the following, we describe a strategy to determine composition and thermal state of Earth's mantle directly from inversion of geophysical data. The theoretical framework we rely on is based on the probabilistic approach to inverse problems, which provides the tools to integrate information from different sources in a consistent algorithm as outlined in more detail by *Mosegaard and Hansen* [2016], *Bosch* [2016], and *Hansen et al.* [2016] (Chapters 2, 3, and 6 *this volume*). While our approach will be that of the inverse problem, we illustrate and discuss the connection between temperature, composition, mineral properties, resultant Earth structure, and seismic data. As an inverse application we discuss results from inversion of surface-wave phase-velocity data [*Khan et al.*, 2013] for thermochemical structure of the mantle beneath continental Australia. As an example of posterior filtering (e.g., see *Khan et al.* [2011]), we compute full waveform synthetics for a set of solution models and compare these to observed waveforms from selected seismic stations across Australia. Finally, the study of *Afonso et al.* [2016] (Chapter 10, *this volume*) presents a complementary study to this contribution and emphasizes the growing popularity of integrated studies.

11.2. THE FORWARD PROBLEM: FROM THERMO-CHEMICAL MODELS TO SEISMIC DATA

The study of Earth's deep interior is highly challenging because of our inability to directly probe the Earth by exploration and lack of representative samples for most of the deep Earth. Instead, models of the interior rely entirely on theory and laboratory measurements performed on mineral analogs. Comparison of these laboratory-based models with radial profiles of seismic wave speeds (e.g., PREM, AK135) or tomographic images, for example, enables qualitative understanding to be gained. This is illustrated in the approach of Birch, who used mostly experimental shock-wave data and compared these to seismological models of the day. This led to the recognition that seismic discontinuities are most likely related to the transformation of major minerals—for example, the transformation of olivine to its high-pressure phases [*Birch*, 1952]. This type of procedure for estimating constitution and structure of the Earth, although more advanced now, is still practiced. Most attempts rely on a two-step "inversion" procedure: Smooth velocity models resulting from linearized seismic tomography are "converted" to properties of interest, typically temperature, using parameterized phase-diagram approaches. However, from a thermodynamic perspective, such a procedure is not self-consistent, because physical properties and phase equilibria are derived independently. As a consequence, physical properties will not satisfy fundamental thermodynamic relations, which can lead to inconsistencies (e.g., see *Cammarano et al.* [2003], *Hacker et al.* [2003], *Goes et al.* [2004], *Shapiro and Ritzwoller* [2004], and *Trampert et al.* [2004]).

Since our aim here is to compute geophysical data (e.g., seismic waveforms) from arbitrary models of Earth's mantle composition and thermal state and to compare these quantitatively with observations, a more general framework is required to properly reconcile the various aspects involved. Our strategy is to develop a sequential algorithm that computes, based on a mineral physics database (compilation of experimental and theoretical results for thermodynamic properties of mantle minerals), the stable mineral modes and their physical properties. These are subsequently employed as input for seismic waveform modeling and comparison to observations. This sequential approach ensures that the data computed are consistent with theory, laboratory-data, and any other information needed in the process.

Figure 11.1 illustrates schematically the various steps involved in the procedure outlined above. The first step involves computing the stable minerals, their proportions, and properties as a function of temperature, pressure, and chemical composition. The second step consists in

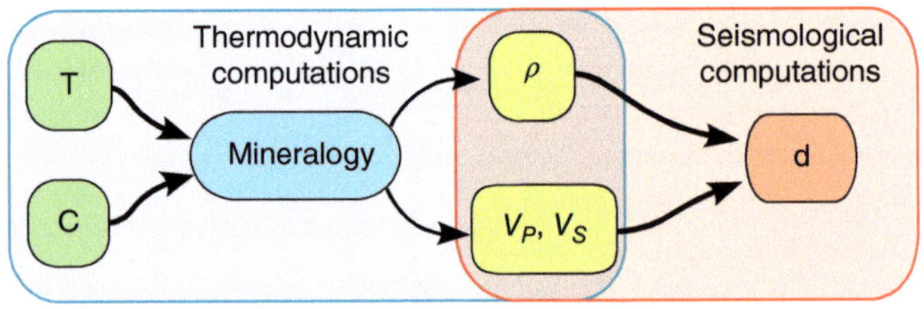

Figure 11.1 Illustration of the joint geophysical–thermodynamic forward problem for computing seismic data from thermochemical parameters (C, T). Mineralogy refers to modal mineralogy and physical properties include density (ρ) and isotropic P- and S-wave speeds (V_P and V_S).

computing bulk rock properties from properties of single minerals through use of the theory of composites. Finally, in the third step, seismic waves, for example, are propagated through the (an)elastic model obtained in steps 1 and 2 to compute seismograms at selected locations on the surface. Although we concentrate on seismology and full-waveform solvers, geophysical data from other disciplines such gravity and electromagnetic sounding could easily be employed [*Verhoeven et al.*, 2009; *Cammarano et al.*, 2011; *Khan et al.*, 2011, 2013; *Fullea et al.*, 2012; *Khan and Shankland*, 2012; *Afonso et al.*, 2013a,b; *Kaban et al.*, 2014].

Table 11.1 Model End-Member Bulk Compositions in Weight Percent[a]

Oxide	Basalt (X_B)	Harzburgite (X_H)
Na_2O	1.87	0.01
CaO	13.05	0.50
FeO	7.68	7.83
MgO	10.49	46.36
Al_2O_3	16.08	0.65
SiO_2	50.39	43.64

[a] Basalt and harzburgite compositions are taken from *Khan et al.* [2009].

11.2.1. Thermodynamics of Mantle Minerals

11.2.1.1. Mantle Compositional Models

Recycling of subducted lithospheric material that originates from differentiation and melting of mantle peridotites at mid-ocean ridges (producing a basaltic crust overlying a depleted an olivine-rich residuum, harzburgite) is believed to be the major mechanism responsible for the production and maintenance of mantle heterogeneities [*Hofmann*, 1997; *Helffrich and Wood*, 2001; *Tackley et al.*, 2005; *Anderson*, 2007]. Based on this compositional end-member model, *Xu et al.* [2008] proposed a simple parameterization to investigate the effect of composition and temperature on seismic properties of the mantle. Mantle composition was parameterized in terms of a single depth-dependent variable that represents the weight fraction of basalt in a basalt-harzburgite mixture. In addition, *Xu et al.* [2008] considered two physical models for defining the "make-up" of the mantle: (1) equilibrium assemblage (EA) and (2) mechanical mixture (MM). In the EA model, the mantle is assumed to have reached chemical equilibrium, whereas in the MM model the bulk rock is composed of a solid–solid mixture of the two end-members.

With the above definitions, phase assemblages for the two models are given by

$$\mathrm{EA}: \Phi\left[\mathrm{C}X_{\mathrm{B}} + (1-\mathrm{C})X_{\mathrm{H}}\right] \tag{11.1}$$

and

$$\mathrm{MM}: \mathrm{C}\Phi\left[X_{\mathrm{B}}\right] + (1-\mathrm{C})\Phi\left[X_{\mathrm{H}}\right], \tag{11.2}$$

where Φ represents the vector of phase proportions, $X_{(\cdot)}$ the vector defining basalt and harzburgite compositions, respectively, and C the basalt fraction (ranging from 0 to 1). In the equilibrium model the properties of interest are calculated from the stable mineralogy for a given composition, whereas in the mechanical mixing model the bulk properties of the rock are obtained by averaging the properties of the equilibrium mineralogies of the two end-member lithologies. Table 11.1 summarizes the basalt and harzburgite compositions in the NCFMAS system, comprising oxides of the elements Na_2O, CaO, FeO, MgO, Al_2O_3, and SiO_2. This chemical model accounts for ~98% of the mass of Earth's mantle (e.g., *Irifune* [1994]).

The advantage of using basalt fraction as compositional variable is its simplicity relative to more complicated parameterizations and, not least, its relation to the dynamical processes that are responsible for generating and maintaining heterogeneities in the mantle as discussed earlier. Note, however, that if the real Earth is not well mixed chemically, it would very likely differ from the model invoked presently, calling for more complex scenarios. In the following, we concentrate on the MM model.

11.2.1.2. Phase Equilibrium Computations and Bulk Physical Properties

Our goal is to estimate density and elastic properties of minerals as a function of pressure, temperature, and composition. For this, two elements are required: (I) calculation of the stable mineral phases, their proportions, and their properties at the conditions of interest and (II) determination of bulk rock properties from single mineral properties. The determination of stable mineral phases as a function of pressure and temperature translates into a search for the amounts and compositions of mineral phases that minimize Gibbs free energy (a nonlinear function of composition), with the additional constraint of mass conservation. The input is represented by a number of chemical components that characterize a particular system in addition to pressure and temperature. The availability of comprehensive thermodynamic data sets for minerals and chemical compounds permits the computation of the geophysical properties of interest for the stable minerals. The computation of phase equilibrium is cast as an optimization problem, where the equations that need to be solved are of the form [*Connolly*, 2005]

$$G^{\text{sys}} = \sum_{i}^{N} \alpha_i G_i \quad \text{with} \quad \alpha_i \geq 0, \tag{11.3}$$

where G^{sys} is the Gibbs free energy of the system, G_i the Gibbs free energy of a certain amount (e.g., a mole) of the *i*th phase, and α_i the amount of the *i*th phase. A mass balance constraint is also taken into account, since the sum of the amount of the components within the phases must equal the corresponding amount in the system. Combination of thermodynamic data measured experimentally and the solution to the above minimization problem allows us to calculate which phases are stable at given conditions and their properties of interest. All free energy minimization computations were done using the Perple_X software package [*Connolly*, 2005] (www.perplex.ethz.ch).

Once stable mineralogy is obtained from Gibbs free energy minimization, physical properties (density and elasticity) of single minerals are determined by equation-of-state modeling, whereas properties of the aggregate are computed using averaging schemes that derive from composites theory (e.g., see *Hill* [1952], *Watt et al.* [1976], *Guéguen and Victor* [1994], and *Berryman* [1995]). A commonly employed scheme is the Voigt–Reuss–Hill (VRH) average [*Hills* 1952], which is constructed as the arithmetic mean from the Voigt (V) and the Reuss (R) bounds (corresponding to upper and lower bounds), respectively:

$$X_{\text{V}} = \sum_{i=1}^{N} f_i p_i, \quad X_{\text{R}} = \left(\sum_{i=1}^{N} \frac{f_i}{p_i} \right)^{-1}, \tag{11.4}$$

$$X_{\text{VRH}} = \frac{X_{\text{V}} + X_{\text{R}}}{2}, \tag{11.5}$$

where p_i and f_i are property and volume fraction of the *i*th mineral, respectively. Other averaging schemes exist—for example, Hashin–Shtrikman bounds [*Hashin and Shtrikman*, 1963], which are narrower than the Voigt and Reuss bounds—but entail making additional assumptions (e.g., *Watt et al.* [1976] and *Berryman* [1995]). Although not rigorously justified as a best estimate of the actual effective moduli [*Stixrude and Jeanloz*, 2007], we nonetheless follow common practice in the field (e.g., see *Connolly and Kerrick*, [2002], *Connolly* [2009], and *Stixrude and Lithgow-Bertelloni* [2005b, 2011]) and consider the VRH average as an estimator. For further discussion, the interested reader is referred, for example, to *Berryman* [1995], *Stixrude and jeanloz* [2007], and *Jones et al.* [2009]. Here we compute only isotropic seismic wave speeds; however, anisotropic schemes have been developed [*Stixrude and Lithgow-Bertelloni*, 2011] but still lack a sufficient support from mineral physics data. To compute pressure at a point inside the Earth, we integrate the load from the surface. Finally, implicit assumptions of the Gibbs free energy minimisation approach include isotropy of the mineral aggregate, homogeneity of bulk composition, absence of elastic wave dispersion, and subsolidus conditions (no partial melt).

As an illustration of the thermodynamic approach, we computed phase equilibria and shear wave speeds (Figure 11.2) for the two end-member compositions discussed in Section 11.2.1.1 (Table 11.1) along the adiabat of *Brown and Shankland* [1981]. We observe clear differences in (1) stable mineralogy and computed S-wave speed profiles between the end-member models and (2) location and size of seismic discontinuities: olivine→wadsleyite (410 km) and ringwoodite→bridgemanite+ferropericlase (660 km) vary considerably and depend strongly on the particular mineralogy and thus thermochemical setting. As shown and discussed in more detail in the next section, variations in physical properties strongly influence the seismic response, which, as a consequence, can be used to distinguish the various profiles and by inference the thermochemical state of the mantle.

The present thermodynamic model has limitations that may have consequences for its validity to Earth's mantle. First, the NCFMAS model system precludes consideration of redox effects that might be expected to be important if native or ferric iron is present in the Earth's mantle. We neglect such effects both out of necessity, in that the thermodynamic data required to model redox is

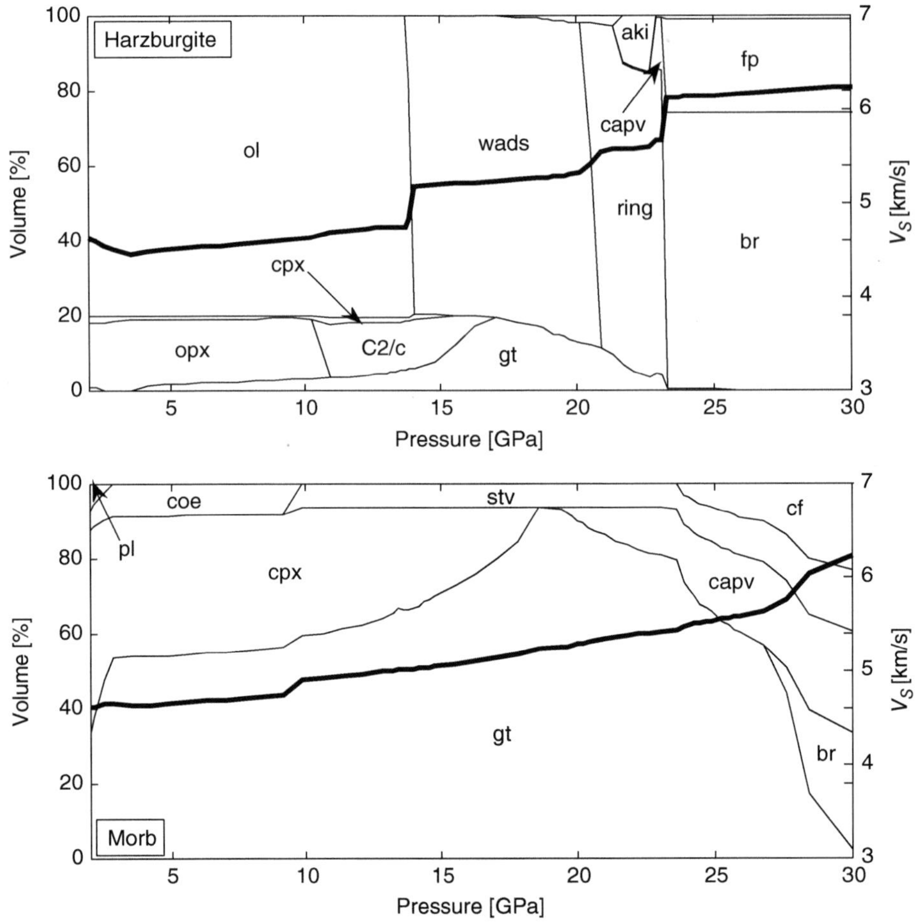

Figure 11.2 Variations in phase proportions (modal mineralogy) and seismic shear-wave speed (V_S) in the upper mantle, transition zone, and uppermost lower mantle (pressure range 2–30 GPa corresponding to 50 to 750 km depth). Isotropic shear-wave speed (*bold solid lines*) and phase equilibria are calculated for two different mantle compositions in the NCFMAS system (comprising oxides of the elements Na_2O, CaO, FeO, MgO, Al_2O_3, SiO_2): harzburgite and morb, along the mantle adiabat of Brown and Shankland (1981) (not shown). Phases are: olivine (ol), orthopyroxene (opx), clinopyroxene (cpx), pl (plagioclase), coe (coesite), stv (stishovite), high-pressure Mg-rich cpx (C2/c), garnet (gt), wadsleyite (wad), ringwoodite (ring), akimotoite (aki), calcium silicate perovskite (ca-pv), ferropericlase (fp), bridgemanite (br; formerly perovskite), and calcium ferrite (cf).

lacking, and simplicity, in that we are unaware of any work that suggests that ferrous iron is not the dominant form of iron in Earth's mantle. Components and phases not considered include H_2O, TiO_2, Cr_2O_3, and partial melt because of lack of thermodynamic data. Uncertainties associated with thermodynamic modeling are difficult to assess on account of the nonlinearity of the free-energy minimization problem and correlations among thermodynamic parameters—for example, bulk modulus and its pressure derivative. We conducted a Monte Carlo analysis of this uncertainty by considering the variance in a single parameter successively—that is, independent of the other parameters. The results indicate uncertainties in density and isotropic P- and S-wave speeds on the order of ~0.1%, ~0.5–1%, and ~1–3%, respectively (unpublished data). This particular type of statistical assessment tends to underestimate the overall uncertainty (cf. *Connolly and Kerrick* [2002], who estimate an uncertainty in seismic velocities of ~1–3% based on thermodynamic data). However, absence of the covariance between the various thermodynamic parameters leaves little alternative for now. Based on a different thermodynamic database than the one employed presently, *Kuskov et al.* [2014], in a study of the thermochemical structure of the lithospheric mantle beneath the Siberian craton find that the temperature at the lithosphere–asthenosphere boundary (LAB) are only accurate to within to ±100°C. An alternative method for investigating this uncertainty is found in the study of *Afonso et al.* [2013a], who consider the problem of determining thermochemical structure of the lithospheric mantle using a diverse set of geophysical data. Afonso and co-workers assess the uncertainty by comparing predicted and observed modal compositions for a set of xenoliths and generally find that differences between computed and observed values are within experimental error. Differences in physical properties between computed and laboratory measurements are also deemed negligible.

11.2.1.3. Precalculated Tables: An Efficient Tool for Computing Properties

The phase equilibria calculations hold a large part of the algorithm. This can be prohibitive, in particular in the framework of an inverse algorithm where typically many thousands of such calculations are required. The free energy minimization can be carried out in two ways: by static or by dynamic implementation. Static implementation means that given composition of rocks, we construct tables representing the requested properties (e.g., P- and S-wave speed, density) on the pressure-temperature space. In this way it is possible to precalculate the seismic properties of the mantle for several given compositions as a function of pressure and temperature and store these in multidimensional tables as done in *Zunino et al.* [2011]. Any property can then be obtained by simple table look-up and interpolation. The other approach is the dynamical implementation: Given a particular value for pressure, temperature, and composition, we compute the resulting properties minimizing Gibbs free energy during the inversion process. Here we will focus on the static implementation (called Phemgp) and construct tables of properties for the two end-member compositions listed in Table 11.1.

With this technique, computation of bulk rock properties, such as density and seismic velocities, becomes very fast and reliable. Once the tables are computed, only interpolation is involved. The only caveat is that the range of variation in pressure, temperature and compositional parameters need to be anticipated in advance in order to construct the look-up tables. In principle, this could have a significant impact on memory requirement depending on the resolution needed, since, for the sake of speed, tables have to be kept in memory during the process; however, we verified that at the resolution necessary for our applications, the amount of memory used is well within the standard capacity found on a modern personal computer. Precalculated look-up tables for the two end-member composition considered here and programs to extract properties from these are available from http://www.perplex.ethz.ch/phemgp/phemgp.html.

11.2.1.4. Effects of Compositional and Thermal Variations on Elastic Properties

Examples of mantle properties obtained by Phemgp for basalt and harzburgite—that is, for basalt fractions of 0 and 1, respectively—are shown in Figure 11.3 and illustrate the influence of phase equilibria on mantle seismic and density structure. Properties extracted with Phemgp along representative mantle geotherms reveal prominent features associated with the major mineral phase transformations and the major seismic discontinuities at 410-, 520-, and 660-km depth. These transitions, which cause strong seismic velocity gradients, clearly depend on temperature, pressure, and composition.

To further illustrate the effect of temperature and compositional variations on mantle properties, we calculated properties as resulting from perturbations applied to a reference state. As reference geotherm we rely on the continental geotherm of *Turcotte and Schubert* [2002] and as reference composition we employ a basalt fraction of 0.2. Reference continental and modified geotherms are shown in Figure 11.4.

Properties extracted from the tables along the various geotherms and compositions and mixed according to Eq. (11.2) are summarized in Figure 11.5. Perturbations on the major phase transitions, particularly olivine →wadsleyite and wadsleyite→ringwoodite, are strongly influenced by temperature variations. Both the "410" and the "520" move deeper as temperature increases, in line with theoretical considerations [*Bina and Helffrich*, 1994] and observations [*Helffrich*, 2000]. The "660" appears to be less influenced. Changes in properties with temperature (i.e., size of discontinuity) across the transitions are almost negligible. Variations in properties due to compositional changes, on the other hand, are clearly more complicated and include relatively strong perturbations in all seismic discontinuities in both location and size. For a basalt fraction of 1, for example, the "410" is entirely absent, while the "660" moves well below 700 km depth (cf. Figure 11.5).

These observations have also been discussed in a recent seismic tomography study where surface-wave dispersion data across the Australian continent were inverted for mantle thermochemical structure using the same methodology proposed here and the parameterization shown in Figure 11.6 [*Khan et al.*, 2013] (to be discussed further in Section 11.3). Analyzing transition zone topography—that is, variations in depth to the major discontinuities in and around the transition zone (Figure 11.7)—Khan and co-workers found that the "410" correlated strongly with temperature, while the behavior of the "520" was somewhat diminished relative to the "410". The "660", on the other hand, was clearly found to correlate negatively with temperature. Variations with composition were, as indicated above, more complicated with the "660" correlating positively with Fe# (Fe/(Fe + Mg)) and "410" and "520" showing negative correlation, while all three discontinuities were observed to correlate positively with (Mg + Fe)/Si (these features are discussed in more detail in *Khan et al.* [2009]). Although surface waves are not ideal for imaging seismic discontinuities, the important point to

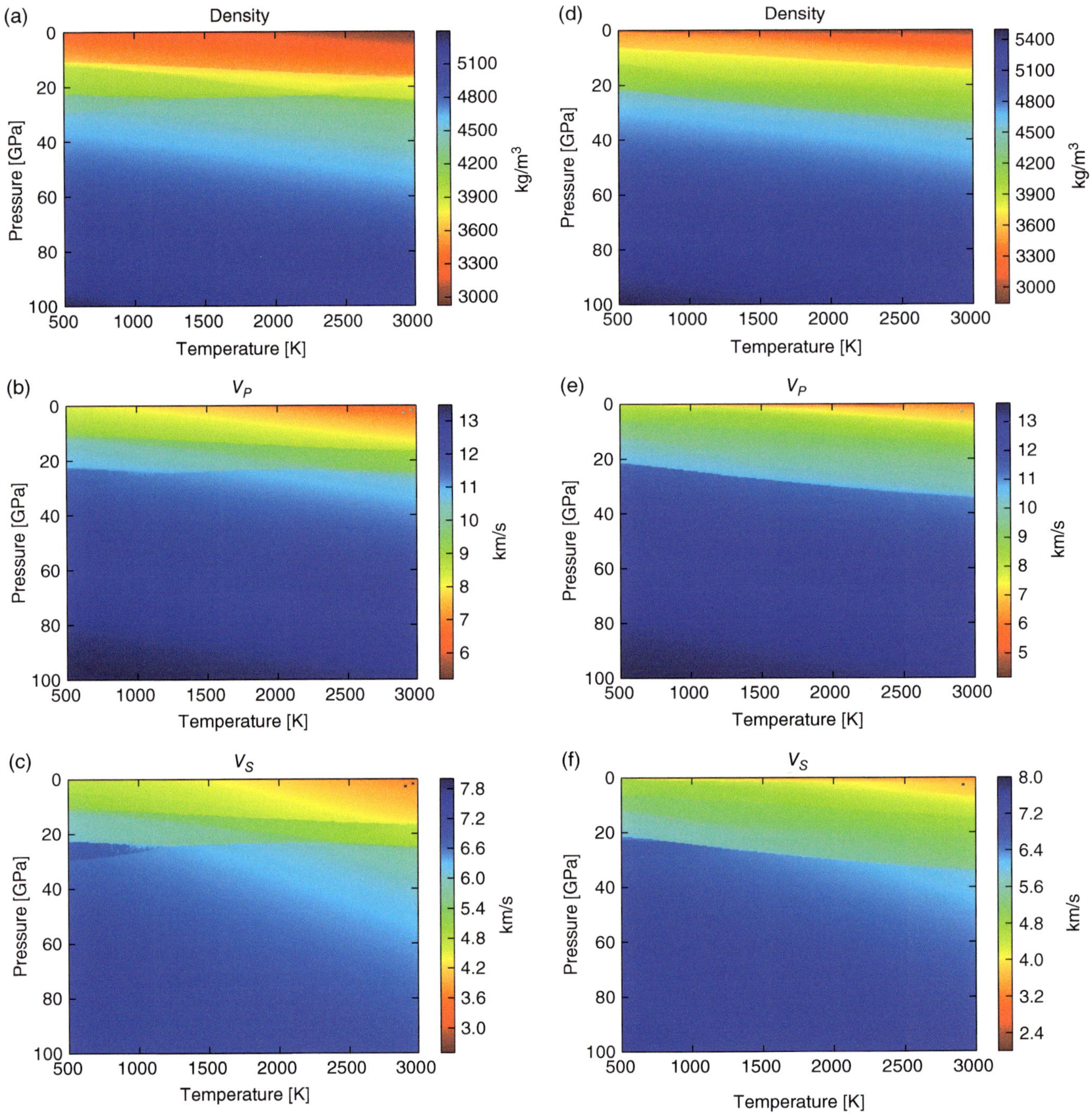

Figure 11.3 Density and isotropic P- and S-wave speeds (V_p and V_s) of the mantle computed for harzburgite (plots **a–c**) and basalt (plots **d–f**) compositions (see Table 11.1).

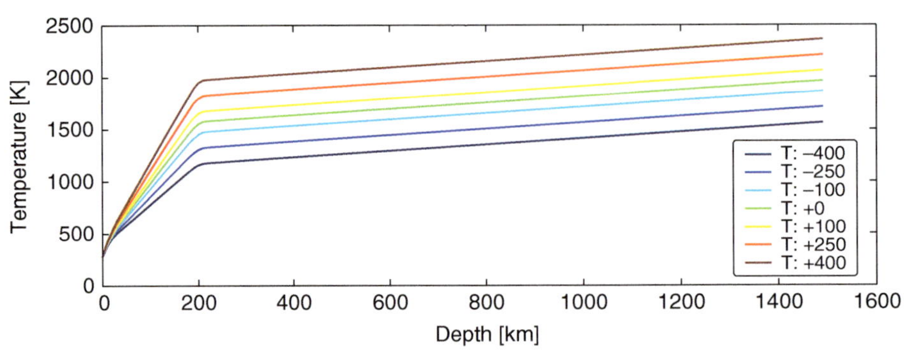

Figure 11.4 Continental reference and modified geotherms. "T" indicates relative temperature perturbation (in K) with respect to the reference continental geotherm of *Turcotte and Schubert* [2002] (T:+0). See main text for details.

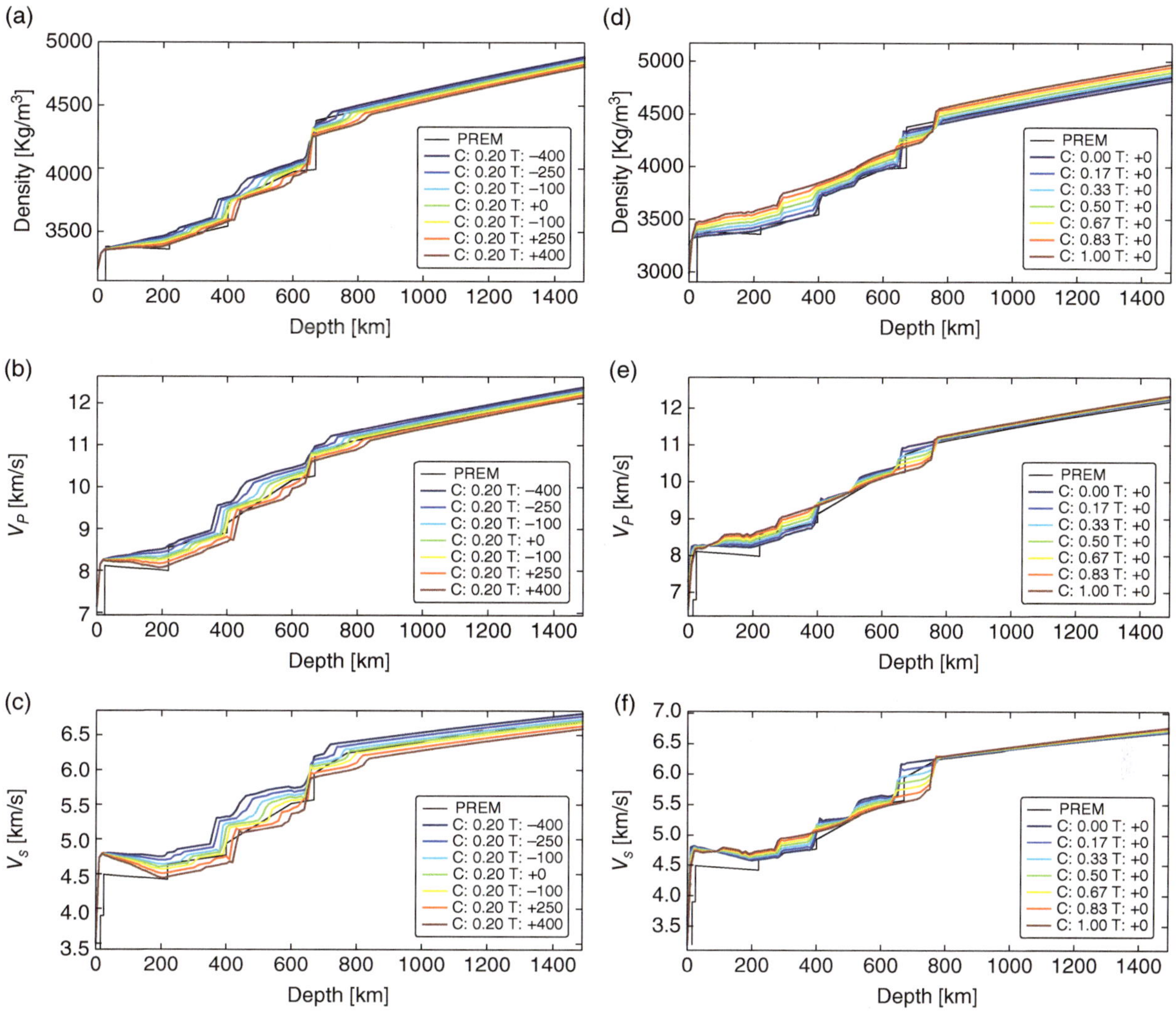

Figure 11.5 Profiles of density and isotropic seismic wave speeds V_P and V_S as a function of depth for the geotherms (a-c) shown in Figure 11.4 and various basalt fractions (d-f). "T" indicates relative temperature perturbation with respect to reference geotherm (T:+0) and "C" variation in basalt fraction. PREM profile is shown in black.

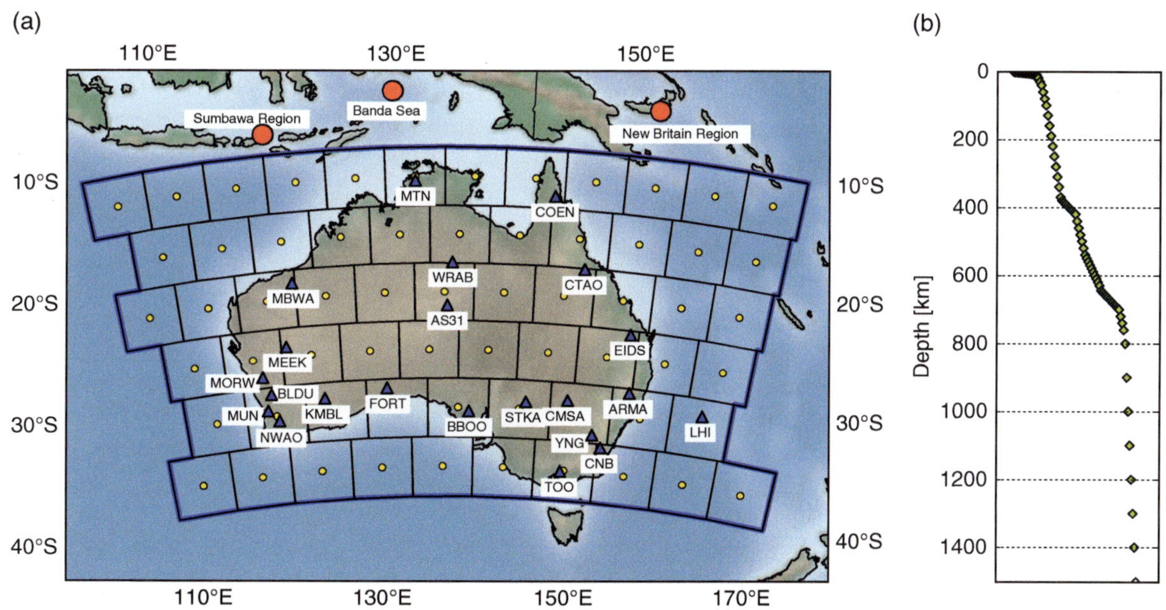

Figure 11.6 Lateral **(a)** and vertical **(b)** parameterization of the area used in the surface-wave-based tomography study. *Yellow circles* represent the centers of "pixels" **(a)**, and *yellow diamonds* represent depths of nodes **(b)**. *Blue triangles* indicate seismic stations, and *red circles* indicate earthquake epicenters. See main text for details.

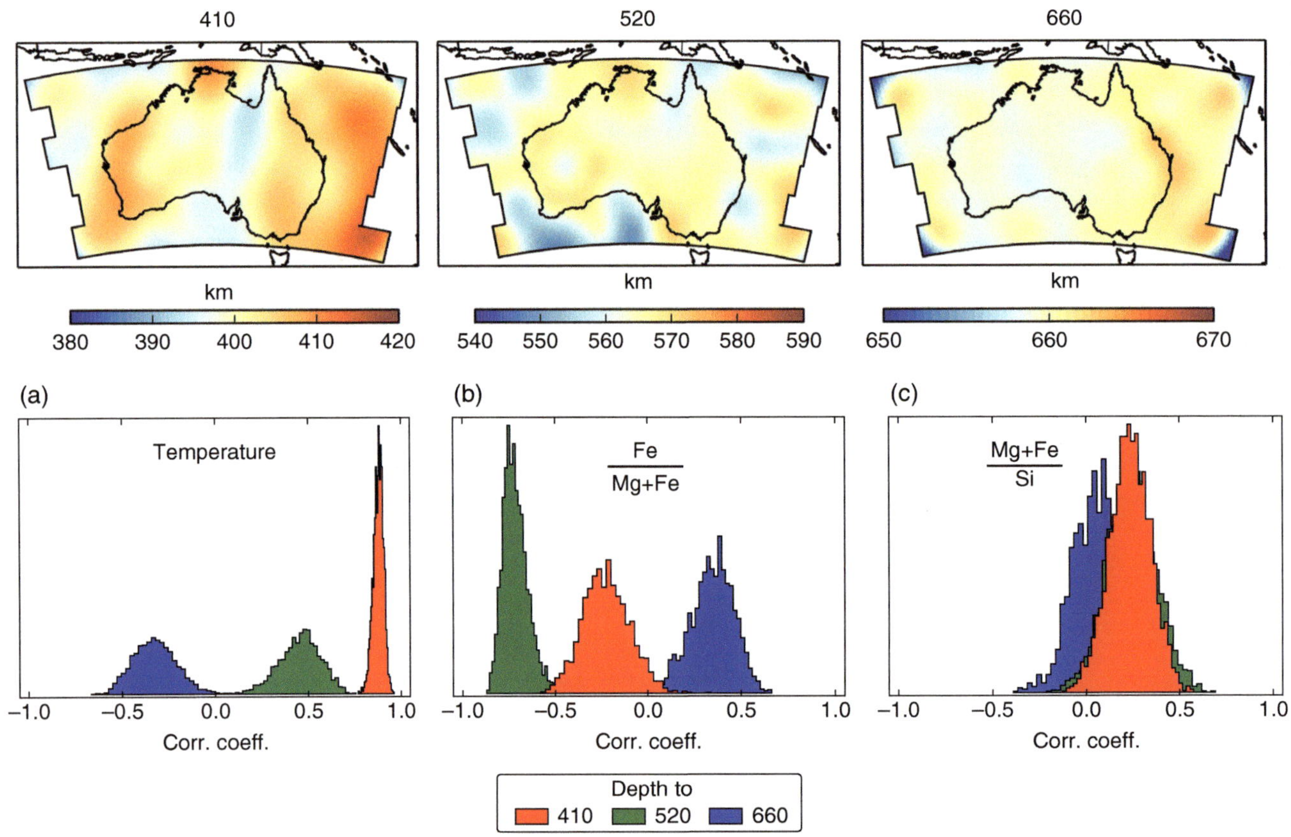

Figure 11.7 Mean depths to "410"-, "520"-, and "660"-km seismic discontinuities beneath continental Australia marking the transition between upper mantle and transition zone, the mid-transition zone discontinuity, and the transition to the lower mantle as defined by the mineral phase transformations: olivine→wadsleyite, wadsleyite→ringwoodite, and ringwoodite→ferripericlase+bridgemanite, respectively (see Figure 11.2 for location and notation of mineral phases). The images were computed from an ensemble of ~10^4 models obtained from stochastic inversion (see *Khan et al.* [2013] for details) and represent the statistical mean of these models. Plots **a–c** show correlation between transition zone depth and thermochemical parameters (composition is shown as variations of Fe/(Fe + Mg) and (Mg + Fe)/Si). Distributions are normalized to unity for the highest peak in each plot and are independent between the various properties shown.

note is that much of the structure visible in the images at the base of the upper mantle, for example, derives from lateral variations in phase transitions associated with the olivine→wadsleyite transformation. The corresponding lateral variations in velocity were observed to be much larger than those associated with purely lateral variations in temperature and composition immediately above and below the transition (see also *Stixrude and Lithgow-Bertelloni* [2007]). Neglect of this contribution to lateral structure very likely results in lateral variations in temperature, or structure in general, that vary too rapidly and probably also too strongly as apparent in some studies (e.g., see *Goes et al.* [2005] and *Trampert et al.* [2004]). Extension of the methodology to include seismic data sensitive to discontinuities and transition-zone structure such as PP and SS precursors and receiver functions (e.g., see *Schmerr and Garnero* [2006], *Tauzin et al.* [2008, 2013], *Shmandt et al.* [2014], *Schmerr* [2015], and *Thompson et al.* [2015]) will be considered in the future.

11.2.2. Seismic Wavefield Modeling

The availability of modern broadband catalogs of seismic data, together with improved numerical modeling of seismic waveforms, has opened up the opportunity to explore Earth's interior more accurately and in more detail than previously possible (e.g., see *Komatitsch et al.* [2002], *Fichtner et al.* [2009], and *Cupillard et al.* [2012]). The possibility to invert entire waveforms has opened a new chapter in seismic tomography. Dealing with waveforms means exploiting all the information contained in seismograms, which in principle leads to a significant improvement in the amount of information we can extract from the subsurface.

11.2.2.1. Seismological Tools

The seismic data used to image the Earth's interior range from body-wave travel times and surface-wave dispersion curves to entire waveforms that include both phase and amplitude information. Correspondingly, the methods used to compute synthetic data can be divided in a few major classes: ray-based, normal-mode summation, direct solvers, finite differences and spectral-element methods [*Liu and Gu*, (2012)]. Ray-based methods rely on the assumption of infinite frequency, which collapses the sensitivity to a ray and allows travel times to be computed by solving the eikonal equation. These methods are computationally very fast, but become inappropriate when the size of heterogeneities in the Earth's model is close to the wavelength of the seismic wavefield. The normal-mode-based approach instead is capable of very accurate calculations, but the computational load increases very quickly with the order, making this approach intractable for probabilistic tomography.

Thanks to theoretical advances, a novel technique termed finite-frequency tomography [*Dahlen et al.*, 2000] has gained popularity among seismologists. This technique takes into account the frequency dependence of the sensitivity of seismic waves. It is used to compute finite-frequency sensitivity kernels, which can be employed for solving the inverse problem. These kernels are characterized by a sensitivity which spans a certain volume around the ray path, such that substantial portions of the model have influence on the seismic response at the surface (e.g., see *Montelli et al.* [2004]).

Eventually, the method that has emerged as the best compromise between accuracy and computational cost is the Spectral Element Method (SEM) (e.g., *Komatitsch and Vilotte* [1998]). It is able to solve the wave equation to very high precision, as it provides an exponential decrease of the aliasing error with the increase in the order of the polynomial basis functions. This numerical property is the so-called spectral precision, from which the name SEM derives.

The elastic wave equation describing the displacement $\mathbf{u}$ can be written as

$$\rho \frac{\partial^2 \mathbf{u}}{\partial t^2} = \nabla \cdot \mathbf{T} + \mathbf{f}, \quad (11.6)$$

where ρ is the density, t the time, $\mathbf{T}$ the stress tensor, and $\mathbf{f}$ the body force. This equation together with the linear constitutive stress–strain relation (Hooke's law)

$$\mathbf{T} = \mathbf{C} : \varepsilon, \quad (11.7)$$

where $\mathbf{C}$ is the fourth-order stiffness tensor and ε the strain tensor ($\varepsilon = \frac{1}{2}\left(\nabla \mathbf{u} + (\nabla \mathbf{u})^t\right)$, forms the elastodynamic problem. Usually, at the surface of the Earth, a homogeneous Neumann boundary condition is imposed ($\hat{\mathbf{n}} \cdot \mathbf{T} = \mathbf{0}$, where $\hat{\mathbf{n}}$ is the normal to the surface), but other conditions can be used, such as absorbing boundaries at the edges of the domain in the case of regional-scale studies. The SEM solves the elastodynamic problem in the weak form and yields the full seismic wavefield, including all seismic phases. The spatial domain of the model is decomposed into hexaedral elements, and high-order Lagrange polynomials defined on Gauss–Lobatto–Legendre (GLL) points are employed for the functional discretization. These polynomials, in addition to a GLL quadrature for evaluating the integrals, make the mass matrix diagonal, which significantly reduces the computational cost. One ends up with a discretized linear ordinary differential equation in time:

$$\mathbf{M}\ddot{\mathbf{U}}(t) + \mathbf{K}\mathbf{U}(t) = \mathbf{F}(t), \quad (11.8)$$

where $\mathbf{M}$ is the mass matrix, $\mathbf{K}$ the stiffness matrix, and $\mathbf{F}(t)$ the source term. Solving Eq. (11.8), with the use of a classic finite-difference scheme in time, yields the entire seismic wavefield for a given source term. The latter is usually in the form of a source time function and a moment tensor that can be retrieved from available earthquake catalogs such as the centroid moment tensor (CMT) catalog [*Dziewonski et al.*, 1981; *Ekström et al.* 2012]. Note that CMT solutions are the result of an inverse problem that relies on waveform data. In principle, therefore, when modifying the seismic velocity model, changes in the moment tensor should also be taken into account. However, the trade-off between source mechanism and Earth structure is beyond the scope of this paper. For simplicity, we assume the CMT solution retrieved from the CMT catalog to be the correct source mechanism.

In this work we rely on RegSEM [*Cupillard et al.*, 2012], an implementation of the SEM for regional-scale simulations. The workflow consists of two main processes: (1) meshing (i.e., defining the three-dimensional grid where the wavefield will be propagated) and (2) solving the wave equation on the mesh. RegSEM uses hexaedral meshes where the size of the elements depends on the minimum wavelength that has to be propagated. Thus, the higher the frequency to be simulated, the higher the computational cost. This becomes prohibitive for Monte Carlo-like methods, where many forward iterations are required. This is the main reason why, at present, we have to limit our study to long-period seismic data (see Section 11.2.2.2).

11.2.2.2. Parameterization, Meshing, Homogenization

To simulate the propagation of seismic waves through the Earth, we have to define a parameterization of our model of elastic properties which accommodates the structures we are interested in and follows the requirements of the chosen strategy for forward modeling. Generally, parameterizations can be divided in two categories: relying on global bases such as spherical harmonics or local bases in the form of cell, node or spline-based grids. The choice of the parameterization has to be adapted to the forward solver and the specific problem under study, as no single approach can be considered universally superior to all others.

The first step to perform seismic forward modeling with the SEM is to define a mesh upon which density and P- and S-wave velocities (and possibly additional parameters such as anisotropic parameters) are assigned. The values of those properties are obtained from the thermodynamic calculations described above. Typically, seismic solvers for the SEM require to pick values of the needed parameters at several points (GLL points) within each mesh element, in a possibly nonsimple spatial arrangement, so, practically speaking, it is required to be able to provide values of properties at any given point in the domain and because of this some form of interpolation is needed.

Meshing the Earth can be a very complex task because of the spherical shape and the various discontinuities and inhomogeneities that should be accommodated. For the spectral element method, structured hexaedral meshes are generally used, which discretize the wave equation and use a functional representation on these grids. The choice of the grid spacing is not arbitrary, but must satisfy some conditions to ensure proper convergence of the solution and thus adapt to the characteristics of the wavefield to be propagated. This translates in a set of rules to constrain the maximum grid spacing allowed for propagating a certain maximum frequency of the wavefield and the time stepping of the algorithm. For instance, for RegSEM, two conditions must be satisfied [*Cupillard et al.*, 2012],

$$d \geq \frac{N}{5}\lambda_{\min} \quad \text{and} \quad \Delta t \leq C\left[\frac{\Delta x}{V_P}\right]_{\min}, \tag{11.9}$$

where d is the size of the elements in the mesh, N is the polynomial order, $\lambda_{\min}$ the minimum wavelength, Δt the time step, C the Courant number (usually chosen between 0.3 and 0.4), Δx the grid spacing, and V_P the P-wave speed. The first condition dictates the maximum size of the elements of the mesh allowed to propagate a wavefield with a minimum wavelength of $\lambda_{\min}$ and is empirically derived. The second condition, the Courant–Friedrichs–Lewy (CFL) condition, ensures stability of the time-marching by setting the time step Δt. These conditions have a strong impact on the computational burden because the smaller the size of the elements of the mesh and the time step, the greater the amount of time it takes to compute the solution.

It can be shown that the band-limited nature of the observed seismic data has the consequence that the heterogeneities whose size is smaller than the minimum wavelength only affect the wavefield in an effective way; in other words, what can be "seen" from the point of view of the observations is an upscaled version of the medium (e.g., see *Backus* [1962] and *Capdeville et al.* [2013]). Theoretical and technical advances [*Capdeville et al.*, 2010a,b; *Guillot et al.*, 2010; *Capdeville et al.*, 2015; *Cupillard et al.*, 2015], permit us to calculate such upscaled medium in terms of elastic parameters provided the shortest wavelength traveling in the medium. The resulting homogenized model, being a smoothed version of the original, can be used to alleviate the computational burden deriving from performing complex and high-resolution meshing in the presence of sharp discontinuities and small heterogeneities. Further, in the case of models parameterized in terms of large blocks, as could be the case for long-period data, it may help decreasing the computation time by allowing coarser meshes to be used.

11.2.2.3. Effects of Thermal and Compositional Variations on Seismic Data

As shown in Section 11.2.1, temperature and composition are observed to strongly impact Earth's elastic properties. To assess the effects of such variations on the seismic response of the Earth, we performed numerical simulations where we computed full waveforms using the models shown in Figure 11.5. For illustration, we calculated the seismic response at station STKA (Stephens Creek, SA, Australia) (see Figure 11.6) for a 6.5-magnitude earthquake that occurred on November 25, 2007 in Indonesia. For the purpose of computing seismograms based on one-dimensional models, we employed the waveform code AxiSEM [*Nissen-Meyer et al.*, 2014], where attenuation was taken into account by interpolating PREM [*Dziewonski and Anderson*, 1981] values. Figure 11.8 shows the resulting vertical-component seismograms for the various models, which were computed down to 10-s period. Temperature-induced variations of the calculated ground displacement appear relatively smooth and are mostly characterized by phase shifts and only a modest impact on amplitude. In contrast, composition has a more pronounced influence, changing both amplitude and phase, which reflects the more complex variation in physical properties incurred by compositional changes (cf. Figure 11.5).

11.3. POSTERIOR FILTERING OF TOMOGRAPHIC MODELS—THE AUSTRALIAN SUBCONTINENTAL MANTLE REVISITED

In this section, we illustrate how actual thermochemical models of the mantle (beneath continental Australia) obtained from inversion of surface-wave dispersion data

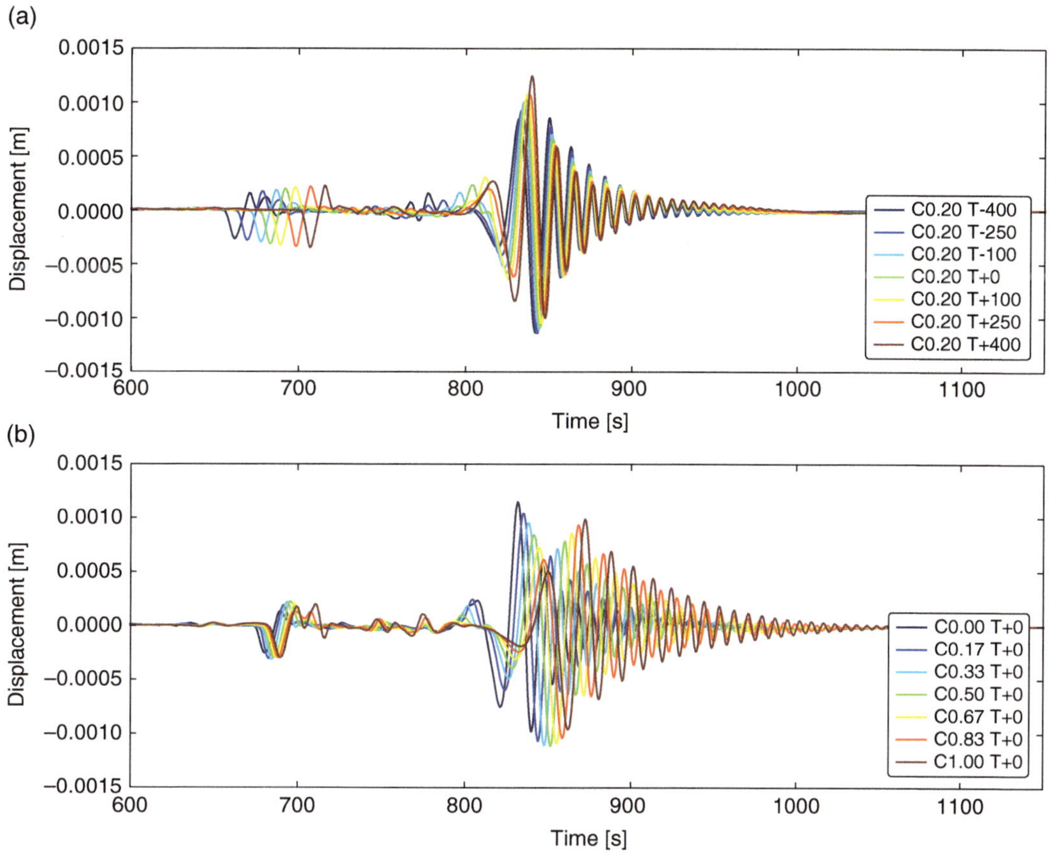

Figure 11.8 Seismograms computed from the radial profiles shown in Figure 11.5 down to 10-s period as recorded on the vertical component at station STKA (see Figure 11.6) at a distance of about 3500 km. The source is a magnitude-6.5 earthquake that occurred on November 25, 2007 in Indonesia. "T" indicates the relative temperature perturbation in Kelvin with respect to the reference geotherm (T + 0), while "C" refers to basalt fraction.

can be further refined by using additional data (waveforms). For completeness, we first briefly summarize the approach used in inverting surface-wave phase-velocity dispersion curves and describe the results (for details the reader is referred to *Khan et al.* [2013]), after which follows computation of waveform data using the surface-wave based tomographic models and comparison to observations.

11.3.1. Inversion of Surface-Wave Phase Velocity Dispersion Curves

The measured data consist of the azimuthally averaged isotropic part of the global azimuthally anisotropic phase-velocity maps of fundamental and higher-mode Love (to fifth overtone) and Rayleigh (to sixth overtone) waves of *Visser et al.* [2008]. Overtones have enhanced sensitivity in depth relative to fundamental-mode data and thus help to constrain regions below the upper mantle. The model was parameterized laterally (longitude and latitude) as an unstructured grid of "pixels" each spanning an area of about 5×5 degrees, whereas in depth a series of non-evenly spaced nodes was used. Figure 11.6a shows the grid of "pixels" along with the pixels' centers where phase-velocity dispersion curves were extracted from the phase-velocity maps of *Visser et al.* [2008]. The depth parameterization consists of a series of non-evenly spaced nodes where parameters (e.g., crustal structure, Moho depth, composition, temperature, anisotropy, and anelasticity) are defined. Values for all these parameters are interpolated to the depth nodes shown in Figure 11.6b. To take into account crustal effects, we use CRUST2.0 [*Bassin et al.*, 2000] as initial model and apply perturbations laterally to both structure and Moho depth.

The prediction of Rayleigh- and Love-wave dispersion data from thermochemical parameters is shown in Figure 11.9, which summarizes the entire forward problem. The g_n's correspond to different physical theories and the various symbols to model and data parameters (see Table 11.2). Computation of anisotropic and anelastic parameters are described in Appendices B and C.

We employed a Markov chain Monte Carlo method to invert the data (see Appendix A). We considered Rayleigh- and Love-wave dispersion curves for each "pixel" independently and inverted these for radial models of the parameters that are summarized in Table 11.2.

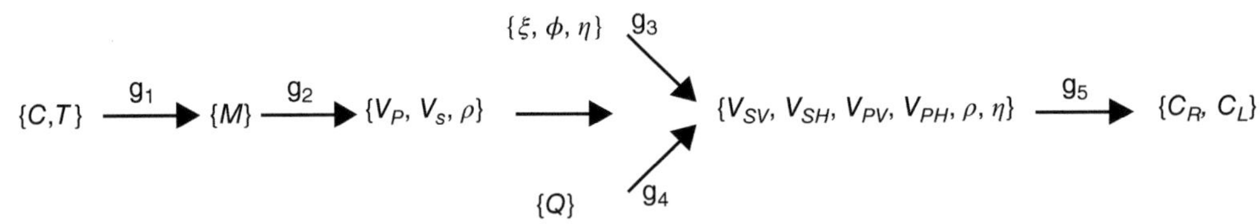

Figure 11.9 A schematic illustration of the forward problem and the different model parameters (c, T, ξ, φ, ...), physical theories (g_1, g_2, ...), and data (C_R, C_L) used to describe it. Symbols are described in Table 11.2.

Table 11.2 Model and Data Parameters and Connections Between the Various Parameters (Physical Laws)[a]

Model and Data Parameter	Description
C	Composition (primary)
T	Temperature (primary)
M	Mineralogy (secondary)
V_P, V_S, ρ	P- and S-wave speed, density (secondary)
ξ, φ, η	Anisotropic parameters (primary)
Q	Attenuation (primary)
V_{PV}, V_{PH}, V_{SV}, V_{SH}	Anisotropic anelastic parameters (secondary)
C_R, C_L	Rayleigh- and Love-wave dispersion data
Physical Laws	
g_1	Thermodynamic modeling
g_2	Equation-of-state modeling
g_3	Anisotropic correction
g_4	Anelastic correction
g_5	Prediction of surface-wave dispersion

[a] Note that we only invert for primary parameters; secondary parameters depend on primary parameters (see *Khan et al.* [2007] for a detailed discussion).

11.3.2. The Posterior Movie

In comparison to deterministic inverse methods where the output is a single best-fit, typically smooth model, the probabilistic approach provides a whole ensemble of models that fit data within uncertainties (~5000 have been selected for analysis here). All of these models constitute the output, represented by the posterior distribution $\sigma_M(\mathbf{m})$ [see Equation (11.A1)]. The posterior distribution is usually a high-dimensional probability density function that is notoriously difficult to visualize in a plot. Depending on the characteristics of the posterior distribution (e.g., multimodality), central estimators such as the mean value, which are easy to plot, do not provide a meaningful representation of the actual models. As an alternative strategy to single estimators, *Tarantola* [2005] has advocated the construction of a "movie" by sequentially showing samples from the posterior distribution. General features that are characteristic of the models, such as those that are well-resolved, will tend to be recurring in the posterior images, whereas less well-resolved features will appear much more scattered. Data-related structural patterns are easily separable from those that appear randomly and in a noncoherent and nonrecurring fashion. Although posterior models can differ significantly, they are nonetheless models with high likelihood values that predict observed data within uncertainties.

For present purposes, we created a movie showing several properties of the mantle using 250 randomly chosen models from the posterior. The movie is available online (www.nbi.dk/~zunino/publsm.html) as supplemental material and depicts the following properties between 100- and 800-km depth: S-wave velocity, density, temperature, compositional parameters [(Mg + Fe)/Si and Fe/(Mg + Fe)], and mineralogy [px/(ol + px) and fp/(br + fp)]. A single model from the posterior distribution is shown in Figure 11.10.

Strong correlation between shear-wave speed and density anomalies is observed in the upper mantle, which also outlines the continental boundaries of the region. The continental regions appear to have a distinct thermochemical signature and are divided into slow [corresponding to "hot" with relatively low (Mg + Fe)/Si and high Fe/(Mg + Fe) values] and fast [corresponding to "cold" with relatively high (Mg + Fe)/Si and low Fe/(Mg + Fe) values] patterns that extend well into the upper mantle. Toward the bottom of the upper mantle, we observe that thermal effects are becoming progressively less important and give way to more complex structures that appear to be compositionally driven. This is most apparent in the case of density at 300-km depth, which reveals a relatively strong compositional signature that correlates positively with Fe/(Mg + Fe). The behavior of the shear-wave speed anomalies, in contrast, are mostly governed by thermal effects.

11.3.3. Comparison with Other Tomographic Models

The shear-wave speed variations found here in the upper mantle have also been reported elsewhere (e.g., see *Fichtner et al.* [2010] and *Fishwick and Rawlison* [2012]). These models are sensitive to about 200- to 250-km depth. The model of *Fichtner et al.* [2010] (hereafter FCT) was obtained using full waveform tomography, whereas

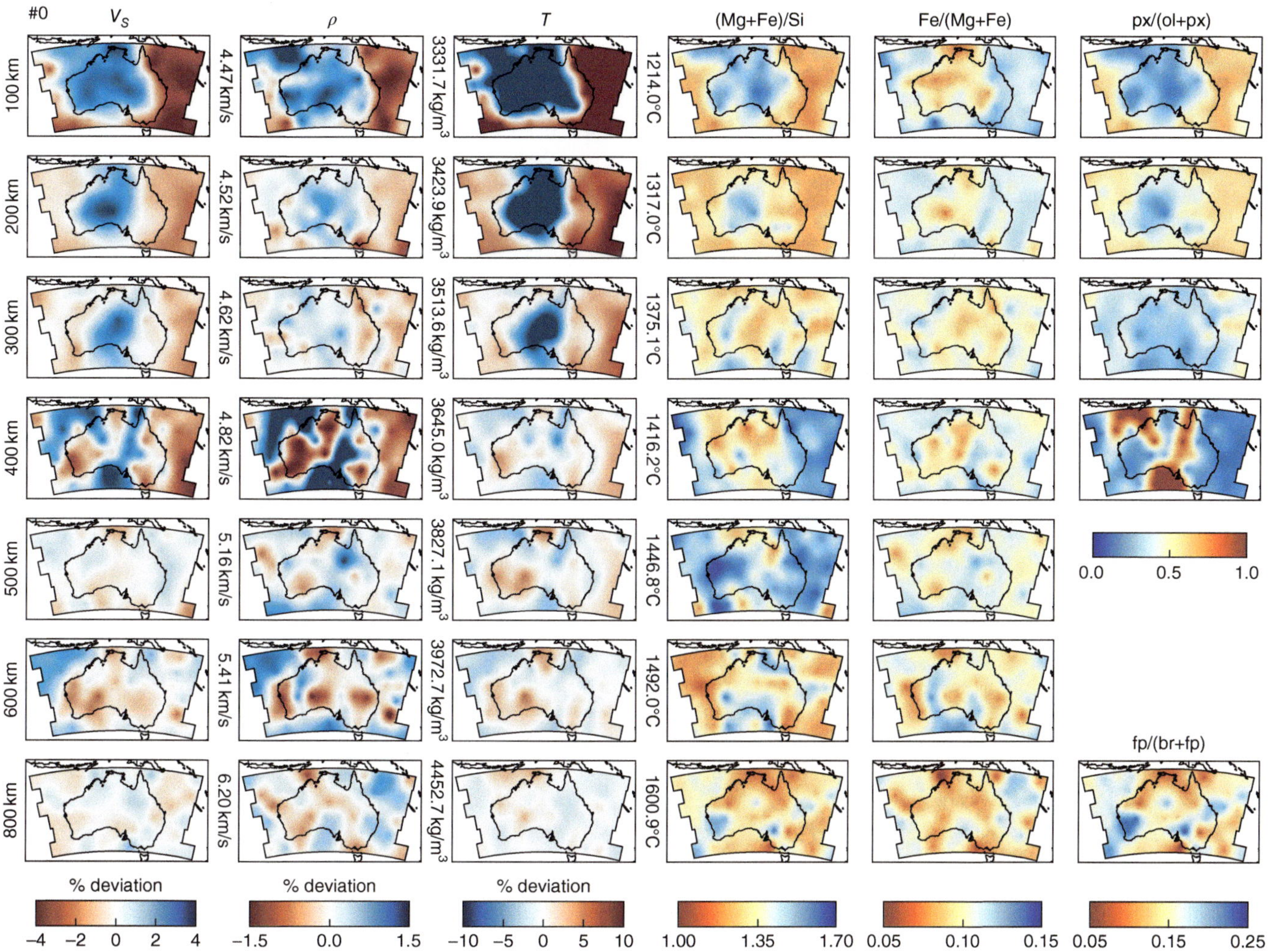

Figure 11.10 A model taken randomly from the posterior distribution showing shear-wave speed (V_S), density (ρ), temperature (T), compositional parameters [(Mg + Fe)/Si and Fe/(Mg + Fe)], and mineralogy [px/(ol + px) and fp/(br + fp)]. For phase names see Figure 11.2. See main text for details.

Fishwick and Rawlinson [2012] (hereafter FSW) considered fundamental-mode surface-wave data. In Figure 11.11 the S-wave models of FCT and FSW are compared with our results using the 95% credible interval [*Chen and Shao*, 1998], the mean model, and the maximum likelihood model. These models are interpolated on a common grid by using radial basis functions. The lateral resolution of our models, compared to FCT and FSW, is lower, as is apparent from their smoothness. The most distinct features of these tomographic models include, much as observed here, separation of structure into slow and fast tectonic regimes.

To provide a more quantitative comparison of the models, we computed the correlation coefficient for each sampled posterior S-wave model and FCT and FSW, respectively, at depths of 100 km and 200 km. For this purpose, FCT and FSW were interpolated on the grid shown in Figure 11.6 using radial basis functions. The resulting correlation coefficients are shown in Figure 11.12. At 100-km depth, the correlation between our models and FCT is ~0.79, whereas for FSW the correlation increases to ~0.92. At 200-km depth, the correlation coefficients are slightly lower than at 100-km depth, but remain relatively high, with coefficients of 0.76 (FCT) and 0.89 (FSW), respectively. Interpolation on alternative common grids with different resolution did not significantly change the distribution of correlation coefficients. Thus, and in spite of differing methodologies, data, and parameterizations, we find good agreement between our ensemble of models and the tomographic models FCT and FSW. The overall level of agreement with other regional models is encouraging and we consider this evidence in support of the joint thermodynamic analysis.

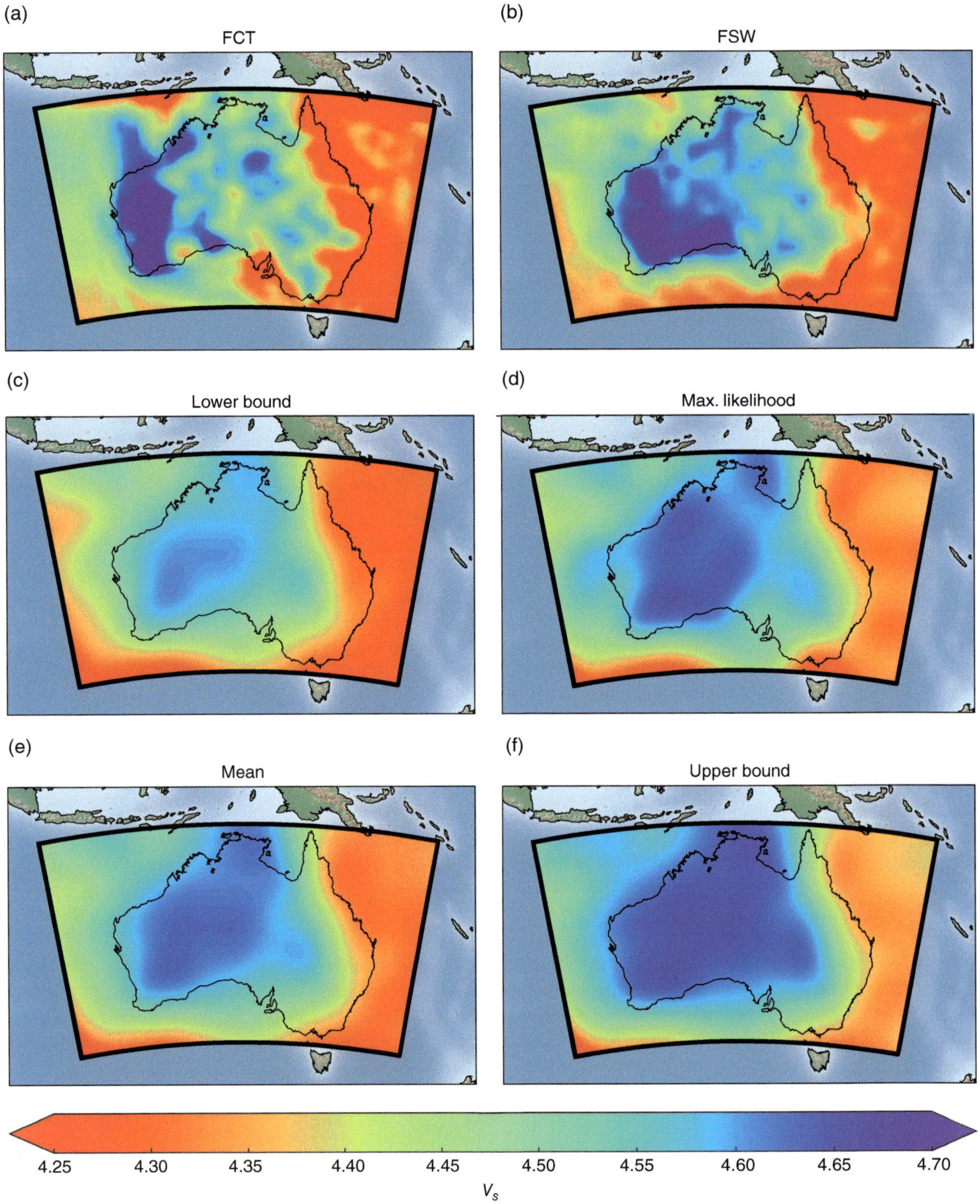

Figure 11.11 Comparison of regional tomographic models in the form of isotropic shear-wave models at 100-km depth from **(a)** *Fichtner et al.* [2010] (FCT), **(b)** *Fishwick and Rawlinson* [2012] (FSW), and models obtained from inversion of surface-wave dispersion data (*Khan et al.* [2013]) (**c–f**). Plots c and f indicate lower and upper credible intervals (95%), respectively, plot e indicates mean model, and plot d indicates the maximum likelihood model.

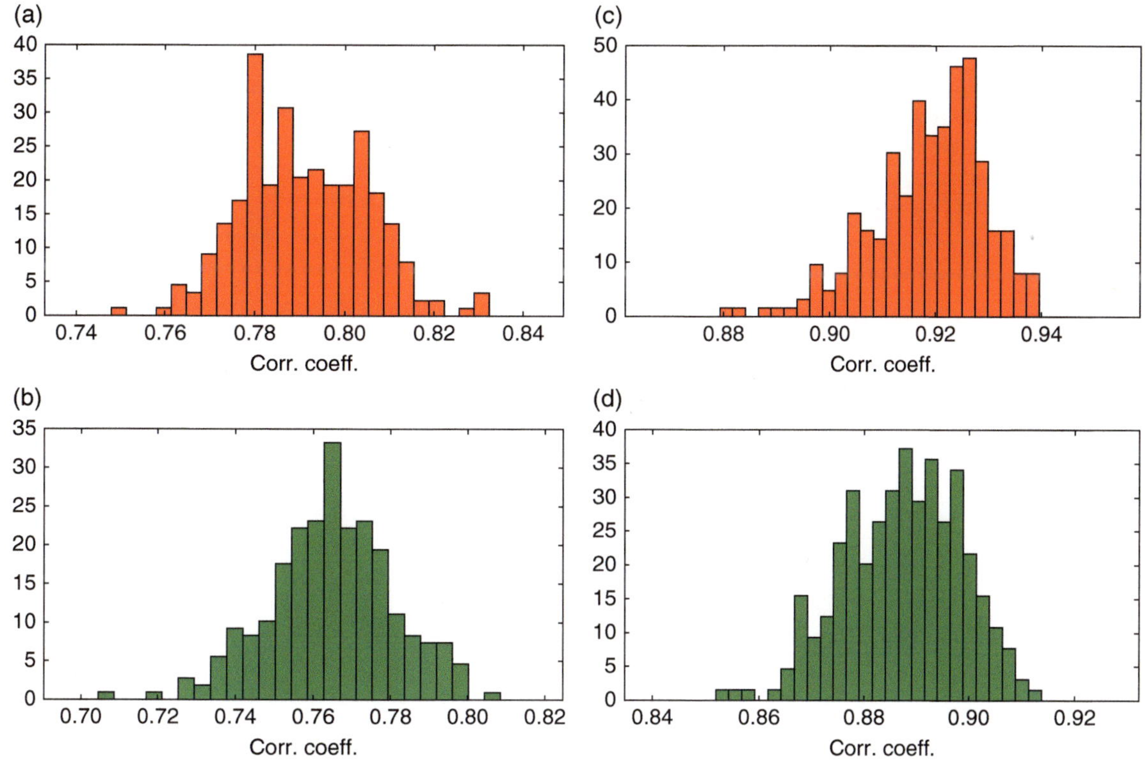

Figure 11.12 Computed correlation between surface-wave based models shown in plots c–f of Figure 11.11 and the regional tomography models of *Fichtner et al.* [2010] **(a,b)** and *Fishwick and Rawlinson* [2012] **(c,d)**. First row shows correlation of models at 100 km of depth, and second row shows correlation at 200 km of depth. See main text for details.

11.3.4. Full Waveform Calculations

To further refine our collection of posterior models, we compute waveforms at 20-s period for a set of our surface-wave-based tomography models. We picked six models randomly from the posterior distribution and computed seismograms using RegSEM.

To reduce the computational costs of running RegSEM down to 20-s period and to prevent propagation of spurious reflections related to the relatively low-resolution input models, we utilized vertical homogenization, based on the concept of an "effective medium" (e.g., see *Capdeville et al.* [2013]) (cf. Section 11.2.2.2). As sources, we used three earthquakes that occurred north of Australia. These include the 6.4-, 6.5-, and 6.4-magnitude events in the regions of Sumbawa, New Britain, and the Banda Sea, respectively (see Figure 11.6). The moment tensor data were retrieved from the global CMT catalogue [*Dziewonski et al.*, 1981; *Ekström et al.*, 2012]. The models are meshed using a structured grid that spans an area of about 46 degrees in longitude and 48 degrees in latitude (centered around 136° W, 23° S). Additionally, perfectly matched layer (PML) boundary conditions were applied to the edges of our grid.

Computed and observed seismograms are shown in Figure 11.13. Observed data have been processed to remove instrument response by deconvolution and filtered according to the parameters used for the source-time function employed in our simulations (the source time spectrum is defined by four corner frequencies: 0.002, 0.005, 0.03, and 0.05 Hz). We generally observe that the phase of the computed seismograms is reasonably well fit for most stations, but that the amplitude is off, which is to be expected given the nature of the input models. Nevertheless, this comparison reveals that much is to be gained by this type of analysis. In particular, the inversion of seismic waveforms has proved to be a strong tool to image Earth's interior, and wedding the methods proposed here with stochastic waveform inversion is an important next step. As shown by *Käufl et al.* [2013], and in spite of the huge computational requirements, stochastic long-period waveform inversion is feasible. While this evidently limits us to large-scale structures only, stochastic waveform inversion provides a powerful means of rigorously assessing measures of uncertainty, accuracy, and resolution—properties of tomographic models that are currently not available.

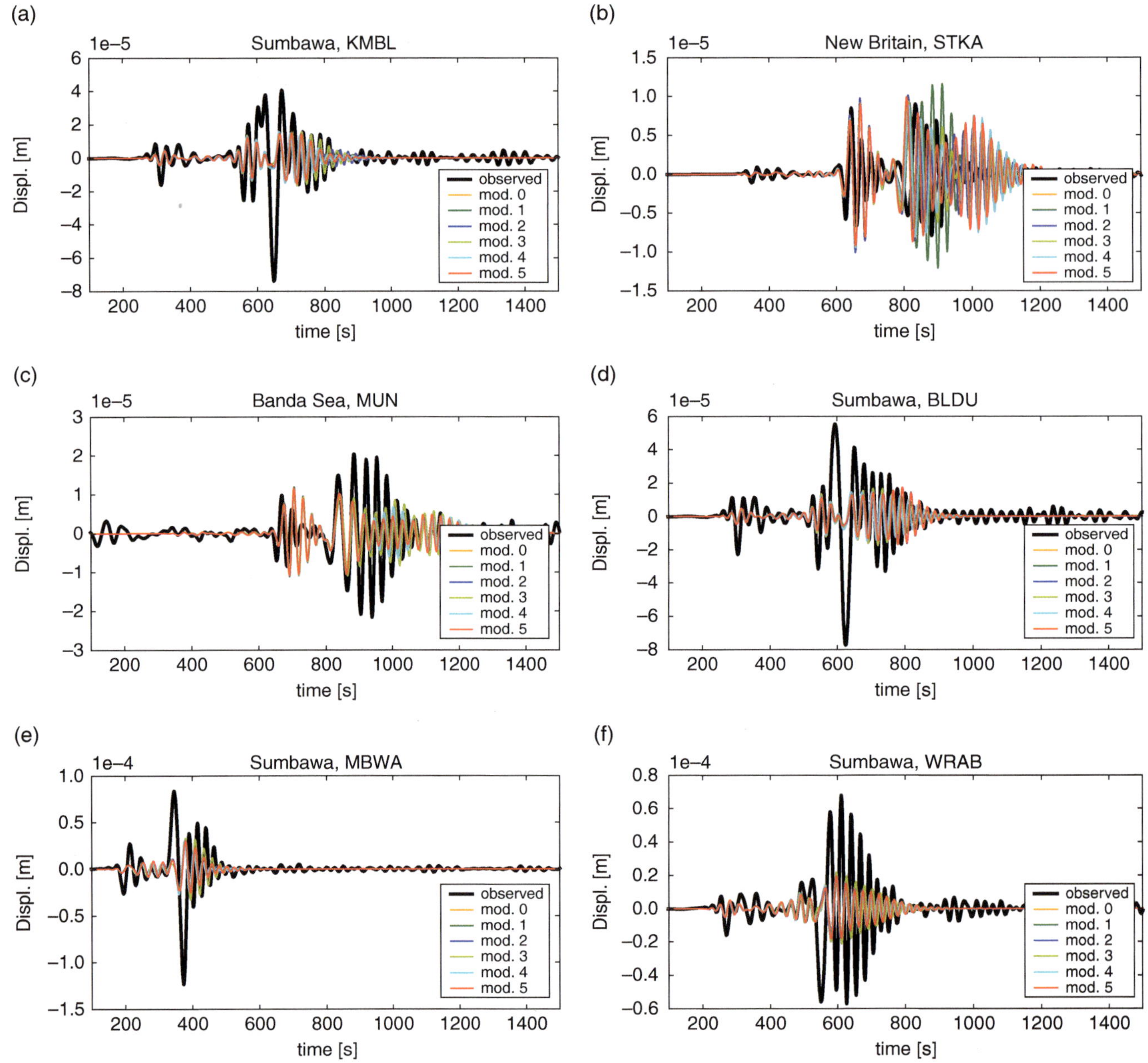

Figure 11.13 Seismograms (vertical component) computed from six regional Australian tomography models (mod. 0–5) from *Khan et al.* [2013], and compared with observed seismograms ("observed") at a minimum period of 20 s. Instrument response was removed and original waveforms were filtered according to the source time spectrum, defined by four corner frequencies: 0.002, 0.005, 0.03, and 0.05 Hz. The title above each subplot designates the name of the event (e.g., Sumbawa) followed by the name of the seismic station (e.g., KMBL). For geographical information on station and event location see Figure 11.6.

11.4. SUMMARY REMARKS

In this study we have described a methodology to estimate Earth's mantle composition and thermal state directly from geophysical data using a joint geophysical–thermodynamical approach. To connect geophysical observables (here in the form of seismic waveforms) with mantle temperature and composition, we rely on a self-consistent thermodynamic modeling scheme of mantle mineral phase equilibria that depends only on composition, temperature, and pressure. What underlies the procedure is a large body of mineral physics-related data that derives from many experimental measurements and supplemented with results from first-principles calculations (e.g., see *Stixrude and Lithgow-Bertelloni* [2011]). In this manner we produce profiles of physical properties to obtain models of mantle conditions that simultaneously combine features of both laboratory and geophysical data. An obvious advantage of the thermodynamic approach is that geophysical–petrological knowledge of, for example, location, size, and sharpness of mantle discontinuities is an inherent part of the data that would otherwise be difficult to resolve.

Increasing effort is being spent in trying to unravel the physical causes of the observed velocity variations that are seen in seismic tomography images. There is a growing consensus that the cause of the imaged wave-speed anomalies not only relates to variations in temperature, but also bears a strong compositional component. However, separation of thermal and chemical effects from seismic wave speeds alone is difficult and is further complicated by the general insensitivity of seismic wave speeds to the density contrasts that are ultimately responsible for driving mantle convection. Thus, placing better constraints on the density structure is key to improve our understanding of interior dynamical processes. However, density–velocity relationships are nonlinear and complex (e.g., see *Khan et al.* [2015]), as a result of which it is common practice to make use of ad hoc scaling relationships to link e.g., S-wave speed and density. In contrast, because composition and temperature act as fundamental parameters, the present set-up provides a natural means of determining all properties and their variations simultaneously and obviates the need for scaling relationships. For example, electrical conductivity, can enhance pictures of mineral composition and density in mantle models (e.g., see *Khan et al.* [2015]).

In summary, interpretation of the anomalies that are being uncovered at increasingly finer scales (e.g., see *Ritsema et al.* [2011] and *Schaeffer and Lebedev* [2013]) with a view to placing constraints on Earth's bulk composition and thermal state and ultimately its origin and evolution, is very likely not possible from a single vantage point. It requires integrated multidisciplinary efforts where data and models from a broad range of disciplines (e.g., geophysics, petrology, geochemistry, mineralogy, and geology where possible) are jointly processed. The foundations and tools for such interdisciplinary approaches already exist and it is not without reason that an increasing effort toward integrated studies is being undertaken given the potential improvements over more traditional approaches (e.g., see *Fullea et al.* [2009], *Khan et al.* [2009], *Verhoeven et al.* [2009], *Fullea et al.* [2012], *Khan and Shankland* [2012], *Afonso et al.* [2013a,b], *Drilleau et al.* [2013], and *Khan et al.* [2015]).

APPENDIX A: THE INVERSE PROBLEM: MARKOV CHAIN MONTE CARLO METHODS

Here we briefly summarize the essentials of solving the inverse problem probabilistically. For more detailed discussions, the reader is referred to the contributions by *Mosegaard and Hansen* [2016], *Hansen et al.* [2016], and *Bosch* [2016] in this volume.

11.A1. A Quick Look at the Probabilistic Approach

The framework that allows us to integrate thermodynamical computations with seismology is the probabilistic approach to inverse problem [*Mosegaard and Tarantola*, 1995]. In the probabilistic approach the target of the inversion is the posterior probability density function, which is given by [*Tarantola*, 2005]

$$\sigma_M(\mathbf{m}) = k\rho_M(\mathbf{m})L(\mathbf{m}), \tag{11.A1}$$

where $\mathbf{m}$ are model parameters, $\rho_M(\mathbf{m})$ the prior probability distribution, $L(\mathbf{m})$ the likelihood function, k a normalization constant, and $\sigma_M(\mathbf{m})$ the posterior probability distribution. The latter expresses the probability related to the space of all conceivable Earth models M. The prior represents the knowledge we have on model parameters independently of the observed data, whereas the likelihood function measures the degree of fit between calculated and observed data and requires computation of the forward model $g(\mathbf{m})$, which may consist of several steps as is the case here. The intention is not to obtain a single "best" fit model as solution to the inverse problem, but a collection of models from the posterior in order to address issues related to uncertainty, nonuniqueness, resolution, and, not least, to accommodate the nonlinearity of the forward model.

In general, prior information plays a fundamental role in solving the inverse problem; in some cases, the availability of information independent of observed data may strongly impact the feasibility of the inversion. Using complex a priori information adapted to a specific problem, including correlations among model parameters, can help make a Monte Carlo search more effective (see, e.g., *Zunino et al.* [2015]).

With the goal of characterizing the posterior distribution, we can employ the "Extended Metropolis algorithm" [*Mosegaard and Tarantola*, 1995] to sample $\sigma_M(\mathbf{m})$ and obtain a (possibly large) collection of models as solution. The drawback of this algorithm is that a very large number of iterations are usually required to sample the posterior, which grows considerably with the number of model parameters. Once the ensemble of solutions is obtained, we use the following relationship (thanks to the *importance sampling* property of the algorithm) to obtain information about Earth's interior, since the posterior collection of models represents the complete solution to the inverse problem:

$$\int_M \Phi(\mathbf{m})\sigma_M(\mathbf{m})\mathrm{d}\mathbf{m} \approx \frac{1}{N}\sum_{i=1}^{N}\Phi\left(\mathbf{m}^{(i)}\right), \tag{11.A2}$$

where $\Phi(\mathbf{m})$ represents a specific feature of interest and N represents the number of posterior models.

11.A2. Data Uncertainty

Observed seismic measurements are affected by noise, which is formally included in the definition of the likelihood function. In case of Gaussian noise, the likelihood function takes the form

$$L(\mathbf{m}) = k \exp\left[-\frac{1}{2}\left(g(\mathbf{m})-\mathbf{d}_{obs}\right)^T \mathbf{C}_D^{-1}\left(g(\mathbf{m})-\mathbf{d}_{obs}\right)\right], \tag{11.A3}$$

where $\mathbf{C}_D$ is the covariance matrix, $\mathbf{d}_{obs}$ the observed data, $g(\mathbf{m})$ the forward model, and k a normalization constant. More generally, under Gaussian assumptions, the covariance matrix $\mathbf{C}_D$ can be expanded to include, if available, any theoretical uncertainty, such that $\mathbf{C}_D = \mathbf{C}_d + \mathbf{C}_T$, where $\mathbf{C}_d$ is the covariance of observed data and $\mathbf{C}_T$ the covariance describing uncertainty of the various steps of the forward modeling scheme.

APPENDIX B: ANELASTICITY

In modeling attenuation we compute shear quality factor Q_S using the laboratory-based model suggested by *Jackson et al.* [2002] and employed in our previous work [*Khan et al.*, 2011]. Specifically, we compute Q_S using the following expression (e.g., see *Jackson et al.* [2002]):

$$Q_S = Q_0 \exp\left[\frac{\alpha\left(E_a + pV_a\right)}{\mathrm{R}T}\right], \tag{11.B1}$$

where Q_0 is a constant, E_a the activation energy, V_a the activation volume, p the pressure, T the temperature, R the gas constant, and α an exponent, which has been determined experimentally to be between 0.15 and 0.25 (e.g., see *Jackson et al.* [2002]). The compressional quality factor is obtained from $Q_P^{-1} = \left(4V_S^2/3V_P^2\right)Q_S^{-1}$. Assuming Q to be weakly frequency dependent (described by α), anelastically corrected (anharmonic) S-wave velocities V_S can be obtained from the expression (e.g., see *Anderson* [1989])

$$V_S(p,T,C) = v_S(p,T,C)\left[1-\frac{2Q_S^{-1}}{\tan(\alpha\pi/2)}\right], \tag{11.B2}$$

where v_S is isotropic anharmonic S-wave velocity as a function of p, T, and composition C obtained using Gibbs free energy minimization as explained in Section 11.2.1. Anelastic corrections to P-wave velocities V_P are obtained by simply substituting v_P and Q_P for v_S and Q_S in the above expression. Based on recent measurements, more complex rheological models (e.g., Andrade, extended Burgers, pseudo-period) have been proposed (e.g., see *Jackson et al.* [2010], *McCarthy et al.* [2011], and *Takei et al.* [2014]), but were not considered.

APPENDIX C: ANISOTROPY

We make the assumption of transverse anisotropy (symmetry axis in vertical direction), in which case the number of independent parameters of the elastic tensor reduce to 5. For present purposes we use the following anisotropic parameters V_{SV}, V_{SH}, V_{PV}, V_{PH}, where the subscripts denote vertically (V) and horizontally (H) polarized S-waves and vertically and horizontally propagating P-waves, respectively. We further introduce the three anisotropy parameters ξ, ϕ, and η, which are related to V_{SV}, V_{SH}, V_{PV}, V_{PH} through

$$\xi = \frac{V_{SH}^2}{V_{SV}^2}, \quad \phi = \frac{V_{PV}^2}{V_{PH}^2}, \quad \eta = \frac{F}{A-2L}, \tag{11.C1}$$

where ξ and ϕ quantify S- and P-wave anisotropy, respectively, η describes the dependence of velocity on the incidence angle of a propagating wave [*Dziewonski and Anderson*, 1981; *Anderson and Dziewonski*, 1982], and F, A, and L are three of the Love coefficients. Reparameterizing in terms of the Voigt average of isotropic P- and S-wave velocities [*Babuska and Cara* 1991], we obtain

$$V_P^2 = \frac{V_{PV}^2 + 4V_{PH}^2}{5} \tag{11.C2}$$

$$V_S^2 = \frac{2V_{SV}^2 + V_{SH}^2}{3}, \tag{11.C3}$$

where V_S^2 and V_P^2 are isotropic (anelastically corrected) P- and S-wave velocities, respectively.

Summarizing, from the set of parameters $\{V_P, V_S, \xi, \phi, \eta\}$, anisotropic velocities are computed using the above equations, which are subsequently employed to determine Love and Rayleigh-wave dispersion curves. Note that because surface-wave overtone data are considered, sensitivity includes both S- and P-wave velocity.

ACKNOWLEDGMENTS

We thank two reviewers for constructive comments which helped improve the clarity of the manuscript. This work was supported by Swiss National Science Foundation grant IZK0Z2_154511. Waveform data used in this study were obtained from the IRIS Data Management Center (http://www.iris.edu).

REFERENCES

Afonso, J. C., J. Fullea, W. L. Griffin, Y. Yang, A. G. Jones, J. A. D. Connolly, and S. Y. O'Reilly (2013a), 3-D multiobservable probabilistic inversion for the compositional and thermal structure of the lithosphere and upper mantle. I: A priori petrological information and geophysical observables, *J. Geophys. Res. (Solid Earth)*, *118*, 2586–2617, doi:10.1002/jgrb.50124.

Afonso, J. C., J. Fullea, Y. Yang, J. A. D. Connolly, and A. G. Jones (2013b), 3-D multi-observable probabilistic inversion for the compositional and thermal structure of the lithosphere and upper mantle. II: General methodology and resolution analysis, *J. Geophys. Res. (Solid Earth)*, *118*, 1650–1676, doi:10.1002/jgrb.50123.

Afonso, J. C., M. Moorkamp, and J. Fullea (2016), Imaging the lithosphere and upper mantle: Where we are at and where we are going, in Integrated Imaging of the Earth: Theory and Applications, Geophysical Monograph 218, M. Moorkamp, P. Lelievre, N. Linde, and A. Khan, eds., AGU, published by John Wiley & Sons, Hoboken, NJ, pp. 191–218.

Anderson, D. L. (1989), *Theory of the Earth*, Blackwell Scientific Publications, London.

Anderson, D. L. (2007), *New Theory of the Earth*, Cambridge University Press, Cambridge, UK.

Anderson, D. L., and A. M. Dziewonski (1982), Upper mantle anisotropy—Evidence from free oscillations, *Geophys. J.*, *69*, 383–404.

Babuska, V., and M. Cara (1991), *Seismic Anisotropy in the Earth (Modern Approaches in Geophysics)*, Springer, New York.

Backus, G. (1962), Long-wave elastic anisotropy produced by horizontal layering, *J. Geophys. Res.*, *67*(11), 4427–4440.

Bassin, C., G. Laske, and G. Masters (2000), The current limits of resolution for surface wave tomography in North America, *EOS Trans., AGU, 81, F897*.

Berryman, J. G. (1995), Mixture theories for rock properties, in *Rock Physics and Phase Relations: A Handbook of Physical Constants*, AGU, Washington D.C., 205–228, doi:10.1029/RF003p0205.

Bina, C. R. (1998), Free energy minimization by simulated annealing with applications to lithospheric slabs and mantle plumes, *Pure Appl. Geophys.*, *151*, 605–618, doi:10.1007/s000240050132.

Bina, C. R., and G. Helffrich (1994), Phase transition Clapeyron slopes and transition zone seismic discontinuity topography, *J. Geophys. Res.*, *99*, 15,853, doi:10.1029/94JB00462.

Birch, F. (1952), Elasticity and constitution of the Earth's interior, *J. Geophys. Res.*, *57*, 227–286, doi:10.1029/JZ057i002p00227.

Bosch, M. (2016), Inference networks in earth models with multiple components and data, in Integrated Imaging of the Earth: Theory and Applications, Geophysical Monograph 218, M. Moorkamp, P. Lelievre, N. Linde, and A. Khan, eds., AGU, published by John Wiley & Sons, Hoboken, NJ, pp. 29–47.

Brown, J. M., and T. J. Shankland (1981), Thermodynamic parameters in the Earth as determined from seismic profiles, *Geophys. J.*, *66*, 579–596, doi:10.1111/j.1365-246X.1981.tb04891.x.

Cammarano, F., S. Goes, P. Vacher, and D. Giardini (2003), Inferring upper-mantle temperatures from seismic velocities, *Phys. Earth Planet. Interiors*, *138*, 197–222, doi:10.1016/S0031-9201(03)00156-0.

Cammarano, F., S. Goes, A. Deuss, and D. Giardini (2005), Is a pyrolitic adiabatic mantle compatible with seismic data?, *Earth Planet. Sci. Lett.*, *232*(3–4), 227–243, doi:10.1016/j.epsl.2005.01.031.

Cammarano, F., B. Romanowicz, L. Stixrude, C. Lithgow-Bertelloni, and W. Xu (2009), Inferring the thermochemical structure of the upper mantle from seismic data, *Geophys. J. Int.*, *179*, 1169–1185, doi:10.1111/j.1365-246X.2009.04338.x.

Cammarano, F., P. Tackley, and L. Boschi (2011), Seismic, petrological and geodynamical constraints on thermal and compositional structure of the upper mantle: Global thermochemical models, *Geophys. J. Int.*, *187*(3), 1301–1318.

Capdeville, Y., L. Guillot, and J.-J. Marigo (2010a), 1-D non-periodic homogenization for the seismic wave equation, *Geophys. J. Int.*, *181*, 897–910, doi:10.1111/j.1365-246X.2010.04529.x.

Capdeville, Y., L. Guillot, and J.-J. Marigo (2010b), 2-D non-periodic homogenization to upscale elastic media for P-SV waves, *Geophys. J. Int.*, *182*, 903–922, doi:10.1111/j.1365–246X.2010.04,636.x.

Capdeville, Y., É. Stutzmann, N. Wang, and J.-P. Montagner (2013), Residual homogenization for seismic forward and inverse problems in layered media, *Geophys. J. Int.*, *194*(1), 470–487.

Capdeville, Y., M. Zhao, and P. Cupillard (2015), Fast Fourier homogenization for elastic wave propagation in complex media, *Wave Motion*, *54*, 170–186, doi:10.1016/j.wavemoti.2014.12.006.

Chen, M., and Q. Shao (1998), Monte Carlo estimation of bayesian credible and hpd intervals, *J. Comput. Graph. Stat.*, *8*, 69–92.

Cobden, L., S. Goes, F. Cammarano, and J. A. D. Connolly (2008), Thermochemical interpretation of one-dimensional seismic reference models for the upper mantle: Evidence for bias due to heterogeneity, *Geophys. J. Int.*, *175*, 627–648, doi:10.1111/j.1365-246X.2008.03903.x.

Connolly, J. (2005), Computation of phase equilibria by linear programming: A tool for geodynamic modeling and its application to subduction zone decarbonation, *Earth and Planet. Sci. Lett.*, *236*(1), 524–541.

Connolly, J. A. D. (2009), The geodynamic equation of state: What and how, *Geochem. Geophys., Geosyst.*, *10*, Q10014, doi:10.1029/2009GC002540.

Connolly, J. A. D., and D. M. Kerrick (2002), Metamorphic controls on seismic velocity of subducted oceanic crust at 100–250 km depth, *Earth Planet. Sci. Lett.*, *204*, 61–74, doi:10.1016/S0012-821X(02)00957-3.

Cupillard, P., Capdeville, Y. and Botella, A. (2015) Homogenization of 3d geological models for seismic wave propagation, in SEG Technical Program Expanded Abstracts, pp. 3656–3660.

Cupillard, P., E. Delavaud, G. Burgos, G. Festa, J.-P. Vilotte, Y. Capdeville, and J.-P. Montagner (2012), RegSEM: A versatile code based on the spectral element method to compute seismic wave propagation at the regional scale, *Geophys. J. Int.*, *188*(3), 1203–1220.

da Silva, C. R. S., R. M. Wentzcovitch, A. Patel, G. D. Price, and S. I. Karato (2000), The composition and geotherm of the lower mantle: constraints from the elasticity of silicate perovskite, *Physics Earth Planet. Interiors*, *118*, 103–109, doi:10.1016/S0031-9201(99)00133-8.

Dahlen, F., S.-H. Hung, and G. Nolet (2000), Frechet kernels for finite-frequency traveltimes—I. Theory, *Geophys. J. Int.*, *141*, 157–174.

Deschamps, F., R. Snieder, and J. Trampert (2001), The relative density-to-shear velocity scaling in the uppermost mantle, *Physics Earth Planet. Int.*, *124*, 193–212, doi:10.1016/S0031-9201(01)00199-6.

Dobson, D. P., and J. P. Brodholt (2000), The electrical conductivity and thermal profile of the Earth's mid-mantle, *Geophys. Res. Lett.*, *27*, 2325–2328, doi:10.1029/1999GL008409.

Drilleau, M., É. Beucler, A. Mocquet, O. Verhoeven, G. Moebs, G. Burgos, J.-P. Montagner, and P. Vacher (2013), A Bayesian approach to infer radial models of temperature and anisotropy in the transition zone from surface wave dispersion curves, *Geophys. J. Int.*, *195*, 1165–1183, doi:10.1093/gji/ggt284.

Dziewonski, A. M., and D. L. Anderson (1981), Preliminary reference Earth model, *Phys. Earth Planet. Interiors*, *25*, 297–356, doi:10.1016/0031-9201(81)90046-7.

Dziewonski, A. M., T.-A. Chou, and J. H. Woodhouse (1981), Determination of earthquake source parameters from waveform data for studies of global and regional seismicity, *J. Geophys. Res.*, *86*, 2825–2852, doi:10.1029/JB086iB04p02825.

Ekström, G., M. Nettles, and A. M. Dziewoński (2012), The global CMT project 2004–2010: Centroid-moment tensors for 13,017 earthquakes, *Phys. Earth Planet. Interiors*, *200*, 1–9, doi:10.1016/j.pepi.2012.04.002.

Emry, E., A. Nyblade, J. Julià, S. Anandakrishnan, R. Aster, D. Wiens, A. Huerta, and T. Wilson (2015), The mantle transition zone beneath west antarctica: Seismic evidence for hydration and thermal upwellings, *Geochem., Geophys., Geosyst.*, *16*, 40–58.

Fabrichnaya, O. (1999), The phase relations in the feo-mgo-al2o3-sio2 system: Assessment of thermodynamic properties and phase equilibria at pressures up to 30 {GPa}, *Calphad*, *23*(1), 19–67, doi:10.1016/S0364-5916(99)00013-9.

Fichtner, A., H. Igel, H.-P. Bunge, and B. Kennett (2009), Simulation and inversion of seismic wave propagation on continental scales based on a spectral-element method, *J. Numer. Anal., Indust. Appl. Math.*, *4*, 11–22.

Fichtner, A., B. L. N. Kennett, H. Igel, and H.-P. Bunge (2010), Full waveform tomography for radially anisotropic structure: New insights into present and past states of the Australasian upper mantle, *Earth Planet. Sci. Lett.*, *290*, 270–280, doi:10.1016/j.epsl.2009.12.003.

Fishwick, S., and N. Rawlinson (2012), 3-D structure of the Australian lithosphere from evolving seismic datasets, *Aust. J. Earth Sci.*, *59*, 809–826, doi:10.1080/08120099.2012.702319.

Forte, A. M., and H. C. Perry (2000), Geodynamic evidence for a chemically depleted continental tectosphere, *Science*, *290*(5498), 1940–1944.

Fullea, J., J. C. Afonso, J. A. D. Connolly, M. Fernãndez, D. Garca-Castellanos, and H. Zeyen (2009), LitMod3D: An interactive 3-D software to model the thermal, compositional, density, seismological, and rheological structure of the lithosphere and sublithospheric upper mantle, *Geochem., Geophys., Geosyst.*, *10*, Q08019, doi:10.1029/2009GC002391.

Fullea, J., M. R. Muller, and A. G. Jones (2011), Electrical conductivity of continental lithospheric mantle from integrated geophysical and petrological modeling: Application to the Kaapvaal Craton and Rehoboth Terrane, southern Africa, *J. Geophys. Res. (Solid Earth)*, *116*(B15), B10202, doi:10.1029/2011JB008544.

Fullea, J., S. Lebedev, M. R. Agius, A. G. Jones, and J. C. Afonso (2012), Lithospheric structure in the Baikal-central Mongolia region from integrated geophysical–petrological inversion of surface-wave data and topographic elevation, *Geochem., Geophys., Geosyst.*, *13*, Q0AK09, doi:10.1029/2012GC004138.

Goes, S., R. Govers, and P. Vacher (2000), Shallow mantle temperatures under Europe from P and S wave tomography, *J. Geophys. Res.*, *105*, 11,153, doi:10.1029/1999JB900300.

Goes, S., F. Cammarano, and U. Hansen (2004), Synthetic seismic signature of thermal mantle plumes, *Earth Planet. Sci. Lett.*, *218*, 403–419, doi:10.1016/S0012-821X(03)00680-0.

Goes, S., F. J. Simons, and K. Yoshizawa (2005), Seismic constraints on temperature of the australian uppermost mantle, *Earth Planet. Sci. Lett.*, *236*(1), 227–237.

Guéguen, Y., and P. Victor (1994), *Introduction to the Physics of Rocks*, Princeton University Press, Princeton, NJ.

Guillot, L., Y. Capdeville, and J. Marigo (2010), 2-D nonperiodic homogenization of the elastic wave equation: SH case, *Geophys. J. Int.*, *182*, 1438–1454, doi:10.1111/j.1365–246X.2010.04,688.x.

Hacker, B. R., G. A. Abers, and S. M. Peacock (2003), Subduction factory 1. Theoretical mineralogy, densities, seismic wave speeds, and H_2O contents, *J. Geophys. Res. (Solid Earth)*, *108*, 2029, doi:10.1029/2001JB001127.

Hansen, T. M., K. S. Cordua, A. Zunino, and K. Mosegaard (2015), Probabilistic integration of geo-information, in Integrated Imaging of the Earth: Theory and Applications, Geophysical Monograph 218, M. Moorkamp, P. Lelievre, N. Linde, and A. Khan, eds., AGU, published by John Wiley & Sons, Hoboken, NJ, pp. 93–116.

Hashin, Z., and S. Shtrikman (1963), A variational approach to the theory of the elastic behaviour of multiphase materials, *J. Mech. Phys. Solids*, *11*(2), 127–140.

Helffrich, G. (2000), Topography of the transition zone seismic discontinuities, *Rev. Geophys.*, *38*, 141–158, doi:10.1029/1999RG000060.

Helffrich, G. R., and B. J. Wood (2001), The earth's mantle, *Nature*, *412*(6846), 501–507.

Hill, R. (1952), The elastic behaviour of a crystalline aggregate, *Proc. Phys. Soc. A*, *65*, 349–354, doi:10.1088/0370-1298/65/5/307.

Hofmann, A. (1997), Mantle geochemistry: The message from oceanic volcanism, *Nature*, *385*(6613), 219–229.

Holland, T. J. B., and R. Powell (1998), An internally consistent thermodynamic data set for phases of petrological interest, *J. Metamorph. Geol.*, *16*(3), 309–343.

Irifune, T. (1994), Absence of an aluminous phase in the upper part of the Earth's lower mantle, *Nature*, *370*, 131–133, doi:10.1038/370131a0.

Ishii, M., and J. Tromp (2004), Constraining large-scale mantle heterogeneity using mantle and inner-core sensitive normal modes, *Phys. Earth Planet. Interiors*, *146*(1), 113–124.

Jackson, I. (1998), Elasticity, composition and temperature of the Earth's lower mantle: A reappraisal, *Geophys. J. Int.*, *134*, 291–311, doi:10.1046/j.1365-246X.1998.00560.x.

Jackson, I., J. D. Fitz Gerald, U. H. Faul, and B. H. Tan (2002), Grain-size-sensitive seismic wave attenuation in polycrystalline olivine, *J. Geophys. Res. (Solid Earth)*, *107*, 2360, doi:10.1029/2001JB001225.

Jackson, I., U. H. Faul, D. Suetsugu, C. Bina, T. Inoue, and M. Jellinek (2010), Grainsize-sensitive viscoelastic relaxation in olivine: Towards a robust laboratory-based model for seismological application, *Phys. Earth Planet. Interiors*, *183*, 151–163, doi:10.1016/j.pepi.2010.09.005.

Jones, A. G., R. L. Evans, and D. W. Eaton (2009), Velocity–conductivity relationships for mantle mineral assemblages in Archean cratonic lithosphere based on a review of laboratory data and Hashin–Shtrikman extremal bounds, *Lithos*, *109*, 131–143, doi:10.1016/j.lithos.2008.10.014.

Kaban, M. K., M. Tesauro, W. D. Mooney, and S. A. P. L. Cloetingh (2014), Density, temperature, and composition of the North American lithosphere—New insights from a joint analysis of seismic, gravity, and mineral physics data: 1. Density structure of the crust and upper mantle, *Geochem., Geophys., Geosyst.*, *15*, 4781–4807, doi:10.1002/2014GC005483.

Karato, S. (2013), *Physics and Chemistry of the Deep Earth*, Wiley-Blackwell, Hoboken, NJ.

Käufl, P., A. Fichtner, and H. Igel (2013), Probabilistic full waveform inversion based on tectonic regionalization – development and application to the Australian upper mantle, *Geophys. J. Int.*, *193*, 437–451, doi:10.1093/gji/ggs131.

Kennett, B., S. Widiyantoro, and R. van der Hilst (1998), Joint seismic tomography for bulk sound and shear wave speed in the earth's mantle, *J. Geophys. Res. Solid Earth (1978–2012)*, *103*(B6), 12,469–12,493.

Khan, A., and F. Deschamps, eds. (2015), *The Earth's Heterogeneous Mantle—A Geophysical, Geodynamical, and Geochemical Perspective*, Springer Geophysics, New York.

Khan, A., and T. J. Shankland (2012), A geophysical perspective on mantle water content and melting: Inverting electromagnetic sounding data using laboratory-based electrical conductivity profiles, *Earth Planet. Sci. Lett.*, *317*, 27–43, doi:10.1016/j.epsl.2011.11.031.

Khan, A., J. A. D. Connolly, and N. Olsen (2006), Constraining the composition and thermal state of the mantle beneath Europe from inversion of long-period electromagnetic sounding data, *Journal of Geophysical Research (Solid Earth)*, *111*(B10), B10102, doi:10.1029/2006JB004270.

Khan, A., J. A. D. Connolly, J. Maclennan, and K. Mosegaard (2007), Joint inversion of seismic and gravity data for lunar composition and thermal state, *Geophys. J. Int.*, *168*, 243–258, doi:10.1111/j.1365-246X.2006.03200.x.

Khan, A., L. Boschi, and J. A. D. Connolly (2009), On mantle chemical and thermal heterogeneities and anisotropy as mapped by inversion of global surface wave data, *J. Geophys. Res. (Solid Earth)*, *114*, B09305, doi:10.1029/2009JB006399.

Khan, A., A. Zunino, and F. Deschamps (2011), The thermochemical and physical structure beneath the North American continent from Bayesian inversion of surface-wave phase velocities, *J. Geophys. Res. (Solid Earth)*, *116*, B09304, doi:10.1029/2011JB008380.

Khan, A., A. Zunino, and F. Deschamps (2013), Upper mantle compositional variations and discontinuity topography imaged beneath australia from bayesian inversion of surface-wave phase velocities and thermochemical modeling, *J. Geophys. Res. Solid Earth*, *118*(10), 5285–5306.

Khan, A., S. Koch, T. J. Shankland, A. Zunino, and J. A. D. Connolly (2015), Relationships between seismic wave-speed, density, and electrical conductivity beneath australia from seismology, mineralogy, and laboratory-based conductivity profiles, in *The Earth's Heterogeneous Mantle—A Geophysical, Geodynamical, and Geochemical Perspective*, A. Khan and F. D., eds., Springer Geophysics, New York.

Komatitsch, D., and J.-P. Vilotte (1998), The spectral element method: An efficient tool to simulate the seismic response of 2d and 3d geological structures, *Bull. Seismol. Soc. Am.*, *88*(2), 368–392.

Komatitsch, D., J. Ritsema, and J. Tromp (2002), The spectral-element method, *Beowulf computing, and global seismology, Science*, *298*, 1737–1742, doi:10.1126/science.1076024.

Koyama, T., A. Khan, and A. Kuvshinov (2014), Three-dimensional electrical conductivity structure beneath Australia from inversion of geomagnetic observatory data: Evidence for lateral variations in transition-zone temperature, water content and melt, *Geophys. J. Int.*, *196*, 1330–1350, doi:10.1093/gji/ggt455.

Kuo, C., and B. Romanowicz (2002), On the resolution of density anomalies in the Earth's mantle using spectral fitting of normal-mode data, *Geophys. J. Int.*, *150*, 162–179, doi:10.1046/j.1365-246X.2002.01698.x.

Kuskov, O. L., and A. B. Panferov (1991), Phase diagrams of the FeO–MgO–SiO_2 system and the structure of the mantle discontinuities, *Phys. Chem. Minerals*, *17*, 642–653, doi:10.1007/BF00203845.

Kuskov, O. L., V. A. Kronrod, and A. A. Prokofyev (2011), Thermal structure and thickness of the lithospheric mantle underlying the Siberian Craton from the kraton and kimberlit superlong seismic profiles, *Izv. Phys. Solid Earth*, *47*, 155–175, doi:10.1134/S1069351310111011.

Kuskov, O. L., V. A. Kronrod, A. A. Prokofyev, and N. I. Pavlenkova (2014), Thermochemical structure of the lithospheric mantle underneath the Siberian craton inferred from long-range seismic profiles, *Tectonophysics*, *615*, 154–166, doi:10.1016/j.tecto.2014.01.006.

Lee, S.-H., J. Rhie, Y. Park, and K.-H. Kim (2014), Topography of the 410 and 660 km discontinuities beneath the Korean Peninsula and southwestern Japan using teleseismic receiver functions, *J. Geophys. Res. (Solid Earth)*, *119*, 7245–7257, doi:10.1002/2014JB011149.

Lekic, V., and B. Romanowicz (2011), Tectonic regionalization without a priori information: A cluster analysis of upper mantle tomography, *Earth Planet. Sci. Lett.*, *308*, 151–160, doi:10.1016/j.epsl.2011.05.050.

Liu, Q., and Y. J. Gu (2012), Seismic imaging: From classical to adjoint tomography, *Tectonophysics*, *566*, 31–66, doi:10.1016/j.tecto.2012.07.006.

Marton, F. C., and R. E. Cohen (2002), Constraints on lower mantle composition from molecular dynamics simulations of $MgSiO_3$ perovskite, *Phys. Earth Planet. Interiors*, *134*, 239–252, doi:10.1016/S0031-9201(02)00189-9.

Matas, J., J. Bass, Y. Ricard, E. Mattern, and M. S. T. Bukowinski (2007), On the bulk composition of the lower mantle: Predictions and limitations from generalized inversion of radial seismic profiles, *Geophys. J. Int.*, *170*, 764–780, doi:10.1111/j.1365-246X.2007.03454.x.

Mattern, E., J. Matas, Y. Ricard, and J. Bass (2005), Lower mantle composition and temperature from mineral physics and thermodynamic modelling, *Geophys. J. Int.*, *160*, 973–990, doi:10.1111/j.1365-246X.2004.02549.x.

McCarthy, C., Y. Takei, and T. Hiraga (2011), Experimental study of attenuation and dispersion over a broad frequency range: 2. The universal scaling of polycrystalline materials, *J. Geophys. Res. (Solid Earth)*, *116*, B09207, doi:10.1029/2011JB008384.

Montelli, R., G. Nolet, F. A. Dahlen, G. Masters, E. R. Engdahl, and S.-H. Hung (2004), Finite-frequency tomography reveals a variety of plumes in the mantle, *Science*, *303*, 338–343, doi:10.1126/science.1092485.

Mosca, I., L. Cobden, A. Deuss, J. Ritsema, and J. Trampert (2012), Seismic and mineralogical structures of the lower mantle from probabilistic tomography, *J. Geophys. Res. Solid Earth*, *117*, B06304.

Mosegaard, K., and T. M. Hansen (2016), Inverse methods: Problem formulation and probabilistic solutions, in Integrated Imaging of the Earth: Theory and Applications, Geophysical Monograph 218, M. Moorkamp, P. Lelievre, N. Linde, and A. Khan, eds., AGU, published by John Wiley & Sons, Hoboken, NJ, pp. 9–27.

Mosegaard, K., and A. Tarantola (1995), Monte Carlo sampling of solutions to inverse problems, *J. Geophys. Res. Solid Earth (1978–2012)*, *100*(B7), 12,431–12,447.

Nissen-Meyer, T., M. van Driel, S. C. Stähler, K. Hosseini, S. Hempel, L. Auer, A. Colombi, and A. Fournier (2014), AxiSEM: Broadband 3-D seismic wavefields in axisymmetric media, *Solid Earth*, *5*, 425–445, doi:10.5194/se-5-425-2014.

Piazzoni, A. S., G. Steinle-Neumann, H.-P. Bunge, and D. Dolejš (2007), A mineralogical model for density and elasticity of the Earth's mantle, *Geochem. Geophys. Geosyst.*, *8*, Q11010, doi:1029/2007GC001697.

Resovsky, J. S., and M. H. Ritzwoller (1999), A degree 8 mantle shear velocity model from normal mode observations below 3 mHz, *J. Geophys. Res.*, *104*, 993–1014, doi:10.1029/1998JB900025.

Ricard, Y., E. Mattern, and J. Matas (2005), Synthetic tomographic images of slabs from mineral physics, *Washington DC Am. Geophys. Union Geophys. Monogr. Ser.*, *160*, 283–300, doi:10.1029/160GM17.

Ritsema, J., W. Xu, L. Stixrude, and C. Lithgow-Bertelloni (2009), Estimates of the transition zone temperature in a mechanically mixed upper mantle, *Earth Planet. Sci. Lett.*, *277*(1–2), 244–252, doi:10.1016/j.epsl.2008.10.024.

Ritsema, J., A. Deuss, H. J. van Heijst, and J. H. Woodhouse (2011), S40RTS: A degree-40 shear-velocity model for the mantle from new Rayleigh wave dispersion, teleseismic traveltime and normal-mode splitting function measurements, *Geophys. J. Int.*, *184*, 1223–1236, doi:10.1111/j.1365-246X.2010.04884.x.

Romanowicz, B. (2001), Can we resolve 3D density heterogeneity in the lower mantle?, *J. Geophys. Res.*, *28*, 1107–1110, doi:10.1029/2000GL012278.

Saxena, S. K., and G. Eriksson (1983), Low- to medium-temperature phase equilibria in a gas of solar composition, *Earth Planet. Sci. Lett.*, *65*, 7–16, doi:10.1016/0012-821X(83)90185-1.

Schaeffer, A. J., and S. Lebedev (2013), Global shear speed structure of the upper mantle and transition zone, *Geophys. J. Int.*, *194*, 417–449, doi:10.1093/gji/ggt095.

Schmandt, B., S. D. Jacobsen, T. W. Becker, Z. Liu, and K. G. Dueker (2014), Dehydration melting at the top of the lower mantle, *Science*, *344*, 1265–1268, doi:10.1126/science.1253358.

Schmerr, N. (2015), Imaging mantle heterogeneity with upper mantle seismic discontinuities, in *The Earth's Heterogeneous Mantle—A Geophysical, Geodynamical, and Geochemical Perspective*, A. Khan and F. Deschamps, eds., Springer Geophysics, New York.

Schmerr, N., and E. Garnero (2006), Investigation of upper mantle discontinuity structure beneath the central Pacific using SS precursors, *J. Geophys. Res. (Solid Earth)*, *111*, B08305, doi:10.1029/2005JB004197.

Schuberth, B. S. A., C. Zaroli, and G. Nolet (2012), Synthetic seismograms for a synthetic earth: long-period p- and s-wave traveltime variations can be explained by temperature alone, *Geophys. J. Int.*, *188*(3), 1393–1412, doi:10.1111/j.1365-246X.2011.05333.x.

Shapiro, N., and M. Ritzwoller (2004), Thermodynamic constraints on seismic inversions, *Geophys. J. Int.*, *157*(3), 1175–1188.

Simmons, N. A., A. M. Forte, L. Boschi, and S. P. Grand (2010), GyPSuM: A joint tomographic model of mantle density and seismic wave speeds, *J. Geophys. Res. (Solid Earth)*, *115*(B14), B12310, doi:10.1029/2010JB007631.

Sobolev, S. V., and A. Y. Babeyko (1994), Modeling of mineralogical composition, density and elastic wave velocities in anhydrous magmatic rocks, *Surv. Geophys.*, *15*, 515–544, doi:10.1007/BF00690173.

Stixrude, L., and R. Jeanloz (2007), *Constraints on Seismic Models from Other Disciplines—Constraints from Mineral Physics on Seismological Models*, Vol. *1*, Elsevier, Amsterdam, Chapter 22, pp. 775–803.

Stixrude, L., and C. Lithgow-Bertelloni (2005a), Thermodynamics of mantle minerals—I. Physical properties, *Geophys. J. Int.*, *162*(2), 610–632.

Stixrude, L., and C. Lithgow-Bertelloni (2005b), Mineralogy and elasticity of the oceanic upper mantle: Origin of the low-velocity zone, *J. Geophys. Res. (Solid Earth)*, *110*, B03204, doi:10.1029/2004JB002965.

Stixrude, L., and C. Lithgow-Bertelloni (2007), Influence of phase transformations on lateral heterogeneity and dynamics in Earth's mantle, *Earth Planet. Sci. Lett.*, *263*, 45–55, doi:10.1016/j.epsl.2007.08.027.

Stixrude, L., and C. Lithgow-Bertelloni (2011), Thermodynamics of mantle minerals—II. Phase equilibria, *Geophys. J. Int.*, *184*(3), 1180–1213.

Tackley, P. J., S. Xie, T. Nakagawa, and J. W. Hernlund (2005), Numerical and laboratory studies of mantle convection: Philosophy, accomplishments, and thermochemical structure and evolution, *Washington DC Am. Geophys. Union Geophys. Monogr. Ser.*, *160*, 83–99, doi:10.1029/160GM07.

Takei, Y., F. Karasawa, and H. Yamauchi (2014), Temperature, grain size, and chemical controls on polycrystal anelasticity over a broad frequency range extending into the seismic range, *J. Geophys. Res. (Solid Earth)*, *119*, 5414–5443, doi:10.1002/2014JB011146.

Takeuchi, N., H. Kawakatsu, S. Tanaka, M. Obayashi, Y. J. Chen, J. Ning, S. P. Grand, F. Niu, J. Ni, R. Iritani, K. Idehara, and T. Tonegawa (2014), Upper mantle tomography in the northwestern Pacific region using triplicated P waves, *J. Geophys. Res. (Solid Earth)*, *119*, 7667–7685, doi:10.1002/2014JB011161.

Tarantola, A. (2005), *Inverse Problem Theory and Methods for Model Parameter Estimation*, SIAM, Philadelphia.

Tauzin, B., E. Debayle, and G. Wittlinger (2008), The mantle transition zone as seen by global Pds phases: No clear evidence for a thin transition zone beneath hotspots, *J. Geophys. Res. (Solid Earth)*, *113*, B08309, doi:10.1029/2007JB005364.

Tauzin, B., R. D. van der Hilst, G. Wittlinger, and Y. Ricard (2013), Multiple transition zone seismic discontinuities and low velocity layers below western United States, *J. Geophys. Res. (Solid Earth)*, *118*, 2307–2322, doi:10.1002/jgrb.50182.

Thompson, D. A., J. O. S. Hammond, J.-M. Kendall, G. W. Stuart, G. R. Helffrich, D. Keir, A. Ayele, and B. Goitom (2015), Hydrous upwelling across the mantle transition zone beneath the Afar Triple Junction, *Geochem., Geophys., Geosyst.*, *16*, 834–846, doi:10.1002/2014GC005648.

Trampert, J., F. Deschamps, J. Resovsky, and D. Yuen (2004), Probabilistic tomography maps chemical heterogeneities throughout the lower mantle, *Science*, *306*(5697), 853–856.

Turcotte, D. L., and G. Schubert (2002), *Geodynamics*, 2nd ed., Cambridge University Press, Cambridge, UK, doi:10.2277/0521661862.

Vacher, P., A. Mocquet, and C. Sotin (1998), Computation of seismic profiles from mineral physics: the importance of the non-olivine components for explaining the 660 km depth discontinuity, *Phys. Earth Planet. Interiors*, *106*, 275–298, doi:10.1016/S0031-9201(98)00076-4.

van der Hilst, R. D., J. D. Bass, J. Matas, and J. E. Trampert (2005), Earth's deep mantle: Structure, composition, and evolution, *AGU, Geophys. Monogr. Ser.*, *160*, 344 pp., doi:10.1029/GM160.

van Gerven, L., F. Deschamps, and R. D. van der Hilst (2004), Geophysical evidence for chemical variations in the Australian Continental Mantle, *Geophys. Res. Lett.*, *31*, L17607, doi:10.1029/2004GL020307.

Verhoeven, O., A. Mocquet, P. Vacher, A. Rivoldini, M. Menvielle, P.-A. Arrial, G. Choblet, P. Tarits, V. Dehant, and T. van Hoolst (2009), Constraints on thermal state and composition of the Earth's lower mantle from electromagnetic impedances and seismic data, *J. Geophys. Res. (Solid Earth)*, *114*, B03302, doi:10.1029/2008JB005678.

Visser, K., J. Trampert, and B. L. N. Kennett (2008), Global anisotropic phase velocity maps for higher mode Love and Rayleigh waves, *Geophys. J. Int.*, *172*, 1016–1032, doi:10.1111/j.1365-246X.2007.03685.x.

Watt, J. P., G. F. Davies, and R. J. O'Connell (1976), The elastic properties of composite materials, *Rev. Geophys. Space Phys.*, *14*, 541, doi:10.1029/RG014i004p00541.

Wood, B. J., and J. R. Holloway (1984), A thermodynamic model for subsolidus equilibria in the system CaO–MgO–Al2O_3–SiO_2, *Geochim. Cosmochim. Acta*, *48*, 159–176, doi:10.1016/0016-7037(84)90358-2.

Xu, W., C. Lithgow-Bertelloni, L. Stixrude, and J. Ritsema (2008), The effect of bulk composition and temperature on mantle seismic structure, *Earth Planet. Sci. Lett.*, *275*, 70–79, doi:10.1016/j.epsl.2008.08.012.

Xu, Y., T. J. Shankland, and B. T. Poe (2000), Laboratory-based electrical conductivity in the Earth's mantle, *J. Geophys. Res.*, *105*, 27,865, doi:10.1029/2000JB900299.

Zunino, A., J. A. D. Connolly, and A. Khan (2011), Precalculated phase equilibrium models for geophysical properties of the crust and mantle as a function of composition, *Geochem., Geophys., Geosyst.*, *12*, Q04001, doi:10.1029/2010GC003304.

Zunino, A., K. Mosegaard, K. Lange, Y. Melnikova, and T. Mejer Hansen (2015), Monte Carlo reservoir analysis combining seismic reflection data and informed priors, *Geophysics*, *80*, 31, doi:10.1190/geo2014-0052.1.

INDEX

Integrated Imaging of the Earth: Theory and Applications, Geophysical Monograph 218, First Edition.
Edited by Max Moorkamp, Peter G. Lelièvre, Niklas Linde, and Amir Khan.